Life Histories of North American Thrushes, Kinglets, and Their Allies

by Arthur Cleveland Bent

Dover Publications, Inc., New York

ADVERTISEMENT

The scientific publications of the National Museum include two series known, respectively, as *Proceedings* and *Bulletin*.

The *Proceedings* series, begun in 1878, is intended primarily as a medium for the publication of original papers, based on the collections of the National Museum, that set forth newly acquired facts in biology, anthropology, and geology, with descriptions of new forms and revisions of limited groups. Copies of each paper, in pamphlet form, are distributed as published to libraries and scientific organizations and to specialists and others interested in the different subjects. The dates at which these separate papers are published are recorded in the table of contents of each of the volumes.

The series of *Bulletins*, the first of which was issued in 1875, contains separate publications comprising monographs of large zoological groups and other general systematic treatises (occasionally in several volumes), faunal works, reports of expeditions, catalogs of type specimens, special collections, and other material of similar nature. The majority of the volumes are octavo in size, but a quarto size has been adopted in a few instances in which large plates were regarded as indispensable. In the *Bulletin* series appear volumes under the heading *Contributions from the United States National Herbarium*, in octavo form, published by the National Museum since 1902, which contain papers relating to the botanical collections of the Museum.

The present work forms No. 196 of the *Bulletin* series.

ALEXANDER WETMORE,
Secretary, Smithsonian Institution.

Published in the United Kingdom by Constable and Company Limited, 10 Orange Street, London W. C. 2.

This Dover edition, first published in 1964, is an unabridged and unaltered republication of the work first published in 1949 by the United States Government Printing Office, as Smithsonian Institution United States National Museum *Bulletin 196*.

International Standard Book Number: 0-486-21086-3
Library of Congress Catalog Card Number: 64-24414

Manufactured in the United States of America

Dover Publications, Inc.
180 Varick Street
New York 14, N. Y.

CONTENTS

INTRODUCTION

This is the seventeenth in a series of bulletins of the United States National Museum on the life histories of North American birds. Previous numbers have been issued as follows:

107. Life Histories of North American Diving Birds, August 1, 1919.
113. Life Histories of North American Gulls and Terns, August 27, 1921.
121. Life Histories of North American Petrels and Pelicans and Their Allies, October 19, 1922.
126. Life Histories of North American Wild Fowl (part), May 25, 1923.
130. Life Histories of North American Wild Fowl (part), June 27, 1925.
135. Life Histories of North American Marsh Birds, March 11, 1927.
142. Life Histories of North American Shore Birds (pt. 1), December 31, 1927.
146. Life Histories of North American Shore Birds (pt. 2), March 24, 1929.
162. Life Histories of North American Gallinaceous Birds, May 25, 1932.
167. Life Histories of North American Birds of Prey (pt. 1), May 3, 1937.
170. Life Histories of North American Birds of Prey (pt. 2), August 8, 1938.
174. Life Histories of North American Woodpeckers, May 23, 1939.
176. Life Histories of North American Cuckoos, Goatsuckers, Hummingbirds, and Their Allies, July 20, 1940.
179. Life Histories of North American Flycatchers, Larks, Swallows, and Their Allies, May 8, 1942.
191. Life Histories of North American Jays, Crows, and Titmice, January 27, 1947.
195. Life Histories of North American Nuthatches, Wrens, Thrashers, and Their Allies. July 7, 1948.

The same general plan has been followed, as explained in previous bulletins, and the same sources of information have been utilized. The nomenclature of the 1931 Check-list of the American Ornithologists' Union has been followed.

An attempt has been made to give as full a life history as possible of the best-known subspecies of each species and to avoid duplication by writing briefly of the other subspecies. In many cases certain habits, probably common to the species as a whole, have been recorded for only one subspecies; as such habits are mentioned only under the subspecies on which the observations were made, it would be well to read all the accounts in order to get the whole story for the species. The distribution deals with the species as a whole, with only rough outlines of the ranges of the subspecies, which in many cases cannot be accurately defined.

The egg dates are the condensed results of a mass of records taken from the data in a large number of the best egg collections, as well as from contributed notes and from a few published sources. They show the dates on which eggs have been actually found in various parts of the country, giving the earliest and latest dates and the limits between which half the dates fall, indicating the height of the season.

The plumages are described in only enough detail to enable the reader to trace the sequence of molts and plumages from birth to maturity and to recognize the birds in the different stages and at the different seasons. No attempt has been made to describe fully the adult plumages; this has been well done already in the many manuals and State books. Partial or complete albinism is liable to occur in almost any species; for this reason, and because it is practically impossible to locate all such cases, it has seemed best not to attempt to treat this subject at all. The names of colors, when in quotation marks, are taken from Ridgway's Color Standards and Color Nomenclature (1912). In the measurements of eggs, the four extremes are printed in boldface type.

Many who have contributed material for previous volumes have continued to cooperate. Receipt of material from over 500 contributors has been acknowledged previously. In addition to these, our thanks are due to the following new contributors: G. D. Alcorn, Amelia S. Allen, A. E. Allin, James Bond. W. H. Carrick, C. D. Carter, H. L. Cogswell, Fred Evenden, Jr., Mae Halliday, F. V. Hebard, H. W. Higman, E. W. Jameson, Jr., G. F. Knowlton, H. C. Kyllingstad, R. E. Lawrence, D. J. Magoon. A. H. Miller, R. F. Miller, R. H. Mills, Tilford Moore, James Murdock, J. W. Musgrove, H. R. Myers, L. N. Nichols, G. A. Petrides, A. L. Pickens, W. T. Shaw, W. E. Shore, J. C. Tracy, H. O. Wagner, G. J. Wallace, Florence G. Weaver, George Willett, and William Youngworth.

If any contributor fails to find his or her name in this or some previous volume, the author would be glad to be advised. As the demand for these bulletins is much greater than the supply, the names of those who have not contributed to the work during the previous 10 years will be dropped from the author's mailing list.

Photographs for this bulletin were contributed by the Museum of Vetebrate Zoology, University of California, and by the University of Minnesota.

This has been, indeed, a cooperative volume. Bernard W. Tucker has contributed eight complete life histories and Dr. Winsor M. Tyler has written three, besides rendering valuable assistance in reading and indexing a large part of the literature on North American birds. Dr. Alfred O. Gross, Dr. George J. Wallace, Mrs. Florence G. Weaver, Francis M. Weston, and Robert S. Woods have each contributed one complete life history.

Egg measurements were furnished especially for this volume by Dean Amadon (for the American Museum of Natural History), James Bond, John R. Cruttenden, Charles E. Doe, Wilson C. Hanna, Wm. George F. Harris (for the Museum of Comparative Zoology), Ed N. Harrison, A. D. Henderson, Turner E. McMullen, Dr. Robert T. Orr (for the California Academy of Sciences), Lawrence Stevens, George H. Stuart, 3d, and Margaret W. Wythe (for the Museum of Vertebrate Zoology). The greater part of the measurements were taken from the register sheets of the United States National Museum by Wm. George F. Harris, who also relieved the author of a vast amount of detail work by collecting and figuring hundreds of egg measurements and by collecting, sorting, and arranging several thousand nesting records to make up the "egg dates" paragraphs.

Through the courtesy of the Fish and Wildlife Service, Miss May T. Cooke has compiled the distribution and migration paragraphs. The author claims no credit and assumes no responsibility for these data, which are taken from the great mass of records on file in Washington.

The manuscript for this bulletin was completed in 1943. Contributions received since then will be acknowledged later. Only information of great importance could be added. The reader is reminded again that this is a cooperative work; if he fails to find in these volumes anything that he knows about the birds, he can blame himself for not having sent the information to—

THE AUTHOR.

LIFE HISTORIES OF NORTH AMERICAN THRUSHES, KINGLETS, AND THEIR ALLIES

ORDER PASSERIFORMES (FAMILIES TURDIDAE AND SYLVIIDAE)

By Arthur Cleveland Bent

Taunton, Mass.

Order PASSERIFORMES

Family TURDIDAE: Thrushes, Bluebirds, Stonechats, and Solitaires

TURDUS MUSICUS COBURNI Sharpe

ICELAND RED-WINGED THRUSH

Contributed by Bernard William Tucker

HABITS

The redwing of British ornithologists is an accidental visitor to Greenland. Schiøler (1926), the eminent Danish authority, in the annotated list of Greenland birds in his great work "Danmarks Fuglo," lists the Greenland visitors as definitely of the Iceland race, *Arceuthornis musicus coburni* (Sharpe), and on geographical grounds this is what would be expected. Dr. J. Reinhardt (1861) recorded two examples, one of which was sent to Dr. Paulsen in 1845, while the other was shot at Frederikshaab, on the west coast, on October 20 of that year and was sent to the Copenhagen Museum, where no doubt it still is. The species is known also from the east coast. Helms (1926) states that Johan Petersen, superintendent of the colony of Angmagsalik, who for many years made observations on the birds occurring there, received three specimens during the period 1894 to 1915. On October 20, 1904, two were shot by the colony. They were flying from one ice floe to another down by the beach, looking for food, and every now and then they made a trip to the shore, where they doubtless caught sandskippers, small snails, and other prey. They were only slightly shy and were easy to shoot. On October 31, 1906, he received one from a Greenlander who had shot it on the beach.

In the British Isles, where both the Iceland and typical forms occur, the red-winged thrush is a familiar winter visitor of markedly gregarious habits, feeding in scattered flocks on open pastures and grasslands or sometimes in stubble and root fields, but perching freely in hedgerows and trees. In northern Europe and Asia it is one of the characteristic birds of the Arctic and sub-Arctic birch forests, though it is also found more sparingly about scrub growth of birch and willow even beyond the tree limit. In Iceland, where woodland is almost absent, the typical haunts of the species are necessarily different from those of the European mainland; it is found in broken, often rock-strewn, country, most commonly where birch scrub exists, but also even in areas almost devoid of scrub. The Iceland race is rather darker above than the typical form, with the underparts more heavily marked and the breast and flanks more washed with olive-brown. The Iceland birds are also generally a trifle larger. Witherby (1938, vol. 2) gives the wing measurements of males as 119-128 mm. (one 117) against 113–119 (one 122) for the typical form; and of females as 120–128 mm. against 113–119 mm. Ticehurst (1925) in a considerable series found that 75 percent of Icelandic birds had the wing 122 mm. or over, this length, as will be seen from the figures just given, being very rare in the typical form.

Courtship.—Unfortunately, nothing whatever has been recorded about the courtship or display of this species.

Nesting.—While its relative and frequent associate the fieldfare is gregarious both at its nesting places and in winter, the redwing is gregarious in winter only, nesting in scattered pairs and being evidently territorial in its habits, for Hantzsch (1905) states that in Iceland the males are quarrelsome and pugnacious and drive off neighboring males from their domains. In the case of the typical form, however, it is not unusual to find a pair or two nesting in or close to a fieldfare colony. In fact, this is so frequent that it can hardly be considered fortuitous; probably the redwings gain protection by nesting near their larger and more aggressive relatives. In Europe the breeding haunts are chiefly the woodlands of birch and alder, though the birds may also be found nesting in pines where the trees are rather small, and they seem to have a distinct partiality for the vicinity of swampy ground. The nest may be placed at a height of 20 feet or so in a tree, but more often lower, and very commonly on a low stump or actually on the ground at the foot of a tree, beneath a bush, or in a bank. Although in northern Europe and Asia the redwing is primarily a bird of wooded country, it nevertheless ranges well beyond the forest limit and then habitually nests on the ground among scrub or in a bank or similar site. Seebohm (1879) found it still common in the Yenisei Valley as far as latitude 71°, but not farther north. Ground sites among rocks and scrub are also typical in Iceland. The Iceland form will also nest on buildings and has established itself as a regular breeder in the town of Reykjavik. The nest is of typical thrush type, built of dry grasses, bents, and fine twigs, lined with a thin but hard layer of mud, which is covered by an inner lining of grass.

Eggs.—The eggs have a greenish ground finely mottled with reddish brown. They are, in fact, extremely like very small eggs of the European blackbird. Jourdain (1938, vol. 2) gives the following measurements based on 116 eggs of the typical form: Average, 25.8 by 19.2 mm.; maximum, 29.5 by 20.3 and 27.6 by 20.6 mm.; minimum, 21.9 by 18.5 and 24.6 by 17.5 mm. His figures for the Iceland race are: Average of 38 eggs, 26.0 by 19.4 mm.; maximum, 29 by 19.2 and 26 by 21 mm.; minimum, 23.5 by 19.5 and 25.2 by 18 mm. In Europe the usual clutch is five or six, but two, three, four, and seven are all recorded. In Iceland the usual number is four to six, but there are records of three and seven. In Europe most eggs are laid in June and July, but some in the latter half of May, while in Iceland they are found from mid-May onward, though also in June and July (Jourdain).

Young.—According to Jourdain, incubation by birds of the typical race may begin with the first egg or the last but one. In the case of the Iceland race it has been stated to begin with the third (G. Timmermann, 1934). The period has been given as about 13 days for the typical form and 14–15 days for the Iceland one, but more data are needed; there is no reason to suppose that there is really any difference between the two. The male appears to take some real share in the task of incubation, but here again fuller observation is needed. A. H. Daukes (1932), observing a pair that bred in Scotland, "noticed that both birds incubated, although there did not appear to be any definite period when the cock took over from the hen," and Timmermann has stated that in Iceland the hen is relieved by the cock at midday in the earlier stages only. The fledgling period is given as about 12–14 days for *musicus* and 11–14 days for *coburni*. The latter is on the authority of B. Hantzsch, whose "Beitrag zur Kenntnis der Vogelwelt Islands" (1905) is still one of the chief sources of information on Icelandic ornithology. Both sexes feed the young, and Timmermann states of the Iceland race that the feces are regularly swallowed by the parents after feeding. The species is double-brooded in Europe and probably at times in Iceland.

Plumages.—The plumages are fully described by H. F. Witherby (1938, vol. 2) in the "Handbook of British Birds." In the nestling the fawn-colored down is distributed on the inner supraorbital, occipital, spinal, humeral, and ulnar tracts, and is long and plentiful on all but the first of these, where it is scanty. The interior of the mouth is gamboge yellow with no spots, and externally the flanges are ivory-colored. In the juvenal plumage the upperparts are less uniform than in the adult, the feathers having darker tips and pale-buffish, drop-shaped central streaks, while the rufous on the flanks is scarcely developed. At this stage it is extremely like the young song thrush, except for the prominent light eye stripe, which at once distinguishes it. The first winter plumage can only be distinguished from that of the adult by white tips to the innermost secondaries and pale buff tips to the greater coverts. In the subsequent summer these distinctive tips often wear off.

Food.—Jourdain (1938, vol. 2) states that the food of the typical race in winter is as follows: "Worms, Mollusca (snails, *Helix aspersa* and *nemoralis*, slugs, *Agriolimax* and even *Arion*, lacustrine univalves and bivalves, *Sphaerium*, etc.). Insects: Coleoptera (*Melolontha* larvae, weevils, etc.), Lepidoptera (Noctuidae, larvae and pupae), Diptera (larvae of Tipulidae), Orthoptera, etc. On sea-shore Crustacea (small crabs, *Talitrus* and *Orchestia*). Berries of hawthorn, yew, rowan, holly, *Vaccinium*."

No exact details on food in the breeding season seem to be available for the typical race. In Iceland the food is recorded as consisting of insects (Coleoptera and larvae of Lepidoptera), worms, and many berries and skins of berries of the previous year, and the young are fed on caterpillars, small Lepidoptera, Coleoptera, and worms (Hantzsch). In British-killed examples of the Iceland race Dr. J. W. Campbell has recorded Mollusca (*Cochliocopa lubrica* and *Succinea* sp.), Coleoptera (larvae of Carabidae and *Barynotus*), and Diptera (larvae of Bibionidae and ? Tabanidae).

In general, the redwing tends to take more animal food and is less addicted to feeding on berries in winter than other thrushes, though, as noted above, it does take berries to some extent.

Behavior.—To observers in the British Isles the arrival of the redwings is one of the notable ornithological events of autumn, for this species and its relative the fieldfare are the two most generally distributed and common winter visitors. One day, probably in October, redwings will be found once again feeding in the fields, or perhaps the first notice of their arrival will be given by their unmistakable thin flight call as flocks pass over at night. Outside the breeding season the redwing is thoroughly gregarious. The birds get most of their food on open grassland, and the delicately built, dark forms widely scattered over the fields, and often mingled with the larger and more robust-looking fieldfares, provide a pleasant picture. Their food, which, as more fully described in the relevant section, consists mainly of worms, insects, and other invertebrates, is obtained from among the grass and from the surface soil and often by rooting about among dead leaves when the birds are feeding near trees or hedgerows or on the outskirts of a wood. On the ground their carriage and gait are those of a typical thrush; they move by a short run or succession of hops followed by a pause and then repeated. When disturbed or resting and when preparing to roost they perch freely in trees and hedges, and it is from the branches of the still bare trees that parties may be heard uttering their low babbling subsong on genial days early in spring while still in their winter quarters. The flight is direct or only slightly undulating.

Voice.—The call note, used especially in flight, is a soft, thin *seeh.* Feeding flocks have also a soft *chup*, and another, quite different but highly characteristic note, much heard at roosts and to some extent at other times, is a very hard-sounding *chittuck*, or *chittick*, with variations. This is also the regular alarm note of breeding birds as heard in northern Europe, where I have also noted the call of fledged young birds as a husky *chucc.*

With regard to the song of the redwing there has been some difference of opinion. In Scandinavia it has been called the "northern nightingale," and various observers, among them the great Linnaeus, have praised its song in terms which others have thought quite extravagant. At least a partial explanation of this divergence of opinion is that the redwing thrush has two perfectly distinct songs. The one by far the most frequently heard is a simple refrain consisting of several, commonly three or four, clear fluty notes generally, but not always, followed by a poor, low, chuckling warble. This song is usually delivered from a pine or other tree, sometimes exposed to view on the top, but often in cover. An interesting feature is its remarkable proneness to local dialects. Thus in Lapland in 1937 I found all or nearly all the birds in one district singing the same very simple stereotyped phrase, which could be rendered as *trūi-trūi-trūi*, while in another district not 30 miles away they were a trifle more ambitious and sang *tee-* (or sometimes *tee-tū*) *tí ruppi-tí ruppi-tí ruppi* with only insignificant variations. The fluty notes are quite musical and pleasing, but the unvarying repetition is monotonous and any comparison with the nightingale would be absurd to anyone familiar with that bird.

There is, however, what must be regarded as a premating song, which appears really to merit the praise bestowed on it. It is thus described by H. W. Wheelwright (1871):

> Of all the northern songsters, perhaps the redwing stands first on the list, and is with justice called the northern nightingale, for a sweeter song I never wish to listen to when this rich gush of melody is poured out from the thick covert of a fir in the "silence of twilight's contemplative hour," or often in the still hour of

midnight when all else in nature is at rest. But as soon as the breeding season commences this beautiful song ceases, and is now changed to a kind of call—"Twee, twee, twee, twee, tweet," ending with a little trill.

The "kind of call" is unmistakably the simple form of song described above, which was all I ever heard in the latter part of June, but in Swedish Lapland two friends and correspondents whom I have quoted in the "Handbook of British Birds" heard, even when most birds were incubating, a few individuals singing what was clearly a form, even if an inferior one, of the full song, which they describe as much like that of a song thrush, but rather lower-pitched with many phrases each repeated several times.

Yet a third performance is the subsong, which can be regularly heard from flocks in their winter quarters on sunny days in February and March. This I have described in the "Handbook of British Birds" (1938), already quoted, as "a low babbling affair of twittering and warbling notes not infrequently punctuated by the fluty 'trui'," and "at times song not apparently differing in any material respect from the simpler phase of true song may be heard from individual birds." The chorus may even be varied with what sound like genuine snatches of full song, but with one recent exception there is no clear record of the full song ever being heard before the return to the breeding ground.

Field marks.—The redwing is an obvious thrush, a little larger than a wood thrush, with the usual spotted, or in this case more strictly streaked, breast of the group, but at once distinguished from any other American or European thrush by its conspicuous pale eye stripe and chestnut-red flanks. The Iceland form is rather darker than the typical one, and in certain of the Scottish islands, where both forms occur regularly, at least two competent and experienced observers have stated that the Icelanders, or at any rate the more strongly characterized specimens, are recognizable even in the field by their darker coloration, an opinion they have confirmed by shooting. But for observers without their exceptional experience and opportunities of comparison the races cannot be considered separable in the field.

Enemies.—There is not much that can be said about enemies of the redwing beyond the kind of general statement that can be made about all the thrushes and other birds of similar size and habits. Like other Turdidae it is subject to the attacks of hawks and perhaps more so than some on account of its being mainly a bird of open ground. It figures in the dietary of the (European) sparrow hawk, goshawk, and at times of the peregrine (duck hawk), and even the little merlin (pigeon hawk). The last-named must be presumed to be the chief enemy of the Iceland form on its breeding ground. Roosting birds are sometimes taken by owls. It is also recorded that hawks may accompany the migrating flocks and take a toll of the wanderers.

A list of parasites recorded from the species is given by Niethammer (1937).

Fall and winter.—Something of the general habits of the red-winged thrush at this period of the year has been indicated under "Behavior." In spite of its breeding in the far north it is not a very hardy bird, and during hard weather in its winter quarters it suffers severely. At such times "weather movements" to milder regions take place, but many birds remain in the frost-bound districts and many pay the penalty of this deficiency in adaptation, perishing from starvation. It is chiefly under stress of hard weather that redwings may be found feeding on berries in the hedgerows if such are available. At other times they are much less addicted to this diet than are fieldfares and some other thrushes, though it is not true—as has been asserted—that they never resort to it except when forced by hunger.

Thoroughly gregarious in its winter quarters by day, the red-winged thrush is, in a sense, even more so at night, when the constituents of many flocks that have been scattered over the farmlands during the daytime will often gather together to roost in the shrubberies of parks and large gardens or in suitable plantations and thickets. Such roosts are very commonly shared with fieldfares, blackbirds, and a few song thrushes. Smaller numbers roost in old, untrimmed hedgerows.

In some of the treeless Scottish islands the winter haunts are necessarily somewhat different from those of the agricultural districts of England, and it is of interest that in these islands, where the Iceland race occurs regularly, as well as the typical one, a distinct difference in habits has been observed. Thus in Fair Isle, between Shetland and Orkney, the very capable observer George Stout (Witherby et al., 1938), has stated that the Iceland birds tend to keep to crops, while those of the typical race prefer the cliffs and bare hillsides. In the Outer Hebrides, where from November to February most of the birds seem to be of the Iceland race, they frequent arable grassland, heather country, and rocky shores, but also the vicinity of habitations, stackyards, etc., and roost in such places as stone dikes and peat stacks or in long heather and willow scrub (J. W. Campbell, Witherby et al., 1938).

DISTRIBUTION

Breeding range.—The breeding range of the typical race is described as extending in Europe north to 70° in Norway, North Sweden, Finland, Russia (Archangel Government), and south to Gotland in the Baltic, northeast Poland, the Baltic States, and in Russia to the Minsk, Chernigov, Kaluga, Tula, Ryazan, Nizhni Novgorod, Kazan, Ufa, and Orenburg Governments. Also in Siberia, north to the tree limit, east to the River Kolyma and south to about 54° in the west, but in the east only to Yakutsk (about 63°). It also breeds in small numbers in Germany and exceptionally in Belgium, and a pair has been recorded nesting in two successive years in Scotland. The race *T. m. coburni* breeds in Iceland and sparingly in the Faeroes, and a pair believed to be of this race has once been recorded nesting in the Shetland Islands.

Winter range.—British Isles and southern Europe to the Mediterranean and Black Seas; also Asia Minor, Syria, and Iran, rarely reaching the Mediterranean islands and Palestine. The Iceland form has been identified on passage and in winter in the Faeroes, Holland, and France. (Jourdain.)

Spring migration.—The migratory movements of the typical race are essentially the same as those of the fieldfare (pp. 13–14), but departures from the British Isles appear to average a fortnight earlier and to finish soon after mid-April. May records are exceptional, and the only three available for the period May 4–25 come from the Scottish islands. Three June dates are recorded for England and Wales, viz, June 9, 13, and 27 (N. F. Ticehurst, 1938, vol. 2). Alexander (1927) records his latest date for the hill country near Rome as March 21. In Germany the passage extends from the second, or more rarely the first, half of March to about mid-April or in East Prussia regularly to the beginning of May (Niethammer, 1937). Arrival is recorded at Vadsö on the Arctic coast of Norway on May 16 (Blair, 1936), at Ust Zylma in Arctic Russia on May 17, and on the Arctic Circle in the Yenisei Valley on June 5 (Seebohm, 1901, 1879). In Iceland, however, according to Hantzsch (1905), it arrives at the end of March and beginning of April (earliest date March 20), which seems surprisingly early for that bleak land.

Fall migration.—Small numbers begin to reach the British Isles at the end of September and the beginning of October, but the main arrival is in October and November. Three July records (earliest July 5) are available (Ticehurst), but these are highly exceptional, and even arrivals in August and the first half of September are very unusual. Niethammer describes the beginning of the passage as usually in the second half of October in central Germany, but at the end of September in East Prussia. In the Mediterranean region Alexander found the arrival in the Rome district very regular about November 1.

Casual records.—Typical race: Spitsbergen, Bear Island, Faeroes, Madeira, Canaries. Iceland race: Bear Island, Greenland, Jan Mayen.

TURDUS PILARIS Linnaeus

FIELDFARE

CONTRIBUTED BY BERNARD WILLIAM TUCKER

HABITS

The fieldfare has been added to the American fauna since the publication of the 1931 A. O. U. list. In *The Auk* for 1940 (vol. 57) P. A. Taverner described a specimen from the southeastern coast of Jens Munk Island, at the head of Foxe Basin, Arctic America, taken during the summer of 1939 and received by the National Museum of Canada from Graham Rowley. "It is a roughly made, semi-mummified skin but quite complete and recognizable. Mr. Rowley found it in the possession of an old Eskimo woman who recognized it as unusual and was keeping it as a curiosity." Mr. Taverner believed this to be the first record for the American list, but there is in fact an earlier authentic record from Greenland. This refers to a male shot on November 24, 1925, at Fiskenaesse on the southwest coast and sent by K. H. Petersen, of Godthaab, along with the skins of several other rarities, to H. Scheel (1927), by whom it was recorded. This specimen was evidently subsequently set up and returned to the Greenland Museum at Godthaab, for in a later article by K. Oldendow (1933) it is recorded that a fieldfare, with the same data as above, is preserved there and constitutes the first record for Greenland. A photograph of a group of birds in the Museum is given on page 187, including an unmistakable fieldfare, which is obviously the specimen in question. A third record can now be added. C. G. and E. G. Bird (1941) record that an adult male was obtained by a norwegian trapper, Magne Räum, near Cape Humboldt on Ymer Island, northeast Greenland, on January 20, 1937. This specimen is preserved in the British Museum (Natural History). The records justly observe that "this is a rather remarkable record, as January is, of course, one of the dark months. The weather for three weeks previously had been one long succession of snowstorms." They add the further noteworthy point that several fieldfares appeared on the island of Jan Mayen on the same date.

The fieldfare, like its relative and frequent associate the red-winged thrush, is a characteristic bird of the forest belt of northern Europe and Asia. But although it ranges quite as far north as the redwing—and indeed, in Europe at any rate, even a little farther—it also extends considerably farther south, breeding locally in parts of central Europe.

Courtship.—The only observer who has described the courtship behavior of the fieldfare appears to be E. J. M. Buxton, who contributed a note on the subject to the "Handbook of British Birds" (1938, vol. 2). The display as he describes it resembles the European blackbird's. The hen stands motionless on the ground with the head on one side, while the cock struts round her many times with the tail fanned and pressed on the ground, the feathers on the rump ruffled up and the bill pointing to the ground. Although the song itself is of the feeblest description, it may be associated with a special display flight, with a peculiar wing action in which the wings are held stiffly out, not fully extended, between the beats, conveying the impression of abnormally slow progress forward.

Nesting.—In its typical northern haunts the fieldfare is first and foremost a bird of the birch forests, in which, unlike the other European thrushes, it breeds in colonies. But it also breeds freely in pine forest, as well as in alders, and in the more southern parts of its range in a variety of other trees. Sometimes colonies may be found in parks, orchards, and gardens, even in towns, and I recall such a colony in trees in a public square in the little town of Kristiansund in Norway. In my experience the nests have generally been about 8 to 18 or 20 feet up, but they may be either higher or lower and occasionally even on the ground or in such other situations as sheds or woodstacks. They are generally fairly widely scattered, and in the Scandinavian birch woods, at any rate, there is not ordinarily more than one in a tree. But in larger trees it is recorded that several nests in one tree are not unusual. Quite small colonies and even isolated nests may be found, especially, though by no means only, toward the limits of forest growth, as well as large ones with scores of nests in a comparatively small area. The species may also be found breeding sparingly even on the high fells above the forests and on open tundra and barren ground beyond the tree limit. Here it is the rule for only an odd pair or two to be found here and there and the nests are built in low scrub or among rocks or on the ground.

The nest is much like that of the European blackbird, built of dry grasses and bents mixed with mud and with a layer of mud beneath the inner lining of fine grass, occasionally with a few twigs or a little moss in the foundation. When built in a tree it is usually placed in a fork or at the base of a branch against the trunk, but it may be placed some way out on a side branch. Whether the male takes any part in building does not appear to be recorded.

Eggs.—The eggs in general closely resemble those of the European blackbird, that is, they are closely freckled with reddish brown on a pale bluish-green ground, but they are much more variable than blackbirds', being often much more richly and boldly marked, sometimes with a well-marked cap or zone at the broad end. There is a type with a bright blue ground more or less prominently spotted and blotched, and unmarked blue eggs are also recorded. The usual clutch is five or six, but seven and even eight occur and occasionally full sets of only three or four are found. Jourdain (1938) gives the measurements of 100 eggs as: Average 28.8 by 20.9 mm.; maximum, 33.5 by 23.4 mm.; minimum, 25.5 by 21 and 29.5 by 19 mm. Eggs may be found from April in Poland, from the last third of April, but usually in May or June, in Germany, and from May to July in Scandinavia.

Young.—According to Jourdain incubation often begins with the first egg and is performed by the hen chiefly if not entirely. The period is given by Armberg as 13–14 days. Both sexes feed the young and the fledgling period is given by Jourdain as 14 days. He states that probably a second brood is sometimes reared. The parent birds are bold and noisy in defense of the nest, and I have seen a pair mobbing a hooded crow with considerable effect. Undoubtedly the social breeding habits are an added protection, for intending marauders will be attacked by a number of birds.

Plumages.—The plumages of the fieldfare are fully described by H. F. Witherby (1938, vol. 2) in the "Handbook of British Birds." The nestling has fairly long and plentiful buff-colored down distributed on the outer and inner supraorbital, occipital, spinal, humeral, and ulnar tracts; it is short on the first of these. The juvenal plumage is not unlike the adult's but a good deal duller, with the head, neck, and rump, which are gray in the adult, washed with brown and with pale shaft streaks to the feathers of the upperparts. After the autumn molt the young birds resemble the adults, though the males have the grays rather browner, and in both sexes the greater coverts generally have whitish tips, which the adults lack.

Food.—Though largely insectivorous in the broad sense, which generally means feeding on a variety of small invertebrates, the fieldfare is a great eater of berries in fall and winter. The flocks may often be seen feasting on berries in the hedgerows, and in places where berry-bearing shrubs are numerous fieldfares may be expected to congregate.

More precisely, the dietary is summarized as follows by Jourdain (1938, vol. 2): "In winter varied, animal and vegetable: Mollusca (slugs and small land-shells), Annelida (earthworms); insects; Coleoptera (*Sitona*, *Otiorrhynchus*, *Megasternum*, *Homalota*, *Quedius*, *Elater*, Curculionidae, *Agriotes* and larvae), Diptera (larvae of Tipulidae, etc.). Also spiders. Many kinds of berries (hawthorn, holly, rowan, yew, juniper, dog-rose, *Pyracantha*, etc.). Swedes attacked in hard weather; fallen apples, grain, and some seeds."

Behavior.—It has been mentioned that the fieldfare is gregarious both in and out of the breeding season. In its winter quarters the flocks lead a wandering existence, roving the fields and hedgerows, often in company with red-winged thrushes. When feeding on grassland both species scatter fairly widely, the fieldfares standing out among their companions by their larger size and bolder, more assertive bearing. But for all this they are wary and alert and not easy to approach very closely. When disturbed from the ground they fly up into any trees that may be at hand, and indeed when not actually feeding they perch freely enough in trees, usually all facing in the same direction with a unanimity appropriate to a social species. The carriage on the ground is much like that of any large thrush. When standing still the position is rather upright, head well up and tail down, and it moves in the usual manner of thrushes in short runs or a succession of hops with pauses between. The flight is fairly direct, with a perceptible but not very noticeable closure of the wings every few beats, and the flocks fly in a rather loose and straggling formation. Except in the breeding season they are characteristically birds of the open. They will resort at times to open woodland if there is a good crop of berries there; indeed on the Continent this would seem to be a more regular habit than it is as a rule in England. But they are not by nature woodland birds and never normally take shelter in bushes or cover of any kind.

Voice.—The ordinary note of the fieldfare is the rather harsh *cha-cha-cha-chack* already mentioned, and the more subdued, conversational chatter of parties in the trees is a variation of this. In the breeding colonies the birds have other more or less similar harsh or chattering notes of alarm or anger. A quite distinct note, a call that with minor variations from species to species is common to a number of thrushes, is a soft prolonged *seeh.* The only note I have heard from young birds on the breeding ground is a shrill *chizzeek.* In contrast to those of many thrushes the song—it really hardly qualifies to rank as more than a subsong—is a remarkably unimpressive performance, consisting of some not unmusical warbling notes mingled with chuckling, whistling, and harsh squeaky sounds and variations of the harsh call. It is more often uttered in flight than from a perch. It is natural to correlate the lack of any highly developed song with the colonial nesting habits of the species. In spring, while the birds are still in their winter quarters, a low guttural warbling subsong may sometimes be heard from parties in the trees, as well as at roosting places, and this seems to be merely a subdued version of the breeding-season song.

Field marks.—The fieldfare is a large, robustly built thrush about 10 inches long, with slate-gray head, nape, and rump contrasted with chestnut back and black tail. The female is like the male, but a little duller. It is largely a bird of open ground, getting much of its food in the fields, but perching freely in trees and hedgerows. It is strongly gregarious and generally seen in parties or flocks, which may be of considerable size, though single birds or little groups of two or three may sometimes be met with. The rather harsh note, *cha-cha-cha-chack*, helps to identify flocks on the wing.

Enemies.—Fieldfares not uncommonly fall victims to the (European) sparrow hawk, goshawk, peregrine falcon, and other birds of prey. In the breeding season their chief enemies are egg-robbing Corvidae, but their bold disposition and colonial habits help to afford protection to the nests and young.

Fall and winter.—Something of the general habits of fieldfares in their winter quarters has been indicated under "Behavior." In hard weather, when other food is difficult to come by, the movements of the flocks are much influenced by the supply of berries, and thick old hedgerows are much frequented. At such times they will also resort to root fields and feed upon swedes or turnips. At dusk numbers gather together to roost in company. It has been asserted that they always or normally roost on the ground and only roost off it in severe weather, or again in an equally sweeping manner that they only roost on the ground of necessity when suitable shrubberies or thickets are not available. Neither is correct. Both types of site are resorted to regularly, and the birds may be found roosting, to quote the summary given by the present writer in the "Handbook of British Birds" (1938), in "rank grass amongst bushes or in young plantations, amongst rushes and marsh plants or other ground vegetation in the open or in woods, in stubble, and in furrows of ploughed fields; but also regularly in tall, thick hedgerows and shrubberies, and even in trees, especially pines and evergreens."

DISTRIBUTION

Breeding range.—The European breeding range is given by Jourdain (1938, vol. 2): "Norway to 71°, N. Sweden, Finland, Russia to Petchora and east to Perm, south to Poltava, Kiev, Voronezh, Saratov and Orenburg Govts., and south to Gotland and Öland, in Germany south to Upper Bavaria, Switzerland, Austria, Czecho-Slovakia, Hungary, and Poland. Has bred Faeroes: also occasionally in Holland and E. France." In Siberia north to the mouth of the river Ob and to 70½° on the river Yenisei, east to lake Baikal and the river Aldan, south to Semipalatinsk and the Altai and Sayan Mountains.

Winter range.—British Isles, central and southern Europe, but rare in the Mediterranean region, though sometimes occurring in Egypt and northwest Africa. Also in Persia, Transcaspia, Kirghiz Steppes, and Turkestan and rarely in Asia Minor, Palestine, and northwest India.

Spring migration.—Leaves winter quarters rather late, parties in May in the British Isles being not uncommon. Departure of winter residents and passage migrants takes place throughout April to early, or sometimes mid, May and even later. A number of late dates are recorded for England and Scotland for the latter part of May, June 2, 3, 6, 10, and 29, and even July 11 and 29 (Ticehurst, 1938). The July dates suggest failure to migrate altogether. Birds reach the German breeding places in March and April and up to the third week of May (Niethammer, 1937). Blair (1936) records first arrivals on the Arctic coast of Norway on May 23, and Seebohm (1901) noted the first appearance of birds on the Arctic Circle in the valleys of the Petchora and Yenisei on May 17 and June 8, respectively.

Fall migration.—Arrives British Isles in small numbers last week of September and early October, followed by large numbers till third week of November (Ticehurst); earliest dates: Fair Isle, August 5; mainland of Scotland, August 10; England (east coast) August 10; Ireland, September 7. In central Germany passage generally from second week of October, in east Prussia from end of September, lasting till first third of November (Niethammer). In Italy from end of October (Arrigoni, 1929). In Rumania from end of October (Dombrowski, 1903). In Greece recorded from November 8; Bulgaria, from mid-November, but one record as early as October 19 (Reiser, 1905, 1894). Recorded in Egypt from November 13 (Meinertzhagen, 1930).

Casual records.—In addition to the Greenland occurrences, Jourdain (1938) mentions: Jan Mayen, Iceland, Spitsbergen, Canaries, Madeira, Balearic Island, Corsica, Sardinia, Sicily, Malta, Cyprus.

TURDUS MIGRATORIUS MIGRATORIUS Linnaeus

EASTERN ROBIN

PLATES 1–5

CONTRIBUTED BY WINSOR MARRETT TYLER

HABITS

The robin, the largest thrush in North America, is widely and familiarly known in the United States and Canada. To millions of people it is as well known as the crow, and far more popular.

The early English colonists gave it its name, doubtless because it resembled in coloration the robin redbreast of England, but they failed to notice the close relationship between our robin and their blackbird, which is a true thrush, *Turdus*, the two birds being very similar in habits, general deportment, and voice, although different in plumage.

H. C. Kyllingstad writes to us from Mountain Village, Alaska: "The robin here is not the confiding creature that it is in the States.

Most frequently it nests away from the village as the native children like robin eggs to eat as well as those of any other bird. The old birds do come about the cabins while feeding or hunting food for the young, but the young are almost never seen, and the old birds keep a sharp watch and will not allow one to approach closely.

"Robins are fairly common in the willow and alder thickets along the Yukon and its branches, but in two years I have found no nests in these places. The only nests I have seen were on the edge of the village under a lean-to attached to a small warehouse. I was able to keep the children away from one nest and the three young left the day after I banded them. The robins are very suspicious of my banding traps; not one has been trapped in two years."

Spring.—From the warm Southern States the robin starts northward early in the year, often in flights of impressive magnitude. George H. Mackay (1897) reports an enormous flight of robins in Florida on February 14, 15, and 16, 1897, observed by James K. Knowlton about 100 miles south of St. Augustine. He says: "They came from a southerly direction, and were continually passing, alighting and repassing, on the above dates, the general movement being in a northerly direction. The air was full of them, and their numbers beyond estimate, reminding him of bees. Mr. Knowlton heard that this movement of Robins had been noted for a distance of *ten* miles away, *across* the flight." And Peter A. Brannon (1921) writes from Alabama: "The annual migration of Robins through the city of Montgomery, took place this year, during the latter part of February, and for ten days thousands were observed on the city streets."

As the robins move northward, they follow very closely the advance of the average daily temperature of 37°, and we may look for them in eastern Massachusetts soon after March 10. They take their place at this time in the opening scene of the grand, dramatic pageant of the long spring migration that follows our bleak, and often comparatively birdless, New England winter.

The robin, however, does not play a leading part in this initial scene; he is a minor character, not at his best so early in the spring. The main actors in the play are the blackbirds, streaming onto the stage in murky, clattering clouds; the bluebirds, mated already, warbling their charming songs to their ladyloves; the song sparrows, filling every acre with their tinkling music.

Wendell Taber and I watched a typical arrival of robins on the morning of March 15, 1936, a day when there was a general influx of the birds into Massachusetts. Looking southward across a broad meadow, we saw them coming toward us, the first we had seen, a flock of a dozen or more, flying in open order, but rather evenly spaced, not closely packed like blackbirds. When they came to the northern edge of the meadow and caught sight of a patch of greensward, they checked their flight and settled on the grass, joining other robins that were running about there, and, after feeding a little while, passed on again to the north. All through the day, spent between Boston and Newburyport, the robin was a prominent bird, chiefly during the morning hours, mostly in small flocks, but sometimes collected in dozens, spread over the open fields. This day's observation is characteristic of the early spring robin flight. It is not spectacular; the great gatherings of the South have thinned out before reaching New England, leaving only small flocks of wild, wary male birds, which wander restlessly about the country, perching in high trees, or feeding in neglected fields or, more commonly, in the cedar pastures where they pluck off the berries. The birds are not in song at this season. They are comparatively silent (i. e., compared to their noisy companions in the migration), expressing themselves only in nervous exclamations.

Early in April we note a sudden, marked change in the behavior of the robins we see about us. We meet many of the birds now in the settled districts of the towns, in our gardens, running familiarly over the lawns. They are tamer than the first migrants and act as if they were our local birds returned to their last year's homes.

The arrival of the female birds at this time precipitates a period of noisy activity. For days our lawns and dooryards become the scene of countless combats and shrieking pursuits full of liveliness and excitement. A male bird will often run at another, seeming to jostle him, and both may then jump into the air against each other, suggesting a fight between gamecocks, or one bird may fly off pursued by the other.

When the noisy pursuits are in full swing, early in April, we sometimes see two robins dash past us, one bird following the other, a hand's breadth apart, sweeping along not far above the ground at a speed so reckless, with lightninglike twists and turns, that collision seems inevitable. Yet they continue on without mishap and pass out of our sight so rapidly that we cannot be sure of their respective sex, and we are left in doubt whether the pursuits are amatory or hostile. The special feature of these pursuits is that only two birds engage in them, and that the flights are maintained for a long distance.

At this season there is still only fitful singing, chiefly in the morning, but all day we hear the long, giggling laugh, *he-he-he-he,* and the scream of attack.

The ground is softening now, and the earthworms, near the surface, are available as food for the next generation of robins.

Courtship.—John Burroughs (1894) ably describes a phase of robin activity, familiar to us all, in which the noisy pursuits assume an element of true courtship. He says: "In the latter half of April we pass through what I call the 'robin racket'—trains of three or four birds rushing pell-mell over the lawn and fetching up in a tree or bush, or occasionally upon the ground, all piping and screaming at the top of their voices, but whether in mirth or anger it is hard to tell. The nucleus of the train is a female. One cannot see that the males in pursuit of her are rivals; it seems rather as if they had united to hustle her out of the place. But somehow the matches are no doubt made and sealed during these mad rushes."

Bradford Torrey (1885) speaks of a quieter courtship:

How gently he approaches his beloved! How carefully he avoids ever coming disrespectfully near! No sparrow-like screaming, no dancing about, no melodramatic gesticulation. If she moves from one side of the tree to the other, or to the tree adjoining, he follows in silence. Yet every movement is a petition, an assurance that his heart is hers and ever must be. * * * On one occasion, at least, I saw him holding himself absolutely motionless, in a horizontal posture, staring at his sweetheart as if he would charm her with his gaze, and emitting all the while a subdued hissing sound. The significance of this conduct I do not profess to have understood; it ended with his suddenly darting at the female, who took wing and was pursued.

It is not uncommon to hear a robin give this hissing note when it is, apparently, alone—standing motionless, as Torrey says, and with its bill pointing slightly upward and the tail expanded. Sometimes, also, a male will utter the hissing sound in phrases much like his song, suggested by the whispered syllables *hissilly, hissilly*. I heard the note once, given in this form when the bird was on the wing.

Audubon (1841) describes what is evidently the culmination of courtship: "During the pairing season, the male pays his addresses to the female of his choice frequently on the ground, and with a fervour evincing the strongest attachment. I have often seen him, at the earliest dawn of a May morning, strutting around her with all the pomposity of a pigeon. Sometimes along a space of ten or twelve yards, he is seen with his tail fully spread, his wings shaking, and his throat inflated, running over the grass and brushing it, as it were, until he has neared his mate, when he moves round her several times without once rising from the ground. She then receives his caresses."

Nesting.—The robin's nest appears as a rather large heap of coarse materials. It is rough on the outside, even unkempt sometimes, because many of the loose ends of grass stalks, twigs, and bits of string or cloth of which the nest is made are not tucked in or neatly woven into the body of the nest, but protrude or hang down from the outer wall. At the top is a deep depression like a round, smooth cup formed by a thick layer of mud, which extends upward to a firm rim, the cup being lined with a little fine, dry grass.

T. Gilbert Pearson (1910) describes thus the structure of a nest built in a balsam: "In its building, a framework of slender balsam twigs had first been used. There were sixty-three of these, some of which were as much as a foot in length. Intertwined with these were twenty fragments of weed stalks and grass stems. The yellow clay cup, which came next inside, varied in thickness from a quarter of an inch at the rim to an inch at the bottom. Grass worked in with the clay while it was yet soft aided in holding it together, and now, last of all, came the smooth, dry carpet of fine grass. The whole structure measured eight inches across the top; inside it was three inches in width, and one and a half deep."

Reginald Heber Howe, Jr. (1898), describes the bird's method of building the nest:

After the site has been chosen the building of a substantial foundation of twigs, grasses, string, etc., is begun; this finished, finer grasses are brought and the bird standing in the centre of the foundation draws them round. After the sides of the nest have been fairly well made the bird by turning around in the nest shapes it to the exact contour of its body, and by pushing its breast far down into the nest and raising the primaries, it presses the nest with the wrist of the wing into a compact and perfect mass. The next work is the plastering with mud; a rainy day is generally chosen for this work; the bird brings the mud in its bill and, placing it on the inside of the nest, flattens it into shape by exactly the methods just described. All that remains now is the lining, which is made of fine grasses and which adheres to the mud, making a substantial though not a particularly beautiful nest.

The average measurements of nest are; depth, outside, 3 inches; depth, inside, 2½ inches; breadth, outside, 6½ inches; breadth, inside, 4 inches.

J. H. Rohrbach (1915) points out that robins may use worm casts as a mud lining for their nests. He says: "A heavy rain of fourteen hours' duration came just at plastering-time. Mud was abundant. Then I observed what was new to me—the Robins passed by all kinds of mud except the castings of earthworms, which they gathered and used for nest-building."

Katharine S. Parsons (1906) describes a nest from which hung "two fringed white satin badges, fastened by mud and sticks" and near them "a knot of coarse white lace" and "two white chicken feathers," and Henry Mousley (1916) states that "Robins here [Hatley, Quebec] are particularly fond of using pearly everlasting (*Anaphalis margaritacea*) in the foundations of their nests."

In early days, before the forests were cleared away, robins presumably built mainly on horizontal limbs of trees or in crotches between the branches as many robins build now in the wilder, heavily wooded parts of the country, but when man felled the trees and replaced them by buildings, he supplied the bird with countless additional sites which afford an ample support, the chief requirement for a robin's nest. Concealment, it seems, is of minor importance to the robin, perhaps because it is difficult to hide so large a nest, perhaps because the bird is well able to defend it. In response to the change in conditions the robin has not only adopted many man-made structures as a site for its nest, but has also accepted man as a neighbor, breeding freely even in large cities in an environment completely changed from that of long ago. At the present time there are probably many times as many robins breeding in the United States as there were in Colonial days.

Frank L. Farley (MS.) makes an interesting comment on this subject in a letter to Mr. Bent: "During the last half century the robin has increased in Alberta at least 100 percent. This is in about the same ratio as the country has become settled. When the hard prairie lands were broken up, it was noted that earthworms were absent, but with the arrival of the settlers, it was not long before the worms began to appear, especially in the gardens surrounding the buildings. The birds increased in numbers at about the same rate as the growth of garden space. It is believed that the settlers inadvertently introduced the worms with the potted plants and shrubs which they brought with them."

There are many records in the literature of robins nesting in various situations which were not available years ago, such as on a rail fence, a fence post, a gate post, or a clothes-line post; Stanley Tess (1926) reports a nest "on the top of a gate-post which forms part of the gate itself. This is not a rarely used gate but, on the contrary, one in the public stockyards where it shuts off the runway leading to the loading platform." On buildings, nests have been placed on the ledge of a window, on blinds, on rain pipes or gutters under the eaves, on a rolled-up porch curtain, on a fire-escape, on beams inside or outside of buildings, piazzas, or porches, sometimes several old nests showing previous occupancy, and even on a lamp bracket in a dance hall. H. P. Severson (1921) tells of a nest that was placed on a trolley wire; "cars passed under this nest every few minutes, their trolley being only a few inches below it. On each occasion the Robin stood up, then settled back on the nest." A nest on a railroad signal gate was observed by Ward W. Adair (1920): "This gate is swung from one position to another perhaps fifty times in twenty-four hours. * * * At night when the red light was placed in position, the signalman's hands were always within a few inches of the bird." A nest may be placed on top of a bird house, or on any open shelf, but Gilbert H. Trafton (1907) tells of one that was actually *in* a bird-house. Wilbur F. Smith (1920) reports three nests inside a blacksmith's shop, respectively, on a wheel hub, on a smoke pipe, and "on some iron used to re-tire wheels, and within eight feet of the anvil before which the blacksmith worked most of the day." Access was provided for the bird by removal of a windowpane.

A. D. Du Bois refers in his notes to a nest in a cemetery, "about 5 feet from the ground, on top of a plain stone base, which supported the sculptured figure of a standing woman."

Other vagaries in nesting sites are: On a last year's hornet's nest, in a vacated nest of a catbird, on a last year's oriole's nest, on a shelf of rock in a cave, and in an old rotted-out woodpecker's hole in which a mud nest was built. Edward C. Raney (1939) tells of a robin sharing a nest with a mourning dove; "the birds shared the duties of incubation and * * * the eggs were hatched and the young were fed and brooded for eight days." The two species had shared a nest the previous year. Mr. Bent once found an occupied nest entirely inside an eel trap on Marthas Vineyard, Mass.; the trap was lying on open ground, and the eggs could be plainly seen through the netting (pl. 4).

Several cases have been reported in the literature where robins have built a series of nests, placed on a row along a beam.

Edward A. Preble (MS.) points out that the robin, when trees are not available, occasionally builds a nest on low cliffs. In Appendix G, by Seton and Preble, in Seton's "The Arctic Prairies" (1912, p. 405) is this record: "The bird was not common on Pike's Portage, between Great Slave Lake and Artillery Lake, but a deserted nest was seen near Toura Lake, near the summit of the divide, where nearly Barren Ground conditions prevail. There being no trees suitable for nesting, the bird had placed its home in a cranny on the face of a low cliff, where it was protected from the elements." A similar observation was later made near the camp at the "Last Woods" on the east side of Artillery Lake, early in the same year, 1907, when Mr. Preble saw a typical robin's nest, then deserted, on a low cliff, 5 or 6 feet from the ground and at least a mile from the nearest grove of spruces, where several deserted nests were observed in normal situations.

There are several records of robins building their nests on the ground, but the following is even more remarkable. Craig S. Thoms (1929) says: "The Robin had actually laid its clutch of eggs on the dry leaves beside a bush which was close to the house, as shown in the photograph. There was no sign of a nest, or even of an attempt to make one."

The nest is built chiefly by the female bird, although her mate aids by bringing in material. Berners B. Kelly (1913) says of a pair which he watched for hours: "On every journey, practically, the female brought larger loads than the male, and twenty-two more of them. The actual shaping of the nest was done entirely by the female, the male usually dropping his load haphazard on the edge of the structure."

Incubation, too, is performed mainly, if not wholly, by the female, the male meanwhile standing guard. Hervey Brackbill (MS.) states that he observed two pairs of robins marked with colored bands and that "every time that I could determine the sex of the incubating bird, it was the female. On one day of combined incubation and brooding all of 13 consecutive sittings were made by that bird." Ora W. Knight (1908), however, says that "the male also takes short turns at incubating, more often helping in this work towards the end of the incubation period." He remarks also: "I have known of a nest being completed and the first egg laid in six days from the time when it was commenced, while other nests have required even up to twenty days from time of beginning to completion, but the longer time required was due to a spell of prolonged rainy weather."

Mr. Preble (MS.) states: "On a morning early in June, about 1886 at my boyhood home in Wilmington, Mass., I happened to see the first few weed stalks deposited on the sloping branch of a medium-sized white oak in our grove, about 8 feet from the ground. At intervals through the day I observed the pair, busily engaged, and taking a look at the site just before dark I was surprised to find the nest virtually finished, the cup of mud fully formed but still wet. The next morning when I went out about breakfast time the earth cup was furnished with the usual lining of dry grass, and an egg had been laid. The clutch was completed promptly and the brood successfully raised."

The nest is kept scrupulously clean while the nestlings are in it, the parents seizing the fecal sacs as they are voided and frequently swallowing them. The male parent takes practically full charge of the fledglings, enabling his mate to prepare at once for another brood. In a nest I had under observation, four fertilized eggs were laid in a nest six days after the young of the first brood had left it.

Thomas D. Burleigh (1931), speaking of the robin in Pennsylvania, remarks: "Two and possibly three broods are reared each year," and he gives the normal height of the nest above the ground as "varying here from five to thirty feet."

Mr. Preble (MS.) submits notes on nesting robins received from W. A. Brown, of Aylesford, Nova Scotia, under date of February 16, 1948: "On my place last year, in an 8-inch-diameter maple, a pair of robins built three nests. The male had a pure-white feather in middle of upper tail coverts. The same year I had a robin's nest in which two broods were raised. A neighbor had a blue spruce in which, three years ago, a pair of robins raised three broods in one nest. Last year I found a robin's nest on the ground, and two years ago one on the ground."

Robins show persistency in their nesting habits, often returning to the same nest or situation year after year. The following quotations illustrate this habit. Dr. Charles W. Townsend (1909) says: "In the 'Birds of Essex County,' page 313, I recorded a Robin's nest that was built under the porch, on the lintel of the front door of my summer house, at Ipswich, Mass., and, at the time the book was published, had been occupied, presumably by the same pair, for four successive seasons. Since then it was used for two more summers, or six in all, but in the winter following the last, i. e., the winter of 1906–7, it was blown down, and the spot has not been built on since. I think, however, that the same pair have since built in a bush close to the front door. This nest over the door was repaired and built a little higher each year, so that in the summer of 1906, when it was last occupied, it had attained a height of eight inches, and was practically a six-storied nest." John H. Sage (1885) reports that "a Robin built her nest five consecutive years in a woodbine that was trained up and over a piazza. We knew her by a white mark on one side of her head."

Hugh M. Halliday, of Toronto, Canada, has sent us a photograph of a very tall nest (pl. 2), in which at least two broods a year had been raised during six successive seasons.

Eggs.—[AUTHOR'S NOTE: Four eggs comprise the usual set for the robin, but often only three are laid; five eggs in a set are rare, and I have taken one set of six, and sets of seven have been reported. The eggs vary greatly in size and shape; the usual shape is typical-ovate, but some are rounded-ovate, elliptical-ovate, or even elongate-ovate. Some are quite glossy after they have been sat upon, but usually they have only a slight luster. Robin's-egg blue seems to be commonly accepted as a standard color and well known; more specifically this means either "Nile blue" or "pale Nile blue," as the eggs appear in collections; some freshly laid eggs may be as dark as "beryl green." I have seen some pure-white eggs. Almost invariably they are unmarked, but I have seen one set that was sparingly marked with a few small spots and dots of very dark brown; and I have heard of a number of other spotted sets, some faintly dotted with pale brown.

The measurements of 50 eggs in the United States National Museum average 28.1 by 20.0 millimeters; the eggs showing the four extremes measure **31.6** by 20.3, 28.5 by **23.1**, **23.8** by 18.8, and 27.9 by **16.8** millimeters.]

Young.—Franklin L. Burns (1915), from the records of several observers, gives the incubation period of the robin as 11 to 14 days. William Edward Schantz (1939), who made an intensive study of three broods of robins, spending "from one to 16 hours each day in direct observation," found that "incubation began in all nests the evening following the deposit of the second egg and lasted for 12½ to 13 days."

Hervey Brackbill writes in his notes: "The incubation period for a marked egg was an hour or two less than 12 days." Of the nestlings he says: "When I lifted them out of the nest to band them, at the age of 7 or 8 days, they clutched the bottom of the nest so tenaciously with their feet that they pulled up a bit of the grass lining. Such a grip must be useful in preventing young birds from being tossed out of the nest during storms."

Schantz (1939) states that one of his broods left the nest 15 and 16 days and another 14 days after hatching. This is about the period of nest life that I noted in a brood in 1912 (Winsor M. Tyler, 1913): These young birds (a second brood) hatched on June 25, or possibly the day before. On the 25th their mouths were just visible above the rim of the nest. On July 1 they filled the nest level full, and tossed about restlessly, apparently preening their feathers. On July 4 they were feathering out fast; they reared up in the nest and flapped their wings, in danger it seemed of falling. On July 7 they were so large that in moving about they overflowed the nest, and one of them stood on a branch of the crotch and moved back and forth between it and the nest, using its wings to steady itself. On July 8 three of the birds, and perhaps the fourth, left the nest.

James Russell Lowell says in his Bigelow Papers that the robins settle down to nesting about the time when the leaves of the horse-chestnut tree begin to unfold. In a normal year we notice this phenomenon in eastern Massachusetts, where Lowell lived, toward the close of April, so, allowing two weeks for the incubation of the eggs, and two weeks more for their life as nestlings, the young birds are ready to fly in early June. At this time a day comes when all the robins in the neighborhood appear to be in the highest pitch of excitement; young birds are blundering about on the ground, and their parents seem distracted for their safety. We also hear a new note on this day, a queer, loud, exclamatory *seech-ook*, which leads us to where the young robins are squatting on the grass, waiting to be fed—plump, innocent-looking birds with spotted breasts and stumpy tails, staring up at the sky with little sign of fear, a choice morsel for the house cat.

They soon become wary, however, and before long are able to avoid attack by running swiftly away, or by flying out of reach. The male parents now take full charge of the broods, and as they scud over the grass plots in search of earthworms, the little birds follow them about expectantly, waiting for them to pull out the worms, shake them, and thrust them into their throats. The fledglings rapidly acquire the manner of adult birds. In a few days they throw off the crouching attitude of the nestling and assume the erect, proud bearing of adult birds, and in less than two weeks are able, but not always willing, to find food for themselves. The male parent is thus free to aid in the care of the next brood, which is almost ready to hatch.

Plumages.—[AUTHOR'S NOTE: Dr. Dwight (1900) says that the natal down of the robin is "mouse-gray." He gives a full account of the juvenal plumage, but I prefer the more concise description by Mr. Ridgway (1907) as follows: "Head as in adults, but the black duller and white orbital markings less sharply defined, sometimes buffy; back and scapulars grayish brown or olive, the feathers with central or mesial spots or streaks of white or pale buff and blackish tips; rump and upper tail-coverts brownish gray or grayish brown, the feathers sometimes narrowly tipped with blackish; wings and tail as in adults, but wing-coverts with terminal wedge-shaped spots or streaks of pale rusty, buff, or whitish; chin and throat white or pale buffy, margined laterally with a stripe of blackish or line of blackish streaks; underparts cinnamon-rufous, ochraceous-tawny, or buffy ochraceous (sometimes the chest and breast much paler, occasionally whitish), conspicuously spotted with black, the lower abdomen white or pale buffy." There is much individual variation in the amount of rufous on the underparts; some juvenals have the sides of the breast largely as bright rufous as in adults, and others have little or none of this color.

A postjuvenal molt, involving all the contour plumage, the wing coverts, and tertials, but not the rest of the wings or the tail, takes place from August to October, the date depending largely on the date of hatching. This produces a first winter plumage which is similar to the winter plumages of the adults of the respective sexes, but the colors are duller and more veiled, browner above, head not so dark, and the white spots on the tail feathers are smaller. The first nuptial plumage is produced by wear; much of the white edging on the breast is lost so that the breast becomes redder; the head becomes blacker and the chin clearer black and white.

Young and old birds become indistinguishable after the next postnuptial molt, which is complete, in August and September. The sexes are alike in the juvenal plumage, but after that the females are always somewhat duller in color, the upperparts lighter and browner, the head not so black, and the breast paler, often edged with whitish.

Dr. Harold B. Wood, who has made a thorough study of the white tail markings of eastern robins, tells me that there is great individual variation in the extent and shape of these markings, which are constant from year to year in individual birds. His studies were based on the examination of 162 robins trapped from 1938 to 1943, and the results will be published.

Albinism is common in the robin. I have seen many partial and some fully albino birds, both in life and in museums. While visiting with Hon. R. M. Barnes, at Lacon, Ill., I saw a beautiful perfect albino robin that had been living in his conservatory for some time.

Melanism, the excess of black pigment in the plumage, is much less common, but it occurs occasionally. Sometimes both phases of abnormal plumage may occur in the same individual, and either may be replaced by normal plumage at the next molt. For further information on albinism, melanism, and other items about robins, the reader is referred to a series of papers by Dr. Earl Brooks, published in the Indiana Audubon Society's Year Books for 1931 to 1935.

Hugh M. Halliday has sent me a beautiful series of photographs (pl. 5) of a pair of nesting robins, one pure white and the other in normal plumage; they have been mated together and nested for three successive years at 78 Broadway Avenue, Toronto; they have raised two broods of three and one brood of four young during the three years, all of which have developed normally colored plumage.]

Food.—Waldo L. McAtee (1926), in his study of the relation of birds to woodlots, makes a distinction between the food of the woodland robins and those which live in our dooryards. He writes the following comprehensive report of the robins' food:

Our knowledge of the feeding habits of the Robin is based mainly of course on studies of the bird as it ordinarily occurs, near to man and his works. We do not have particular information on the mode of life of the woodland Robins. We may, however, be assured on two points, namely that cultivated fruits do not play the part in the diet of these birds that they do in the case of our (in this respect, too familiar) neighbors, and that wild berries therefore are of much greater importance to this fruit-loving bird.

Like the true thrushes the Robin approves of a 60–40 dietary composition, but in a reverse sense, the larger item in its case being vegetable rather than animal food. There is no question about Robins sometimes taking too much cultivated fruit, thus necessitating reduction in their numbers. However, the woodland Robins with which we are here especially concerned have little or no part in these depredations, and their fruit-eating is a benefit rather than an injury because it results in the planting of numerous trees and shrubs. The favorite wild fruits of New York robins are those of red cedar, greenbrier, mulberry, pokeweed, juneberry, blackberry and raspberry, wild cherry, sumac, woodbine, wild grape, dogwood, and blueberry.

Beetles and caterpillars are the items of animal food taken in greatest quantity by the Robin, with bugs, hymenoptera, flies, and grasshoppers of considerably less importance. Spiders, earthworms, millipeds, sowbugs, and snails are additional sorts of animal food worth mentioning.

Various insects which are pests or near pests in woodlots have been identified from stomachs of Robins and we may be sure that a special study of Robins actually living in forests would greatly increase the list. * * *

In the economic court the Robin of the forest, and the Robin of the houseyard, must be adjudged separately, and regardless of the fact that it is differences in opportunities largely, that gives the former a much better character than the latter. The forest Robin has no chance at cultivated fruits and it has much greater opportunities to devour woodland insect pests. As we have seen, it improves these opportunities and should be credited accordingly. In the woodlot the Robin is certainly more beneficial than injurious.

F. E. L. Beal (1915a) in a report of an extensive study of the robins' food carefully weighs the benefit that the robin renders man by consuming harmful insects against the birds' depredations upon the fruit in his orchards. In his summary he says: "While the animal food of the robin includes a rather large percentage of useful beetles, it is not in the consumption of these or any other insect that this bird does harm. A bird whose diet contains so large a percentage of fruit, including so many varieties, may at any time become a pest when its natural food fails and cultivated varieties are accessible. While the robin to-day probably is doing much more good than harm, it must be acknowledged that the bird is potentially harmful."

Professor Beal (1915a) suggests a means by which we can divert the robin's attention from our fruit trees. "For a number of years," he says—

> the writer was engaged in the cultivation of small fruits in Massachusetts, and although robins were abundant about the farm they did no appreciable damage. On the farm where the writer lived when a boy was a fine collection of the choicest varieties of cherries. The fruit first to ripen each year was shared about equally by the birds and the family, but that which matured afterwards did not attract the birds, probably because in that section the woods and swamps abound with many species of wild fruits.
>
> Reports of depredations upon fruit by birds come principally from the prairie region of the West. This is just what might be expected, for but few prairie shrubs produce the wild berries that the birds prefer and for lack of these the birds naturally feed upon the cultivated varieties available. Reports of fruit losses caused by birds in the East are usually from the immediate vicinity of villages or towns where there is no natural fruit-bearing shrubbery. From this it follows that an effective remedy for the ravages of birds upon cultivated fruits is to plant the preferred wild varieties.

The following food-bearing trees, shrubs, and herbs appear on his list: Red cedar, common juniper, bayberry, hackberry, mulberry, pokeberry, sassafras, juneberry (*Amelanchier*), spiceberry (*Benzoin*), mountain-ash, chinaberry, hawthorn, burningbush (*Evonymus*), woodbine, flowering dogwood, and other cornels and viburnums. Professor Beal also gives a list of over 200 species of insects and 7 species of mollusks that have been found in the stomachs of robins.

W. J. Hamilton, Jr. (1935), during a study of four robins' nests, found that the food fed to the nestlings "during late May and early June consisted principally of cutworms." He says: "From the earliest period these larvae form a prominent share of the menu." Dr. Hamilton continues:

> In order to determine the quantity of food eaten by the young birds, the freshly fed cutworm, adult insects, worms, etc., were occasionally removed from the young with blunt forceps, immediately upon being fed by the parent birds, and immediately weighed. This procedure was inaugurated while the birds were but a day or two old, and continued on alternate days until the young left the

nest. By this method it was estimated the birds brought to the young approximately two grams of food at each visit, or a daily feeding of 200 grams of animal matter to the nestlings, be they three, four, or five.

The estimate is high for the early days in the nest and low for the days immediately preceding the time of leaving the nest. It is thought to be fairly accurate and, at least, gives some clue to the amount of food eaten. Robins feed their young, apparently regardless if there be three or five, approximately 3.2 pounds of food during the two weeks while in the nest. The observations were made several weeks before cherries ripened and, because of this, the food consisted almost entirely of animal matter.

In a more recent article Dr. Hamilton (1943) gives the following interesting analysis "of 200 Robin droppings collected between May 1 and June 12, 1942. The figures indicate the percentage of frequency of occurrence of the different food items.

"Plants, 81.5: barberry, 61.0; sumach, 29.0; coral berry, 4.5.

"Animals, 93.5: beetles, *** 82.5; millipedes, 38.5; ants, *** 27.0; cutworms, 9.5; sowbugs, 6.5; wireworms, 4.0; flies, 3.0; cockroaches, 1.5."

A. W. Perrior (1899) writes that the young birds are sometimes fed on hairy caterpillars, the "larvae of *Clisicocampa* (probably *C. americana*)"; Lotta A. Cleveland (1923) says that in 1922 the 17-year locusts on their emergence from the ground were used extensively as food for the young; John C. Phillips (1927) reports a remarkable instance of robins catching trout fry at the State Hatchery at Sutton, Mass.; A. C. Bent (MS.) speaks of the robins' fondness for crab apples; and Floyd Bralliar (1922) tells of the intoxicating effect of the berries of the "umbrella china" tree. "They fall to the ground," he says, "and lie on their side, occasionally feebly fluttering, apparently as happy as any drunkard in his cups."

One of the familiar features of summer to those of us who live in the Northern States within sight of a bit of greensward is the patrol of the robins over the grass in search of earthworms. Almost every little New England village has its common, a level bit of "green" near the town center, and these grass plots, from April, when the worms begin to stir, until the parching droughts of August dry up the grass, become the feeding grounds of all the robins in the neighborhood.

Sometimes half a dozen or more birds, widely scattered, may be seen running over the closely cropped grass, generally in amity, although sometimes one will fly at another and drive him off a little way. The birds take a short, straight run with a quick, tripping gait, then pause to look or listen for their prey. As they run, the back is nearly parallel to the ground, and the head is drawn back and settled between the shoulders, in the position of a decoy duck. When they stop to investigate the grass, they lean forward, turning the head to one side, bringing eye or ear to bear on a suspected spot, resembling the little semipalmated plovers as they feed on the wet sand of the seashore. The robin thrusts his bill deep among the grass blades, prods about the roots and, seizing a worm, leans backward, and bracing his feet against the pull, carefully draws the worm from the ground. Then, looping it up in his bill, he flies off to his nest or perhaps continues his search for another worm.

Robins are not always on the lookout for worms when they course over the grass. Often, early in spring, before the worms are within reach, and late in autumn, after they have retired deep under ground for the winter, robins frequent grassy fields. Here they are seeking smaller game which they see, apparently, above the ground. We may watch them snatching up, over and over again, little bits of food, tiny insects perhaps, which seem very numerous at these seasons among the grass and weeds of the open fields. Sometimes, when the grass is too long for the bird to run over it easily, he hops along with his head high and his primaries lowered, almost sweeping the grass, suggesting the Hylocichlae as they spring over the forest floor.

Tilford Moore writes from St. Paul, Minn., that the robins there seem to be fond of honeysuckle berries and feed them to their young. They "seem to prefer the red berries of the pink honeysuckle to the orange ones of the white honeysuckle. In fact, the yellow ones seem rarely to be touched until all the red berries are gone."

Behavior.—The robin impresses us as a bird of a nervous, highly excitable character, ever on the point of flaring up to an excess of emotion amounting almost to uncontrolled hysterics. For this reason it is a relief to see him in the role described above, quietly feeding on our lawn. The most frequent notes we hear the robin utter, perhaps, are fretful expressions of uneasiness, complaint, or resentment at our presence or at some other distraction, yet it is characteristic of him to break out with a phrase or two of song even in the midst of complaint. He seems always apprehensive, often standing alert and restless, wing tips lowered or twitching, head high, and tail pumping, on the watch for danger, and the least alarm upsets his equilibrium and startles him into vociferous, unrestrained remonstrance. Not an attractive nature, we think. How different the calm preoccupation of the little brown creeper!

Yet the robin has many good qualities: he is robust, confident, a straightforward personality, and no more nervous, perhaps, than many another American. Morning and evening he adds a charming hour to the summer day when he and all his neighbors join in a chorus of singing, in the twilight before the sun rises and after it sets.

It is easy to recognize the robin on the wing, even at a distance. He flies with a very straight back, like a runner with head thrown back, and his breast appears puffed out, expanded, giving a curved outline to the underparts in contrast to the long, straight line of the back and tail. The wings, at the end of a stroke, are not clapped close to the sides, as in the flight of a blackbird or woodpecker. The robin nevertheless accomplishes a full stroke by flipping the tips of the wings well backward so that, at the end of the stroke, the primary feathers of each side are nearly parallel, while the wrist remains out a little way from the body. The wings move rapidly and regularly and there is commonly no soaring or sailing.

A. Dawes DuBois (MS.) sends a note to Mr. Bent describing fearless behavior of the robin. He says: "The robins that nested on my rain pipe became almost entirely fearless. When there were well-grown young in the nest, the male, darting from a tree, struck me a sharp blow on the forehead when I looked out of my window, and one day, when I was at the window, the female flew into the room and grabbed me by the hair with her claws." He adds: "A nest built in a Virginia-creeper was only about 3 feet from a house wren's nesting box. Sometimes the robins drove the wrens away, but usually there seemed to be no friction between the two species."

A. C. Bent (MS.) speaks of the robins' sun bath. "Even on the hottest days," he says, "I often see a robin taking a sun bath on my lawn; he crouches on the grass with wings spread, or lies over on one side, with the wing on the sunny side uplifted, so that the sun penetrates under the fluffed-out feathers of the body. It may remain in this position for several minutes, sometimes for many minutes, as if it enjoyed the warmth of the sun, or derived some hygienic benefit from it. Again in a light, drizzling rain, I have seen them taking rain baths, standing erect for some time, with the bill pointing upward, so that the rain washed the plumage and drained off."

Speaking of territory, Aretas A. Saunders (1938) says: "Robins seem to have territories and to guard them, but they must be small, and probably a large part of the area, where food is found, such as groups of berry-bearing bushes, forms neutral territory. One gets the same impression of neutral territory in this bird, when noting several robins hunting earthworms on a lawn during the nesting season. There seem to be no earthworm hunting tracts here [Allegany State Park], for earthworms are scarce and hard to find. How small the territories are, is shown by finding nests rather close to each other on the school grounds."

This report is in accord with Hervey Brackbill's experience. He states (MS.): "The extreme points at which I saw one pair of color-banded robins that nested in a suburban neighborhood of detached houses indicated a territory extending about sixty yards north and south and sixty yards east and west. Other robins nested closely about on all sides. Both adults defended the territory. Of seven defences which I saw, the male made five and the female two. Strange robins, both adult and immature, were the object of attack five times, a blue jay once, and a gray squirrel once."

There are three records, W. A. Marshall (1921), F. G. McIntosh (1922), and Harry F. Binger (1932), each describing a robin's capture of a small snake, presumably as food for its young.

Robins not infrequently attack their own images reflected in a windowpane, sometimes returning to the attack for days. J. A. Allen (1879) reports a yellow warbler acting in the same manner, but most of the records of this habit refer to the robin, probably because it is the most conspicuous bird of a belligerent nature which breeds about our houses.

J. W. Lippincott (1912) speaks of robins feeding on the ocean beach. He says: "On August 20, 1912, a number of unusually large, dark-colored birds could be seen running along the beach [at Watch Hill, R. I.], which, upon closer inspection, proved to be Robins. They did not mingle with the little shore birds, but followed the retreating waves in much the same manner as these, and evidently ate the same food," and Dr. Charles W. Townsend (1905) says that they frequent "the dry parts of the beaches, the sand dunes, and the salt marshes."

May Thacher Cooke (1937) reports on the age of a bird. A robin, "banded at Philadelphia, Pa., on August 18, 1925, by Dr. William Pepper, was retrapped at the same place on September 25, 1929, and May 5, 1932," and Alexander Wilson (Wilson and Bonaparte, 1832) recounts the following story: "A lady, who resides near Tarrytown, on the banks of the Hudson, informed me that she raised and kept one of these birds for seventeen years; which sung as well, and looked as sprightly, at that age as ever; but was at last unfortunately destroyed by a cat."

Margaret Morse Nice (1933) speaks of a pair of robins "having been mated three years in succession. In 1932 the male arrived February 10; in 1933 on January 25th. * * * His mate never comes till March."

It was not until comparatively recently that the robins' habit of roosting during the breeding season was brought to the attention of ornithologists. The older writers, Wilson, Nuttall, and Audubon, say nothing of the habit.

In 1890 William Brewster published a comprehensive account of the robin roosts in the neighborhood of Cambridge, Mass., and showed that a large number of the breeding birds in the region gathered every evening at a roost and spent the night there during most of the breeding season. He had been aware of the habit for over 20 years, and he traces the history of several roosts during this period. He says (1890):

Our Massachusetts Robin roosts are invariably in low-lying woods which are usually swampy and are composed of such deciduous trees as maples, oaks, chestnuts, and birches, sometimes mixed with white pines. I have never known Robins actually to spend the night, however, in the latter, or indeed in any species of evergreen, except at Falmouth, Mass., where there has been a small gathering, these past two seasons, in a white cedar swamp. The trees in the roost may be tall and old with spreading tops, or crowded saplings only twenty to thirty feet in height, but it is essential that they furnish a dense canopy of foliage of sufficient extent to accommodate the birds which assemble there. As a rule, the woods are remote from buildings, and surrounded by open fields or meadows, but the latter may be hemmed in closely by houses, as is the case with a roost which at present exists in the very heart of Cambridge. A roost once established is resorted to nightly, not only during an entire season, but for many successive seasons. Nevertheless it is sometimes abandoned either with or without obvious cause, as the following account of the movements of the Cambridge Robins during the past twenty odd years will show.

We can form some idea of the multitude of birds that may compose these gatherings from the following quotation from Mr. Brewster's article:

I made no counts at the Maple Swamp roost, but as I remember it, it never contained more than about 2000 birds. Its successor at Little River was not only very much larger, but if my notes and memory can be trusted, was by far the largest gathering that has ever fallen under my observation. Thus I find that on the evening of Aug. 4, 1875, I estimated the Robins which came in on two sides only at 25,000. This estimate was not mere guess work but was based on a count of the birds which passed during an average minute, multiplied by the number of minutes occupied by the passage of the bulk of the flight. Such a method, of course, is far from exact, and it very probably gave exaggerated results, but a deduction of fifty per cent would surely eliminate all possible exaggeration. As the birds were coming in quite as numerously on the two sides opposite to those where my estimate was made, it follows that the total, after making the above deduction, was still 25,000, and this I feel sure was far below the actual number.

Of the dates when the roosts are resorted to, he says:

During the past season Mr. Faxon saw a few Robins going to the Beaver Brook roost as early as June 11, but I have never observed any well marked flights at Cambridge before the 20th of that month. The time probably depends somewhat on the date at which the first broods of young are strong enough to make the necessary effort, for the earlier gatherings are composed chiefly of young birds still in spotted plumage. Perhaps not all of those able to undertake the journey actually perform it at this period, for the movement, at its inception, is slight, and it gains momentum slowly. After July 1 it increases more rapidly, and by the middle of July becomes widespread and general, although it does not usually reach its height until the latter part of that month or early in August. By this time the old birds have brought out their second broods, and old and young of both sexes and all ages and conditions join the general throng. In fact it is nearly certain that during August practically *all* our Robins visit some roost nightly. * * *

After the middle of September the roosting flights diminish rapidly, and by the end of the first week in October the roosts are practically deserted. The latest date in my possession at which any Robins have been actually found in a roost is Oct 20, 1889, when Mr. Faxon noticed a few still lingering at Beaver Brook, but my notes record that on Nov. 6, 1888, I saw a succession of flocks flying, at sunset, into these Beaver Brook woods which, at the time, were "leafless"! About 200 Robins were seen on this occasion. They were in unusually large flocks, one, which passed me closely, containing fully 100 birds. If, as seems probable, they were migrants from further north it is interesting that they should have found their way to this roost; but perhaps enough local birds were with or near them to serve as guides. Mr. Faxon believes that our roosts receive some accessions from the north as early as September.

Continuing, Mr. Brewster adds: "Most of the roosts which I have visited are resorted to by other birds besides Robins." Among these he mentions bronzed grackles, cowbirds, red-winged blackbirds, kingbirds, Baltimore orioles, cedar waxwings, and brown thrashers.

Brewster (1906) also gives an interesting account of the behavior of the robins at a roost in his dooryard in the city of Cambridge. He says:

Late in June, 1902, they began assembling every evening—to my infinite surprise—in some ancient lilacs which form a dense and rather extensive thicket in the garden immediately behind our house. At first there were not more than twenty or thirty birds, but their numbers rapidly increased until by the close of summer we often counted as many as four or five hundred. * * * During the whole of May the roost was frequented nightly by fifty or more birds, all apparently old males. By the middle of June these were joined by the first broods of young, and a month or so later by the old females with their second broods. Thus the number of Robins steadily increased until early in August, when it probably reached its maximum and when we sometimes noted upwards of seven hundred birds in the course of a single evening. The frequent presence of members of my family on the back piazza (which is only a few yards from the lilacs) when the evening flight was coming in, gave the Robins some concern at first, but they soon became perfectly reconciled to it. * * *

As the piazza faces a little opening about which the lilacs are grouped on the remaining three sides, it commands an unobstructed view of the roost and affords rare facilities for watching the birds at close range. I have been interested to learn that a sound resembling the pattering of hail, which is heard when they are fluttering among the foliage and which I had formerly supposed to be caused by their wings striking the leaves, is really made, at least in part, by their bills.

When two or more of them are contesting for possession of the same perch they first threaten one another with wide-opened beaks and then bring their mandibles rapidly and forcibly together, thereby producing the sound above described. After they have quite ceased their calling and fluttering one may pass—even in bright moonlight—within a yard or two of branches where they are roosting by dozens without disturbing them. They invariably begin to leave the roost at daybreak, usually departing singly or in small parties, and scattering in every direction. When the exodus is performed in this manner, it often continues until sunrise. On several occasions, however, I have seen practically the entire body of birds leave simultaneously in the morning twilight, in one immense flock, with a prodigious whirring of wings. The evening flights vary similarly in character but to a less degree. Ordinarily the incoming birds are arriving more or less continuously for half an hour or more, but occasionally the majority of them will appear in the course of ten or twelve minutes, this usually happening when the weather is stormy.

Other references to accounts of the roosting of robins are: A. J. Stover (1912), Arthur R. Abel (1914), William Youngworth (1929), Mrs. J. Frederick Clarke (1930), Joseph C. Howell (1940), and Bradford Torrey (1892).

Tilford Moore writes to us that when some heavy bombing planes were flying over in formation, a robin in his backyard became very much excited, as it would if a cat were about, flitting from one perch to another, with much flicking of wings and tail and worried calls.

Voice.—The robin is at his best when he is singing. In the long choruses at morning and evening, and frequently for shorter periods during the day, he devotes himself to song, and as he stands motionless on a high perch, his head thrown back a little, whistling his happy phrases, his nerves relax, it seems, and a thrushlike calm comes over him: for the time, he seems at peace. *Cheerily, cheery* is a favorite rendering of his song, aptly suggesting by sound and meaning the joyous tenor of the phrases, and the liquid quality of the notes. The song lacks the artistry and poetic quality of the Hylocichlae, and the gentle charm of the bluebird's voice, but it is nevertheless an earnest, pleasing expression of happy contentment. It is generally a long-continued performance made up of paired phrases of two or three syllables each, often alternating up and down in pitch, given with perfect regularity at the rate of about two phrases per second. Close attention, however, will detect, after every few phrases, an almost imperceptible break in the beat, so that an uninterrupted run of a dozen phrases is rare. Frequently in the course of a long period of singing the bird pauses for a longer interval, perhaps for a second's duration, and then continues his song. Often, too, we hear a singing robin raise the pitch of his phrases higher and higher as the song goes on, apparently striving to attain a note beyond his range, until his voice breaks into hissing phrases without tone quality, the acme of his attempt. This peculiarity is characteristic also of the hermit thrush's song.

The robin's song is so characteristic, with its regular beat, its full round tone, and the robust quality of cheerfulness that pervades it, that we recognize it instantly. Yet as we listen to the robins in our dooryards singing day after day, we soon learn to distinguish some of the birds by slight differences in their songs; by a peculiar note recurring in a phrase, by the number of phrases which compose a group, or by a tempo slower or more rapid than the normal rate of the song. Also we notice sometimes that a bird will take a stand to sing his evening chorus on a branch, or perhaps the roof of our house, each night on the same perch, and if we are able to mark down this bird by some peculiarity in his song, we shall find that it is always the same bird that comes to the perch and that he often returns to it to sing during the day.

Aretas A. Saunders (MS.) sends this analysis of the robins' song to Mr. Bent: "The song of the robin is long-continued; made up of phrases with short pauses between them. These phrases are repeated, alternated, or otherwise arranged in groups of two to five, with longer pauses between the groups. Each phrase is composed of one to four notes, but most commonly two or three. The notes are frequently joined by liquid consonant sounds like r or l. I have records of portions of the songs of 49 different robins: in these the pitch varies from A'' to B''', one tone more than an octave. My records are fairly complete for 24 of these birds, and in these the average variation in pitch is about three tones, the least two tones, and the greatest five and a half. The time of the song is regularly rhythmical, the phrases and pauses being of even length. Ordinarily the robin sings at a rate of two phrases per second. In the very early morning they often sing faster and more continuously, the phrases not being broken up into groups. Then the rate is about two and a half phrases per second. Individual robins differ from each other in the phrases they use and the order in which they sing them. While many of the phrases are common to robins in general, nearly every individual will have some peculiar phrase. The average number of phrases used by one individual is about 10, but there is great variation: one bird I listened to for some time had apparently only 2; another had but 3, while a third unusual bird had 26. Two- and three-note phrases are the rule, but a single note used as a phrase is not uncommon. Only twice have I heard a phrase of four notes."

Hervey Brackbill (MS.) writes: "The robin frequently sings on the ground, sometimes for minutes at a stretch while standing at one place, sometimes intermittently between hops or runs in its foraging. I have also noticed a robin singing while on the wing; one sang a three-note phrase during a fifty-foot flight from one tree to another in the early morning."

The robin is apparently the first New England bird to awake in the morning. A few males begin to sing in darkness, at the earliest dim sign of approaching dawn; soon, as the light strengthens, more and more birds awake and join the singing until, gaining in volume, the song swells into a general chorus which lasts all through the morning twilight. I remember that William Brewster was much impressed by the element of drama in the great wave of robins' song which sweeps overhead every morning during the breeding season in the darkness before daylight, and continues on, westward, keeping pace with the sun, but beginning far in advance of its light, as it moves across the continent from the Atlantic to the Pacific.

As July advances, the morning chorus, which the robins have been performing since early in April as an almost formal observance in the hush before dawn, begins to fade out and wane. By the middle of the month, if we listen at our window as the sun approaches the horizon, and its light increases to the degree when robins are accustomed to awake and sing, there is silence—or at most a single robin singing alone, far away; we hear only the birds of night, the killdeer and the nighthawk. But after half an hour of waiting, as day comes nearer, when the gray of night no longer shuts in our vision, and we look out on a green world again, we may see a robin shoot swiftly past our window, then another, and then others, flying to the trees near the house. Soon we hear them singing, rather freely to be sure, but not in the organized chorus of early summer.

This delay in the morning singing is doubtless due to the fact that at this season the male robins do not spend the night near their nesting sites but at a roost to which they escort the young birds of the first brood. If we watch the fading sky at evening, we may see the robins of the neighborhood start off toward the roost, trailing along in loose order, after calling restlessly in the trees for a while, and perhaps singing a little. The evening chorus, too, is over for the season.

Horace W. Wright (1912) and Francis H. Allen (1913) have published the results of careful studies of "The Morning Awakening" to which the reader is referred.

Robins sing freely from early in April to the close of the nesting season late in July. In August and September they sing very infrequently, but later in the autumn and even in winter we hear sporadic songs from the wandering flocks of late migrants and wintering birds.

Albert R. Brand (1938) gives the approximate mean vibration frequency of the robin's song as 2,800, a little lower than that of the red-eyed vireo, 3,600, and of the scarlet tanager, 2,925, birds whose songs resemble somewhat the song of the robin. However, the highest recorded note of the redeye is much higher, 5,850, than the highest note of the robin, 3,300.

The robin has a variety of notes in addition to his familiar song. Some of these, although as well known perhaps as the song, are not easily suggested by syllables. Many observers have their own set of renderings in phrases and syllables, which represent to them the various utterances of the robin, but these renderings, even for the same note, differ from one another in marked degree. Also, a feature that adds to the difficulty in describing robins' notes is that they resemble one another sometimes rather closely, so that it is hard to draw the line between them, to decide whether we are dealing with two different notes or variants of one note.

The following list, it is hoped, will serve to differentiate 10 common notes of the robin. The syllables, of course, are merely approximations of what we hear, and the few words of comment aim to help out the shortcomings that must arise when we attempt to transcribe into letters the voice of a bird. 1. *Seech-ook;* an exclamatory note which the young robin utters soon after leaving the nest. 2. *Pleent, tut-tut-tut;* the first note, which might be written *plint*, and sometimes sounds more like *week*, is usually single, but may be repeated once or twice, and may be given without the *tut* notes. It is a sort of gasp, accented, higher in pitch than the succeeding, more rapid *tuts*. The latter (*huh* suggests the aspirated quality) may be likened to the interjection commonly written "humph," representing a low-spoken exclamation. 3. *Sss, tut-tut-tut;* a sibilant variation of the above, a tremulous, sibilant sound, a shaky squeal, followed by troubled sobbing. 4. *Skeet, skeet;* two or three high screams, uttered as if in haste. 5. *Seech, each-each-each;* a screaming variant of 2 and 3. It may be given *see-seech* with the second note accented and on a higher pitch. A common note, suggesting unrest. 6. *He-he-he-he-he;* a rapid, laughing giggle, suggesting sometimes a note of the red-winged blackbird, or in lighter, more musical form it may run quickly up and down the scale. This is the note which reminded Schuyler Mathews (1921) of the once popular song "Hiawatha." 7. *Chill-ill-ill-ill;* varying from 3 to 8 notes, given in a tinkling voice, the *chill* struck firmly, the *ills* successively losing force and dropping slightly in pitch to the final *ill*. The rhythm strongly suggests the ringing of the kind of bell formerly used on ambulances and police wagons. In tone of voice and in pitch this note resembles the song but differs from it in phrasing. 8. *Hisselly-hisselly;* sibilant, whispered phrases arranged as in song. It is associated with courtship apparently. The hiss may also be given in one long syllable, repeated slowly with downward inflection. 9. *Sssp;* a faint, trembling hiss, a refinement of the shriek (4) often given when a bird starts away in flight, and at the close of the day as it flies to its roost. 10. A low, sobbing note with a deep undertone; a note of trouble. A modification of the *tut* or *huh*, but clearly recognizable in quality and slow delivery as an entity. It is given when a cat is prowling near.

Tilford Moore tells in his notes of June 19, 1941, of a young robin's attempt at song: "He was in our lilac, not three feet from our dining room window, facing us, so we could see his speckled breast moving with his song. The song was a squeaky and quiet effort, much like the baby feeding cry in tone, but definitely a song after the adult morning song pattern."

Enemies.—The three following reports show that snakes are sometimes enemies of the robin: Ethel M. Spindler (1933) states that three young robins were taken from a nest 13 feet from the ground and swallowed by two blacksnakes; Laura Raymond Strickland (1934) saw a blacksnake eat a robin's egg; and Harold B. Wood (1937) writes of a robin strangled by a snake, *Liopeltis vernalis*. "The snake was wound so tightly around the bird's neck, by four complete turns, that it could not be shaken loose."

Ruthven Deane (1878) quotes from a letter written by the granddaughter of Audubon describing a "deadly combat" between a robin and a mole in which, apparently, they killed each other.

C. M. Arnold (1907), at a time when English sparrows were more abundant than they are at present, calls attention to their habit of following a robin about and snatching earthworms away from it.

John Lewis Childs (1913) notes the destruction of robins by "the most severe electric storm I have ever witnessed." It "annihilated the Robins that live in the trees about my lawn. Thirty-six were picked up the next morning on about an acre of ground, and others in the near vicinity brought the total up to about fifty. The English Sparrows were very abundant also but very few were killed; the Starlings escaped uninjured as far as I can learn. * * * The birds were evidently blown out of the trees where they were roosting and perished from the awful wetting they were subjected to on the ground."

Predatory hawks often capture robins. Walter Faxon, years ago, was standing in his garden watching a robin, near at hand, running over the grass. Suddenly, like a thunderbolt, a little sharp-shinned hawk struck the robin, pinning it to the ground and covering it all over with its open wings. Mr. Faxon frightened the hawk away, but the robin was dead, killed in an instant, its life snuffed out by a bird no larger than itself.

The domestic cat is the most destructive enemy of the birds that breed about our houses. It has been estimated that a cat will capture, on an average, 50 birds in a season, and the helpless young robins provide a large part of the kill.

Herbert Friedmann (1929) says of the robin in relation to the cowbird: "Probably an uncommon victim. It is hard to state definitely the extent to which this bird is affected by the Cowbird because the parasitic eggs are practically always thrown out. Half a dozen or more records from New York, Connecticut, Iowa, North Dakota, and Alberta have come to my notice."

Harold S. Peters (1933 and 1936) reports the presence in the plumage of the robin of 17 species of external parasites—lice 6, flies 4, ticks 2, and mites 5.

In former times a great number of robins were shot for food. Audubon (1841) says: "In all the Southern States, * * * their presence is productive of a sort of jubilee among the gunners, and the havoc made among them with bows and arrows, blowpipes, guns, and traps of different sorts, is wonderful. Every gunner brings them home by bagsful, and the markets are supplied with them at a very cheap rate. Several persons may at this season stand round the foot of a tree loaded with berries, and shoot the greater part of the day, so fast do the flocks of Robins succeed each other. They are then fat and juicy, and afford excellent eating."

Fall.—Of the behavior of robins during the late summer and autumn William Brewster (1906) says:

Soon after rearing their second broods of young—most of which are able to shift for themselves before the middle of August—our Robins change not only their haunts but their habits, also. Abandoning their diet of earthworms, and assembling in flocks, they now range widely over the country in search of berries of various kinds, on which they subsist almost wholly during the remainder of the year. It is true that they revisit our city gardens in early September when the rum cherries are ripe, and that even later in the year we occasionally see them running about in the old familiar way over our lawns and flower-beds, but throughout the autumn they spend most of their time in retired fields, pastures and woodlands, or in swampy thickets bordering brooks and meadows. Most if not all of our local-bred birds depart for the south before the close of October. In November their places are taken by migrants from further north, which sometimes appear suddenly in immense flocks and, after literally flooding the country for several successive days, pass on further to the southward. Robins are ordinarily scarcer in December than at any other season, and occasionally they are almost wholly absent during that month.

Francis Beach White (1937), speaking of the "stragglers in the woods" late in summer, says: "It is now that their habits undergo a complete change, for these birds are now like different beings, shy, furtive, wary, excitable. You may hear a rustling in the foliage, a soft 'whut-whut', and all vanish unseen, or you may come on one that assumes the motionless pose of a Hermit Thrush on a branch in a dim thicket."

There are days also in mid-September when a furor of excitement seems to possess the flocks of robins in the woods. They are restless and noisy, moving about high in the trees, and making long flights in companies of half a dozen or more: a businesslike air of migration pervades the gatherings.

On September 4, 1931, Wendell Taber (MS.) saw robins in actual migration. He was on the tableland on Mount Katahdin, Maine, at an elevation of 4,300 feet in a dense fog when 24 robins flew past him, near together, at close range in a southerly direction. He says (MS.): "Visibility was limited to a few yards, and I have no doubt that I saw only a small part of the flight. A deviation of a few miles to either side would have avoided passing over the high range."

Winter.—Most of the robins pass southward in fall to spend the winter in the milder climate of the Middle Atlantic and Gulf States, but occasionally flocks of considerable size remain in the Northern States and eastern Canada where they are exposed to very low temperatures. They have been reported as present during the winter in the Province of Quebec, Canada, by Napoleon A. Comeau (1891), in southern Maine by Nathan Clifford Brown (1911), in the Upper Mississippi Valley by Miss Althea R. Sherman (1912), and in Nova Scotia by Harrison F. Lewis (1919).

In the Southern States robins gather in almost incredible numbers. Mrs. Lotta T. Melcher (MS.) writes to Mr. Bent of watching robins flying to a winter roost in Florida. She estimated that no fewer than 50,000 birds assembled to spend the night "in low evergreen bushes, in a cypress swamp." She says: "I could think of nothing but being out in a snowstorm whose giant flakes never came to the ground."

Lester W. Smith (MS.) also writes of the invasions of robins during the winter. "When a cold snap descends into the Florida peninsula," he says, "with real truck-killing effect, there may come an invasion of robins. A multitude of robins appears suddenly on the lawns, and particularly in and under the cabbage palmetto trees, for it is on the abundant, wild-cherrylike fruit of this native palm that the robins feed, regardless of the protestations of the resident mockingbirds. When the robins arrive here in vast numbers, the cabbage palms of the entire district are soon stripped of their fruit."

Julian D. Corrington (1922), speaking of the bird in winter in Mississippi, says: "The Robin here is by no means a bird of the lawns and gardens as in the north in summer, but is as wild as the wildest and frequents only remote districts for feeding and roosting."

Otto Widmann (1895) gives this interesting account of a winter robin roost in Missouri, a contrast to the summer roosts of the north:

> The lower parts of the marsh, with the exception of the slough itself, are overgrown with reeds five feet high, bending over in all directions. These reeds are matted into a regular thicket which is not easily penetrated. In the fall the reeds are dry and yellow, some cinnamon and even dark chestnut brown.
>
> It is in these reeds that the Robin finds a safe retreat for the night, sheltered equally well from wind and cold, rain and snow, and comparatively safe from prowling enemies. During the day nothing betrays the roost. Not a Robin is seen in the neighborhood all forenoon and for several hours of the afternoon. An hour or two before sunset a few may arrive and stay in the trees along King's Lake, but nobody would suspect anything extraordinary until half an hour before sunset when the great influx begins.
>
> The new arrivals no more fly to the trees but alight on the ground, some in the wheat field, some in the meadows, some on the corn and hay stacks, but the majority flies directly into the reeds, while the others shift from place to place

until they, too, disappear. They do not come in troops like Blackbirds, but the whole air seems for a while to be filled with them, and standing in the marsh, one can easily see that they come from all points of the compass, all aiming toward a certain tract of reeds, a piece of about forty acres on some of the lowest ground where the last remains of water are now vanishing, leaving heaps of dead and dying fishes in the puddles (mostly dog, cat, and buffalo fishes).

When unmolested the Robins are not long in settling down and out of sight amongst the high and thickly matted reeds, and it is not nearly dark when the last has disappeared and nothing indicates the presence of so many thousand Robins but an occasional clatter, soon to give way to entire silence. If one enters their domain at night, they start with a scold, one by one, and not until one approaches very closely, to drop down again at no great distance.

Associating with them in the roost sleep a goodly number of Rusty Blackbirds, while the Bronzed Grackles keep somewhat apart. They arrive in troops with the last Robins and leave also a little later in the morning.

DISTRIBUTION

Range.—From extreme northern continental America to Guatemala.

Breeding range.—The robin breeds **north** to Alaska (Cape Prince of Wales, rarely, the Jade Mountains, Alatna, Fort Yukon, and the Porcupine River); northern Yukon (Old Crow River and Lapierre House); northern Mackenzie (east branch of the Mackenzie Delta, possibly the Arctic coast at Kittigazuit, Fort Anderson, Horton River, Coppermine River at latitude 67° 20′ N., and the Thelon River); northern Manitoba (Cochrane River, Churchill, and York Factory); northern Ontario (Fort Severn and Moose Factory); and northern Quebec (Great Whale River, Chimo, and Port Burwell, rarely). **East** to northern Quebec (Port Burwell); the coast of Labrador (Okkak, Nain, Hopedale, Rigolet, and Henley Harbor); Newfoundland (St. Anthony, Humber River, and St. John's); Nova Scotia (Sidney, Halifax, and Yarmouth, occasionally Sable Island); the Atlantic Coast States south to North Carolina (Raleigh, and has occurred in summer near Cape Fear). **South** to North Carolina (Raleigh and Charlotte); northern South Carolina (Rock Hill, Spartanburg, and Greenville; rarely Columbia); northern Georgia (Brasstown Bald); northern Alabama (Anniston and Birmingham; rarely Montgomery); northern Mississippi (Aberdeen); central and western Arkansas (Helena, Hot Springs, Arkadelphia, and Delight); eastern Texas (Tyler, Waco, Houston, and Somerset); western Tamaulipas (Galindo); western Veracruz (Las Vegas, Jalapa, Córdoba, and Orizaba); and Oaxaca (Totontepec and Mount Zempoaltepec). **West** to Oaxaca (Mount Zempoaltepec); Guerrero (Chilpancingo); Jalisco (Sierra de Nayarit); Nayarit (Santa Teresa); western Durango (Durango and El Salto); western Chihuahua (Pinos Altos); eastern Sonora (Alamos, Mina Abundancis, and Oposura); eastern and central Arizona (Huachuca Mountains, Tucson, Santa Catalina Mountains, and Prescott);

southern California (Redlands, Los Angeles, and Mount Pinos), the mountains and interior valleys and the Pacific coast from Monterey northward; the Coast Range and Willamette Valley to northwestern Oregon (Pinehurst, Fort Klamath, Corvallis, Portland, Tillamook, and Astoria); western Washington (Vancouver, Cape Disappointment, Clallam Bay, Lake Crescent, Seattle, and Blaine); British Columbia (Vancouver Island, Metlakatla, Inverness, and Queen Charlotte Islands); and Alaska (Sitka, Yakutat, Kenai Peninsula, Nushagak, Bethel, Yukon Delta, St. Michael, Nome, and Cape Prince of Wales).

Winter range.—The robin winters with considerable regularity in suitable localities **north** to southern British Columbia (Vancouver Island; Victoria and Comox, Port Moody, and the Okanagan Valley); Washington (Blaine and Spokane); southern Idaho (Meridian), northern Utah (Bear Lake and Utah Lake Valleys); southwestern and eastern Colorado (Durango, Beulah, Colorado Springs, Denver, and Boulder); southeastern Wyoming (Laramie and Wheatland); southeastern Nebraska (Red Cloud, Lincoln, rare, and Nebraska City); central Missouri (Kansas City, Marshall, and St. Charles); southern Illinois (Alton and Olney); southern Indiana (Terre Haute, Bloomington, and Richmond); central Ohio (Columbus); central West Virginia (Parkersburg and Charleston); and central Virginia (New Market, Variety Mills, and Bowers Hill). **East** to the Atlantic coast from Virginia (Bowers Hill) to southern Florida (Royal Palm Park). **South** to southern Florida (Royal Palm Park) and the Gulf coast from Florida to Texas (Galveston, Victoria, and Brownsville); Tamaulipas (Matamoros); southern Veracruz (Tuxla); and Guatemala (Coban and the Sierra Santa Elena). **West** to Guatemala (Sierra Santa Elena); Oaxaca (Coixtlahuaca); Jalisco (Zapotlan and Bolaños); Sonora (San José de Guaymas, Sonoyta, and El Doctor); northern Lower California (Rosario and Ensenada); and the Pacific coast of California (Los Angeles, Watsonville, and San Francisco); Oregon (Fort Klamath, Corvallis, Salem, Portland, and Astoria); Washington (Grays Harbor, Olympia, Seattle, and Everett); and southwestern British Columbia (Vancouver Island).

In addition, the robin sometimes occurs in winter north to northern Idaho (Coeur d'Alene); Montana (Kalispell and Billings); southeastern South Dakota (Yankton); southern Minnesota (Minneapolis and Red Wing); southern Wisconsin (La Crosse, Madison, and Milwaukee); southern Michigan (Ann Arbor and Detroit); southern Ontario (London, Toronto, and Ottawa); southern Quebec (Quebec, Kamouraska, Godbout, and Bonaventure Island); and Newfoundland (St. John's).

The distribution as outlined is for the entire species, which has been separated into several subspecies or geographic races. The typical race, the eastern robin (*T. m. migratorius*), breeds from western Alaska, northern Mackenzie, northern Manitoba, southern Quebec, New Brunswick, and Nova Scotia south to southern Alaska, central British Columbia, central Alberta, central Oklahoma, northern Arkansas, central Illinois, Ohio, Pennsylvania, and New Jersey, and south in the mountains to northern Georgia. The black-backed robin (*T. m. nigrideus*) breeds in northeastern Quebec, Labrador, and Newfoundland; the southern robin (*T. m. achrusterus*) breeds from southern Illinois and Maryland south to northern Mississippi, central Alabama, and northern South Carolina except in the higher mountains; the northwestern robin (*T. m. caurinus*) breeds from Glacier Bay, southeastern Alaska, in the humid, coastal belt south to northern and western Washington. The western robin (*T. m. propinquus*) breeds from southeastern British Columbia to the eastern Rocky Mountains to extreme western Texas, western Chihuahua, and eastern Sonora. Other races occur in Mexico.

Migration.—Some late dates of spring departure from the winter home are: Florida—Pensacola, April 13. Georgia—Macon, April 21. South Carolina—Aiken, April 6. Louisiana—New Orleans, April 21. Mississippi—Oakvale, April 5. Texas—San Antonio, April 10. Oklahoma—Kenton, April 13.

Some early dates of spring arrival are: District of Columbia—Washington, February 25. West Virginia—Charleston, February 26. Pennsylvania—State College, February 22. New York—Plattsburg, March 16. Massachusetts—Stockbridge, March 10. Vermont—Burlington, March 11. Maine—Machias, March 14. Nova Scotia—Halifax, March 26. New Brunswick—Chatham, March 25. Quebec—Kamouraska, March 20. Prince Edward Island—North Bedeque, March 31. Newfoundland—St. John's, April 6. Kentucky—Versailles, February 20. Illinois, Chicago, February 21. Ohio—Toledo, February 7. Michigan—Grand Rapids, March 3. Ontario—Toronto, March 8. Missouri—Kansas City, February 28. Iowa—Sioux City, February 28. Wisconsin—Madison, March 5. Minnesota—Red Wing, March 1. Manitoba—Winnipeg, March 13. Kansas—Onaga, March 5. Nebraska—Omaha, February 10. South Dakota—Yankton, March 1. North Dakota—Fargo, March 22. Saskatchewan—McLean, March 31. Mackenzie—East Branch Mackenzie River Delta, May 15. Colorado—Denver, February 16. Utah—Salt Lake City, March 14. Wyoming—Yellowstone Park, March 11. Idaho—Rathdrum, February 19. Montana—Great Falls, March 11. Alberta—Banff, March 25. Yukon—Dawson, May 9. Alaska—Kobuk River, May 20.

Some late dates of fall departure are: Alaska—Kobuk River, September 7. Mackenzie—Simpson, November 17. Alberta—

Belvedere, October 28. Montana—Missoula, November 21. Wyoming—Laramie, November 20. Colorado—Boulder, November 2. Saskatchewan—Eastend, October 24. Manitoba—Aweme, November 8. North Dakota—Charlson, November 2. South Dakota—Sioux Falls, November 26. Nebraska—Red Cloud, October 17. Kansas—Hays, November 6. Minnesota—St. Paul, November 15. Wisconsin—La Crosse, November 28. Iowa—National, November 14. Missouri—Independence, November 13. Ontario—Ottawa, November 21. Michigan—Detroit, October 31. Ohio—Oberlin, November 28. Prince Edward Island—North River, November 15. Quebec—Quebec, November 10. New Brunswick—St. John, November 15. Nova Scotia—Wolfville, November 9. Maine—Portland, November 8. Massachusetts—Boston, November 23. New York—Rochester, November 27. Pennsylvania—Pittsburgh, November 7. District of Columbia—Washington, November 12.

Some early dates of fall arrival are: South Carolina—Charleston, October 20. Georgia—Savannah, October 15. Alabama—Prattville, October 19. Arkansas—Winslow, October 1. Louisiana—New Orleans, September 12. Mississippi—Biloxi, October 13.

Records from banded robins show that the migration is by no means a strictly north-and-south movement but that individuals, at least, deviate considerably from that line. Records of 340 individuals banded on their breeding grounds and recovered the following winter give the following results: In **Virginia** the 3 recoveries include 1 from Nova Scotia, 1 from Massachusetts, and 1 from South Dakota. In **North Carolina** the 23 recoveries include 5 from Massachusetts, 3 from New York, 4 from New Jersey, 6 from Pennsylvania, 1 from the District of Columbia, 1 from Ontario, 1 from Ohio, 1 from Illinois, and 1 from Wisconsin. The 23 recoveries in **South Carolina** include 1 from Nova Scotia, 1 from Quebec, 2 from Massachusetts, 4 from New York, 2 from New Jersey, 5 from Pennsylvania, 3 from Ontario, 2 from Ohio, 1 from Michigan, 1 from Indiana, and 1 from Tennessee. The 38 birds recovered in **Georgia** include 1 from Nova Scotia, 2 from Quebec, 1 from Connecticut, 3 from NewYork, 3 from New Jersey, 6 from Pennsylvania, 1 from the District of Columbia, 3 from Ontario, 4 from Ohio, 3 from Michigan, 3 from Indiana, 4 from Illinois, 1 from Wisconsin, 1 from North Dakota, 1 from Kentucky, and 1 from Tennessee. **Florida** has 31 recoveries including 4 from Massachusetts, 1 from Rhode Island, 2 from New York, 1 from New Jersey, 9 from Pennsylvania, 2 from Ohio, 2 from Michigan, 1 from Indiana, 7 from Illinois, and 2 from Wisconsin. The 28 recoveries in **Alabama** include 1 from Massachusetts, 1 from New York, 1 from New Jersey, 3 from Pennsylvania, 3 from Ohio, 4 from Michigan, 3 from Indiana, 5 from Illinois, 4 from Wisconsin, 1 from

Minnesota, and 2 from North Dakota. **Mississippi** has 29 recoveries including 1 from New York, 2 from Pennsylvania, 3 from Ohio, 5 from Indiana, 2 from Michigan, 8 from Illinois, 1 from Wisconsin, 1 from North Dakota, 2 from Kentucky, 1 from Iowa, 2 from Saskatchewan, and 1 from British Columbia. The largest number of recoveries, 86, is from **Louisiana** and includes 2 from Massachusetts, 2 from New York, 1 from Ontario, 8 from Ohio, 8 from Indiana, 6 from Michigan, 24 from Illinois, 10 from Wisconsin, 6 from Minnesota, 3 from South Dakota, 7 from North Dakota, 2 from Manitoba, 1 from Saskatchewan, 3 from Iowa, 1 from Missouri, and 2 from Tennessee. The 25 recoveries in **Arkansas** include 1 from Ohio, 2 from Michigan, 2 from Indiana, 8 from Illinois, 3 from Wisconsin, 4 from Minnesota, 1 from South Dakota, 1 from North Dakota, 1 from Iowa, and 2 from Missouri. The 43 recoveries in **Texas** include 1 from Ohio, 2 from Indiana, 2 from Michigan, 9 from Illinois, 3 from Wisconsin, 1 from Minnesota, 3 from South Dakota, 6 from North Dakota, 4 from Manitoba, 1 from Saskatchewan, 10 from Iowa, and 1 from Missouri. The 5 recoveries in **Oklahoma** include 1 from Saskatchewan, 1 from South Dakota, 2 from Iowa, and 1 from Missouri. The 2 recoveries in **California** are 1 each from Alberta and British Columbia. Many other records that show longer elapsed time between the dates of banding and recovery serve to confirm the evidence of those cited.

Several records are sufficiently interesting to warrant detailed citation. A robin banded at Groton, Mass., on October 24, 1940, was found in Bladen County, N. C., on November 24, 1940; one banded at Germantown, Pa., on March 25, 1928, was recovered at Torquay, near Selbys Cove, Newfoundland; one banded at Summerville, S. C., on March 23, 1934, was killed on May 26, 1934, at Fond du Lac, Wis.; one banded at Nashville, Tenn., on March 25, 1940, was found dead about June 19, 1940, at Gowanda, N. Y.; one banded at Blue Island, Cook County, Ill., on October 8, 1938, was found dead April 14, 1939, at Salisbury, Somerset County, Pa.; a young bird banded September 3, 1935, at Aberdeen, S. Dak., was found dead May 24, 1936, at Plentywood, Mont.; one banded at Modesto, Calif., on February 26, 1939, was found April 25, 1939, at Vernonia, Oreg.; one banded at Pasadena, Calif., on February 23, 1933, was killed about June 22, 1934, at West Jordan, Utah; one banded in Yosemite National Park, Calif., on February 21, 1934, was killed by a hawk on May 25, 1934, at Sandpoint, Idaho; one banded at Crystal Bay, Lake Minnetonka, Minn., on July 7, 1924, was killed December 17, 1925, at Pachuca, Hidalgo, Mexico; one banded at Fargo, N. Dak., on September 27, 1937, was shot on January 31, 1940, at Villa Ocampa, Coahuila, Mexico; and one banded at Barkerville, British Columbia, on May

27, 1929, was found February 1, 1930, at Chunky, Miss. In the case of the last record the band was forwarded to the Biological Survey.

Casual records.—The robin has been recorded three times on St. Paul Island, Alaska, from 1872 to September 15, 1919. It has been collected twice at Point Barrow, Alaska, on May 14, 1930, and in the summer of 1931. One was reported at Herschel Island, Yukon, many years ago. One was reported at Warren Point, Mackenzie, 200 miles from Pearce Point, on June 19, 1917; and a specimen was collected near the mouth of the Kogaryuak River, Coronation Gulf, on June 19, 1911.

Four specimens have been recorded from Greenland: Specimens taken at Qôrnoq near Godthaab about 1865; at Sukkertoppen about 1881; at Graedefjord on September 26, 1899; and at Kangek, Godthaab Fjord, between October 7 and 21, 1944. The last specimen was identified as *nigrideus,* and it is quite probable that the others were also of this race. The robin is a straggler to Cuba, three records; and it has been collected in Bermuda in five different years. There are several European records, in Ireland, England, and Germany.

Egg dates.—Alaska: 8 records, May 26 to June 15.

California: 46 records, April 6 to July 14; 24 records, May 13 to June 17, indicating the height of the season.

Colorado: 21 records, May 10 to July 15; 11 records, May 24 to June 10.

Illinois: 29 records, April 18 to July 20; 15 records, April 29 to May 17.

Massachusetts: 50 records, April 28 to July 1; 25 records, May 17 to May 30.

TURDUS MIGRATORIUS ACHRUSTERUS (Batchelder)

SOUTHERN ROBIN

HABITS

Charles F. Batchelder (1900) in naming this subspecies gave as its characters: "Size considerably less than in *M. migratoria.* Colors in general much lighter and duller." Then follows a detailed description. Ridgway (1907) describes it more concisely as follows: "Adult male with black of head broken by more or less broad grayish margins to feathers; gray of back duller and browner, rarely, if ever, with blackish centers to feathers; color of breast, etc., tawny-ochraceous to tawny cinnamon-rufous. Adult female with grayish margins to feathers of head broader, sometimes nearly concealing the central dusky areas, and color of breast, etc., yellowish ochraceous-buff to tawny-ochraceous. Young paler in color than that of *P. m. migratorius,* with under parts largely (sometimes mostly) whitish and less heavy spotted."

At the time Batchelder described this form, the details of its distribution had not been worked out, but he was safe in stating that "probably all the robins breeding in the Carolinas and Georgia, outside of the mountain region of these States, will prove to belong to the new form, while those that pass the summer among the mountains, and in the low country of the adjacent region to the north, may be expected to be variously intermediate between it and true *migratoria*."

The 1931 Check-list gives it as breeding "from southern Illinois and Maryland to northern Mississippi, central Alabama, northern Georgia, and upper South Carolina." Recent investigations (Wetmore, 1937 and 1940) have shown that the southern robin breeds at the lower elevations at least as far north as West Virginia and Kentucky, where it begins to intergrade with the northern form.

Although generally considered to be a bird of the lower levels, it occurs on the tops of some of the higher southern mountains, notably Mount Mitchell, which rises to a height of 6,684 feet in western North Carolina. There, according to Thomas D. Burleigh (1941), it is—

a fairly plentiful breeding bird in the fir and spruce woods at the top of the mountain where, to one familiar with this species about the lawns in towns and cities, it seems at first rather out of place. Its arrival in the spring is influenced to a certain extent by the weather, and while it invariably appears by the latter part of March a relatively mild winter, as in 1933–34, has seen its return as early as March 8. It is rarely observed after the last brood of young are fully grown, the one exception being a flock of twenty birds noted October 28, 1932. It is possible that two broods are reared for a nest found June 3, 1930, held three well-incubated eggs, while on August 10, 1931, young barely able to fly were seen being fed by the two adult birds. There are no records for the occurrence of the northern race here, all specimens taken both in the spring and in the fall being clearly referable to *T. m. achrusterus*.

A. L. Pickens writes to me from Paducah, Ky.:

One of the most remarkable extensions of range that I have observed is that made in recent decades by the southern robin. Early in the present century the robin, in the South Carolina Piedmont, was regarded as a harbinger of cold weather. They descended from the mountains and the more northern areas to feed on chinaberries especially; and some were reputed to have become intoxicated from eating the fermented fruit, a condition which I personally never observed. As the smaller towns installed civic waterworks and water was available for lawns, and incidentally for earthworms, robins apparently began to spread, as inhabitants of cities and towns, until they may now be found in summer, even far down on the coast plain. In wet summers, when pastures are lawnlike, the birds may be found even out among the farms six and seven miles from town; but let a dry summer succeed and they yield the areas of farmland held the season before.

The reader is also referred to an extensive paper by Odum and Burleigh (1946) on this general subject, which is too long to be quoted here.

Nesting.—M. G. Vaiden, of Rosedale, Miss., writes to me that he finds the southern robin a very common nesting bird within the city limits. There are at least 50 or 75 pairs nesting in the town. They seem to be content to nest on limbs that reach out over the streets, and the passing cars, trucks, and other vehicles do not seem to disturb them. Their territory of defense seems to be the side of the tree in which their nest is constructed. His first nest was located in a cedar tree some 8 feet up and within 3 feet of the end of a limb. Another, near his house, was 25 feet from the ground and within 5 feet from the end of a limb that was about 15 feet long. This nest was on the top of the limb with only a small twig to support it on the south side and no support whatever on the north side.

He tells of another pair that started to build their nest on March 16 but did no more work on it after the next morning, apparently having given up the idea of nesting. On March 28 there was a heavy rain; and on the next day the robins worked hard carrying mud and completed the nest on April 1. There was no mud available in the vicinity until the rain came, and the birds had to wait for it. He says that the robins nest there in April, May, June, and the early part of July, but mainly in May and June. He is quite positive that two broods are raised there in a season.

Margaret Morse Nice (1930–31) says that in Norman, Okla., the southern robin is an "increasingly common summer resident. We have records of 48 nests. In 1921 there were eight nests with complete sets before the end of March, the earliest being March 23; but most seasons the first eggs are found during the first week in April. * * * In four cases there have been 4 eggs, in eleven cases 3. Seventeen elms have been chosen, two maples, two walnuts, one box elder and one apple tree. One nest was built at a height of 3 feet, two at 8, three at 10, one at 12, five at 15, three at 20 and three at 25, the average being 15 feet."

Eggs.—The eggs of the southern robin, usually three or four to the set, are practically indistinguishable from those of its northern relative. The measurements of 21 eggs average 27.9 by 20.0 millimeters; the eggs showing the four extremes measure **29.5** by 20.9, 29.2 by **21.3**, and **25.7** by **18.5** millimeters.

Young.—Mr. Vaiden tells me that the young in a nest near his house hatched on June 29, 1942; "on July 10, a frisky youngster dropped from the nest to the ground and the parents fed it for two days, along with the one in the nest, until it moved out of our yard. The second bird left the nest by coaxing from the parents on July 13, at 1:33 p. m. The date of hatching was accepted as the first day of actual feeding which we observed. Only two birds seemed to have reached maturity in this nest, as no others were observed being fed as we watched them through binoculars."

Food.—Mr. Vaiden says in his notes: "We have several chinaberry trees within the town limits, yet I do not find the robins that migrate through the Delta as fond of the berry of the chinaberry tree as I observed years ago in the hill section. I have seen the robin gorge himself with berries until he would fall to the ground; and we were told at that time that they were drunk from eating the berries of the chinaberry tree.

"I find great concentrations, such as 200 to 300 robins, feeding on the levee after it has been burned over late in the winter; probably the levee has a certain type of soft soil where they find plenty of worms to feed upon. Robins are not great seed-eaters, but I have found many weed and grass seeds in the stomachs of the birds dissected. I have found the seeds of Johnson grass, coco-grass, Bermuda grass, giant ragweed, and dwarf ragweed on many occasions."

Enemies.—Dr. Friedmann (1934) records two instances in which the southern robin had served as a host for the eastern cowbird.

TURDUS MIGRATORIUS CAURINUS (Grinnell)

NORTHWESTERN ROBIN

HABITS

In describing this subspecies, Dr. Grinnell (1909) says that the "full-plumaged male resembles *Planesticus migratorius migratorius* of corresponding plumage in the matter of size and darkness of coloration, the latter being excessive, but lacks the extended white patch on inner web of outer tail feathers; resembles *Planesticus migratorius propinquus* in the extremely narrow white tippings of the outer tail feathers, but coloration much darker and size smaller. In other words, this new form shares some characters of both, but presents in addition an extreme darkness of coloration seldom or never found in even *migratorius.* Young very much darker than in either *migratorius* or *propinquus.*"

Dr. Grinnell's specimens all came from southern Alaska, but the race is now known to extend its range southward in the humid coast region through British Columbia and Washington. According to some extensive notes on western robins received from Samuel F. Rathbun, the paler race, *propinquus*, would seem to be the common breeding form in western Washington, at least in the older and more settled regions of the interior, the darker race, *caurinus*, occurring there mainly in fall, winter, and spring. "But no line of demarcation can be drawn between these two forms of robins as to their distribution; they intermingle wherever found, although in some localities one or the other may predominate in numbers."

Based on his 20 years of observation, Mr. Rathbun sums up the status of the two forms (MS.) as follows: "I feel safe in saying that the robins so commonly seen from early spring until well into October, in and about the long reclaimed and older settled sections of the region, almost always represent *propinquus;* but associated with this form during the rest of the year will be seen numbers of what can be regarded as *caurinus*, for both are common residents of the region, although apparently each differs to some degree in its distribution.

"Ordinarily, the robins found in the wilder parts of western Washington, and in and about the tracts of heavy coniferous forest, particularly if such have more or less of a growth of spruce, can be regarded as *caurinus*. In particular, this appears to be the case within the Olympic Peninsula, where what seems to me to be *caurinus* is the prevailing form throughout the year; and, although I have found it quite well distributed here, it seemed as if the height of its abundance was in the spruce, or so designated 'coastal belt,' along the Pacific Ocean, where also the varied thrush is found to be so common. By 'coastal belt' is meant a rather wide strip extending inland from the Pacific coast, a section of heavy rainfall; the U. S. Weather Bureau records show that this strip has an annual precipitation of 75 inches or more; in fact, there are records of 150 inches at Clearwater, not far north of Lake Quinault.

"Then, at the approach of autumn, *caurinus* commences to scatter widely to the Sound region and adjacent sections. But, at its nesting period, the farther east from the coast, the less common is *caurinus*. One never sees it in summer in the backyard, for then the robin is *propinquus*."

Spring.—The northwestern robin does not seem to be permanently resident in southern Alaska, but to be, at least partially, migratory. The only specimens definitely recorded in Mr. Rathbun's Washington notes were taken in spring, March 19 and April 16. He says that it is "not uncommon in spring" and that it "is quite often seen in the forest, from the west end of Lake Crescent to the Pacific Ocean. No matter how dense the forest, or how far distant one may be from any clearing or habitation, at times robins will be seen, and as a general thing they resemble this race."

George Willett writes to me that this robin arrives in southeastern Alaska mostly in April and leaves in October; but he has seen it occasionally during the winter months; he has seen it at Craig on January 29 and March 16, 1923, on December 11, 1924, and on February 23 and 24, 1925, only a single bird in each case.

Alfred M. Bailey (1927) writes: "Robins are very common throughout the summer, and were first noted at Wrangell April 13, when half a dozen were seen feeding in a garden; they were abundant by April 26. Mrs. Bailey recorded her first Robins at Juneau April 14, and they were common a week later. * * * Mr. Gray tells me that Robins have wintered, occasionally, at Wrangell."

Nesting.—Mr. Willett (MS.) took a set of four slightly incubated eggs at Ketchikan on May 30, 1925; the nest was placed 7 feet up against the trunk of a young spruce tree on a hillside; the nest was made of grass and twigs and lined with fine grass; its external measurements were 150 by 85 millimeters, and the inner cavity measured 90 by 55 millimeters.

Mr. Rathbun (MS.) tells me that he has found "more than a few nests of *caurinus*" in the coastal strip in western Washington, as described above, and says: "From the first one I discovered, I noted that without an exception, the nest proper always rested on a platform or base of twigs, similar to the nest of Steller's jay or the varied thrush, and in this respect it differed from that of *propinquus.*"

Eggs.—The northwestern robin usually lays three or four eggs, which are similar in every way to those of the eastern bird. The measurements of 40 eggs average 29.5 by 21.1 millimeters; the eggs showing the four extremes measure **32.5** by 20.2, 29.9 by **22.5, 27.9** by 22.2, and 28.5 by **19.5** millimeters.

Food.—The food of the northwestern robin is evidently of the same general character as that of other robins. Where it lives in settled communities it may be seen grubbing for worms on the lawn, catching various kinds of noxious insects and their larvae, or taking what berries and fruit are available.

Wild fruits and berries of various kinds form most of the food in fall and winter. Dr. Bailey (1927) writes: "At Hooniah Sound, May 8–24, they were exceedingly plentiful, being the most common bird of the vicinity. They fed along the beaches exclusively, none being seen back in the woods, or on the muskegs; while droves worked the beaches like so many Sandpipers, in fact, we considered them as 'shore-birds' for the time being."

I. McT. Cowan (1942) lists the northwestern robin as one of the species that feeds on the flying termites (*Zootermopsis angusticollis*). "In extreme southwestern British Columbia the extensive areas of deforested land, strewn with decaying logs and stumps, provides ideal habitat for termites." The robins and other birds "have been observed catching *Z. angusticollis* close to or on the ground."

G. D. Sprot (1926), of Cobble Hill, Vancouver Island, tells the following story: "On the 13th of June 1926 a Robin (*Planesticus migratorius* subsp. ?) slipped from its perch on a seat in my garden into the nearby shrubbery, returning to the lawn with a dead field mouse which it proceeded to beat upon the ground, endeavoring also, so it appeared, to crush it in its bill. Every now and then it would pick it up, run a short distance and repeat the motions. For five of ten minutes it kept this up, then lifting up the battered remains in its bill it swallowed them head first. It then remained rigid for a few seconds with the tail of the mouse still protruding. Having apparently discovered that it had not overdone it, it gave a flick with its bill, when the tail disappeared down its throat." He thought that his terrier might have caused the death of the mouse, rather than the robin!

Theed Pearse (MS.) adds fallen apples and pears, honeysuckle berries, and ripe seeds of dogwood to the food of this robin on Vancouver Island.

Behavior.—Mr. Pearse has sent me the following account of a female robin that he saw "anting" on October 5, 1942, at Courtenay, Vancouver Island: It was "standing on top of a nest of red ants and kept picking up something, presumably an ant; it placed it on a primary, generally halfway up the feather, impressing the ant onto the feather as though trying to make it stick there. Occasionally the ant would be pushed into the feathers of the anal region. It was never seen to place the ant under the wing. The actions of the bird suggested that the ant was distasteful and the desire was to dispose of it as quickly as possible. Much of the time the bird held its wings quite loose from the body and, at times, was practically reclining on the surface of the ants' nest. Sometimes it appeared as if the bird were ruffling its feathers, as though bathing. It was watched for ten minutes until disturbed, and all this time it was 'anting.' The day was very dull after a rain. An examination of the ant hill showed no disturbance, except where pecked at. There were only a few ants working then; later it became a seething mass, as the day became warmer, and probably any bird would hesitate to venture there then."

Fall.—In his notes from western Washington Mr. Rathbun says: "In September, the first evidence of a tendency to gather together may be seen, and during October flocks will be noted. Among the later birds are individuals that may be regarded as *caurinus*, but there is no difficulty in distinguishing *propinquus*. These flocks roam about the country, evidently being first attracted to the localities that have food of the nature of the fruit borne by trees and shrubs, such as the mountain-ash, the dogwood, madrona, etc., these during the winter months being stripped clean. In sections lacking such food, robins will be missing to an extent; and, as the plants named vary in fruiting each year, this is reflected in the numbers of the birds seen."

Mr. Pearse tells me that, on Vancouver Island, there is an early southward migration from early in July until the middle of August. Late in December, there is, almost yearly, a flight of robins arriving weeks after all the resident robins and the earlier migrants have moved on; these are the birds that spend the winter. The regular migrants stay as long as berries are abundant. On January 23, 1942, there was a migration from the south; this was a very open winter. He heard robins singing as early as February 11, 1925, but does not usually hear them until about March 10. He has also heard them singing in October and November and as late as December 14.

TURDUS MIGRATORIUS PROPINQUUS Ridgway

WESTERN ROBIN

HABITS

According to the 1931 Check-list, the western robin "breeds mainly in the Canadian and Transition zones from southeastern British Columbia and Montana south to southern California, Jalisco, Oaxaca, and Vera Cruz, and from the Pacific coast east to the border of the Great Plains. Winters from southern British Columbia and Wyoming south to middle Lower California and to the highlands of Guatemala."

It probably intergrades with the eastern races somewhere near the western edge of the Great Plains, but all the robins that we collected in the Maple Creek region of southwestern Saskatchewan were referable to *propinquus*, though Professor Macoun referred the birds of that region to *migratorius*. North of the range outlined above, typical *migratorius* is the breeding form.

In Washington, the breeding ranges of *propinquus* and *caurinus* are more or less mixed; this distribution of the two races has been referred to under *caurinus*. It appears from Mr. Rathbun's (MS.) notes that, although *caurinus* breeds in the coniferous forests, especially on the Olympic Peninsula, the pale form of the robin, *propinquus*, is found everywhere throughout western Washington, from the Cascades to the Pacific, and seems to be the only breeding form in the older, more settled and more open regions.

The western robin is slightly paler both above and below than the eastern robin and decidedly paler than the northwestern robin, but the most conspicuous difference is that the white tips of the lateral tail feathers are entirely lacking, or reduced to a very narrow edge.

Before the prairies of the Middle West were settled and when the bison roamed in vast herds over the boundless grassy plains, the eastern robins bred in the northern woods of Michigan, Wisconsin, and Minnesota; but, as civilization moved westward and trees were planted about the ranches, the robins adapted themselves to the new and welcome conditions and made their summer homes near human dwellings in regions they had formerly passed over on migrations. Robins prefer to build their nests in trees or on suitable ledges to be found on human structures. Furthermore, they must have short grassy areas in which to forage. The treeless plains covered with long grass were not to their liking.

A similar extension of the breeding range of the western robin has taken place in California and other parts of the West and is apparently still continuing in the drier lowlands, where irrigation is reclaiming arid lands and where more lawns are being developed and planted with trees and shrubbery.

Under primitive conditions, while the lowlands were too dry to suit the robins, the summer haunts of the western robins were in the mountains, from 5,000 feet up to 12,000 feet, even to timberline; and in many of the wilder sections of the West such is still the case, especially in the mountain ranges of California and Arizona. Grinnell and Storer (1924) write: "Summer travellers in the Sierra Nevada recognize the Western Robin at once as characteristic of the mountains, inhabiting the small meadows which floor the openings in the coniferous forests; people who live in the foothills and valleys of California know the bird as a winter visitor to their orchards, fields, and gardens. Upon the establishment of towns within either its winter or summer range, the robin quickly becomes a dooryard bird, regardless of whether the dooryards are those of permanent houses or those of the ephemeral tent cities which, as in Yosemite Valley, grow and vanish with the passage of each summer." Similar primitive conditions were noted by Taylor and Shaw (1927) in Mount Rainier National Park, where "one is likely to find robins on open grass-covered areas, whether clearings in the thick timber, extensive alpine parks, or high ridges nearly at timberline. The robin's preference, however, seems to be for burns, where berry vines, decaying logs full of insects, and a wealth of other food-furnishing material are generously abundant."

Such were evidently the original breeding haunts of the western robin, but civilization has been encouraging changes and extension of range, which have been taking place even during the present century. Dr. Tracy I. Storer (1926) has published an extensive paper on this subject and states that up to 1915 "there were no known breeding records for Marin County, the San Francisco peninsula, the adjacent Bay region, or the Transition Zone of Monterey County. It seems very unlikely that the presence of the Robin as a nesting species could have escaped the attention of the numerous keen-eyed observers who have worked these areas during the preceding three decades."

He then goes on to cite a number of localities in California where the robin had extended its breeding range during the previous ten years, and explains some of the reasons for the changes.

> The general summer range of the Robin (as a species, as well as of the western subspecies, *propinquus*) everywhere includes territory where there is moist grassland (or its equivalent) in which this "soft-billed" bird can find soft-bodied insect larvae or earthworms as food for itself and young during nesting time. This seems to be a prime requirement of the Robin. The original "natural" range of the Western Robin in California included only those parts of the State where damp

meadows, with short grass in which the adults might seek their forage, persisted during the summer months. These areas vary from a few hundred square feet of grassland, as along the banks of small creeks, to large level tracts in the high Sierra Nevada, sometimes embracing several square miles of continuous grassland. The number of birds present in any given place usually seems to be proportional to the amount of such forage surface available.

Such favorable conditions had not previously prevailed in the regions to which the robin has recently extended its breeding range. "But with the development of lawns, with continued moisture supply and 'green feed,' various species of insects are able to persist there as larvae during the summer season. With irrigation, earthworms also are able to live up near the surface of the soil when normally they would be aestivating in deep burrows to avoid desiccation."

A. J. van Rossem (1942) was prompted by recent reports of robins nesting in the vicinity of Pasadena to state that a pair first nested on the grounds adjoining the residence of Donald R. Dickey in 1923; and he remarks significantly: "Generally speaking, it may be said that the transition of Pasadena from a small farming community to a residential city took place in the late 1890's and the early 1900's. It was thus about twenty-five years from the establishment of suitable territory until the robins first made use of it, although the species has always been common in summer in the Transition Zone in the immediately adjacent mountains."

Migration.—As many western robins spend the winter as far north as southern British Columbia, the migrations in the northern part of their range are not well marked as north and south migrations. At any time during late fall and early winter, large flocks of robins may be seen moving about from one locality to another in search of suitable feeding grounds, their presence or absence in any one place being dependent on the supply of berries or other food. Thus their migrations are mainly local wanderings, coupled with a downward movement from the mountains in the fall and a return to the higher levels when the snows disappear in spring. Mr. Rathbun (MS.) tells me that, in western Washington, "about the close of winter, or sometime during the month of February, single robins or perhaps pairs of the birds will be seen again around the residence districts of the cities and towns."

In California conditions are somewhat similar, with great variations from year to year in the winter population of robins, depending on the food supply. But there the migration, especially in spring, is well marked. The robins that winter in Mexico and Guatemala have a long way to go to reach their breeding grounds, and large numbers are often seen flying north.

On February 11, 1929, I saw large flocks migrating over Pasadena flying high and headed northward. On March 6 and 7 large numbers of robins gathered in the camphortrees in front of my house; the trees were fairly alive with them. Others were seen flying about in loose flocks and were probably migrants.

Nesting.—The nesting habits of the western robin are, in the main, very similar to those of its eastern relative. The nest may be placed anywhere from on the ground up to 75 feet in a tree, in bushes or in trees of many kinds, but most of the nests are not over 12 feet above the ground. Nests on structures erected by human beings seem to be less common than with the eastern bird. They are usually typical robin nests, made of the usual materials, including a liberal supply of mud, and firmly built. Mr. Rathbun (MS.) describes a poorly constructed nest that "was placed at a moderate height in the fork of a young alder, this crotch being filled with fresh leaves of the vanillaleaf (*Achlys triphylla*), with which were a few twigs. Next was a thin coating of mud, and its lining was an abundance of green grasses and a few dried ones. Very little skill was shown in the construction of this nest, my attention being attracted to it simply by seeing a mass of green leaves piled in the fork." Mrs. Wheelock (1904) says that the nests that she has "found have been somewhat different from those of the Eastern bird and very much prettier, being decorated with moss woven in the mud instead of straw, and carefully lined with moss. It is really a beautiful structure, with the mud practically concealed from view."

Dr. Walter P. Taylor (1912) says that, in northern Nevada, "nests were found on the ground and at various heights up to six feet above it, and were located in willow thickets, wild-rose bushes, sage-brush, quaking aspens, poplars (at Big Creek Ranch) and limber pines." Various other species of pines, spruces, and firs, as well as a variety of deciduous trees, have been occupied as nesting sites in other sections.

Joseph Mailliard (1930) had an opportunity to watch all the happenings at a robin's nest within 10 feet of his office window and published a full account of what took place from the building of the nest until the young left, to which the reader is referred. The nest was built in six days.

Eggs.—The western robin is said to lay three to six eggs; three seems to be a commoner number than four, and the larger numbers are very rare. The eggs are indistinguishable from those of the eastern robin. The measurements of 40 eggs in the United States National Museum average 29.2 by 20.7 millimeters; the eggs showing the four extremes measure **32.5** by 20.3, 30.9 by **22.4**, and **23.4** by **17.3** millimeters.

Young.—In the nest watched by Mr. Mailliard (1930) incubation was performed entirely by the female; this lasted for 14 days, during which time she left the nest only occasionally for 15 or 20 minutes at a time. The young were fed by both sexes, but mainly by the female, and left the nest in about two weeks. The young seemed to be troubled by some insect pest, so Mr. Mailliard tied a can to a long stick and sprinkled insect powder over the nest, after which the young seemed to be quieter.

Mrs. Wheelock (1904) asserts that the young are fed by regurgitation for the first four days and that by the fifth day "earthworms are given the nestlings after being broken into small mouthfuls, and, as the days go by, these worms as well as large insects are given whole."

James L. Ortega (1926) saw a robin apparently carrying water to its young on a very hot day, 100° in the shade. "It took a few swallows of water, then suddenly dipped its bill in the water and flew up into an acacia tree nearby. There its nest was situated, containing young robins. It didn't pause in its flight but flew straight to the nest, and I believe that it was carrying water to its young. It made repeated trips from the nest to the water pan, always flying rapidly and straight to the nest. However, on returning to the water it flew more slowly."

Grinnell and Storer (1924) relate the following incident:

> A robin was seen to fly away from its nest nearby carrying in its bill something which looked like a mouse dangling by the tail. The bird happened to drop the object within the camp precincts and it proved to be a juvenile robin (with feathers still in the sheaths). The old robin had obtained a large piece of liver from a pile of discarded mammal bodies and had carried this material to the youngster as food. When the young bird had swallowed as much of the liver as it could hold, a portion still protruding [sic] from its mouth. The parent, in haste to clean the nest, had picked up the free end of the piece of liver, not appreciating the fact that the youngster had swallowed the other end, and had carried both the liver and the young robin out of the nest.

Bailey and Niedrach (1936) report two instances of western robins and house finches using the same nest. "In May, 1934, we were informed that House Finches were feeding young robins in a nest on a front porch in east Denver, Colorado. On investigation we found four half-grown robins, two newly hatched finches and four finch eggs. There were two female finches apparently with the same mate, and the three finches and the two adult robins fed the young regularly. Unfortunately, however, the large robins smothered their small nest mates. We did not determine whether the four remaining eggs hatched. All three adult House Finches fed the young robins in the nest, and after the young had left the nest." In the other instance, the nest was on the back porch of Dr. Bailey's house, and here, too, the adult robins and the adult finches fed the young robins, though there was no evidence that the pair of finches had laid eggs in the nest.

Like the eastern robin, the western bird probably raises two broods in a season, and perhaps often three. J. Hooper Bowles (1927) has on two occasions known this robin to raise three broods; and once he saw three broods raised in the same nest, near Tacoma, Wash. This nest contained four well-grown young on May 20, three well-incubated eggs on June 12, and two new eggs on July 10. The three sets of young were apparently raised successfully, as the nest was frequently visited. The male always used the same singing perch near the nest; and the female, at first wild and noisy, became so tame that she had to be lifted off the nest.

Food.—The western robin eats the same *kinds* of food as the eastern robin, but naturally it includes many different *species* of insects, berries, and fruits. Professor Beal (1907), in his study of the food of California robins, had the stomachs of only 71 birds, collected in the winter months from November to April, inclusive. He found that, for the three winter months, the eastern robin eats 18 percent of animal food and 82 percent of vegetable; whereas the western bird eats 22 percent animal and 78 percent vegetable food during the same period, more insects being available on the Pacific coast than in the East at that season. Beetles, which amounted to 54 percent of the whole food in April, amounted to 13 percent for the six months. Caterpillars came next in importance, over 4 percent, and the remainder consisted of various insects and a few angleworms.

E. R. Kalmbach (1914) gives a better idea of the summer food of the robin, based on the examination of many stomachs collected in Utah during the months of April, May, June, and July. A large share of the food (14 percent) consisted of the destructive alfalfa weevil. Out of 45 April birds, 28 had eaten adults of this weevil and three others showed traces of it. "Caterpillars, many of which were cutworms, were taken with almost as great avidity as the weevil, occurring in 27 stomachs, but the larger size of these insects resulted in a much higher percentage, 23.24. One stomach contained at least 90 young caterpillars. Click beetles (Elateridae) and their larvae, wireworms, were found in 18 stomachs and amounted to 11.10 percent of the contents. One bird had eaten no less than 5 adults and 40 larvae of *Limonius occidentalis*. The other important elements of the animal food were earthworms (8.68 percent), flies (5.97), dung beetles (Scarabaeidae) (5.70), and ground beetles (3.97)."

During June, 17 stomachs contained 23.77 percent alfalfa weevils. One bird "destroyed 2 adults and 253 larvae, and the other 3 adults and about 241 larvae; the latter composed 80 percent of the food."

The examination of 18 robins collected in July showed a falling off in the number of these weevils eaten, but one stomach contained 2 adults and about 220 larvae. Caterpillars amounted to 37.72 percent, and earthworms made up nearly a fifth of the food.

I have often been asked whether a robin sees, or hears, or feels the worm, as we see it cocking its head to one side, then taking a few steps forward and extracting the unsuspecting worm from its burrow. Probably all three senses are used at different times, but I should think that eyesight might be the most important one. Claude T. Barnes tells me that he watched a robin feeding on a lawn where there was an incessant din of street cars and automobiles passing nearby, and it seemed as if the bird could not possibly hear the slight noise made by the worm.

Other items of animal food have been reported. Mrs. Bailey (1902) adds crickets and grasshoppers to the list of insects taken. Aretas A. Saunders (1916) saw a robin eating butterflies, which it swallowed wings and all; there were two species at a wet place, one yellow and black and the other cream color and black; he watched it for some time and "noticed that the yellow butterflies were the only ones eaten, although the others outnumbered them almost three to one." Charles W. Michael (1934) saw numbers of robins feeding on stranded fish at Mirror Lake, Yosemite, and says: "I saw the long isolated arm of the lake go dry, and I saw thousands of trout fry perish. * * * Scattered along the margin of the brown pool, feeding on the mud flats like a company of sandpipers, were at times as many as nineteen robins. Occasionally a spotted robin would plunge in belly-deep to capture a fish. The old birds were content to stand on the shore and to pluck their fish when they came into shallow water. The fish taken by the robins were about two inches long. These fish they would toss out on the beach, mangle with their bill, beat on the ground, and otherwise soften before attempting to swallow. One robin was seen to capture and to consume four fish."

Mr. Rathbun mentions in his notes that the western robin eats the coddling moth and its larva, locusts, spiders, and snails.

During fall and winter the greater part of the robin's food consists of wild fruits and berries. Dawson (1923) writes: "The madrona tree (*Arbutus menziesii*) often fruits in such abundance that hordes of Robins can thrive upon it throughout the winter. Christmas berries (*Heteromeles arbutifolia*) are another staple of winter fare, while haws, service berries, cascara berries, and all available representatives of the genera *Rhus*, *Prunus*, *Cornus*, *Pyrus*, *Celtis*, *Juniperus*, and a dozen others, furnish their quota." On March 7, 1929, the camphor-trees in front of my house in Pasadena were alive with robins feasting on the profusion of black berries.

Claude T. Barnes writes to me: "Last year I planted along my back fence that vigorous climber known as the wild mockcucumber, or wild balsamapple (*Echinocystis lobata*). Its dried leaves and egg-shaped, prickly fruit still drape the fence in midwinter. Today (February 9, 1939), in the midst of one of the worst blizzards Salt Lake City has had in three years, robins were feeding on the vine. Each would fly to a seed pod, hover over it, like a hummingbird, peck until the seeds began to run, and then flutter below to pick the fallen seeds from the snow. No bird that I observed hovered in midair for more than 4 seconds. One bird performed the same feat in pecking something from an English ivy growing on the house wall."

Mr. Mailliard (1930) found seeds of the English ivy in the nest he was watching and others have reported them as eaten by the robin. Other berries reported by others are blueberries, elderberries, coffee berries, mistletoe berries, manzanita berries and the berries of the peppertree, and chokecherries. Though the eastern robins eat a few seeds of the poison-ivy, and thus help to spread that noxious vine, Professor Beal (1915a) says that the seeds of the California poison-oak (*Rhus diversiloba*) were not found in the stomachs of west-coast robins, which is much to their credit; this is rather remarkable, as the poison-oak is one of the most abundant shrubs in California and as the robins feed freely on other species of *Rhus*.

Professor Beal (1907) says that from November onward the bulk of the vegetable food was cultivated fruit, "grapes in 5 stomachs, figs in 3, prunes in 2, pear, apple, and black berries in 1 each." These were, at that season, waste fruit that had not been gathered. He then goes on to say: "From the foregoing the robin would not appear to do much damage, or at least not more than is amply paid for by the insects it destroys. But, unfortunately, more is to be said about its food habits, which does not redound so much to its credit. In certain years when their customary food is scarce, robins appear in the valleys in immense numbers, and wherever there are olives they eat them so eagerly and persistently that the loss is often serious and occasionally disastrous. Sometimes, indeed, it is only by the most strenuous efforts, with considerable outlay of labor and money, that any part of the crop can be saved. Fortunately, such extensive damage is not done every year, although here and there the olive crop may suffer." In some cases it was necessary to employ men with shotguns and keep them constantly firing, in order to save more than 50 percent of the olives. Some of the birds shot had as many as six olives in the crop.

Howard L. Cogswell tells me that early in spring the robins "take toll from the many red decorative berry bushes (*Pyracantha*, *Cotoneaster*, *Eugenia*, etc.), which, however, remain practically untouched as long as the camphor berries last."

Behavior.—There does not seem to be anything in the behavior of the western robin that is peculiar to this subspecies. Mr. Rathbun tells me that he has, on numerous occasions, tested the speed of this robin in ordinary undisturbed flight, and found it to vary from 25 to 28 miles an hour; once a test showed 30.

Harold S. Gilbert (MS.) tells the following interesting story: "Nothing more startling in a bird way ever happened to me than during a drill by the Hawaiian police (June 17, 1936), when one of their number came out and did a whistling stunt. There was an audience of some 25,000 that witnessed the show in the Multnomah Stadium; and soon after the man began to whistle, about 11 P. M., a robin came down out of the darkness onto the field within a few feet of the whistler (the field was lighted by high-powered flood-lights), and sang as long as the whistling continued. Many of us thought it was a prearranged stunt, but as soon as the whistling was over, the robin flew away into the darkness."

Mr. Rathbun (MS.) relates the following: "While standing in the road our attention was drawn to the actions of a male robin that was springing about on the roadway. While we watched, the bird suddenly took wing, holding in its beak some object which proved to be a small snake. We could easily see the snake writhe about, and at times it appeared to have part of its length around the body of its captor. The robin, not meeting with success in its attempt to carry off the snake, dropped it on the road, then eyed it for a moment, meanwhile cocking its head first on one side and then on the other, as if puzzled by the actions of its prey. But the instant the snake attempted to crawl away, the robin again seized it with the same results as before. This action on the part of the bird took place four times; it then gave up its attempt to take the snake and flew away. We picked up the snake, which proved to be a western garter, about 8 inches in length. It was to all appearances uninjured, none the worse for its experience."

Robin roosts occur in the West as well as in the East and under similar circumstances. L. Ph. Bolander, Jr. (1932), gives an interesting account of a large winter roost in Lakeside Park, Oakland, Calif., in which he estimated that there were 165,000 birds. Space will not permit including much of his account here, but the following paragraph is too interesting to omit:

> Another interesting observation connected with worm pulling by the robins is the action of the gulls. I observed a Glaucous-winged Gull, three California Gulls and one Ring-billed Gull standing on the grass plot amid about eighty robins. Every time a robin would start pulling out a worm a gull would make a run toward him. Of course the robin would let go of the worm and then the gull would gobble it up! This was repeated again and again; but I could not determine whether the Ring-billed Gull followed this practice, as it left soon after I arrived on the scene. Sometimes the worm would come out quickly enough for the robin to get it down before the gull could get on the job. If the worm was too big for the robin to swallow immediately the gull would pursue it, but the robin usually dived under a protecting oak tree or madrone. The gull would not follow there.

Voice.—The song of the western robin is evidently no different from that of its eastern relative, but the song period seems to be of somewhat shorter duration. Mr. Rathbun writes to me: "By about the middle of July the robin no longer sings near the close of day, except on rare occasions. And this evening song began to shorten in the latter part of June. It commences to come into full song early in March, but snatches of song are given on sunny days in February." Mr. Saunders (MS.) says that his notes show that the period of song in Montana is shorter than in the East, "the birds beginning to sing early in April, or the last few days of March, and ceasing to sing late in July, rather than August."

Enemies.—J. K. Jensen (1925) saw young robins robbed of the their food and tells this story about it:

During the latter half of May, 1925 a pair of Robins built a nest in a locust tree in front of my house. Four eggs were laid and in due season four young appeared. The parent birds have since been busy feeding the young. A pair of English Sparrows discovered the Robin's nest and saw the process of feeding. Now for about two weeks the Sparrows have been watching the Robins closely, and whenever one of them flies down on the lawn in search of food for the young the Sparrows will follow it. As soon as the Robin captures a grasshopper or a worm and flies to the nest, the Sparrows will follow and alight on the rim opposite the Robin. As soon as the Robin has placed the food in the open bill of one of the youngsters, one of the Sparrows reaches over and pulls the food out and flies away to a quiet place to devour it.

Jays are among the worst enemies of robins, as well as of other birds, as they craftily and persistently rob the nests of eggs or young. Susan M. Kane (1924) gives the following account of a spirited battle in which the jay was the loser:

For several days a Steller's Jay had been pestering a Robin sitting on her nest in a bush against the corner of the house. The nest was in full view of a window at which I often worked. The Jay had employed every ruse to get the bird's eggs. He watched for her absence; slipped upon her to frighten her off; sounded alarms; engaged the male in skirmishes. These were but a few of his pernicious tactics to further his aims. I missed his final move but the Robin did not. There was a cry of distress from the Robin and when I looked up the Jay's toes were already in the air and contracting. The Robin had made a master thrust. Its beak had penetrated the Jay's head in a vulnerable spot, causing instant death.

But death for the villain did not satisfy the Robin. She shrilled for her mate again and again as repeatedly she pounced upon the fallen bird and pommeled him with beak and claw. The mate must have been gallivanting about the country for it seemed every other bird on the campus was at the scene before he arrived. When he did come it was in hot haste and with wild cries. He leaped into the fray. At times the birds fought by leaping into the air striking with beak and wing and pouncing with feet as a barn-yard cock fights. More often the attack was made from low branches of trees to which they flew and then struck with a flying dash. The battleground was sloping. Up and down the incline they kicked and tossed that brilliant brigand until his plumage was in sad array.

Fall and winter.—Some of the fall and winter movements and habits have been referred to above. Grinnell and Storer (1924) write: "After the young are grown, family parties are to be seen for a while. As soon as the young are capable of getting their living independently they gather into flocks. Meanwhile the adults go off by themselves and remain sequestered until completion of their annual molt. Then, in late September, the robins, without regard to sex or age, gather into mixed flocks and, for the most part, spend the winter in such gatherings. * * * Only a few venturesome robins continue in the mountains above the 3000-foot level during the Sierran winter."

W. E. D. Scott (1888) says of the winter range of the western robin in Arizona: "This form of the Robin I found to be a regular fall, winter, and early spring resident in the Catalinas, altitude 3500 to 6000 feet. They arrive here in the fall about November 1, and are soon quite common in small flocks or companies. All through the winter they are more or less common, but towards spring their numbers seem to be very considerably increased, and they are quite common until late in March, and are to be seen sparingly during the first week in April."

There is plenty of evidence that western robins congregate in enormous numbers in winter on favorable feeding grounds, as well as in winter roosts, such as that mentioned by Mr. Bolander (1932) at Oakland, which he estimated to contain 165,000 birds. John B. Price (1933) made some interesting observations on the winter feeding territories of western robins, of which he writes:

Two semi-albino robins were observed during the winter season at Stanford University, California. One was observed daily on the same lawn from January 19 to February 18 with the exception of three days. The other was observed on another lawn from February 12 to February 18. Each night they flew away (in all probability four or five miles) to roost and returned to the same small areas before sunrise the next day. This suggests that each individual robin in a flock may have its own individual territory during the winter season. [The bird that he called White-head] was always on the Jordan Hall lawn, and during the month of observation it was never once seen on the neighboring lawn in front of the Psychology Building although about fifty other robins regularly foraged there. Furthermore, it was always seen in the middle portion of the lawn, occasionally going into the bordering bushes. This feeding territory had an area of about 400 square yards and the bird was never observed to feed elsewhere. * * *

The White-headed robin did not have exclusive possession of its portion of the lawn. A few other robins fed there but they were never very close together. If another robin approached too closely, White-head would drive it a few yards farther on. On February 12 instead of the dozen or so there before, over fifty robins were seen on the Jordan Hall east lawn. The newcomers may have been previously feeding on berries in the nearby oval and moved to the lawn when the berries were exhausted. Many of the newcomers were in White-head's territory and it was very vigorous in combating them. During a three-minute interval in the late afternoon it was observed to combat ten times. Usually the opponent would retreat a short distance as soon as White-head rushed at it; sometimes both flew up in combat; but in every case White-head was successful. In a few days the number of robins on the Jordan Hall east lawn was once more only about a dozen.

His experience with the other robin, "White-tail," was similar. It was not observed until February 12 but may have been there before that; it had a smaller territory, about 300 square yards; it was driven out of another small lawn that was being defended by a normal robin and forced to return to its own territory.

Howard L. Cogswell, of Pasadena, Calif., writes to me: "Over much of the valley area robins flock with cedar waxwings, which seem to prefer much the same food; a dozen or so robins to a hundred waxwings is about the usual proportion. In some localities, though, there are regular robin roosts. One such in a eucalyptus grove at the base of the Santa Monica Mountains, near North Hollywood, was frequented by hundreds of birds each night during the winter of 1940–41, according to my friend Arthur Berry. On December 27, 1942, when I visited this spot at dusk, small flocks of robins came flying over at a height of about 250 feet. As they were directly over the trees, several groups half closed their wings and tumbled precipitously into the thickest of the topmost branches, immediately ceasing their call notes, which had been given by the whole flock flying over."

Dr. Helmuth O. Wagner tells me that the robins from the north arrive in the vicinity of Mexico City during the first part of October, mostly in flocks of 10 to 30 birds. They frequent the bushy forests of oaks, pines, and cypresses, preferring the open forests. The flocks are not compact, and, if they are frightened, one after another of the birds flies away. Sometimes, they are in berry-bearing trees, together with *Ptilogonys cinereus*, which are living in fixed flocks and are coming to the same trees for berries. In December, or later, if it is very dry, you will see more single birds, or flocks of three or four, in the cornfields near the borders of the forests, looking for insects or other food. At all times they are very shy, and if they see anyone they fly into the bushes on the borders, or into the high pines of the forests. So far as it is possible to identify them, the same flocks remain all winter within a fixed area. In the summer of 1935 he observed a flock of 10 birds, more or less, in a forest of pines and liquidambar at 1,700 meters in the mountains of Chiapas.

TURDUS MIGRATORIUS NIGRIDEUS Aldrich and Nutt

BLACK-BACKED ROBIN

HABITS

In naming this northeastern subspecies, Aldrich and Nutt (1939) give its subspecific characters as "nearest *Turdus migratorius migratorius*, but darker throughout. Upper parts: gray areas darker, more blackish and black areas more extensive; wings and tail more blackish, back much darker, more blackish mouse gray, in males gray more or less completely obscured by an extension posteriorly of the black of the head. Lower parts: More deeply colored, hazel rather than cinnamon rufous, with white areas less extensive and black areas more extensive; in male, black streaks of throat tend to coalesce laterally and posteriorly; gray areas of under tail-coverts and under surface of tail darker; black spots on breast of juvenile specimens larger, tending to coalesce anteriorly."

Of its geographical distribution, they say: "Breeds in Newfoundland. South in winter to eastern Canada and the eastern United States." Specimens have been taken from Nova Scotia, Wolfville, April 20; New York, Shelter Island, March 28; Ohio, Geauga County, March 22 and April 18. "The robin is apparently partially a permanent resident in Newfoundland since natives report them to be common about St. John's in the winter months."

In a later note, Dr. Aldrich (1945) writes: "In view of the recent extension of the known breeding range of *Turdus migratorius nigrideus* across the Straits of Belle Isle from Newfoundland to the coast of Labrador (Peters and Burleigh, Auk, 61: 472, 1944) it would seem to be of interest to put on record additional material that has recently come to my attention. In the United States National Museum there are two adult male breeding specimens from Chimo, northern Quebec. These birds, taken by L. M. Turner on May 27 and June 8, 1884, are almost typical *nigrideus* and extend the breeding range of the Black-backed Robin considerably to the northwest. This discovery makes less surprising the occurrence of migrants from as far west as Illinois and Michigan." In the same note he records identified specimens from as far south as New Jersey, Pennsylvania, Virginia, North Carolina, and South Carolina.

Aldrich and Nutt say that this robin is an abundant breeding bird on the Avalon Peninsula in eastern Newfoundland, where the type was taken, "but is exceedingly wary as compared with its Ohio relatives. The noisy and precipitous departure of robins while the observer is still as much as 100 yards away is characteristic of birds of that region."

On the west coast of Newfoundland, Ludlow Griscom (1926) found this robin "almost ubiquitous, but common only near dwellings and cleared land." In his unpublished manuscript on the birds of Labrador and Ungava, Lucien M. Turner reports it as common or abundant all along the east coast of Labrador, but not beyond Nakvak, and especially numerous about Fort Chimo, Ungava.

Dr. Oliver L. Austin, Jr. (1932), calls it "a common summer resident north to the tree line, occasionally straying farther north shortly after the breeding season." He says that it "is seldom seen north of Nain, where it breeds commonly about the settlement. It is usually to be found in the small-tree growth, but comes out into the barren coastal zone to build its nest in abandoned dwellings and under the cod-flakes."

Spring.—According to L. M. Turner (MS.), the robins arrive at Fort Chimo from the 9th to the 13th of May. "The first individual is always a male who sings suspiciously low, as if afraid he had come too early. In a day or two after the arrival of the first male will be seen a few females and as many males. I have reason to suspect that they have already paired before reaching this locality, as the labor of nidification begins immediately, and the first nest was obtained June 5, containing two eggs. * * * At the date of arrival the birds frequently find that several inches of snow have yet to fall and cover the ground for three or four days at a time, or that a cold spell comes and freezes the ground for several days and thus prevents the birds from procuring mud with which to stiffen their nests."

Nesting.—He says: "The nest of the robin is placed at various distances from the ground and even in the midst of an elevated mass of sphagnum rising round a clump of bushes. Many of the nests are remarkable for their great bulk and when just secured have a great weight from the thick mud walls."

Townsend and Allen (1907) found a nest at Rigolet on July 18, containing three eggs. "It was placed about seven feet up in a spruce, near the houses of the Hudson's Bay company's post, and was constructed of twigs, lichens, and mud, lined with finer material."

Dr. Austin (1932) says: "The Robins nest persistently under the Battle Harbor flake. I have found nests there on numerous occasions, but the birds are seldom successful with their broods, for the combination of small boys and husky dogs is deadly."

Mr. Griscom (1926) says that in the fishing villages along the Straits of Belle Isle, "it nested in the racks for drying fish, on top of fences, under the wharves, and other unlikely places. A few individuals nest in the stunted spruces on the Blomidon tableland."

Evidently the nests, and to some extent the nesting sites, are similar to those of our familiar eastern robin, with due allowance made for the difference in environment. The eggs of the two races are apparently practically indistinguishable. Mr. Turner (MS.) inferred that incubation begins before the set is complete.

He says further: "I found nothing in their habits to differ from their actions in other localities. Their food consists of insects during the breeding season; and in the earliest days of their arrival they subsist principally on the berries of *Empetrum nigrum* and *Vaccinium*, which were preserved by the frost during the winter. When the berries ripen in the fall, these birds apparently eat nothing else. During the early days of June and before it was possible for young birds to be hatched, I frequently observed male robins searching near my house for worms and other food. During these times I never saw a female, yet the male birds secured their beaks full of food and flew away with it, leading me to conclude that the food was destined for the females which were sitting at that particular time."

He observed this robin at Fort Chimo as late as October 17 during the fall of 1882.

TURDUS CONFINIS Baird

SAN LUCAS ROBIN

HABITS

The San Lucas robin is a beautiful pale edition of our familiar robins, clad in the softest, blended colors. The upperparts are plain "smoke gray" or "mouse gray"; there is no black or even blackish on the head, and the breast is creamy buff or creamy white, instead of the rich "cinnamon-rufous" so characteristic of our northern birds. We may miss the rich colors, but there is no mistaking it as a robin.

The type specimen was collected by Xantus at Todos Santos, in the Cape region of Lower California, during the summer of 1860. This specimen, still in the United States National Museum, remained unique for over 20 years. During the winter of 1882 and 1883, Lyman Belding (1884) explored the mountains of the Cape region and obtained two more specimens of this robin, which were deposited in the National Museum. He writes:

The most important localities visited were in the Victoria Mountains [now known as the Sierra de la Laguna], which were probably never previously explored by any collector. I ascended these mountains by three different trails on as many different spurs. The trail leading to Laguna is the longest, highest, and possibly the worst; however, I suppose either of them would be considered impassable in any other country than Mexico. On this trail an altitude of about 5,000 feet was reached. From an altitude of about 3,500 feet and upward the flora was partly that of the temperate zone.

This region is well watered and well timbered with medium-sized oaks and pines, the latter constituting about a tenth of the forest, being distributed unevenly among the oaks. Bunch grass was everywhere abundant. * * * Upon meeting the first pines, I discovered almost simultaneously the long sought Cape

Robin (*Merula confinis* Baird), the beautiful new Snowbird (*Junco bairdi*), and other interesting species. * * * Only about a dozen Cape Robins were seen, and these were all on the Laguna trail. About half were found singly, one as low as 2,500 feet above sea level.

Mr. Cipriano Fisher, an American, who had often hunted deer at Laguna, informed me that Robins were sometimes abundant there. This may be the case when the berries of the California Holly (*Heteromeles*), which grows abundantly in the neighborhood, are ripe. * * * The type specimen, shot by Xantus at Todos Santos in summer, may have been a straggler from the mountains.

During 1887 M. Abbott Frazar spent about nine months in Lower California, collecting for William Brewster, and sent him over 150 specimens of this hitherto rare robin. Mr. Brewster (1902) writes:

Mr. Frazar was the next to meet the St. Lucas Robin in its native haunts. He found it first on the Sierra de la Laguna, during his ascent of this mountain on April 26, 1887. It was common at this date, and by the end of May, exceedingly abundant, for its numbers continued to increase during nearly the whole of Mr. Frazar's stay, but up to the time of his departure (June 9), it was invariably seen in flocks, and none of the many specimens examined showed any indications that their breeding season was at hand. The people living on the mountain asserted that the birds do not lay before July. * * *

During his second visit to La Laguna, Mr. Frazar saw in all only ten St. Lucas Robins,—one on November 28, two on November 30, one on December 1, and six on December 2. This led him to conclude that most of them leave the mountains in winter, a supposition speedily confirmed, for about two weeks later (December 18–25) he found them abundant at San Jose del Rancho. At this place a few breed, also, for three were seen during July, and one of them, a female, shot on the 27th, was incubating, and must have had a nest and eggs somewhere in the immediate neighborhood.

"The St. Lucas Robin," Brewster continues, "is evidently one of the most characteristic species of the Cape Fauna, for it does not range even so far to the northward as La Paz, and, according to Mr. Bryant, is unknown to the people living in the central and northern portions of the Peninsula." This statement is correct, so far as I know today, but Brewster then goes on to cite the record of a specimen, supposed to have been taken by W. Otto Emerson at Haywards, Calif. This record had long stood unchallenged in the literature, until the curiosity of that critical student of California ornithology, Dr. Joseph Grinnell, was sufficiently aroused to prompt him to examine the specimen. After a critical examination and comparison with pertinent material, he and Mr. Emerson both agreed that it was, in all probability, merely an extremely pale individual of a female western robin (see Condor, vol. 10, pp. 238–239).

Nesting.—"Mr. Frazar found a number of old nests which were constructed precisely like those of the common Robin, and placed in similar situations" (Brewster, 1902). J. Stuart Rowley was in the Sierra de la Laguna during the last few days of May and the first part of June 1933, and says in his notes that "these pale-colored robins were reasonably abundant throughout the higher mountain area. At this time they were just commencing to nest and only one set of eggs was taken, consisting of three eggs."

Col. John E. Thayer (1911) seems to have published the first description of the nest and eggs of the San Lucas robin. His collector, Wilmot W. Brown, sent him two sets of three eggs each with the nests. They were taken in the Sierra de la Laguna on July 5, 1910. One was placed "in an Oak tree at the juncture of a limb with the trunk, about 40 feet from the ground." The other was placed "in an Oak, on a horizontal limb, about thirty feet from the ground. * * * The two nests are fine specimens. They are built of dried grass, weed stalks and lichens, neatly held together with mud. * * * Both these nests are much better built than any Robin's nest I have ever seen."

I have examined these two nests and can find no evidence of mud in their construction, except in the bases where there is some muddy moss and mud picked up with the decayed rubbish used as foundations; there is no mud visible in the sides or rims, as usually the case with northern robins' nests. The same is true of the two nests referred to below.

The dimensions of Thayer's nests are approximately as follows: Height, 4 inches; outside diameter, 5 inches; inner diameter, 3¼ inches; inside depth, 2¼ inches.

Another nest, in the Museum of Comparative Zoology, was taken by Mr. Brown in the same locality on June 28, 1913; it was placed near the end of an oak branch 40 feet above ground; this is a beautiful nest, mainly like the others in construction, but larger and more elaborate; the foundation is a great mass of coarse and fine lichens, coarse and woody weed stems, and the flower stalks of everlasting, which are carried up into the rim of the nest; it is neatly lined, as are the others, with very fine yellow grasses; it measures 6½ by 7 inches in outside diameter, and the inner cavity is 3½ inches in diameter.

There is a nest of the San Lucas robin, in my collection in Washington, that was taken by Mr. Brown on June 13, 1912, at 6,000 feet altitude in the Sierra de la Laguna; it was placed near the end of a branch of a mountain oak, about 20 feet from the ground; it is similar in construction to those described above.

Eggs.—All the nests referred to above contained three eggs, which seems to be the usual number for the San Lucas robin. These are much like the eggs of the eastern robin, varying in shape from ovate to elongate-ovate, with a tendency to be somewhat pointed; they are only slightly glossy.

The color does not vary much from "pale Nile blue," and there are no signs of markings. The measurements of 19 eggs average 30.3 by

20.5 millimeters; the eggs showing the four extremes measure **33.0** by 20.8, **28.3** by **22.1**, and 28.5 by **19.1** millimeters.

Plumages.—Strangely enough, neither Mr. Frazar nor Mr. Brown ever collected any young birds, though they took close to 200 adults. It is a pity that collectors neglect to take immature or molting birds, which are always scarce in collections and are exceedingly interesting to students of plumages and often indicate relationships. Fortunately, I have in my collection a young male San Lucas robin in juvenal plumage, taken by Chester C. Lamb in the Laguna Valley, at 6,000 feet, on July 29, 1929, 69 years after the discovery of the species. It is about half grown, but fully feathered on the body, with a very short tail. The upperparts are similar to those of the eastern robin, but paler; the pileum is "olive-brown," and the back is only slightly paler; the light spots on the back are larger and a paler buff than in the eastern bird; there is much more white on the underparts, where the extensive rufous of our familiar robin is replaced by a very limited suffusion of "pinkish buff" on the chest, sides, and flanks; the dusky spotting on the underparts is about the same as in our bird, but the white tips of the outer tail feathers are very narrow.

In all the large series I have examined I could not find *one* molting bird. Most of the specimens are in the pale, faded nuptial plumage, in which the underparts are pale "cream-buff" or paler, mixed with a large amount of dull white, sometimes nearly all white; I find birds in this plumage through August and up to September 2, mostly in worn condition.

There are no birds in the series that were taken late in September or in October. But from November 10 and through December we find birds in fresh autumnal plumage, in which the underparts are clear, rich "ochraceous-buff" or "chamois"; these are probably fall adults. The inference is that the postnuptial molt of adults, and probably the postjuvenal molt of young birds, are accomplished in September and October.

In the series are many fall birds and some spring birds that show more or less ashy clouding or obscure spotting on the chest, in some cases forming an almost solid pectoral band. Mr. Brewster (1902) was probably correct in suggesting, in his extensive remarks on their plumages, that this and the dark bill, which is not always correlated with the ashy clouding, are signs of immaturity. If this idea is the correct one, it means that young birds can usually be recognized by these characters all through their first year and do not assume the fully adult plumage until their second fall.

Voice.—Mr. Brewster (1902) quotes from Mr. Frazar's notes: "The song resembles that of the eastern robin, but is weaker and less distinct, reminding one of the efforts of a young bird just learning to sing. I did not hear a single loud, clear note."

Mr. Rowley says in his notes: "This species has the same pleasing habit of singing at dusk, as do the robins of the United States, and it is a toss-up between this species and the San Lucas western flycatcher as to which bird sings the latest into the evening darkness."

DISTRIBUTION

Range.—The Cape region of Lower California.

Breeding range.—The San Lucas robin is nonmigratory and is confined to the Cape region of Lower California, chiefly in the mountains but found also in the lowlands. The range extends **north** to Todos Santos and Las Lagunas on the coasts and possibly a little farther in the interior as one record reads "road to Triunfo."

Egg dates.—Lower California: 15 records, April 6 to August 6; 8 records, June 13 to June 29, indicating the height of the season.

TURDUS MERULA MERULA Linnaeus

EUROPEAN BLACKBIRD

CONTRIBUTED BY BERNARD WILLIAM TUCKER

HABITS

The claim of the European blackbird to a place on the American list rested for some years on a solitary specimen from Sydproven near the southern tip of Greenland, a short distance up the west coast (Helms and Schiøler, 1917). It was a young male in the first winter plumage, with faint gray-brown edgings to many of the feathers and as yet showing no yellow on the bill. Since the date of this occurrence the species has been met with on the northeast coast. Pedersen (1930) records that he was brought a blackbird by a Greenlander, who had shot it from a flight of snow buntings at Cape Tobin on April 8, 1928, and C. G. and E. G. Bird (1941) record that no less than six were seen by K. Knudsen on Bass Rock on November 22, 1922, following a northeasterly blizzard which lasted six days. He shot two and skinned them, but they were lost by shipwreck.

There is also extant a specimen of the blackbird taken many years ago in California, but there is reason to believe that this was an introduced example. Its somewhat curious history was recorded by Tracy I. Storer in *The Condor* (1923):

For a number of years there has reposed in the collection of the Museum of Vertebrate Zoology a dark plumaged thrush which was thought by some people to be merely a melanistic example of the Western Robin. In fact, the writer had so accepted the bird, and had used it on two or three occasions in demonstrating color abnormalities to classes in vertebrate zoology, contrasting it with an almost complete albino Robin of undoubted identity. But a recent critical study, made at the suggestion of Mr. H. S. Swarth, showed that the bird was not a Western Robin at all. On the presumption that the bird in question was an individual which had strayed out of its normal path of migration, the descriptions and illustrations of dark-colored thrushes in Central America and Eastern Asia contained in Seebohm's Monograph of the Turdidae were examined, but without revealing any species with which the specimen in hand might be linked. The bird was then submitted to Dr. Charles W. Richmond for comparison with the National Museum material and he identified it as a female English Blackbird, *Planesticus merula*, (Linnaeus).

The specimen in question was collected by F. O. Johnson at Oakland, California, on December 6, 1891. It came with the rest of the Johnson collection to the Museum of Vertebrate Zoology and is now number 10688 of the bird collection. In an article published soon after its capture (Zoe, III, 1892, pp. 115–116), Johnson described the bird, identifying it as a melanistic Robin (*Merula migratoria propinqua*). He also gave the circumstances of capture and they are worth recording in the present connection.

"* * * While pursuing a Townsend's Sparrow which had flown to the top of a tall growth of jasmine, I noticed on the opposite side of the bush a strange bird moping in the shade. It observed me just as I saw it, and hopped sluggishly to another branch putting a bough between us. * * * My first impression was that it might be a catbird which had strayed from his rightful home. I crept up * * * and easily approached within twenty feet. It made no note and did not pay the least attention to my maneuvers. When I killed it, I was still more puzzled, for it was totally different from anything I had ever seen. It appeared much like some European thrush."

Dr. Storer goes on to point out that a short time prior to the capture of this bird there had been some activity in the importation of European song birds, including blackbirds, on the Pacific coast. Such an importation and release of 16 pairs of "black thrushes (*Turdus merula*)" is recorded at Portland, Oreg., in May 1889. It seems not impossible that the bird in question was one of these, which had migrated southward, or perhaps more probably it was the result of some other, unrecorded, introduction.

In its habits and behavior, though not in its coloring, the blackbird is in many ways a close counterpart of what an English writer may perhaps be pardoned for calling the *American* robin, and the reader will probably find in the description that follows a number of passages that could be applied without much alteration to the latter species.

In England, the blackbird is one of the most familiar birds, well known owing to the distinctive appearance of the male, his fine song, and the partiality of the species for gardens. The male, with his striking glossy black plumage and orange-yellow bill, is one of the very few birds which probably the least ornithologically inclined country dweller knows by sight, though the brown female is less generally recognized and often confused with the song thrush by the unobservant. It is a typical garden and shrubbery bird, and to a bird lover an English lawn would seem hardly complete without its pair or more of blackbirds foraging for earthworms, for the species is one that spends much of its time, and gets much of its food, in the open. Unfortunately, however, it does not confine its attentions to worms, and fruit growers would perhaps prefer to be without it. Apart from gardens it is found plentifully in open woodland where there are some bushes and undergrowth (it has been described as perhaps the most typical British woodland bird), in thickets and about hedgerows, as well as in more open localities with some cover, such as rough hillsides, bushy commons, and the like. It is no doubt commonest in cultivated, but not too intensively cultivated, districts where there are scattered woods, copses, and bushy places, but it is found also in quite uncultivated country and, being an adaptable species, may be met with at times even on relatively small treeless islands, nesting on the ground in the shelter of rocks or in the meager cover of brambles or some stunted bush. In the Orkney Islands to the north of Scotland, a change of habitat on a more considerable scale has recently been shown to have taken place. Here it is not only found about woods and haunts of the more usual type, where these exist, but is "also characteristic of open fields and low-lying moorland away from all trees, bushes, gardens and rocks" (Lack, 1942). This has made it possible to colonize considerable areas in these northern isles which would otherwise have been closed to it. At the other extreme it has also adapted itself successfully to life in the more sophisticated surroundings of the parks and residential districts of the larger towns, both of Britain and of the continent of Europe. Here, it must be admitted, its garden haunts, except for being interspersed among houses, are not really so very different from its normal ones, but by no means all garden birds of the country can accommodate themselves to town life.

Being, as already remarked, a well-known and common European bird, the blackbird has become the subject of a not inconsiderable literature, yet there is still room for a more intensive study of its life history and behavior, by no means all points of which are fully understood. Reference may be made here to a valuable recent study (Lack, 1943) of the age attained by wild blackbirds, based on an analysis of the British banding returns. The author points out that strictly all that the returns show is the average age of the very small percentage of banded birds found dead by human beings. "But there is no particular reason to think that those adult birds found dead by human beings are on the average either older or younger than the adults in the population as a whole." Leaving out of account banded birds found dead in the first two months after leaving the nest, which are considered for good reasons not to provide a fair sample of the population, it is found that the expectation of life of a first-year bird on August 1 is about 1.6 years. At the next August 1, it is about 1.9 years, and, in contrast to the situation in a modern human population, it remains about the same in subsequent years up to the fifth, after which the records are too few to be of value. The average life of wild blackbirds is thus only about 8 percent of the potential age of about 20 years indicated by records in captivity. This result may be compared with the average span of 2½ years under favorable conditions established by Mrs. Nice for the song sparrow and serves to emphasize how few wild birds—or at any rate small passerines—survive to die of old age.

Territory.—The activities of the blackbird in spring present a number of features of particular interest, but in spite of some notable recent observations it is still not possible to give a completely rounded and satisfactory picture of the early stages of the breeding cycle of this common bird. The species is certainly territorial, though perhaps not equally strongly so under all conditions. Lack and Light (1941) in a valuable, though incomplete, study, interrupted by war service, found that in Devonshire territories varied from about 1½ to 2½ acres. If an attempt was made to drive one out of its territory, it behaved in the manner characteristic of strongly territorial species, flying in front of the observer till it reached the boundary, then refusing to go farther, and eventually flying back past the observer into the center of the territory.

In February and March, males regularly patrolled their territories, taking short flights with intervals for feeding and perching quietly, and usually the observer was not long in the territory before the male came by on his round, the female often being in attendance.

The owning male at once attacks any other male Blackbird trespassing in the territory, and does not desist till the trespasser leaves. Probably the male also drives out trespassing females, as some violent male-female chases were seen * * *. But females were not attacked nearly so often as trespassing males, and sometimes were apparently ignored.

The authors found that "females certainly take much less part than males in the defense of the spring territory," though chases, and even fights, between females do occur and there are certainly temperamental differences between individuals.

As with other territorial species, most encounters between males are settled by threat-display * * *. On seeing an intruding male, the owning male flies towards it and, if the intruder flies off pursues it out of the territory. But if the intruder stays put, the male does not usually attack at once, but perches some feet away and, with lowered and retracted head, approaches gradually and indirectly in a series of hops, runs or very short flights. This occurs both on the ground and in the trees * * *. By the time the attacker is within a few inches, the intruder usually departs. On three occasions when the intruder did not retreat, the attacking male repeatedly snapped its beak open and closed. The bright orange-yellow of the beak and the inside of the mouth, and also of the eyelid, are then in contrast with the black plumage, and can perhaps be regarded as threat-colours. Except for this, the attacking male does not usually posture. But on two occasions the wings were flicked open and closed and the body jerked. Nearly always the bird keeps silent, but occasionally give a sibilant "seep" note and on one occasion faint "chucking" accompanied the beak-opening. Comparatively rarely does the intruder wait to be attacked, but this was seen occasionally.

Threat-display is also common between two resident males along the common boundary of their territories, and is usually remarkably formalised and unexcited * * *.

Of course territorial encounters are occasionally more serious, and there are records of one male killing another * * *. But * * * serious fights are rare, and probably occur chiefly when one male is trying to dispossess another of its territory.

The statement in Niethammer (1937) that choice of territory and pairing up take place at the same time as the beginning of the male's song or soon afterward is certainly not correct in England. It is one of the peculiarities of the blackbird that it does not start to sing until the majority have been paired and in occupation of territories for some time. Just how long is at present a matter of uncertainty. Lack found pairs already formed in January and February and even late in December, and the late T. A. Coward, an excellent observer and author of by far the best of the smaller works on British birds (1920), stated that pairing begins in October and November. He probably had good grounds for this assertion, but unfortunately gives no details. The manner of staking out of territories and of pair formation still awaits elucidation. It will be understood that the observations quoted in the present section refer to resident birds, but in the northern parts of its range the species is migratory.

Courtship.—Although the blackbird is such a common European species and so regularly frequents gardens it is nevertheless a shy bird, and it is but seldom that any courtship display or the act of coition is observed. The most striking single feature of display, though not a constant one, is the erection of the feathers of the rump, which gives the bird a most peculiar appearance as it runs about in a somewhat crouching posture with the tail usually fanned and depressed, an expression of excitement which may also be observed in other types of display and is common to both sexes. Several writers (Kirkman, 1911, et al.) have described this odd erection of the rump feathers and the present writer has also observed it. For the rest, the recorded display actions show wide variation and conform to no well-defined pattern. This inconstancy of pattern is observable in the display actions of a number of passerine species, which seem to be much less stereotyped than in some types of birds. Perhaps such individually variable and erratic posturings under the stress of nervous excitement represent the kind of raw material out of which the more set displays have been evolved.

Geyr (1933) saw the male after coition assume a striking posture with the head stretched very steeply, but not quite vertically, upward and the tail nearly vertical. The wings were strongly drooped and fluttered with a quick rhythm. Garnier (1934) observed a female on a low wall surrounded by bushes, where she was joined by a male, which settled on the wall singing softly, circled several times round her with tail fanned and depressed, wings slightly dropped and 'ruffled back feathers', no doubt meaning the erection of the rump feathers already mentioned. Coition followed. König (1938) watched a male which was being followed and solicited by a female. The male had the head and body feathers ruffled and the bill wide open. Suddenly he turned round so that both birds were facing each other. The female's bill was now wide open, but not so wide as the male's. Both then approached closer and remained thus until after a brief interval coition took place. Throughout a soft twittering was maintained. This seems a usual accompaniment of coition and of any preliminary posturing. It is described also by Antonius (1937), another observer who has published a note on coition. In this case the female crouched with quivering wings in the usual solicitation posture of passerines, but had the bill directed almost vertically upward and wide open. The male too had the bill wide open and seemed to fondle the female's head feathers with it. Otherwise there seem to have been no special preliminaries. A. W. Boyd (1941) has recorded coition as early as February 8, after the male and female had followed each other in small circles with fanned tails sweeping the ground and wings half spread and depressed.

Others have described display actions which, at any rate on the occasions observed, did not lead to coition. J. M. Boraston (1905), in the third week of March, saw a cock blackbird creeping up a ditch bank in what he describes as a most unusual manner, with body low and the fully fanned tail trailed along the ground. The head was stretched forward on the stiffly extended neck, and from the slightly open bill came a continuous flow of small squeakings and pipings which first attracted the observer's attention. There proved to be two females present, both of which regarded the male intently, paying no attention to one another. "When the cock bird, facing the two females, had serenaded them in this fashion for about a minute, he again turned tail upon them, fanning it and trailing it as before, and as he wormed along they followed him silently, appearing fascinated by his wild skirling. He, stopping abruptly, and with his back still turned toward them, drew himself up, flung the spread tail askew on one side and jerked his head awry on the other, as if he were set on a crooked wire." Unfortunately the proceedings were disturbed, as so often happens in such cases, by one of the females noticing the observer and becoming alarmed. Again, Miss Averil Morley (1938, vol. 2) has seen a male with head sunk between the shoulders and the feathers of shoulders, neck, breast, etc., puffed out so that the yellow bill appeared to form a conspicuous center to a round black shield, and bowing by both sexes also occurs. Other variations have been observed, but the cases described serve to illustrate the general type of display behavior and to emphasize how far from stereotyped the posturings are in spite of the fact that certain features tend to recur.

Nesting.—Some account of the territorial behavior of the blackbird has been given under the section "territory." The nest is usually built in hedges or bushes or brambles, or in ivy on trees, walls, or buildings. It may also be situated low down in a tree or, less frequently, at a fair height, though a record of one 30 feet up in a tree is exceptional. Not rarely it is built on banks or on the ground in woods, in quarries or among rocks, and on islands with no trees and not much cover such situations may be normal. Occasional sites are the insides of barns and outbuildings, hollows in trees, and so on. The nest is normally built by the hen, though the cock frequently escorts her and sometimes assists by bringing material. Exceptionally he may even work some of the material into the nest. It is a solidly constructed cup of dry grasses, straws, rootlets, and moss solidified with mud and with a solid mud lining, but unlike that of its relative the song thrush the mud is concealed by an inner lining of fine dry grass and rootlets.

Eggs.—The ground-color of the eggs is a pale bluish green or greenish blue closely freckled all over with reddish brown. Coarser spots or markings of the same color may also be present and in some eggs the markings may be concentrated in a cap or zone at the broad end. Not rarely a blackish streak is present. Varieties with a clear blue ground color, sometimes without markings, occur and a rare erythristic variety is white with red-brown spots (Jourdain, 1938, vol. 2). Jourdain gives the measurements of 100 British eggs as: Average, 29.4 by 21.5 millimeters; maximum, 35 by 21.5 and 34 by 24; minimum, 24.2 by 19 millimeters. Niethammer's (1937) average for 56 German eggs agrees closely with the above, namely 29.1 by 21.7 millimeters. As two or three broods are normally reared, and exceptionally even four, eggs may be found over a considerable period. In England some nests with eggs are regularly to be found in March and even in February, but laying is not general until April and the egg season extends to July. Exceptionally eggs have been recorded in January and December. In central Europe, according to Niethammer, the normal egg season begins in April, and eggs even in the last half of March are considered exceptional. The normal clutch in England is four or five, but often there are only three, rarely six, while seven, eight, and even nine are recorded (Jourdain). Five or six is considered typical in Germany (Niethammer).

Young.—Incubation is performed by the female only. A good many observers have flushed male blackbirds off eggs, and this has led to the assertion that the male occasionally takes part in incubation. But Lt. Col. B. H. Ryves (1943) has pointed out that such an assertion in this and analogous cases is probably not justified. There is reason to believe that in such instances the male is doing no more than "brood" the eggs and is incapable of incubating them in the proper sense of bringing sufficient warmth to bear on them for development to proceed. Colonel Ryves found that eggs which had been brooded for 35 minutes by a male blackbird, whose mate was for some reason losing interest in them, were still almost cold when he left them.

According to Jourdain incubation usually begins on completion of the clutch. The incubation period ranges from 12 to 15 days, but is usually 13 to 14. C. and D. Nethersole-Thompson (1942), in a detailed survey of eggshell disposal by British birds, mention a female having been seen with a shell in her bill, and shells may often be found in the vicinity of nests. The young are assiduously fed by both parents and usually leave the nest in 13 to 14 days, though 12 and 15 days have also been recorded. Observers contributing to a recent investigation on nest-sanitation (Blair and Tucker, 1941) found that the feces of the nestlings are generally swallowed by the parents for about the first week after hatching and sometimes longer, while in the later stages they are usually carried away. The droppings are deposited by the young on the rim of the nest from about the eighth day. "Injury-feigning" by birds with young has been recorded, but is rare: It has also been recorded at least twice by a bird off eggs.

Plumages.—The plumages of the blackbird are fully described by H. F. Witherby (1938, vol. 2) in the "Handbook of British Birds." The nestling has fairly long, but rather scanty down of a pale buffish-gray color, distributed on the inner supraorbital, occipital, humeral, ulnar, and spinal tracts. The flanges of the bill are pale yellowish white externally, and the mouth inside is deep yellow without spots.

The juvenal plumage is not unlike that of the adult female, being umber-brown above, but not so dark as most females and more inclined to rufous, with rufous shaft streaks on the feathers, rather noticeable on the wing coverts. The underparts present a more definitely mottled or spotted effect than in the female, having a paler, rufous-buff ground color with dark brown markings. The young females tend to be less dark than the males, but there is much individual variation in both sexes. In their first winter and summer males are decidedly browner, less jet black than adults, and the yellow bill is not acquired until spring.

Food.—The Rev. F. C. R. Jourdain (1938) in the "Handbook of British Birds" summarizes the food of the blackbird as follows:

> Largely vegetable as well as animal. Very destructive to fruit, especially during drought (apples, pears, gooseberries, strawberries, raspberries, currants, cherries, figs, etc.). Also takes berries of many species (holly, rowan, rose, hawthorn, ivy, yew, bird-cherry, blackberry, guelder-rose, *Cotoneaster*, etc.), seeds of many plants and grain. Besides earthworms, insects, Coleoptera (*Aphodius, Otiorrhynchus, Sitona, Agriotes, Carabus*, etc.), Diptera (larvae of *Tipulidae*), Lepidoptera (*Noctuidae, Bombyx*, etc. and larvae), Hymenoptera (*Bombus*, ants, *Ichneumonidae*, gall-insects (*Neuroterus*), etc.), and Trichoptera. Spiders, millipedes, and small Mollusca (*Helix, Zonites, Cochlicopa*, and occasionally slugs), and small frog and stranded minnow recorded as brought to nest.

Tadpoles have also been recorded and in one case were even seen fed to the young.

Partly on account of the damage that, it must be admitted, it does in fruit gardens the food of the blackbird has been studied a good deal. Collinge (1927) in England carried out an extensive investigation involving the examination of the stomach contents of 285 adults. His results showed that 39 percent (by bulk) of the total food was of an animal nature and 61 percent vegetable. The former figure was made of 31 percent insects (subdivided, perhaps a little arbitrarily, into injurious 22 percent, beneficial 3.5 percent, and neutral 5.5 percent), 4 percent earthworms, 2.5 percent slugs and snails, and 1.5 percent miscellaneous animal matter; the latter of 25.5 percent cultivated fruits, 2.5 percent wheat, 2.5 percent roots, and 24.5 percent wild fruits and seeds, and 6 percent miscellaneous vegetable matter.

Collinge's conclusion was that the blackbird was too numerous in Britain and should be reduced. More recently (1941) he has published the results of a further study based on material obtained from two areas in successive years. Since the date of his earlier work the blackbird, though still plentiful, appears to have undergone a distinct decrease, partly as a result of some exceptionally severe weather in winter, and the analysis shows, at any rate for the areas in question, a quite considerable change as compared with the earlier results. The quantity of cultivated fruit and fruit pulp taken shows a drop from 25.5 percent to 15.2 percent and injurious insects a rise from 22 percent to 30.5 percent.

Behavior.—The blackbird is an alert and wary species rather than really shy, though away from habitations where it has had little opportunity of growing accustomed to human beings even the latter adjective might often be used with justice. It is also highly excitable. At the slightest alarm it dashes off with a shrill volley of alarm notes, startling all the birds in the vicinity, and no other species is so much addicted to noisy vocal assaults on owls. The gait on the ground consists of short runs or a succession of quick hops followed by a pause and then repeated. On alighting the tail is elevated with an easy and graceful motion, and the same movement may be seen in a lesser degree when it halts on the ground. Though much of its food is secured in the open it is seldom far from hedges or other shelter and readily retires into or underneath bushes and similar cover. It also regularly obtains food by turning over dead leaves in woods, hedge bottoms, etc., chiefly with the bill, but also at times with the feet. Generally speaking, it is not much given to long or high flights and when merely flying across a lawn or from one bit of cover to another close at hand the action often appears rather flitting and weak, but over longer distances this is less noticeable and the flight, as I have described it in the "Handbook of British Birds" (1938, vol. 2), is "direct or only slightly undulating, with occasional closure of wings often too rapid to be conspicuous, but at times more marked."

Though in its communal displays and at roosts it shows a definite social tendency, the blackbird is never really gregarious except when migrating.

Voice.—The blackbird is one of the most pleasing songsters among European birds, and American observers in the Old World have remarked a distinct family similarity to the song of the robin. It has a rich mellow quality and is delivered with an easy, unhurried fluency which is a part of its charm—the very spirit of a peaceful English garden. It is uttered in short warbling phrases in which separate notes cannot usually be distinguished, generally of about 2 to 4 seconds' duration, but occasionally 5 to 8 seconds or even more (Nicholson, 1936), with pauses of several seconds between each. Though the phrases are of the same general type, there is much variation in detail; there are no set forms, though the same phrases recur in each bird's repertoire. A weakness is a tendency for some of them to tail off into a feeble and rather ignominious ending of subdued, creaky, chuckling notes, which, however, are not audible for very far. The song post selected is usually a fairly elevated one, commonly a tree, not rarely the roof of a house or some other building, occasionally a mere bush or fence. It is not unusual for the bird to utter one of its short phrases as it flies a little way to a fresh perch, and it may even sing on the ground. Occasionally song may be heard at night. As mentioned in an earlier section, the song begins later in the season than in many resident birds. In the south of England and the midlands it is not regular until late in February, though occasional song may be heard earlier in the month, and it continues into July, declining and usually ceasing before the end of the month. The subsong has been described as "a low, continuous sweet but rather tentative stumbling performance, sometimes mellow * * * sometimes including the blackbird's alarm cry 'mocked' in a less spirited tone * * * and often breaking out into * * * repetitions of the same note." It may be used in display and also in an aggressive context in territorial disputes.

On human analogies the blackbird's song might seem to suggest a placid and easy-going temperament, but some of its other vocal performances would not at all bear out such an impression. When disturbed, but not seriously alarmed, the note is a low *tchook, tchook, tchook*. But it is easily startled and on remarkably little provocation it dashes off in a momentary panic with an excited, screaming clamor which is quite as characteristic as the song. This so-called "alarm-rattle"—though it is not really a rattle at all—has a well-defined pattern, though subject to much variation in detail. Very commonly it begins with the *tchook* notes, rises in a crescendo of shriller notes in quick succession, and dies away again in *tchooks*, so that a typical version might be rendered *tchook, tchook, tchook-a, tchwée, tchewée-chewéechewéechewéechewée, tchook, tchook*. But the *tchooks* may also be omitted. Occasionally a slightly quieter version may even be given on the ground. Another note, often uttered with tiresome persistence when the birds are going to roost, is a monotonous *chick-chick-chick* * * *, and it is also this note which is used as a scold against owls, cats, and other objects of its resentment. A thin *tsee* is a less frequently heard but not uncommon call used chiefly by the male—in fact I cannot recall having heard it from a female—and a high, rapid tinkling titter has been heard from the female during courtship. The fledged young have a curious, rather shrill *tsee-tsee-tsee-tseep*, which is retained for some time after they are full grown and is a familiar garden sound in early summer.

Field marks.—The European blackbird is a bird of the size and somewhat the same form and build as the robin of America. The male is quite unmistakable with his smart glossy black plumage and orange-yellow bill. The female and young are less distinctive. They are obviously birds of the thrush group, but considerably darker than any American thrush. The female is umber-brown, the underparts somewhat lighter, usually inclined to rufous, with more or less distinct dark mottlings, and a paler, more whitish, throat. The chief distinctions of the young birds from the adult female have been noted under "plumages."

Enemies.—Although owing to its rather skulking habits it is perhaps less frequently taken than some other species, the blackbird is nevertheless one of the commoner victims of the (European) sparrow hawk (*Accipiter nisus*), and it is recorded in the dietary of other hawks. The nestlings not uncommonly fall victims to jays, magpies, stoats, and other marauders, and in gardens or near habitations cats are a serious menace. Where fruit is grown man must also be reckoned among its enemies.

A list of parasites recorded from blackbirds is given by Niethammer (1937).

Fall and winter.—It must be confessed that although hundreds of ornithologists see blackbirds daily the life of the species during the fall has not yet been studied in the intensive fashion required before it will be possible to speak with the precision desirable on a number of points relating to this period. There is reason to believe that among resident blackbirds the old pairs maintain their association at least to some extent outside the breeding season, but how far this association is constant and regular is a matter of some uncertainty. There is evidence that females, or some of them, in fall and winter show a marked attachment to particular places and during this period of the year may show a stronger territorial sense than the males, driving off intruders of both sexes with considerable determination (Morely, 1937). Again, it has been stated, probably on good grounds, as has already been noticed, that the formation of new pairs takes place, or may do so, as early as October and November, but details are unfortunately lacking. Lack (1941) has noticed a definite pair as early as December 25, but their earlier history is not known. What is certain is that from early winter onward highly interesting, but in some ways puzzling, activities of a communal kind take place. As they continue into early spring they might also have been treated in the section devoted to that season, but since they begin in winter they will be dealt with here. These communal displays, involving much chasing and posturing between males at fixed assembly places on open ground, seem to be of erratic occurrence and very varying intensity. Attention was first drawn to them by Miss Averil Morley, already quoted, but the most spectacular type of performance, which is evidently rare, has been described by H. Lambert Lack (1941):

At first four, later six cock Blackbirds were congregated on a small area of the lawn: one, sometimes two, females were seen feeding some twenty or more yards away from the group. Only the males took part in the display. With wings drooping and slightly extended so that their tips were visible, with tail spread and depressed almost to the ground, head, neck and beak fully extended and the neck feathers fluffed out, one bird would rush rapidly at another and chase it, or run round and round it at a distance of apout 15 inches. Sometimes two birds would circle round each other or round a third bird, or all three would be running in circles. Or again two birds would run straight side by side and some twelve to fifteen inches apart for a distance of three or four yards, then switch round and run back again; often two birds in a similar fashion would chase a third, one on each side of it. On rare occasions these chasings ended in a brief aerial combat, two birds flying up at each other to a height of two or three feet in the air and apparently attacking witn beak, claws and wings. The fights lasted but a second or two and though apparently fierce no damage seemed to result. It was particularly noted that in the chasings the birds always *ran* with very rapid steps, they never hopped: that their beaks remained closed and that they uttered no sounds. Those birds which for the moment were not actively engaged in these performances stood motionless with wings and tail spread and depressed and with feathers fluffed out as above described, but the head, neck and beak stretched upwards and forwards at an angle of about 45 degrees giving them a curiously melevolent expression. This first-seen display was so spectacular that it excited the curiosity of two other people who took little or no interest in bird behavior.

The display was seen on February 10, beginning at about 7.15 A. M. and lasting till about 7.55, with one interruption of some 10 minutes. Regular watching was then instituted in the early morning, and the displays continued until April but were very irregular. Really active displays only took place on a few days. "On a few occasions two or more females appeared on the scene and might chase each other or be chased by single male birds." Some similar evening displays were also observed.

Variations of the formalized drill-like movements which Lack mentions have since been recorded by others. A. W. Boyd (1941) has described three males moving about in a triangular formation, maintaining their relative positions as the triangle turned first one way and then another, and J. Staton (1941) observed four moving about in single file 2 or 3 feet apart in the posture which Lack's birds assumed when not actively performing, with beak elevated at about 45°. It will be apparent that these displays are of no fixed form and that different groups may "improvise," as it were, their own particular version, but the common sort of performance seems to be that consisting of little but indiscriminate, often mechanical-looking, chasings without conspicuous posturing. The displays recorded by Miss Morley were of this type.

It will probably not have escaped the reader that some of the postures mentioned, such as the depressed and fanned tail and the head stretched upward, are much like those described under courtship, but the former at any rate is a generalized excitement posture, and in any case it is no unusual occurrence in bird life for the same posture to be used in more than one type of situation. It seems clear that these communal tourneys are not in any direct sense sexual, though they may be in some vague way related to the reawakening of sexual impulses and the associated (primarily territorial) aggressiveness. But in fact their significance remains problematical.

In fall and winter blackbirds also display a communal tendency in their roosting habits, and may gather in considerable numbers to roost in shrubberies, plantations, thickets, and old overgrown hedgerows, often in company with redwing thrushes and fieldfares. Excitement rises as roosting time approaches and the birds become very vocal, keeping up a persistent chorus of the tiresome *chick-chick-chick* note, which otherwise, oddly enough, is chiefly used as a scold at owls and cats.

In the colder parts of its range the blackbird is a migrant and the milder parts such as the British Isles receive considerable accessions in the fall, though some of the birds are only passing through on their way farther south. On the other hand, a certain number of British-bred birds, especially juveniles and birds from the north of Scotland and exposed places elsewhere, leave their breeding grounds and migrate to the south of England or to Ireland and some few even to the Continent. Banding returns indicate Germany, Sweden, Norway, Denmark, Holland, and Belgium as breeding quarters of some of the immigrants into Britain (Ticehurst, 1938, vol. 2). Birds from the northern regions also winter in central Europe. In Germany about one-third of the breeding population is stated to migrate, ranging in a more or less southwesterly direction to Belgium, France, and northern Italy.

In hard winters, although the blackbird population may suffer considerably, it is not too much to say that it invariably comes through better than its relative the song thrush. This is no doubt due, at least in part, to its propensity for foraging for food among dead leaves in places which are less affected by frost than open ground.

DISTRIBUTION

Range.—The blackbird is a Palearctic species ranging across Europe and Asia and extending to northwest Africa. The range of the present race extends north in the breeding season to about latitude 63° in Scandinavia, 61½° in Finland, the Leningrad, Kostroma, Kazan, and Ufa Governments in Russia, and south to north Portugal, north Spain, Italy, north Yugoslavia, Hungary, south Poland, and central Russia. The winter range extends southward to central Spain, the Balearic Islands, the Crimea, southeast Russia, etc. Allied races are found in the Azores, Madeira, and the Canary Islands, in north Africa east to Tunisia, in Spain, the Balearic Islands, Corsica, southeastern Europe, and Asia Minor with others in southern Asia.

Spring migration.—Continental migrants in the British Isles are on the move from late February to early April, and the departure lasts to the end of April and even the beginning of June in the northern isles (Ticehurst, 1938). A considerable passage of northern-breeding blackbirds takes place in Heligoland in March and April.

Fall migration.—The autumn immigration of blackbirds into the northern isles and down both coasts of Great Britain, as well as in Ireland, takes place from late September to the end of November

(Ticehurst). Immigration into Germany is described as taking place in October and November (Niethammer).

Casual records.—In addition to the Greenland record Jourdain mentions occurrences in Iceland, Jan Mayen, Spitsbergen, Bear Island, Faeroes, Archangel Government, Malta, and Egypt.

IXOREUS NAEVIUS NAEVIUS (Gmelin)

PACIFIC VARIED THRUSH

HABITS

I owe my introduction to this large and elegant thrush to my old friend Samuel F. Rathbun, who first showed it to me in the vicinity of Seattle and who has given me a wealth of information on it in his copious notes. While we were waiting for the good ship *Tahoma* to sail for the Aleutian Islands, in May 1911, he helped our party to locate for two weeks in the then small town of Kirkland across Lake Washington from Seattle. At that time the shores of the lake and the country around the little town were heavily wooded, much of it with a primeval forest of lofty firs, but more of it had been lumbered once and grown up again to dense second growth, with some clearings and little farms scattered through it. The principal forest growth consisted of firs of two or three species, with a considerable mixture of hemlock and cedar; and in some places there was a heavy forest growth of large alders and maples, with an undergrowth of flowering dogwood and wild currant. The favorite haunts of the varied thrushes were in their dark, shady retreats in the dense stands of firs that were often dripping with moisture, for it rained most of the time that we were there. Here we often heard the clear, rich, vibrating notes of the thrushes, uttered without inflection, but with a weird double-toned or arpeggio effect. Mist and rain did not appear to dampen their ardor; their voices seem to be at their best in such gloomy weather and to form a fitting accompaniment to the patter of raindrops on the dripping foliage.

Mr. Rathbun tells me that the varied thrush is a common summer resident in the mountains, and to a less extent in the lowlands, from the summit of the Cascades to the shores of the Pacific and Puget Sound, especially about the mountain lakes where the dense coniferous forest extends down to the water's edge. It is evident everywhere that this thrush loves shade, coolness, and dampness. W. Leon Dawson (1923) writes: "To have earned the right to speak appraisingly of the Varied Thrush as a bird of California, one must have lingered in some deep ravine of Humboldt County, where spruce trees and alders and crowding ferns contend for a footing, and where a dank mist drenches the whole with a fructifying moisture. * * * For the Varied Thrush loves rain as a fish loves water; while as for the eternal drizzle, it is his native element and vital air. Sunshine he bears in stoical silence or else escapes to the depths of the forest glade."

Similar haunts are to be found all along the heavily forested, humid coast belt throughout British Columbia and southern Alaska, as far north as Yakutat Bay, where the dense conferous forests flourish down to the shores, nourished by fogs and frequent rains, and where the Pacific varied thrush feels most at home.

Spring.—Although some varied thrushes may spend the winter as far north as extreme southern Alaska, others are found as far south as southern California, indicating a decided latitudinal migration. Through the central portion of the summer range this migration is not much in evidence, but Mr. Rathbun tells me that in western Washington there is a decided movement in spring toward the higher altitudes and again in fall from the highlands to the lower levels. Probably there is such an altitudinal migration in other parts of the range of the species.

George Willett writes to me that the Pacific varied thrush arrives in southeastern Alaska during the latter half of April and leaves early in October, but sometimes it is plentiful throughout the winter, as it was at Craig in the winter of 1919–20; it was common at Ketchikan after January 18, 1925.

Nesting.—We found no nest with eggs while we were at Kirkland, but on April 30, 1911, I found a new nest, apparently completed and ready for the eggs; it was located about 10 feet from the ground in a small fir on a knoll in the coniferous woods and near a small swampy run; it was built on some small branches against the trunk; it was a bulky nest, made mainly of soft mosses, reinforced with fir twigs and lined with fine grasses. We found several old nests in the vicinity that were similarly located.

In the same general locality, on April 26, 1914, my assistant, F. Seymour Hersey, found two nests similarly located; one contained three young birds, apparently about five days old, and the other held three well-incubated eggs; the female was on the nest in both cases. Mr. Rathbun was with Mr. Hersey when these nests were found and remarked that in the first of these nests the eggs must have been laid early in April; once in a while he finds one of these early-nesting pairs, but as a rule fresh eggs can be looked for about the first of May.

Mr. Rathbun mentions in his notes a number of nests found by him in that same general region. Most of them were in small or medium-sized firs in dense second-growth forests, sometimes alongside a path, and sometimes in heavier fir forest or near the shore of the lake; they were all built against the trunk and supported by small branches;

none was over 14 or 15 feet above the ground. An exceptional nest, containing two heavily incubated eggs, was in a small cedar in low ground along the shore of Lake Washington. Two nests are described in detail. One "was made outwardly of dead twigs, a few of which were 5 inches long; next were many dead leaves of the alder and maple interwoven with fresh green moss, strips of rotten wood and fibrous dead grasses, within which was a thin layer of decayed vegetable matter smoothly molded. Soft dead leaves and fine dry grasses represented the lining." Another was made "outwardly of dead leaves, dead twigs, bits of rotten wood, a very large amount of green mosses stripped from logs and tree trunks, strips of dry inner bark, and coarse dry grasses, interwoven with dry twigs from hemlocks and firs, many of which were covered with lichens. The lining of the nest was wholly of soft dry grasses to a depth of one and one-half inches."

Alaska nests, described by Mr. Willett in his notes, are similar; he says that they are usually placed against the trunk of a young spruce, 10 to 25 feet up, but are sometimes on horizontal limbs of larger trees. One especially large and fine nest measures 225 by 150 millimeters externally, with an inner cavity measuring 110 by 55 millimeters.

Maj. Allan Brooks (1905) published data on five nests in southern British Columbia, as follows: "The birds were found nesting in heavy coniferous forest of very tall timber, with very little undergrowth for the coast district, where dense brush is the rule. The nesting site was usually a small tree heavily draped with the rank growth of green moss which grows in such profusion in these dark woods." One nest was in a small hemlock 9 feet from the ground, one in a moss-covered spruce, one in a leaning cedar, one in a vine maple (*Acer circinatum*), and the other in the pendant branch of a large cedar, 12 feet from the ground. One nest contained four eggs and all the others three. The nests were similar in construction to those described above. "The average dimensions are about six and a half inches for outside diameter, and three and a half across the cavity. * * * In no instance were two pairs of birds found breeding near each other; the nests were about half a mile apart. The proximity of the nest is usually betrayed by the actions of the birds, which flutter from tree to tree uttering a peculiar chatter not heard at other times."

Mr. Dawson (1923) says: "Old nests are common; and groups of half a dozen in the space of a single acre are evidently the consecutive product of a single pair of birds. There is a notable division of territory among these thrushes. As a rule, they maintain a distance of half a mile or so from any other nesting pair. In two instances, however, one observer found nests within three hundred yards of neighbors." He mentions a nest that was placed "full sixty feet up and eight feet out on the first limb of a stately hemlock."

Robert R. Talmadge writes to me: "The nest in my collection is made of moss with a little shredded redwood bark; this was placed on a foundation of mud and small twigs of the Sitka spruce. The nest cup was slightly arched over and placed to one side of the main body. This cup was lined with fine dried grasses and fine bits of shredded bark. The outside diameter varies from 7½ to 8½ inches. The depth is 4½ inches. The cup measures 3½ inches at the top, and widens out to 4 inches at the bottom. The depth of the cup is 2½ inches."

Henry C. Kyllingstad sends me the following note from Alaska: "The varied thrush is one of the most abundant nesting birds at Mountain Village, building its nest on the ground among the Sitka alders (*Alnus sinuata*) or at heights from a few inches to 15 feet above ground in the willow (*Salix* sp.) thickets. Clutches of three, four, and five eggs are about equally common. All nests have in the lining some of the fine, black, hairlike roots of *Equisetum*, which the birds gather from the cutbanks of the river.

"If one approaches the nest too closely, the brooding bird flies off with a harsh squawking cry, and its mate soon joins the noise. Both birds slink through the branches a little way from the nest calling like young robins just off the nest. If the nest is not completed or if incubation has only just begun, the nest is usually deserted after the most casual and brief visit. Closer examination of the nest and handling of the eggs will cause desertion even when incubation is advanced several days.

"These thrushes are among the earliest to arrive in spring of all our passerine birds, and they begin nesting almost immediately. Two years arrival dates are May 7 and May 1. The first nests with full clutches of fresh eggs for the two years were found on May 21 and May 16."

Eggs.—The Pacific varied thrush may incubate on two, three, or four eggs, or even five, but nearly all the sets consist of three. The shape varies from ovate to a rather long-oval. The ground color is a robin's-egg blue, rather paler than the usual robin's egg, or "Nile blue." The eggs are rather sparingly, though more or less evenly, marked with small spots or fine dots of dark brown, "raw umber," "burnt umber," or "seal brown"; and there are often a few underlying shell markings of "ecru-drab" or "Quaker drab."

The measurements of 38 eggs average 30.5 by 21.3 millimeters; the eggs showing the four extremes measure **34.5** by 20.5, 31.0 by **23.0**, **27.9** by 21.5, and 29.2 by **18.5** millimeters.

Young.—Incubation is apparently performed entirely by the female; I can find no record of the male being seen on the nest; he is, however, always present in the neighborhood and perhaps feeds the female on the nest. Nothing definite seems to be known on the length of the incubation period, which is probably not far from two weeks.

Dr. Joseph Grinnell (1898) says that, near Sitka, the first young, scarcely feathered, were "taken on July 2. By August 1, the young began to gather in considerable numbers and together with the Robins and other Thrushes were feeding on the blueberries."

The fact that young have been found in the nests in April and in July would seem to indicate that perhaps, in some cases, two broods are reared in a season.

Plumages.—A small nestling in my collection, in which the plumage is just beginning to show in the feathered tracts, is scantily decorated on the head and on the alar and dorsal tracts with long "vinaceous-buff" down, which is fully an inch long on the back.

The young nestling is scantily covered with grayish down.

Ridgway (1907) describes the young as "much like the adult female, but under parts more yellowish ochraceous, with feathers of breast and lower throat narrowly margined or tipped with olive or dusky (these markings sometimes indistinct), the jugular band sometimes uniform dull olive, oftener with feathers ochraceous centrally broadly margined with olive and dusky." He says in a footnote: "I am not sure whether the sexes differ or not in first plumage." But seven sexed specimens of young birds indicate "a decided sexual difference, the males having nearly the whole of the lower parts ochraceous and the jugular band indistinct, the females having the posterior half of the lower parts mostly white and the jugular band more or less conspicuous."

Dwight (1900), however, says that the sexes are indistinguishable in juvenal plumage; they are olive-brown above, plumbeous on the rump, with very faint whitish shaft streaks; "wings and tail clove brown with ochraceous bands edging the quills and tipping the coverts. Below ochraceous buff, whiter on abdomen, a pectoral band and edgings of throat and breast, olive-brown. Supra auricular line buff."

The first winter plumage is acquired by a partial postjuvenal molt, "which involves the body plumage and wing coverts but not the rest of the wings nor the tail, young and old becoming practically indistinguishable." He describes the male in this plumage as "above, deep plumbeous gray with brownish edgings, darker on the pileum, the wing coverts broadly tipped with deep orange buff, forming two wing bands. Below rich orange buff, the abdomen and crissum chiefly white mixed with buff and olive-gray, the sides with olive-gray edgings. A black pectoral band, somewhat veiled with gray, orange tinged." This is, approximately, the handsome adult plumage, which becomes only slightly grayer by wear before spring; the birds probably breed in this plumage. The female in first winter plumage, as in the adult plumage, is much duller in coloration, the upper parts grayer, the lower parts paler and all the markings paler, duller, and less distinct, the pectoral band being gray instead of black.

The complete first postnuptial molt, the following summer, produces little change, though the colors are somewhat deeper and richer than the spring colors, the feathers of the upperparts are at first indistinctly margined with olive, and the pectoral band of the male may be broader and blacker.

The partial postjuvenal molt of young birds and the complete postnuptial molt of adults is apparently completed in August.

Food.—Professor Beal (1915b) writes: "The varied thrush appears to be a pronounced ground feeder, and the stomachs show an unusual quantity of such food as thousand-legs, sow bugs, snails, and angleworms; but spiders are rarely eaten." Only 58 stomachs were examined, taken in the months from October to April, inclusive; this analysis, therefore, indicates only the winter food. Beetles aggregate 4.46 percent of the food; ants comprise 4.08 percent, and other Hymenoptera (bees and wasps) 2.24 percent; Hemiptera (bugs) amount to 1.09 percent, Diptera (flies) 1.47 percent, Lepidoptera (caterpillars) 2.18 percent, Orthoptera (grasshoppers and crickets) 1.99 percent, and all other insects to 1.18 percent of the food. Spiders amount to only 0.10 percent; myriapods (thousand-legs) are more popular and are taken to the extent of 3.08 percent. Earthworms, snails, and sow bugs collectively amount to 3.97 percent of the food. The total animal food shown in this analysis is only 25.85 percent of the whole, as against 74.15 percent of vegetable food. Probably the percentage of animal food would be higher in the summer months.

"The vegetable food of the varied thrush consists of fruit, weed seed, and mast, with some unidentifiable matter." Mast, mostly acorns, was found in 16 stomachs and amounted to 76.71 percent in November, when acorns were abundant, fresh fruit was not obtainable, and insects were scarce. Unidentified seeds averaged 16.78 percent for each of the winter months, and were found in 10 stomachs. Cultivated fruit amounted to only 3.63 percent, while wild fruits and berries made up 23.21 percent of the food. Snowberries were found in six stomachs, apple in three, California honeysuckle in two, and juniper berries, wheat, amaranth, blackberries or raspberries, filaree, pepperberries, poison-oak, sumac, buckthorn, and black nightshade were identified in one stomach each, in the form of fruit or seeds.

Many years ago, 1899, complaint was made that these thrushes did considerable damage by pulling up sprouting peas.

Much of the food of the varied thrush is found on the ground, where it forages among the masses of fallen leaves, picking up acorns, or casting the leaves aside; many of the lower forms of animal life are found under the damp and thickly matted leaves. Mr. Rathbun describes this action in his notes, as follows: "On various occasions we have watched this thrush feed on the leaf-scattered ground near the edge of an open spot in the forest. Springing forward the bird would seize some of the litter in its beak, and almost at the same instant make a backward spring, scattering the leaves in various directions. A motionless instant would follow before the bird searched for food among the disturbed carpet of leaves; and then, whether or not successful in its quest, it would repeat these actions. Often, when feeding on the ground, it will give its musical notes in a soft, faint tone."

Dawson (1923) has seen a varied thrush hunting for worms on a well-kept lawn, after the manner of a robin, and says that these thrushes search the ground for fallen olives. Ralph Hoffmann (1927) adds madrone berries to the vegetable food and has seen them feeding on the ground under live oaks. And Dr. Grinnell (1898) includes blueberries in the food of young birds.

Behavior.—The varied thrush on its breeding grounds is a rather shy and retiring bird, perhaps more retiring than shy, as it fades away into its dense and shady retreats on the approach of an intruder. But at other times it is often far from timid; Mr. Rathbun has had them come to within 10 feet of him while he sat quietly watching them on the lake shore. He says in his notes that in spite of the striking color pattern these birds are often quite inconspicuous where their colors seem to match their surroundings and help them to fade into the picture. This is especially true of the females; the pale buff shades of the breast are exactly the tint of many of the mosses on the rocks and logs, or the ends of broken branches on the fallen trees; and the pale grayish olive of the back is the color of logs or rocks. If motionless among such surroundings, the bird might easily be overlooked. The male's colors make him more conspicuous, and consequently more shy, but when his back is turned his colors match his surroundings; and even the conspicuous black band on his breast may tend to break up the continuity of his form. Mr. Rathbun (MS.) adds: "More than once I have noticed that when this thrush alights on a branch of an alder, to which yet clings a trace of dead foliage, how perfectly the markings of the bird harmonize with the russet tinge of the dead leaves and the grayish colored bark of the tree."

He found these thrushes very demonstrative and active in the defense of their nest; the female usually slipped quietly off the nest, but soon returned and made a great outcry, which soon brought her mate to the scene. On one occasion, he says: "While I was at the nest, both of the pair made quick dashes at me; and they continued to give their harsh alarm notes as long as I was in the vicinity."

Theed Pearse tells me that the varied thrush "is a very quarrelsome bird, continually driving its own and other species from a feeding station. One bird will adopt a regular dog-in-the-manger attitude, stopping any other bird from taking food, though not feeding itself. Its place may be taken by another, who on driving it off will take up the same position. Males are worse than females."

Voice.—The strange, rich, musical, yet almost melancholy notes of the varied thrush are one of its striking characteristics, most appropriate for the somber glades in which it lives, and to which they add a decided charm. Mr. Rathbun (MS.) writes: "In March and April a return movement of these thrushes toward their breeding habitat takes place, and by the middle of May probably all have reached their destination. It is in this and the following month that the song of this bird is heard at its best, and the character of this is decidedly different from that of any other. The song consists of a succession of long, vibrating notes, generally on five different tones, all of pure quality. When one is very close to the singing bird, a sixth note is often heard, but this is very soft and low; and interjected, at times, in the song will be a harsh rasping note, not to be heard at a distance. There is no regular sequence followed in giving the tone notes; and the song carries a long way through the forest, being sung with deliberation.

"The birds seem to sing at their best on rather warm, misty mornings; I have heard some really remarkable concerts given by about a dozen thrushes in the immediate vicinity, and supplemented by others at a distance. To hear this song at its very best, one should be close to the performer; then there is a quality in the tones that is lost with distance."

On a warm, misty morning in June, he listened for some time to the voice of an exceptionally good songster, and jotted down the varied sequence in which the different tones were heard, as follows: "High—low—medium—low—very low, this followed by a harsh note; high—very high—low—medium—high—low, then a pause as if the bird was reflecting on its performance; high—medium—medium—low—medium — high — medium — medium—high—medium—medium—faint low—medium—high—medium—low, then a pause; medium—high—very low—medium—high—very high—low—medium, pause; medium—very low, this given softly, an exquisite effect, followed by a long pause; then high—very high—medium and a pause; high—high—soft low, followed by a long pause; medium—high—medium—

high—, a pause; low—medium—high—medium—medium—very low, followed by a harsh note; medium—high very high—low—medium—very low, followed by a harsh note; medium—high—very high—low—medium—high—medium—low—medium—low, followed by the harsh note without intermission, low—high—the harsh note—medium—this repeated slightly lower—medium—the harsh note—high—medium—medium—high—slightly lower—medium—a pause; low—medium—high—low—medium—a long pause, and the song ceased from the place where the bird had been singing, a few minutes later beginning again from some distance away."

Until it flew away, the bird had not changed its position during the above recital, which lasted between three and four minutes. It will be noted from the above that the songs varied greatly in length, as well as in the order in which the tones were given. The intervals between the tones were about one second. In order to determine the number of times the different tones, or keys, were used, two typical songs were analyzed as follows:

Very high key, 3 times.	High key, 9 times.
High key, 14 times.	Medium key, 7 times.
Medium key, 21 times.	Low key, 10 times.
Low key, 12 times.	Very low key, once.
Very low key, 4 times.	Lower key, 6 times.
Harsh note, once.	Harsh note, 6 times.
Very high key, once.	

Mrs. Bailey (1902) quotes an excellent description of the song by Louis Agassiz Fuertes, who says that it is—

most unique and mysterious, and may be heard in the deep still spruce forests for a great distance, being very loud and wonderfully penetrating. It is a single long-drawn note, uttered in several different keys, some of the high-pitched ones with a strong vibrant trill. Each note grows out of nothing, swells to a full tone, and then fades away to nothing until one is carried away with the mysterious song. When heard near by, as is seldom possible, the pure yet resonant quality of the note makes one thrill with a strange feeling, and is as perfectly the voice of the cool, dark, peaceful solitude which the bird chooses for its home as could be imagined. The hermit thrush himself is no more serene than this wild dweller in the western spruce forests.

Howard L. Cogswell, of Pasadena, Calif., writes to me: "The only call notes I have heard from varied thrushes in winter are a weak, but very thrushlike *tschoook*, and a vibrant, vocalized whistle *vwoooeeee*, somewhat like the noise produced by blowing on a comb wrapped in paper, though not nearly so loud."

Field marks.—The varied thrush is so conspicuously marked with such striking colors that it could hardly be mistaken for anything else. It is a large, stoutly built thrush, about the size of a robin. The rich, orange-buff throat and breast are separated by a broad black band across the chest; there is an orange-buff stripe above and behind the eye, two orange-brown wing-bars, with a patch of similar color in the open wing, and the outer tail feathers are tipped with white. The color pattern of the female and the young bird is similar, but the colors are much duller.

Fall.—Referring to the fall migration from the mountains to the lowlands, Mr. Rathbun (MS.) says: "This I have seen on occasions in October, when at a considerable altitude, the thrushes passing by our camp in considerable numbers day after day. The birds invariably were to be seen on or near the ground, and they were not at all timid, sometimes coming close to the camp searching for food. On occasions with them would be seen a few Steller's jays; it was amusing to see the jays make unsuccessful attempts to drive away the thrushes, who proved much too active and not at all intimidated; such contests often ended by all the birds feeding together. October is a silent month for bird song in the forests of this region, and in the valleys of the mountains are many sombre and rainy days; so, it gives one particular delight to hear, on rare occasions, the long vibrant song of this bird steal through the forest, mingled with the sound of falling raindrops."

Joseph Mailliard (1908) witnessed a great wave of varied thrushes migrating southward in California on October 20, 1906, of which he writes:

A very strong, and exceedingly warm and dry north wind was blowing, amounting in places to a veritable gale. We drove from the house to the extreme end of the ranch, a distance of about four miles, before sunrise, in the face of the gale, and putting up the team in the barn there, commenced on foot to ascend the range with the purpose in view of looking over the property and, incidentally, seeking for quail in their accustomed haunts. The sun was rising as we began the ascent and the air startlingly clear. We had taken but a few steps when my attention was attracted by the sight of a few birds, about the size of robins, flying high and scattered over the sky. This was so unusual at this time of day that I remarked upon the phenomenon to my companions.

As the sun cleared the horizon and the light increased we realized the fact that the sky was dotted in every direction with birds flying singly, and at quite an elevation, mostly toward the south. As the light grew stronger individuals here and there dropped to a lower altitude and allowed us to discern the fact that they were Varied Thrushes (*Ixoreus naevius*). For some time their numbers increased until there were hundreds in sight at once in any direction one might look, and as we were by this time pretty well up on the range we had quite an extended view. soon we noticed single birds dropping out of the flight and settling in the bushes. These must have been our regular winter residents. Apparently not over one in a hundred dropped out in this way. After nine A. M. the numbers decreased and by ten o'clock the flight was over, with no birds in sight except a few flying from one canyon to another and settling down locally, apparently.

Winter.—Throughout at least half of its breeding range the varied thrush is present in winter in reasonable numbers, but in somewhat different surroundings. And a few remain well up toward the northern limit of the summer range. Alfred M. Bailey (1927) saw one at Juneau, Alaska, on January 26, and says that "not many winter as far north as Wrangell, however, only stragglers being seen during the milder months."

George Willett (MS.) found it plentiful at Craig, Alaska, during the winter of 1919–20, and says that it was common at Ketchikan after January 18, 1925.

In British Columbia, Major Brooks (1905) records the abundance of the varied thrush during the winter of 1904–05: "During other winters a few may be seen, even in the coldest weather, throughout the district west of the Cascades. This winter they fairly swarm; and reports from Okanagan show they are even wintering in the cold interior of the Province."

In western Washington, according to Mr. Rathbun (MS.), varied thrushes are sometimes very common in winter, having moved down from the higher levels into the low country, even down to the sea coast. "On rare occasions and at long intervals, heavy falls of snow have occurred in the lowlands; at such times, this species must suffer severely and sustain a diminution of its numbers. Such an occurrence took place in February 1916, when snow fell to a depth of several feet throughout the Sound region. This caused the varied thrushes to appear in very large numbers in and about the cities and towns, wherever there was a human habitation. To this apparent appeal people generally responded. A large proportion of the yards became feeding stations for the birds, who resorted thereto until the bulk of the snow had disappeared."

Suckley and Cooper (1860), referring to the same general region, say:

> In winter it is a shy bird, not generally becoming noticeable in the open districts until after a fall of snow, when many individuals may be seen along the sand beaches near salt water. * * * I suppose that they are driven out of the woods, during the heavy snows, by hunger. It may then frequently be found in company with the common robin, with which it has many similar habits. * * * At this time of the year it is a very silent bird, quite tame, allowing near approach; flying up when the intruder comes too near, but alighting on the ground again at a short distance in front. It appears to be fond of flying by short stages in a desultory manner, sometimes alighting on the ground; at others on fences, bushes, or trees.

W. A. Kent writes to me: "The winter migration of the varied thrush south into Los Angeles County is a food problem rather than that of climate. Should there be an acorn shortage in Ventura County, north of us, varied thrushes will be here in numbers of from two or three to ten together in the oak groves. They are usually seen in December and January; they were with us during the winter of 1937–38, none in 1939–40, but were here again in 1941–42."

Mr. Cogswell (MS.) referring to the same general region, says: "Here, the varied thrushes seem fondest of the shady areas under the oaks and sycamores of the semiopen canyons of the foothill mesas, but on December 21, 1941, in Mandeville Canyon, Santa Monica Mountains, I saw two fly out of a section of pure chaparral when frightened by a passing sharp-shinned hawk; in this case there were no large trees within 200 yards. I also have a record of one bird seen February 19, 1942, in a willow-cottonwood association along Mill Creek in the Prado Basin near Corona, Calif., a distance of some 6 or 7 miles from the nearest foothill oak-sycamore association, in which they are most typically found."

DISTRIBUTION

Range.—Western North America.

Breeding range.—The varied thrush breeds **north** to northern Alaska (valley of the Kobuk River, probably the vicinity of Alatna, near Fort Yukon, and the Porcupine River below the mouth of the Coleen) and northern Mackenzie (Aklavik). **East** to the Mackenzie River Valley (Aklavik, Fort McPherson, Fort Good Hope, Fort Norman, near mouth of the Nehami River, and Fort Liard); eastern British Columbia (Nelson, Yellowhead Pass, and Kootenay National Park); probably extreme western Alberta (Banff); northwestern Montana (Glacier National Park, Fortine, and Swan Lake); northern Idaho (Clark Fork, Coeur d'Alene, and St. Joe National Forest); eastern Washington (mountains near Spokane and near Dayton); northeastern Oregon (Powder River Mountains) and the Cascades of western Oregon to north-western California (mouth of Redwood Creek, Humboldt County, Eureka, and Peanut, Trinity County). **South** to northwestern California (Trinity and Humboldt Counties). **West** to California (Humboldt and Del Norte Counties); western Oregon (in the mountain forests and at Tillamook Bay and Astoria); western Washington (Grays Harbor, Tacoma, and Seattle); western British Columbia (Chilliwack, Alta Lake); Vancouver Island (Comox); the Queen Charlotte Islands (Rose Spit and Massett); and Alaska (Forrester Island, Sitka, Nushagak, Bethel, Nulato, and the Kobuk River).

Winter range.—In winter the varied thrush is found **north** to southeastern Alaska (Craig, casually to Wrangell and Juneau; a specimen was taken near Flat in late November 1927); **east** to southeastern Alaska (Wrangell); southern British Columbia (Okanagan Landing and Edgewood); western Washington (Bellingham, Edmonds, Seattle, Tacoma, and Camas); western Oregon (Portland, Salem, and Brownsville); extreme western Nevada (Carson); central California (Stockton,

Fresno, and Redlands); and casually northwestern Lower California (Laguna Hanson and Rancho San Pablo). **South** casually to northern Lower California (Rancho San Pablo). **West** to the Pacific coast of Lower California (Guadalupe Island, casually); California (Witch Creek, San Diego County, San Clemente Island, and Santa Barbara, San Francisco); Oregon (Coquille, Newport, and Astoria); Washington (Cape Disappointment and Port Townsend); British Columbia (southern Vancouver Island and Vancouver); and southeastern Alaska (Craig). In migration the varied thrush has been found in southern California in Death Valley and east of the Mojave Desert and may occur in winter farther east than the range as outlined.

The above ranges apply to the entire species, of which two subspecies or geographic races are recognized. The Pacific varied thrush (*I. n. naevius*) occurs in the coastal region from Kodiak Island, Alaska, to northern California and eastward to the Cascades; the northern varied thrush (*I. n. meruloides*) occurs from northern Alaska and Mackenzie through eastern British Columbia to northwestern Montana, northern Idaho, eastern Washington, and northeastern Oregon.

Migration.—Many varied thrushes move only from the mountains to the valleys for the winter, others make longer migrations. Some late dates of spring departure are: Lower California—Guadalupe Island, March 4. California—Berkeley, April 11. Oregon—Portland, April 21. Montana—Missoula, April 16.

Some early dates of spring arrival are: Idaho—Coeur d'Alene, March 8. Montana—Missoula, March 9. Alberta—Banff, March 12. British Columbia—Atlin, April 24. Yukon—Dawson, May 25. Alaska—Fairbanks, April 28.

Some late dates of fall departure are: Alaska—Alatna, September 11. Wrangell, October 19. Yukon—Forty-mile, September 12. Alberta—Edberg, October 10. Montana—Fortine, October 24. Idaho—Priest River, November 8.

Some early dates of fall arrival are: British Columbia—Atlin, September 1. Oregon—Portland, September 18. Nevada—Fallon, November 3. California—Diablo, October 31. Lower California—Laguna Hanson, November 17.

Casual records.—In migration the varied thrush has been found, (usually in company with robins) far east of the normal range: Alberta, Wyoming, Colorado, Nebraska, Kansas, and New Mexico. Farther east the following occurrences have been recorded: Maniconagan, Quebec, a specimen collected on August 28, 1890; a specimen collected near Port Jefferson, Long Island, on December 20, 1889; one at Islip, Long Island, "in fall"; a specimen at Miller's Place, Long Island, on November 19, 1905; one seen almost daily at Richmond, N. Y., from

November 24 to December 6, 1936; a specimen taken at Hoboken, N. J., in December 1851; and one at a feeding station near Clementon, N. J., from November 26, 1936 to March 20, 1937. The easternmost record is of a specimen collected at Ipswich, Mass., in December 1864.

Egg dates.—Alaska: 36 records, May 15 to July 2; 17 records, June 1 to June 12, indicating the height of the season.

British Columbia: 8 records, April 28 to June 20.

California: 17 records, April 11 to July 17; 10 records, April 19 to May 5.

IXOREUS NAEVIUS MERULOIDES (Swainson)

NORTHERN VARIED THRUSH

HABITS

Swainson and Richardson (1831) first described the varied thrush of the northern interior, under the name *Orpheus meruloides,* from a specimen taken at Fort Franklin, in northern Canada. The Pacific varied thrush was described from a specimen taken at Nootka Sound, Vancouver Island. Dr. Joseph Grinnell (1901) discovered that two races of this species should be recognized, the difference between the two being most marked in the female. In explaining his contention, he says:

Well-marked differences exist in the case of the female between the Varied Thrush breeding in the humid Sitkan District and that of the drier interior region of northern Alaska. The Sitkan race is characterized by a predominance of deep browns, restriction of white or light markings, and by a shorter and more rounded wing. The northern and interior race has a much grayer and paler coloration, greater extension of white markings, and a longer and more pointed wing. Unfortunately I have no male birds from Sitka, except juveniles; but three spring males from the Kowak Valley, when compared with late winter males from northern California taken along with females referable to *naevia,* are of a lighter slate color dorsally and slightly paler tawny beneath. The females of this species appear to be much more subject to protective coloration, so-called, than the males, and it is therefore reasonable to expect climatic variations to be more pronounced in the females than in the males, especially when the climate of the *summer* habitat is of an extreme nature. In the winter home of the Varied Thrushes there is also a different distribution of the two races, but their latitudinal relation is reversed. Thirty-five skins from Los Angeles County, California, are all but one referable strictly to *meruloides,* while the majority of the winter skins from the coast region of central and northern California are near *naevia.* So that *meruloides,* although its summer habitat is northernmost, goes farthest south in winter, and its migration route is much the longest.

Evidently this subdivision of the species did not impress Mr. Ridgway very favorably, for he (1907) says: "With a series of one hundred and forty specimens (sixty-seven adult males from the coast district and seven from the interior, forty-two adult females from coast and fourteen from interior localities, besides nine young representing both districts) I have been unable, after very careful comparison, to discover the slightest reason for recognizing two forms of this species."

This northern and inland race, according to the 1931 Check-list, "breeds in the Hudsonian and upper Canadian zones from the Yukon Delta, Kowak Valley, and Mackenzie delta south to Prince William Sound, Alaska, the southern part of the Mackenzie Valley, and south in the mountains through eastern British Columbia to northwestern Montana and northeastern Oregon. Winters mainly in the interior of California south to Los Angeles County and irregularly to northern Lower California." This is the form that has wandered as a straggler as far east as Quebec and Massachusetts.

Spring.—Dr. Grinnell (1900) writes:

The Varied Thrush proved to be an abundant summer resident of the Kowak Valley, and was observed in every tract of spruces visited. * * * Its arrival in the neighborhood of our winter camp was noted on May 21st, when the twanging notes of the males were heard several times in the morning and evening. The next day they had arrived in full force and were to be seen and heard throughout the spruce woods. The snow had by this date nearly all disappeared, though the rivers and lakes were still covered with ice. The food of the Varied Thrushes at this time consisted largely of the cranberries and blueberries which were left from the previous summer's crop, and had been preserved beneath the winter snows. For a few days the birds were quite lively for being of the thrush tribe, which are usually of a quiet demeanor. When not feeding on the ground in one of the fruitful openings in the forest, they would be seen in wild pursuit of one another, either courting or quarrelling. The males were often seen in fierce combat; that is, fierce for a thrush. Of course some female ensconced in a thick evergreen in the vicinity was the cause of the duel. I never saw just how a quarrel would commence. The swift pursuit would follow a tortuous route around and about, twisting among the close-standing trees and across openings, so rapidly as to be difficult to follow with the eye. The *finale* would be a brief scrimmage among the thick foliage of a spruce, with a clatter of fluttering wings and a few sharp squeals like a robin's. They would fall slowly through the branches to the ground, when the contestants would separate, panting and puffing out different parts of their plumage. The greatest apparent injury to either of the belligerents would be the loss of two or three feathers, yet one of them would consider himself fairly beaten and soon retire leaving the victor free to continue his courting.

Nesting.—Dr. Grinnell (1900) found many nests of the northern varied thrush and collected eleven sets of eggs; he writes:

In the Kowak Valley I noticed the first signs of nest-building * * * on the 25th of May, just four days after their arrival, and by the 28th nearly every pair were busy; for the summer is short, and there is no loitering, as is often the case with our southern birds, after their arrival. The female does all the work of constructing the nest, the male accompanying her constantly in her many trips after material, but, as far as my observations go, never proffering any assistance. Many of the nests are built on those of the previous year as a foundation, and I have even found three-storied nests. The old nests are flattened and dilapidated by the heavy August rains and winter snows, with the mud mostly dissolved out of them. During the winter a tour of the woods discloses hundreds of old thrushes' nests in various states of preservation, and in some sections nearly every tree harbors one or more. Where well-protected in dense spruces they may survive many years. Probably the same pair of birds return to a single nesting site for several successive seasons. * * * All the nests of the Varied Thrush observed were in spruces, and varied in height above the ground from 6 to 20 feet, the latter being far above the average height, which I should judge to be 10 feet. Even in the tallest timber the nesting sites are chosen in the lower foliage at a similar elevation. * * * The majority of the nests are situated on the south side of the tree-trunks, as probably being the driest and warmest side, and then, too, strong, cold north winds are of frequent occurrence. All the nests which I examined are very much alike in composition and structure. The foundation is a rather loose and bulky mass of plant stems and dry grasses, but the nest proper is a solid, closely-felted structure. The bottom and sides are substantially formed of a mixture of mud, and wet, partly-decomposed grasses and moss. The amount of mud varies in different nests, and in some there is scarcely any, but the various vegetable materials are always incorporated when wet, so that after the structure dries, the walls and rim are very firm like *papier-maché*. When finished the nest presents a neatly-moulded cup-shaped cavity, with an inner lining of fine dry grasses. The measurements of a typical nest are as follows: Inside diameter, 3.25; depth, 2.25. Outside diameter, 6.50; depth, 4.50. The weights of the dry nests vary from one-half to one pound, depending on the amount of mud in their composition.

Dr. E. W. Nelson (1887) reports a nest found by Dall at Nulato, Alaska, on May 22, 1867, that "was built in the midst of a large heap of rubbish in a group of willows, about 2 feet above the ground, and close to the river bank," a most unusual location!

From near the southern limit of its range in the interior, Thomas D. Burleigh (1923) reports a late nest, found on August 7, in Bonner County, Idaho; it was "fifteen feet from the ground and twenty feet out at the outer end of a limb of a large hemlock at the side of a swift roaring stream in a deep wooded ravine."

Along the shores of the lower Yukon River, during the first few days of July, Herbert Brandt (1943) found a number of nests of the northern varied thrush "in an upright crotch in a small tree, and from three to eight feet above the ground. The bird generally chooses an aspen or a willow but we found one nest in a small spruce tree"; some of the nests were still under construction and some held eggs or young.

Eggs.—Apparently the northern varied thrush lays larger sets of eggs than the Pacific coast race. In the 11 sets collected by Dr. Grinnell (1900), there were two sets of three, seven sets of four, and two sets of five eggs each. His description of the eggs tallies well with the eggs of the other race, but "the eggs of one set show larger blotchy markings of raw umber. One egg is almost without markings, thus resembling a robin's."

He gives the average measurement of 44 eggs as 30.0 by 21.3 millimeters, the largest measuring 31.8 by 22.3 and the smallest 28.0 by 21.1 millimeters.

Food.—The birds that he saw in the Kowak Valley in August were feeding "almost exclusively on cranberries and blueberries." Grinnell and Storer (1924) say of the food of this subspecies in the Yosemite region in winter:

> Like its not distant relative, the Western Robin, the Varied Thrush (sometimes called Oregon Robin) feeds in the winter season chiefly on berries of various sorts, and its local occurrence and relative abundance is governed by the season's crop of these. Two or three of these birds seen among golden oaks near Camp Lost Arrow, November 13, seemed to be feeding on mistletoe berries. On the Big Oak Flat road, about 3 miles out of Yosemite Valley, on December 28, 1914, 8 or more Varied Thrushes were seen feeding on the sweetish berries of a manzanita (*Arctostaphylos mariposa*). On the Wawona Road at Grouse Creek, November 26, 1914, two were apparently feeding on berries of the creek dogwood. In the Upper Sonoran foothill region, the Christmas berry or toyon (*Heteromeles arbutifolia*) furnishes a favorite food as long as the crop lasts.

Voice.—A. Dawes Du Bois sends me the following tribute to the song of the northern varied thrush: "I have reason to remember the day that I heard, for the first time, the voice of a varied thrush. It was on the Middle Fork of the Flathead, in the mountains of northwestern Montana. I had seated myself for a brief rest among the rocks near the bank of the river. All at once a voice of astonishing clearness came from somewhere up the stream—a single tone, long sustained with remarkable constancy of pitch. It was followed by a long rest, then another long note and a silence, and yet others, each note of distinctly different pitch. This mountain piper was sounding his pipes with deliberate precision, swelling each note with matchless truth and purity of tone. His music was radiant with spiritual quality, thrilling in its effect. The murmurings and whisperings of the river supplied the connecting passages. The whole became an imposing largo, in perfect accord with the wildness, the clarity, the beauty of the mountains."

Fall.—"In the fall of '98," in the Kowak Valley, the varied thrushes "remained common until the last of August, though at that season the birds were quiet and of secretive habits. * * * The last Varied Thrushes, two in number, were seen on September 4th," according to Grinnell (1900). The species was first seen on October 24, 1920, and thereafter "large flocks" were observed on various dates throughout November and December, in the Yosemite Valley, as reported to Grinnell and Storer (1924) by C. W. Michael.

HYLOCICHLA MUSTELINA (Gmelin)

WOOD THRUSH

CONTRIBUTED BY FLORENCE GROW WEAVER

HABITS

The nature lover who has missed hearing the musical bell-like notes of the wood thrush, in the quiet woods of early morning or in the twilight, has missed a rare treat. The woods seems to have been transformed into a cathedral where peace and serenity abide. One's spirit seems truly to have been lifted by this experience.

These birds are found in low, cool, damp forests, often near streams. This probably follows because of the need of mud and damp plant material, which are used in the construction of the nest. Undergrowth and the presence of saplings seem to help determine the suitability of an area during the breeding season. I found no nests in conifers, which were numerous in the mixed coniferous and deciduous forests in which my studies of this species were carried on, but short dead branches of these trees were often used as song perches. There are citations in the literature, however, that record nests in conifers. A hemlock was used in New York, cedars in Florida, and recently many nests have been found in coniferous bogs in northern Michigan; two nests were found in balsam firs. Thickets were usually not chosen by the wood thrush, although one pair built in a sapling ash-leaved maple in a dense growth of such saplings.

Dr. A. A. Allen (1934) believes that this bird dislikes bright sunlight, probably because its eyes are so large that too much sunlight makes the bird uncomfortable, so that it keeps to thick woods or ravines where there is plenty of vegetation and resulting shade.

Besides the locations already mentioned, a number of birds choose places near human habitations, or in parks or gardens. Tracing this adaptation of habitat back through the literature, it seems that this change took place during a 20-year period, from 1890 to 1910. Widmann (1922) reports a nest inside the conservatory in Shaw's Garden, St. Louis, Mo. Orchards are seldom chosen as locations for nests.

There is considerable evidence of the gradual northward extension of the breeding range of the wood thrush. In reviewing the literature for this paper, I find a recent breeding record for Montreal, Quebec, Canada, which is a little north of previous records for the vicinity (Cleghorn, 1940).

Root (1942) reports records for Cheboygan County in northern Michigan in Canadian Zone coniferous bogs. Roberts (1932) has noted the northward spread of this bird into the Canadian Zone in Minnesota. Dr. Fred Lord in Hanover, N. H., has witnessed the northward spread in New Hampshire, saying that 30 years ago there were none in the village, while at the time he was writing, 1943, they were common.

M. W. Provost (1939) writes as follows: "Before the advent of the white man, the Wood Thrush was not found farther north than the Lake Region in central New Hampshire and Hanover in the Connecticut Valley. In the two decades 1890–1910 there was a remarkable invasion of the White Mountain valleys by this bird. Today it is by no means rare in the transition valleys throughout the mountains and even up into the forests on slopes up to 2,000 feet. I have found a Wood Thrush on July 8, 1937, in the deep forests of the Mountain Pond region of Chatham, at an elevation of over 1,500 feet and three miles from the nearest settlements in North Chatham."

This statement from Minot (1895), which is a footnote by the editor, William Brewster, will give a basis for the comparison of the distribution then and at the present time: "A summer resident, very common and generally distributed in Connecticut, less numerous and more local in Massachusetts, and rare or accidental north of the latter State, excepting, possibly, near shores of Lake Champlain in western Vermont, where it is said to breed regularly and in some numbers." Goss (1891) also gives habitat as "north to Massachusetts."

Some early records of the northward movement are as follows: A note from F. H. Kennard, written in 1910, states that Horace Wright recorded wood thrushes in Jefferson, N. H., which is in the northern part of State. But Wright (1912) says that in 1904, 1905, and 1908 the wood thrush was as yet a rare bird. He heard it at 1,600 feet. F. H. Kennard also heard these birds at Averill, Vt., in the northeastern tip of the State, June 21, 1912, at 1,850 feet. William Brewster (1938), on May 14, 1896, heard a wood thrush singing at the Pearly White farm, which is on the Maine side of Umbagog Lake.

Spring.—These data are from records in the files of the United States Fish and Wildlife Service, which divides the range into three sections—the Atlantic, the Mississippi, and the Western. I will deal with each in order. Most wood thrushes have spent the winter south of the United States. This will be discussed in more detail in the section called "Winter." Spring migration begins in March when birds have been reported in South Carolina, Alabama, and Georgia, at about the 33d parallel north latitude. Some records show that some birds are still south of the United States at this time; therefore migration is not really in full swing until April. There are reports of birds in North Carolina on April 3, in Maryland on April 6, in Pennsylvania, and in New Jersey on April 11. A few early individuals have been reported for New York on April 9 and 12. They progress northward rapidly and by the end of April are seen in Connecticut, and by May 19 they have already reached the northern limits of their range. They have been reported from Vermont, Nova Scotia, and New Brunswick.

In the Mississippi Valley, in spring, wood thrushes appear in the United States between March 26 and 30. (Audubon seems to be the source of statements saying that the wood thrush winters in southern Louisiana, but more recent studies do not bear him out in this.) Ten days later they have usually advanced northward to Dunklin County, Mo.; in the next 10 days up to April 20 they have advanced to the parallel of St. Louis; by April 30 they have reached Glen Ellyn, Ill.; by May 10 they average northward as far as London, Ontario; and in the next 10 days they reach the northern boundary of their range.

For the western part of their range, we have very few records for the early spring months. The birds reach the southern limits of the western flyway about April 1 and advance northward to the 44th parallel and westward to the 102d meridian by May 10.

In summarizing, we find that wood thrushes appear earliest in the Atlantic Coast States and move northward there earlier than in the Mississippi Valley or west of it. The speed of the advance, however, is greater in the Mississippi section than east or west of it. In the western flyway, the birds are slow in moving northward, but they do not move as far northward as in the other two flyways.

Courtship and territory.—According to my observations, borne out by those of Brackbill (1943), the male arrives in the territory he chooses and stays there waiting for his mate to arrive. In one case, which I was able to watch from the moment of arrival, the male arrived first and the female two days later. These two birds were circling in their courtship performance on this second day. A banded male came back to the same territory for three years in succession. The choice of the actual nesting site seems to be made by the female. She was seen to alight upon a branch, hop to a crotch, stick her head barely above the leaves, and turn around several times. This particular spot was not chosen, however, so this may have been of no significance.

The territory of the wood thrush may be as small as one-fifth of an acre or as large as two acres.

In defense of this territory the wood thrushes I studied seemed to vary in their reactions. To birds smaller than themselves they appeared to pay little attention. Birds that disturbed the thrushes most were other wood thrushes, veeries, and robins, although even then reactions varied greatly. Sometimes other wood thrushes were driven out, but at other times they were not. Sometimes the male would challenge the intruder with a burst of his most complete song, which would last as long as 10 minutes. On other occasions the intruder was merely chased from the territory. Most robins were chased, but some were not. All degrees of alarm and retaliation were displayed from the mildest form, in which the adult sat quietly and merely raised the feathers on the head, to a real bird fight. When the young in one nest were 10 and 11 days old, the male fought the most furious fight I have ever witnessed. He flew upward attacking the robin in the air. The two birds flew at each other with wings beating rapidly and feathers fluffed from the body. They attacked each other over and over again, and peace was not restored until the robin was chased from the territory. A veery chased two wood thrushes from its territory in an area containing five wood-thrush nests. A scarlet tanager was attracted by the call of a young wood thrush when it was just a day out of the nest. The tanager flew away unchallenged while another wood thrush, which alighted on the same branch as had the tanager, was driven out. Brackbill (1943) notes that the territory was defended against other wood thrushes but that the birds seemed very tolerant of other species. He says the only birds toward which they displayed hostility were a blue jay and a purple grackle.

The males arrive first on the breeding grounds, as the records of Cornell University show that the first wood thrushes reported for 12 different seasons were singing birds. In one instance the female of a pair arrived in the territory three days after the male. Early in the morning the male bird sat high up in a leafless tree singing. A low *trrrr*, which I have often heard both male and female give as if in acknowledgment of its presence, could be heard. There was a sudden flight to the ground. This was followed by six or seven swift, circular flights of about 30 feet in diameter, one bird in pursuit of the other. They both alighted contentedly in the same shrub and began feeding among the fresh leaves. This circular flight was accompanied by swift turnings to bank with the curve. A few low notes were uttered during this performance. Four days later the song of the male in this territory was noticeably loud and long. Loud calls of excitement were also heard, leading me to believe that the territory was well established. Another observation of a similar performance was made. The female stood on a low branch and fluffed her feathers and raised her wings. The male chased her in half a dozen circular flights. Between flights both birds fed among the fresh leaves, often biting off pieces which fell to the ground. In this case the female arrived six days after the male. He used the very highest tree in the area as his song perch. This form of courtship was observed even after the nest was built. Dr. A. A. Allen (1934) states that he believes the act of copulation normally takes place on the ground. On one occasion I observed what appeared to be copulation but it took place in flight. The birds made several circular flights after which one went to a small trickling stream for a drink. Then the two met in the air. With a flutter of wings and with wings spread, the male lowered himself to contact the female. This took place about 8 feet above the ground, just over the top of a bushy shrub. They continued to fly around after each other.

Nesting.—The nest of the wood thrush is much like that of the robin. Both contain a middle layer of mud or plant material mixed with mud. The wood thrush's nest can be distinguished from the robin's by the presence of dead leaves and sometimes moss. Each wood thrush nest I examined was lined with brown rootlets, whereas the robin's nests were lined with dried grass.

All except two of 20 nests examined contained paper, cellophane, white cloth, or some white material. One of the two exceptions had long pieces of dried grass hanging from the bottom, yet this was in a position within easy access of paper. The other was away from habitations where such materials were not available. These pieces of paper or rags used in the foundation of the nest would seem to make it more conspicuous. Dr. A. A. Allen (1934) suggests that perhaps enemies do not recognize such large affairs as nests. My interpretation is that its use follows the theory of concealing coloration in that it breaks the nest contour. The bird's white underparts were sometimes used also. The female fluffed her flank feathers so much that they and the under tail coverts were visible, thus breaking the contour lines of the bird. Also, the incubating bird, in holding her head high, shows the white triangle of the chin and throat. It was also noted that when something happened to frighten the incubating bird, she pressed her body more deeply into the nest and held her head back farther, pointing her bill upward. This made the white throat still more conspicuous.

In all cases where rags, cellophane, and paper were used they were white or transparent. In two instances in which it was known that nests had been broken up or new ones were started, pieces of colored paper, cellophane, and tinfoil were scattered about to see if these would be used. In one case the second nest was already built; in the other the birds moved beyond the area of the scattered materials so the experiment came to naught. White paper was used in both these second nests.

The nest is a firm, compact cup of grasses and weed stalks with a middle layer of mud or leaf mold. In a few cases where the nest was saddled on the branch, the nest was cemented to the branch with mud. Every nest observed had dead leaves or leaf skeletons tucked into the bottom. The nest is loosely lined with fine dark rootlets. Inside, the nest measures 2 inches deep and 3½ inches wide.

The nest is usually fixed in a fork in saplings or undergrowth, but some were found saddled on horizontal branches. Several nests were found in shade trees near houses. Nests ranged from 6 to 50 feet from the ground. The average height of 15 nests was 10 feet.

Nesting materials varied considerably with the availability of materials. Where paper was in abundance much was used. In some nests more leaf skeletons were used than in others. One nest contained many weed stalks because they were available. The outside depths of nests varied greatly with their location. On a horizontal branch the nest was shallow, to fill a crotch it was deep.

I was not fortunate enough to observe the entire construction of a nest. In one case building a blind frightened a pair from its chosen site while a nest was under construction. In another instance I found a deserted, partially built nest, so that the progress could be traced. In one instance a pair was observed to start the nest. Five days later it was finished and the first egg was laid.

There seems to be little choice in selecting the kind of tree, shrub, or vine used. The following list is a summary of 14 nests: one nest each in basswood, juneberry, birch, locust, and a grapevine; two nests each in maple, witch-hazel, and hawthorn; and three nests in elms. Many references state that many kinds of trees are used.

Territories used in 1936 and 1937 were used also in 1938. In one case a pair built in the identical spot it had used the year before, since I had removed the old nest after nesting was over.

I have no evidence of this species ever using the same nest a second time or a second season.

The only reference to the fact that the wood thrush nests on the ground is in the "Key for Identifying Bird's Nests," prepared by Helen Blair (1935). Mary H. Benson, a former student at Alleghany State Park Nature Camp, has informed me that a wood thrush's nest was found on the ground there and that Aretas A. Saunders photographed it.

A wood thrush's nest, after being used by a family, is a well-worn and sorry-looking abode. Pieces of loosened lining are removed by the adults so that finally there is little left; bits of the rim break off leaving it quite irregular; bits of feather sheaths from the young are found on the bottom despite the immaculate care given the nest by the adults.

The time of nesting is probably determined by the character of the food of the young and also the concealment of the nesting site and the young out of the nest. In Ithaca, N. Y., the nesting season extends from May 12 until the end of July.

Nesting instincts are very strong in both the male and female birds. The female often gives evidence of this by remaining on the nest while a person walks up to it and stands within 3 feet of her. An attempt was made to reach out and touch an incubating female; she sat quietly until the hand was within a few inches of her before she flew. The female, after banding, returned to the nest within 15 minutes, while it took the male bird an hour to quiet down after being excited. Several times the female returned to the nest while the male bird was still uttering the *quirt, quirt* call of alarm. The least movement in the vicinity of the nest would cause the male to utter a low *trrrr* call.

Eggs.—The eggs of the wood thrush are smooth, ovate in shape, plain in color. According to Ridgway's "Color Standards" (1912), the color is "beryl green" if the egg is dark; "pale sulphite green" to "Nile blue" if it is light; or, in common parlance, greenish blue much like those of the robin. The eggs of the wood thrush are smaller than those of the robin and are more pointed toward the small end.

[AUTHOR'S NOTE: The measurements of 50 eggs in the United States National Museum average 25.4 by 18.6 millimeters; the eggs showing the four extremes measure **28.5** by 19.3, 28.2 by **19.8**, and **22.4** by **16.5** millimeters.]

Authors give varying figures as to the number of eggs in a clutch, varying from two to five. Data from my study show that in no case was a clutch composed of five eggs. A fifth egg was laid in one case after the removal of a punctured egg. Of 16 nests observed in 1937, the average number of eggs per nest was three. In 1938 the average number in 15 nests was four. Together they average three.

In 1938 in 15 nests, 51 eggs were laid out of which 33 young were produced from which 22 survived. This is a survival of 43 percent of the eggs laid, and 66 percent of the young hatched.

Evidence at hand does not show more than two clutches of eggs for any pair of birds. In one case positive evidence was obtained by marking birds that the second clutch was a second brood. This will be discussed later. In many cases when the first nest was broken up a second nest with a second clutch was found. In one instance a pair built three nests, but in the second no eggs were laid.

An egg is laid each day until the clutch is complete. In one case I observed and timed the laying of eggs and found that they were laid about 10:30 o'clock on two successive mornings.

Incubation.—Eggs were marked so that the incubation period could be determined. It was found to be 13 or 14 days. Incubation begins with the laying of either the second or third egg. These conclusions are drawn from observations on two closely watched nests of four eggs each. Eggs were marked. The first two eggs in each case hatched in 14 days and the last two in 13 days.

The brood spot occupies the region of the abdomen. It is deep red in color and is devoid of all feathers. This was found only on female birds and was used to identify the females when banding them.

The eggs were turned by the female by clasping the egg between the angle made by the bill, chin, and throat when the head was pointed downward into the nest. With a raking motion she turned the eggs.

Incubation is performed solely by the female wood thrush. In order to prove this, a unique method of catching nesting birds was employed. William Montagna had learned this in Italy. At the time of this study he was assistant to Dr. G. M. Sutton, of the Laboratory of Ornithology of Cornell University. He employed a piece of orchard grass 2½ feet long. The leaves were stripped from it and a slip knot was tied at the end. The grass was moved up to the bird on the nest without frightening her. It was slipped over her head and tightened, and the nesting bird was captured. These birds were then marked with colored feathers, as well as aluminum bands, for identification purposes. Since no desertions resulted, the method was considered successful. Subsequent visits to the nests of birds marked in this fashion and observations on other banded birds give evidence of the fact that only the female incubates. Several birds were trapped with a drop trap as the young were about to leave the nest. Only one bird of each pair had a brood spot. Examination of the gonads of a bird, believed to be the male of a pair with young in the nest, indicated that the bird on the nest was the female. Brackbill (1943) also states that only the female incubates. His study was made with marked birds.

I should like to describe the hatching of eggs by giving an account of an all-day observation period beginning at 4:10 A. M. on June 22, 1937, in Ithaca, N. Y. The female sat deeply huddled in the nest. The male was singing soon after 4 A. M. From this time until 7:15 A. M. the female would raise herself from the nest, back off, and look in. Sometimes she would peck in the nest, then settle on again with a rolling motion. She repeated this over a period of four hours on the average of once every 15 minutes. At 9:15 A. M. her anxiety was relieved by the hatching of one of her four eggs. Male birds of this species had not been observed to feed the female on the nest, and so it was thought that perhaps the male in this case was acting in anticipation of the coming event when he arrived at the edge of the nest in the absence of the female with a small green caterpillar dangling from his mouth. He stood on the edge of the nest, looked in, ate the caterpillar, then sang his full song. This was four hours before the hatching of the first egg.

It took six minutes for the first egg to hatch. Four minutes later the young was free from the shell and it called and opened its mouth. The female fed it with a small insect she had picked up in the nest. She waited 15 minutes after the hatching before she carried away the shell. After leaving the blind, I found the shell about 50 feet southeast of the nest. She took the small end first, then six minutes later carried away the other piece. After she had disposed of the first piece, the male brought food. The female tried to give the food to the young but gave up and ate it herself. Later the male came again with food. The female took it from him and swallowed it.

The time required for the hatching of an egg varied greatly. In one case it required 22 hours after the shell was pipped. Another egg in the same nest required only 5¼ hours. In another instance the shell was pipped one-half inch at 8 p. m., but the egg did not hatch until 4:25 the next morning. The time probably varies with the vitality of the young bird as well as with the temperature of the egg.

The actual hatching is not a consistent procedure. The movements of the young cause the shells to open differently, so the young must use various means of freeing themselves from the shell. In one case 10 minutes after the shell was completely cut the left wing protruded from the opening. In another 10 minutes the right wing and head were free. The feet were freed last. The head of the bird lies in the large end of the egg. The head is bent downward causing the egg tooth to cut the egg at the large end at a distance of about one-fourth the length of the egg.

The egg tooth is a small whitish dot on the upper mandible near the tip. It is still visible at the time when the young birds are ready to leave the nest.

No one procedure was followed consistently in disposing of the eggshell. In one instance the female carried the two halves separately to a distance of 50 feet. Another time she carried away the larger part but ate the smaller part. In one case she ate the broken bits of the shell as it was being opened by the young.

Progress of a typical wood-thrush nest:

Building nest, 5 days, May 28 to June 1.
Eggs laid, 4 days, June 2 to June 5.
Incubation, 13 days, June 3–4 to June 16–18.
Hatching, 2 days, June 16 to June 18.
Brooding, 12 days, June 16 to June 30.

The number of young hatched from 43 wood-thrush eggs laid in 1937 was 27, or 63 percent. In 1938, of 51 eggs, 33 young were hatched, or 64 percent. In 1938, of the 51 eggs laid, 43 percent of the young survived to leave the nest.

In 1937, of the four nests which contained four eggs each, in only one instance did all four young survive to leave the nest.

In 1938, of eight nests containing four eggs each, in two cases all four young survived, and in one of these two. A cowbird survived with the four young wood thrushes.

Young.—The first plumage of this species is the natal down, which is light gray in color. Young wood thrushes, at hatching, average 5.08 grams in weight. They are approximately 46 millimeters in length; the tarsus averages 17 millimeters; the wing 9 millimeters; and the gape 9 millimeters. On the second day the eye slit breaks through the skin but the eyes are not open. The eyes open between the fifth and seventh days (Brackbill, 1943). On the third day the first feathers, wing primaries, and tail rectrices pierce the skin. The first wing fluttering occurs between the ninth and twelfth days (Brackbill, 1943).

The female alone broods the young. Brackbill (1943) states that there was no progressive daily decrease in brooding at either of the nests he observed. During the cool hours of early morning the female brooded oftener and for longer periods than during the hotter part of the day. Brooding lasted throughout the nest life of the young even on the day they left the nest.

Both females on the nest and young wood thrushes were observed to sleep occasionally. The nesting adult would open her eyes quite frequently only to close them again. The young slept both in the nest and after they had left it. On one occasion a young bird that had been out of the nest four days sat for 10 minutes on a small branch with its feathers all fluffed out and its head under its wing. A young bird in the nest was so relaxed during sleep that its head hung down over the edge of the nest.

Young birds in the nest were marked with small pieces of colored feathers glued to the tops of their heads so that individual records of their activities might be kept. Both male and female birds feed the young in the nest, each bird being fed about every 20 minutes when the birds were seven and eight days old. Young in the nest were fed mulberries and honeysuckle berries as well as animal matter. The food call of nestlings is very weak. It is a single chip in a high-pitched tone. Sometimes the adults gave a squeaky call uttered with the mouth full of food, to get the young to open their mouths. Once the male bird pecked the female on the head to get her to move so he could feed the young. At times the female flew away while the male fed the young, but at other times she stood on the rim of the nest. Both adults disposed of excreta either by swallowing it or by carrying the excretal sac away. The adult that happened to be present at the time the sac was expelled attended to the disposal.

Brackbill's (1943) calculations showed that the male made two-thirds of the feedings while young were in the nest. The feeding day corresponds roughly to the time between sunrise and sunset. He also states that each bird was fed 47 times per day. However, he found that birds of the second brood were fed every 39 minutes in comparison with 19½ minutes for the first brood, which resulted in but 24 feedings per bird per day, or half as many as the first brood received.

The female birds observed by both Brackbill (1943) and myself seemed to use some care in the type of food given the young. She fed them caterpillars and small insects. If the male brought large or hard-shelled or winged insects she ate the head, wings, and other less easily digested parts before offering the remainder to the young. The female was seen to divide food brought by the male among several young instead of allowing one to have it all.

Brackbill states that young birds begin to forage for themselves when 20 to 23 days old, although they may beg for food from adults anywhere up to 32 days.

The adult male birds were usually in the close vicinity of the nest to defend the young. The degree of their attentions varied greatly with the individual bird. There is some indication that the protective instinct is stronger early in the season. In one case the male spent most of the time while the female was off the nest perched above the nest. Other males perched in view of the nest but sometimes as much as 20 feet away. The male at one nest early in July was very inattentive. He did not guard the nest closely, almost never helped in feeding the young.

During the two days before leaving the nest, the young birds engaged in preening their feathers and occasionally beating their wings rapidly. They would also stretch one leg or a wing its full length and at times raise the whole body while standing on and stretching both legs. A few times one was observed to shake the whole body to fluff all the feathers. Sometimes one bird would spend as many as five minutes exercising while the other two in the nest remained quiet. When the one that had been exercising had finished, it would settle down to rest while another would take up the business of exercising. At times all the birds fluffed feathers, preened, and stretched at the same time. In doing so they crowded each other to the edge of the nest so far that a sudden stretching of a wing was necessary to prevent the bird from falling from the nest. Between periods of exercising the young spent a considerable amount of time sleeping.

Few references were found in the literature to the number of broods raised by the wood thrush in a season. Minot (1895) indicates second broods near Boston, saying that first sets were laid last week of May and "those of the second, if any, in the early part of July." Harbaum (1921) observed a pair through two nestings, but no evidence of the birds being marked is given. The writer set out to find definite proof with marked birds (since also found by Hervey Brackbill in Baltimore, Md., 1943). In Ithaca, N. Y., young from a nest under observation were placed in a drop trap just previous to the time for their departure from the nest. First one adult was caught. This one was kept at a distance in a collecting cage. The food calls of the young in the trap attracted the other adult. When caught the adults were marked with colored celluloid bands, as well as aluminum bands. A colored chicken feather was glued to the tail feathers of each bird.

The first egg was laid in the first nest on May 19. The young birds left the nest on June 15. On July 2 the second nest was found. The same birds had mated and two eggs were in the nest. These hatched on July 16 and 17. This second nest was located about 10 feet from the first nest. If a pair of birds has to make two or three trials before being successful in raising a brood, the season is too far advanced for a second brood; if, however, the first nest is successfully raised a second brood may follow.

The young birds unceremoniously leave the nest 12 or 13 days after hatching. I observed no coaxing or "teaching" on the part of the adults. The first in one case flew and alighted on the ground about 20 feet south of the nest. The second, when frightened, stood on the edge and flew, alighting on a small branch near the trunk of a hemlock tree 20 feet above the ground. The third bird in this nest remained quietly resting for half an hour. Without being disturbed it stood on the edge of the nest and flew to the ground alighting about 10 feet away.

The male bird seemed to defend the entire feeding territory although he took charge of feeding certain of the young out of the nest while the female fed certain others. This seems to need further study. Brackbill (1943), in a case of three birds in a brood, says that the male fed two out of the three in each of two successive broods, while the female fed the third bird.

After leaving the nest the young stay in a limited territory near the nest for several days. As their ability to fly increases they move about. Birds were found in the vicinity of nests six to nine days after they had left. They were still being fed by the adults. The adults would return to the spot where the young was fed last. The food-getting was usually confined to an area close to the young. The young bird would often lift and flutter its wings rapidly when approached by the parent with food, at the same time uttering rapid, squeaky calls.

Plumages.—The newly hatched wood thrush is clothed in natal down. The juvenal feathers are a continuation of this down, which is carried out on the tip of the new feathers and is finally rubbed off. The down is still present when the young leave the nest. The loss of the down is the postnatal molt. The juvenal plumage is the first complete plumage of the bird following the natal down and is acquired by the growth of new feathers. Sheaths are short and are lost quickly, as a bird when ready to leave the nest has no sheaths on the short body feathers. Therefore, when sheaths are found approximately six weeks later, they are known to be those of new feathers. Then, at six weeks the juvenal plumage is lost by an incomplete postjuvenal molt in which the body feathers are lost but not the flight feathers, neither in wings nor tail. This molt brings the bird into its first winter plumage.

The first nuptial plumage is supposedly acquired by abrasion or feather wear accomplished by casting the points of the feathers. This takes place on the wintering grounds before spring migration.

The postnuptial molt is complete, both body and flight feathers being lost. Specimens were found in molt at the end of July. These birds were practically "bob-tailed." At this time the birds are secretive in habits, of necessity.

Wetmore (1936) states that a female weighing 60.4 grams had 2,075 feathers, which weighed 3.2 grams.

Brackbill (1943) banded a partially albino female wood thrush. It had a white feather in the crown, some white feathers among the upper tail coverts, and four white rectrices. The eyes were normal. No sign of the inheritance of this character could be noted in any of the two broods of three young.

Food.—Thrushes are insectivorous but are fond of fruit. The U. S. Fish and Wildlife Service supplied the following data on the food of this species, based on 179 stomachs examined. These birds were taken from 19 States, 5 from Ontario, Canada, and 19 from the District of Columbia. They were unequally distributed over nine months of the year, the months of May and July yielding more specimens.

In the 179 examinations, 62.25 percent of the material consisted of animal matter and 37.75 percent vegetable matter. Of the latter, 3.49 percent was cultivated fruit and 31.2 percent wild fruit. In general, this bird is beneficial, although 2.17 percent of its food is composed of useful Coleoptera, 8.38 percent of Arachnida, and 3.49 percent of cultivated fruit, making a total of 14.04 percent of food substances useful to man. This percentage is canceled by its consumption of the following predators: Orthoptera, 2.1 percent; Rhynchophora, 2.16 percent; and Lepidoptera, 11.29 percent, making a total of 15.55 percent of food substances in the harmful class.

During my observations the following foods were fed to young birds: Several species of moths, ants, spiders, caterpillars, mulberries, honeysuckle berries, earthworms, and cankerworms.

Economically, then, the wood thrush is to be encouraged, for its food habits prove that it is a valuable aid in the destruction of many injurious insects and but few beneficial ones.

It was observed that soon after their arrival the birds fed from the foliage of the newly leaved trees. In their feeding they sometimes broke off pieces of leaves which fell to the ground. Later most of the feeding was done on the ground. Their presence in an area can often be detected by noticing them turning over leaves with the bill. A letter from D. J. Nicholson to A. H. Howell contains an interesting incident in which the writer saw a wood thrush eating pokeberries by springing up 18 inches and plucking off a berry. He also took a picture of these birds eating fallen gallberries.

Behavior.—When alarmed, wood thrushes, both adults and young, raise and lower the feathers of the head, giving the appearance of a crest. The young have been observed to do this only just previous to the time for leaving the nest, or after the fear instinct has been developed.

While incubating and brooding, when the temperature was high, the female lifted her wings from her sides and raised the feathers of her back and sides to allow air to circulate beneath them. She and the young in the nest often sat with mouths open when it was hottest.

The wood thrushes' reactions to storm and rain were noted by Brackbill (1943) and by the writer. Brackbill observed a female in a heavy downpour. She sat closely, holding her head at an angle of 60°, presumably to compress the feathers at the nape of the neck. During a late afternoon storm it became very dark and the male bird I was watching, who had been guarding the nest in the absence of the female, settled on the nest and covered the young. It rained and the wind blew quite hard. During a very heavy wind, Brackbill watched an incubating female. At times the eggs seemed perilously near falling from the nest, as it was tilted by the wind, but the female remained calm. A few times she almost lost her balance.

[AUTHOR'S NOTE: When living in our towns and cities, the wood thrush losses much of its natural shyness and timidity. We often see it leave the shelter of the shrubbery or leafy thickets in the more secluded borders of our grounds and come out onto the lawn in search of food, almost as fearless as a robin. It may even visit the bird bath or take a shower bath under the automatic lawn sprinkler, provided that we are careful not to frighten it by too close an approach. It seems to be more trustful of its human neighbors than does its shy relative in the woodlands.]

Voice.—Saunders (MS.) says: "The song of the wood thrush is long continued, made up of a number of different phrases, sung in varied order, with rather long pauses between the phrases. It sounds like *eeohlay——ayolee——ahleelee——ayleahlolah——ilolilee,* etc." Elsewhere (1924) he writes: "Each phrase may have three parts, an introduction of two or three short notes, usually low in pitch and not especially musical; a central phrase of two to five notes, most commonly three, loud, clear, flute-like and extremely musical; and a termination of three or four notes, usually high-pitched, not so loud, and generally the least musical part of the song. Phrases may be sung either with or without either introduction, termination or both, and sometimes, especially late in the season, birds indulge the habit of singing only introductions and terminations, leaving out the beautiful central phrases." Saunders made records of 115 wood-thrush songs, which showed the pitch to range from D'' to D'''', two complete octaves. The average bird has a range of a tone or two over an octave.

The song of the wood thrush early in the season is more elaborate, performed with more vigor, and is of longer duration than songs later in the season.

The calls fell into three classifications discussed here in order of the degree of feeling seemingly expressed by them. When slightly disturbed, or uneasy, apparently to indicate his or her presence, both male and female utter a sound that can be expressed by *trrrrrr, trrrr,* a sort of rattle or trill. The other bird would then often respond with the same call. If they became alarmed they used the *pit, pit, pit* call. When greatly alarmed, as when danger threatened nest or young, the call changed to *quirt, quirt, quirt, quirt,* usually accompanied by swift zooming flights at the intruding person, bird, or object. During this defensive demonstration the bill was snapped and the birds came within a few inches of the object of their fury. This happened once upon the erection of a blind.

Another sound made by both male and female was a squeaky whistle. During the nesting period both male and female were heard to give this clear whistle upon several occasions. This was used by either adult upon arrival at the nest with food, especially when the young did not open their mouths to receive it. Sometimes it was necessary for an adult to repeat this at least four times. It can be described by saying that it seemed to have been produced by inhaling with the bill almost closed. It was often given when the bill was filled with food. The female sometimes gave this sound while she was on the nest. It was also used by the male when he sat at some distance from the nest or when he arrived with food and the female did not leave the nest so that he could feed the young.

Brackbill (1943) describes a "rudimentary or vestigial song"—an explosive one used by the female in defense of territory.

The young birds uttered a faint *chip*, which was the food call. This was not loud enough to be heard at any distance but could be heard easily from the blind. My first record of such a call was made, as mentioned before, four minutes after the young was free from the shell. This call also serves to indicate the location of the birds after they have left the nest. Brackbill (1943) says that the juveniles began using a rudimentary form of the adults' rattle or trill, which consisted of three or four notes, at the age of 21 days. Also when chased a young bird bursts into a series of calls similar to the adult call I have described as *pit, pit.*

Early in the season wood thrushes perched in the tops of the highest trees in their territories to sing their loudest, most complete, and most varied songs of the season. A week later perches were about 15 feet above the ground. Often they chose short, dead branches of hemlocks. Others were known to sing from the ground, from large logs, in the nest tree, or even from the edge of the nest in the absence of the female.

From my observations, wood thrushes begin their morning songs with the break of day, singing at the end of June at 3:45 A. M. in Ithaca, N. Y. At this time it is still quite dark and feeding has not yet begun. This singing continues both through the periods of incubation and brooding. Evening song usually ceases at dark, or about 8:00 P. M. in June in Ithaca. Wright (1912) made a study of morning awakening and evening songs of birds in the White Mountains of New Hampshire in which he recorded the average of early wood-thrush songs as 3:26 A. M. The earliest sunrise during the study was 4:02, and so the wood thrushes sang about half an hour before sunrise.

Song on the breeding grounds begins with the arrival of the first birds; so it is believed that the males do not wait until the females arrive before the song period begins. Males arrive and sing to denote the possession of their territory. There was song in the evening only, after the young had left the nest.

On one occasion Brackbill heard a wood thrush sing a song of good quality while on the wing and not in defense of territory.

Forbush (1929) describes the calls and song as follows: "Notes, a liquid *quirt*, a low *tut tut*, a sharp *pit pit* or *pip pip* and a shrill *tsee tsee*. Song, a pure, clear, sweet, expressive, liquid refrain, often with a bell-like ending; usually composed of a series of triplets, each beginning with a high note, then a low one, then a trill, often highest of all, but the different phrases varying in pitch. It is calm, unhurried, peaceful, and unequaled in both power and beauty by any other woodland songster of New England."

Saunders (1921a) claims that individual male wood thrushes have characteristic songs by which they can be identified. John Burroughs (1880) also makes this statement. Saunders says that a count of pairs merely by singing males is not reliable, but such a count would be possible if individual songs were studied.

Charles W. Townsend (1924) and Francis H. Allen, quoted by him, state that they heard catbirds mimic the song of the wood thrush. The white-eyed vireo, according to Brand, is an imitator of this bird, and Forbush (1929) reports that it has been known to imitate the wood thrush.

About the middle of October 1927, at 2 A. M. on a moonlight night, a large flock of birds alighted in tops of street basswoods in residential Washington, D. C. (Hazen, 1928). Immediately at least 10 wood thrushes burst into full song. They sang continuously for 20 minutes, then one lone bird sang until the flock disappeared at 2:45 A. M. The thrushes were accompanied by small tree-top birds, either vireos or kinglets.

The latest songs of the wood thrush in the autumn were recorded on July 28 and August 10 at Ithaca, N. Y. Brackbill (1943) records August 2 in Baltimore, Md. Saunders (MS.) gives July 29 as his average fall song date. His latest dates are August 8, 1928, and September 7, 1941. As a rule, then, the song period closes about the end of July, and little is seen or heard from the birds from then on. During the postnuptial molt birds were located by listening for the calls of excitement, but no songs were heard.

Field marks.—The members of the genus *Hylocichla* have more or less spotting on the underparts and the young are spotted above and below in the juvenal plumage. The distinguishing characteristics that separate this species, *mustelina*, from the others of the genus are its larger size (over 8 inches in length) and greater sturdiness. Its upperparts are bright cinnamon-brown, being brightest on the *head* and changing gradually to olive on the upper tail coverts and tail. (The hermit thrush has the cinnamon-brown most pronounced on the *tail.*) The underparts are white, thickly marked with large *rounded* dark brown spots, except on the throat and middle of the belly. This species is more strongly marked than others of the genus; the spots are larger, more distinct, more numerous, and more generally dispersed. The spots extend well down on the flanks, more so than on any of the other thrushes. In distinguishing this bird from confusing species other than thrushes, we may eliminate the brown thrasher by the fact that it is longer-tailed and is streaked below rather than spotted. The large dark eye of the thrush is to be compared with the yellow eye of the thrasher (Forbush, 1929). The fox sparrow is reddish-tailed, underparts streaked, not spotted, and the bill is thick, conical, and sparrowlike rather than slender like that of the thrush.

Enemies.—Cats are responsible for the destruction of some wood thrushes. "Causes of death" listed on banding returns in the files of the United States Fish and Wildlife Service show that on 74 returns 8 percent of the deaths were due to cats.

In 1938 three nests were deserted because of the destruction of the entire clutch of eggs. In each case a few very tiny bits of the shells were found in the nests. Those acquainted with the predators of the region offered the suggestion that the red squirrel may have been guilty.

Approximately one-fifth of the nests studied were parasitized by the cowbird. This social parasitism decreases the numerical strength of the species by causing desertion in some cases. A few instances will be cited. In a nest containing one cowbird egg and three wood-thrush eggs, the thrush eggs disappeared, one each day, until only the cowbird egg was left. The nest was deserted after the last egg disappeared. In another case the writer removed three cowbird eggs from one nest. In one nest containing a day-old cowbird and one thrush egg, the thrush egg did not hatch. In contrast, however, a nest contained four wood-thrush eggs and one cowbird egg, all of which hatched and all the young survived to leave the nest.

Friedmann (1929) in his book on the cowbird gives interesting information on its relation to the wood thrush. The wood thrush is larger than the cowbird yet is frequently parasitized and is also seen caring for young cowbirds. Often the cowbird is the only survivor in a thrush nest. Records of such parasitism come from New England, Connecticut, New York, Pennsylvania, and Maryland, Ohio, and Indiana.

Perry (1908) describes a wood-thrush nest in Illinois in which five cowbird eggs were laid. These were laid by at least two cowbirds, since two were deposited on the same day. Despite this, one young wood thrush lived to leave the nest successfully.

In one instance I recorded the feeding times of a young cowbird which was in a nest with one wood thrush. There are many uncontrolled factors that would discount any conclusions drawn, but in this case the feedings were of the same number, averaging about one every 15 minutes.

Wood thrushes react to cowbird eggs in several ways. There was some evidence that the wood thrush tried to imbed the intruder's eggs. Friedmann (1929) said in the majority of cases the eggs and young were tolerated, and that Lynds Jones knew a case in which the wood thrush tried to throw the cowbird egg out.

Wood thrushes seem to be fairly free from external parasites, as none were found on the many birds handled during the study made by the author. There were indications at times that the adults picked them from the nest; one time, after picking in the nest, the adult put her bill into the mouth of a young bird seeming to feed it. Peters (1936) lists two parasites found on this species, one a louse (*Myrsidea incerta* (Kellogg)), on a specimen from New York, and a tick (*Haemaphysalis leporis-palustris* Packard), on a specimen from North Carolina.

A study of the Federal Government's records of bird-banding shows that, up to May 1936, the oldest banded wood thrush was six years old. From my own records, I had a bird three years old, which nested in the same vicinity each of the three years.

Fall.—Records of the United States Fish and Wildlife Service were studied to make this summary of fall migration. In the Atlantic flyway in *September*, wood thrushes are still in their summer range, being found as far north as they were during the summer months, although there is no way of telling in what numbers they are present. The autumn migration begins in *October*. The northernmost records for October are one for Maine, one for central Vermont, and one for New Haven, Conn. They are still numerous in New York, Pennsylvania, New Jersey, and the District of Columbia. In *November* there are about a dozen records of birds in the United States, so that by this time most birds have left our country. The first actual record of a bird south of the United States is one for Almirante, Panama, on October 30.

In the Mississippi flyway in *September*, wood thrushes are still as far north as they were during the summer, being found in Ontario, but *October* records show a southward movement. The last fall record is at Ozark Beach, Mo., on October 27.

There were only six records on file of fall dates in the Western flyway. In *September* a record was made in Sioux Falls, S. Dak. The last fall record was made at Independence, Kans., near the end of *October*.

In comparing these data we find that wood thrushes remain longest in the United States in the Atlantic Coast States.

There is an interesting note by Weston (1935) about the fall migration of wood thrushes at Pensacola, Fla.: "A heavy flight of wood thrushes filled the swamps with birds on October 13." The latest date given by Howell (1932) for Florida is October 14.

In Ithaca, N. Y., the latest fall record is September 18.

A. A. Allen (1934) and Lincoln (1935) state that the thrushes migrate at night because light is less intense, they can then better avoid their enemies, and they can take care of feeding during the daytime.

Fall migration, then, can be described as irregular since some birds are found in the United States (Florida) every month of the year, but most have left the country by the end of October.

A study of migration records leads me to conclude that there is no definite route of migration either in spring or fall. This is contrary to the statement that thrushes prefer the Mississippi Valley flyway.

Winter.—Most wood thrushes have left the United States by the end of October, but there are a few records for November some of which are quite far north: New Hampshire, New York, New Jersey, and Pennsylvania. The scarcity of records, however, would indicate that these are stragglers and that by November most of the birds have left the United States. The United States Fish and Wildlife Service had but a dozen records for November at the time this study was made, and of those, four were from places south of the United States: One for Nicaragua, two for Costa Rica, and one for Panama. In December there are records from New Jersey, South Carolina, Alabama, and Florida, in the United States, and two for places south of our country, one from Costa Rica and one from Barro Colorado Island, Panama. The January records are from Georgia, Florida, and Mexico, only one record in each case. These were the only records in the files for that month. The February records consist of four from Florida, with others from Mexico, Guatemala, Honduras, and Costa Rica.

Roberts (1932) summarizes, saying: "Winters in southern Mexico and Central America and occasionally in Florida. Casual in migration in the Bahamas, Cuba, and Jamaica; accidental in Colorado and Bermuda."

In checking the Christmas bird-census records in Bird-Lore, I found two records of observations of the wood thrush. At Cape May, N. J., 1934, one was seen by McDonald and others. Another record was from Paris, Tenn., where three were seen in 1933.

Howell (1932) lists four records of wintering individuals in Florida.

O. Salvin (1888), in writing of the birds of the islands off the coast of Yucatán, says: "A migratory species from the north, and common in Cozumel Island. It has not been noticed in Northern Yucatan, but it occurs in Cuba, though rarely. It is abundant in the winter months in Southern Mexico and Eastern Guatemala, the southern limit of its range being Northern Honduras."

Alexander F. Skutch, an American ornithologist in Costa Rica, writes me that he observed a wood thrush on Barro Colorado Island, in the Canal Zone, in March 1935. He says it "was singing when I came upon him. Although it is stated by Carriker and others that the North American birds which winter in Central America are 'almost invariably as silent as so many shadows', this is quite untrue. Many of the song birds which pass the winter here begin to sing a short time before their departure for the north."

A note from Mr. Skutch to A. C. Bent summarizes his Central American observations of the wood thrush: "The wood thrush winters in Central America throughout the length of the Caribbean lowlands; but I have found it far from abundant in Guatemala and Honduras, and exceedingly rare in Costa Rica and Panama. During the winter months it does not form flocks, but leads a solitary life, in the undergrowth of the forest, in low moist thickets, or even in banana plantations. On March 21, 1935, I heard a wood thrush sing in the undergrowth of the forest on Barro Colorado Island, Canal Zone. His song was subdued but perfectly distinct, and beautiful as always. The single thrush was in company with antbirds of several kinds. My only records that suggest the time of arrival and departure of this migrant are: Tela, Honduras, October 1, 1930; near Los Amates, Guatemala, April 4, 1932; and Barro Colorado Island, March 21, 1935."

DISTRIBUTION

Range.—Eastern North America from southern Canada to Panama.

Breeding range.—The wood thrush breeds **north** to northern Minnesota (Deer River and Duluth); northern Michigan (Iron County, McMillan, and Mackinac Island); southern Ontario (Lake Nipissing, Algonquin Park, and Ottawa); southern Quebec (Lac Manitou and Montreal, and casually to Gaspé County); central Maine (Phillips, Sidney, and Dover-Foxcroft, occasionally); southern New Brunswick (St. Stephen); and possibly Nova Scotia (Digby). **East** to southwestern New Brunswick (St. Stephen); possibly Nova Scotia (Digby); the Atlantic Coast States south to northern Florida (Orange Park, Middleburg, and Bostwick). **South** to northern Florida (Bostwick, Waukeenah, Whitfield, and Pensacola); Alabama (Spring Hill), Louisiana (Madisonville, Baton Rouge, and Avery Island); and southeastern Texas (Houston). **West** to eastern Texas (Houston, Tyler, and Marshall); central Oklahoma (Fort Reno and Ponca); central Kansas (Wichita, St. John, and Hays); central Nebraska (Red Cloud, North Platte, North Loup, and Neligh); southeastern South Dakota (Yankton and Sioux Falls, casually Pierre); and northern Minnesota (St. Cloud and Deer River).

Winter range.—In winter the wood thrush is found **north** to extreme southern Texas, rarely or casually (Brownsville); northern Florida (Whitfield and Gainesville). **East** to Florida (Gainesville and Fort

Myers); Quintana Roo (Palmul and Xcopen); British Honduras (Orange Walk and Cayo); Honduras (Omoa, Tela, and Ceiba); Nicaragua (Escondido River); Costa Rica (Peralta and Tuis); and Panama (Almirante and Barro Colorado, C. Z.). **South** to Panama. **West** to Costa Rica (Palmar and Miravalles); El Salvador, rarely (Mount Cacaguatique and Lake Olomiga); Guatemala (Godines); Oaxaca (Tehuantepec); western Veracruz (Matzorongo); eastern Puebla (Metlaltoyuca); eastern Tamaulipas (Soto la Marina); and southern Texas (Brownsville). There are late December records from Plainfield, N. J.; Columbus, Ohio; Raleigh, N. C.; Fort Worth and Houston, Tex., which may be delayed migration or accidental wintering.

Migration.—Some late dates of spring departure from the winter home are: Canal Zone—Barro Colorado, March 21. Honduras—Tela, March 7. Guatemala—Quiriquá, April 4. Veracruz—Cerro de Tuxtlá, March 29. Florida—Fort Myers, May 24.

Some early dates of spring arrival are: Florida—Pensacola, March 24. Georgia—Macon, March 28. South Carolina—Aiken, March 17. North Carolina—Waynesville, April 4. West Virginia—Morgantown, March 16. District of Columbia—Washington, April 4. Pennsylvania—Glen Olden, April 26. New York—Syracuse, April 25. Massachusetts—Stockbridge, May 2. Vermont—Rutland, May 4. Louisiana—New Orleans, March 27. Arkansas—Helena, April 4. Tennessee—Nashville, April 5. Indiana—Indianapolis, April 23. Michigan—Grand Rapids, April 12. Ohio—Oberlin, April 10. Ontario—Toronto, April 24. Missouri—St. Louis, April 18. Iowa—Keokuk, April 20.—Wisconsin—Milwaukee, April 30. Minnesota—St. Paul, May 3. Kansas—Independence, April 25. Nebraska—Lincoln, April 25. South Dakota—Yankton, April 27.

Some late dates of fall departure are: South Dakota—Yankton, September 19. Nebraska—Omaha, September 30. Minnesota—Minneapolis, October 10. Wisconsin—Madison, October 12. Iowa—Marshalltown, October 2. Missouri—St. Louis, October 19. Ontario—Ottawa, October 4. Ohio—Columbus, October 5. Michigan—Detroit, October 5. Illinois—Lake Forest, October 28. Kentucky—Danville, October 20. Louisiana—New Orleans, October 18. Vermont—St. Johnsbury, September 29. Massachusetts—Marthas Vineyard, October 2. New York—Rhinebeck, November 4. Pennsylvania—Pittsburgh, October 4. District of Columbia—Washington, November 27. North Carolina—Raleigh, October 16. Georgia—Athens, October 18.

Some early dates of fall arrival in the winter home are: Honduras—Tela, October 1. Nicaragua—Bluefields, November 7. Panama—Almirante, October 30.

Casual records.—In October 1849 several specimens were taken in Bermuda. One was noted on the Mazaruni River, British Guiana, on March 1, 1916. There are three positive records of occurrence in Colorado: a specimen from near Holly, Powers County, on May 12, 1913; two specimens from Dry Willow Creek, Yuma County, on June 24, 1915; one from Boulder, on May 13, 1942, and several sight records.

Egg dates.—Illinois: 39 records, May 3 to July 10; 20 records, May 22 to June 7, indicating the height of the season. Massachusetts: 33 records, May 14 to June 24; 16 records, May 26 to May 30.

New Jersey: 45 records, May 20 to July 18; 26 records, May 23 to May 30.

West Virginia: 62 records, April 25 to May 24; 48 records, May 12 to May 21.

HYLOCICHLA GUTTATA GUTTATA (Pallas)

ALASKA HERMIT THRUSH

HABITS

This, the type race of the species and the first of the several races to be named, is the form that breeds from the Mount McKinley region in Alaska south to Cross Sound, Kodiak Island, and northern British Columbia. It migrates south in winter to Cape San Lucas and northern Mexico.

All the western subspecies of the hermit thrush have the sides and flanks grayish or olivaceous, rather than brown or buffy brown; the bill is relatively smaller or more slender, the tail relatively longer, and the feet relatively smaller than in the eastern subspecies. The Alaska hermit thrush is one of the two smaller races, and its coloration is lighter than the other even smaller race, the dwarf hermit thrush.

Spring.—In his notes made at Lake Crescent, western Washington, for April 25 to 30, 1916, S. F. Rathbun (MS.) thus describes the behavior of the Alaska hermit thrush on its migration: "Each morning at daybreak I hear the songs of several hermit thrushes, coming from the thick fringe of shrubs and young growth along and near the shore of the lake. This song is most delightful and continues for the space of about an hour, with but little intermission; after that it is seldom heard, and through the day not at all; but near sunset the birds sing again for a short time, and the effect of this beautiful song, heard in the waning light of day, is most pleasing.

"These birds are of a retiring nature; they haunt the rather more open forest, or its edges where there may be open spots, but they always seem to remain in close proximity to the forest. In such a locality one morning I had an opportunity to watch one of the hermits for some time; it is a very quiet bird, and when in repose it is perfectly motionless; from time to time it would rapidly move a short distance, looking for its food on the surface or under some of the leaves with which the ground was covered; but it was not continually in motion, and there were frequent intervals of complete repose, though should a winged insect chance to fly past, which happened several times, the bird sprang into the air quicker than a flash and was generally successful in catching it; after that its former motionless position was resumed. During my observation, which lasted for some time, the bird uttered no note and, as I remained perfectly quiet, it came at times within a few feet of where I was.

"By the first of May all the birds seemed to have passed by, and their songs were no longer heard."

Nesting.—The nesting habits of the Alaska hermit thrush seem to vary somewhat in different portions of its range, and for no apparent reason.

In the Stikine River region, in northern British Columbia, Harry S. Swarth (1922) found this thrush in the spruce woods on the mountains at about 3,000 feet altitude. He collected two nests with eggs; the first nest was taken on May 26, with a set of five eggs. "It was in the creek bottom, about two miles north of the town of Telegraph Creek, some three feet from the ground, in a spruce sapling. The nest rested against the trunk and upon some small branches. The outer structure is of twigs, weed stems, rootlets and bark strips; the lining is of fine rootlets and grass, with a good many of the long overhairs of a porcupine. It measures as follows: greatest outside diameter about 160 mm.; outside depth, 90; inside diameter, 60; inside depth, 40 mm."

The second nest was taken on June 4, with four slightly incubated eggs; it was in similar surroundings and was very much like the first one in structure, even to the porcupine hairs in the lining, but this one was "placed between two small spruce trees, thirty inches from the ground. * * * Both were in situations where there was little concealing vegetation, and were easily seen from some distance."

A short distance farther north, in the Atlin region, Mr. Swarth (1926) found three nests: "One, June 13, with three fresh eggs; one June 23, with four fresh eggs; and one July12, with four fresh eggs. All were on the ground, the first in a clump of small willows at the edge of a muskeg, the second in an opening in mixed poplar and spruce woods, and the third in rather dense poplar woods."

Eggs.—The Alaska hermit thrush lays three to five eggs, probably usually four. These are practically indistinguishable from those of the dwarf hermit thrush and differ from those of the other hermit thrushes only in size. The measurements of 18 eggs average 21.9 by 16.2 millimeters; the eggs showing the four extremes measure **23.3** by 16.7, 22.4 by **16.8, 20.8** by **15.8** millimeters.

Food.—The food of all the western races of the hermit thrush may as well be considered here, as several of the forms live or spend the winter in California, and as Professor Beal's (1907) study of the stomach contents of 68 hermit thrushes, taken in California, does not separate the food of the different subspecies. The food of all the races is much the same under similar conditions. Beal's analysis shows the food to consist of 56 percent animal and 44 percent vegetable matter. Of the animal food, Hymenoptera, mostly ants, constitute the largest item, 24 percent; caterpillars come next, 10 percent; beetles, all harmful species and more than two-thirds weevils, form 11 percent of the food; other insects, spiders, and miscellaneous items amount to 12 percent. One stomach contained the bones of a salamander. Beal writes:

The vegetable food is made up of two principal components—fruit and seeds. The former amounts to 29 percent of the whole, and is composed of wild species, or of old fruit left on trees and vines. A few stomachs contained seeds of raspberries, which, of course, must have been old, dried-up fruit. Seeds of the pepper tree and mistletoe were the most abundant and, with some unidentifiable pulp and skins, make up the complement of fruit. * * * Seeds of all kinds amount to 14 percent of the food, but only a few are usually reckoned as weed seeds. The most abundant seed was poison oak (*Rhus diversiloba*), which was found in a number of stomachs. While this plant is not usually classed among weeds, it is really a weed of the worst description, since it is out of place no matter where it is. It is unfortunate that the birds in eating the seeds of this plant do not destroy them, but only aid in their dissemination.

Mr. Dawson (1923) watched them feeding in his yard and says:

They tackle the pepper berries, and rather awkwardly at first. It is evidently new business for some of them, and they make hard work of it. One bird that I particularly observed would fly up to a bunch, hover a moment in midair, snatch a berry, and return to a more secure position. This he did repeatedly, without once endeavoring to alight on the berry cyme itself, or trying to find a place where he might eat his fill unmolested. Another dashed up and fell to eating the berries as they lay strewn upon the ground. He fed very daintily, taking care in each instance to discard the red husk. * * *

One of our garden faucets drips incessantly and this is the favorite drinking place of the Hermit. A bird will alight on the faucet and, stooping over, will pluck the drops one by one as they fall. One morning I saw five birds at a time either waiting their turn or else making suggestive dives at the fellow who seemed to be tarrying too long at the faucet.

Behavior.—Dawson (1923) gives the following good description of a well-known bit of action that is common to all hermit thrushes, and by which they can often be recognized:

Perhaps the most prominent characteristic of the Hermit Thrush, and the one which does most to remove it from the commonplace, is the incessant twinkling of the wings—the action is so rapid and the return to the state of repose so incalculably quick that the general impression or silhouette is not thereby disturbed; but we have an added feeling of mobility of tensity on the part of the bird which gives one the impression of spiritual alertness, a certain high readiness. I tried on a time to count these twinkles, with the compensatory flirt of the tail, as the bird was hopping about on the ground in my rose garden. The movements occurred about once per second, yet oftenest in groups, and so rapidly, that not a twentieth part of the bird's time seemed so consumed. * * *

In one station which the bird occupied, being not over seven feet from me, I could, by closing one eye and focussing the other upon a closely placed background of greenery, note the extreme limit of the wing-motion. The tip, in each instance, travelled at least two inches from the body; yet the return was so instant and the dress so quickly composed that no detail of the readjustment could be traced.

Howard L. Cogswell refers in his notes to the wing-flipping habit, described above, and says that "the hermit's habit of slowly raising its tail after alighting, so often used in identification in the East, does not always take place in western birds, I have found."

Fall.—Mr. Rathbun's notes mention the first arrivals of the Alaska hermit thrushes in Clallam County, Wash., on October 10, 1915, when seven were seen along the shore of Crescent Lake near the beach, where the shore was overhung with bushes growing at the edge of the water.

Referring to the Yosemite region in California, Grinnell and Storer (1924) say: "By the latter part of September, birds which have nested in various parts of southern Alaska begin to arrive, to spend the winter here. In the fall the Dwarf and Alaska hermit thrushes, as the two races from the north are called, occur in considerable numbers at all altitudes below 9,000 feet. The arrival of heavy snow forces most of those in the higher zones to below the 4,000 or 3,500 foot contour."

Winter.—Mr. Cogswell (MS.) says of this species: "The hermit thrush is common throughout the winter in coastal southern California, but most common in the shady oak-sycamore association of the canyons of the foothills, in tall, dense growth of climax chaparral (in small canyons), and in the residential sections of many cities wherever there is plenty of brush and hedge cover and a steady water supply. The highest altitude at which I have seen the hermit thrush in midwinter was about 5,000 feet in the upper Santa Ana Canyon, San Bernardino Mountains, on December 28, 1941. This one was calling the usual *chuck-chuck* note from underneath a canopy of snow-covered chaparral, about 9 A. M., with the temperature at 20° F. Occasional birds are heard singing in soft, detached phrases from mid-March until they leave early in April; but during the rest of their stay in the southern California lowlands, the *chuck* note and a louder, ringing *cheeeeeeeee* (slightly rising pitch) are their only notes."

DISTRIBUTION

Range.—North America from central Alaska and northern Canada to Guatemala.

Breeding range.—The hermit thrush breeds **north** to central Alaska (Lake Aleknagik, Lake Clark, Mount McKinley, and Chitina Moraine); southern Yukon (Donjek River, Little Salmon River, and Watson Lake); southwestern Mackenzie (mouth of the Nahanni River, Fort Simpson, Fort Providence, Fort Resolution, and Hill Island Lake); central Saskatchewan (Snake Lake and Hudson Bay Junction); southern Manitoba (Lake St. Martin, Portage la Prairie, and Hillside Beach); central Ontario (Minaki on Lake of the Woods, Port Arthur, Lake Abitibi, and Ottawa); southern Quebec (Lake Mistassini, Godbout, and Natashquan); and southern Labrador (Mary Harbor and Chateau Bay). **East** to southern Labrador (Chateau Bay); Newfoundland (St. Anthony and St. John's); Nova Scotia (Baddeck, Halifax, and Barrington); New Hampshire (Rye Beach); eastern Massachusetts (Belmont, Roxbury, Cape Cod, and Marthas Vineyard), and Long Island (Yaphank). **South** to Long Island (Yaphank), northern New Jersey (Beaufort Mountain); northeastern Pennsylvania (Lords Valley and Pocono Mountains) and, in the Appalachians south to western Maryland (Grantsville and Mountain Lake Park) and West Virginia (Cheats Bridge); southern Ontario (Dunnville and Plover Mills); northern Michigan (Douglas Lake, Wequetonsing, and Escanaba); northern Wisconsin (Mamie Lake, Rhinelander, and Lake Owen); central Minnesota (Mille Lacs Lake and Otter Tail County); southern Manitoba (Winnipeg and Margaret); southern Saskatchewan (Indian Head); western Montana (Chief Mountain Lake, Great Falls, and Bear Tooth Mountains); and south along the eastern slope of the Rocky Mountains in Wyoming (Laramie); Colorado (Estes Park, Manitou, Wet Mountains, and Fort Garland) to southern New Mexico (Cloudcroft and Silver City); southeastern and central Arizona (Chiricahua Mountains, Tombstone, Santa Catalina Mountains, and San Francisco Mountain); southern Nevada (Charleston Mountains); and southern California (Providence Mountains, San Bernardino Mountains, the Sierra Nevada south to Big Cottonwood Meadows, and the Coast Range to Little Sur River). **West** in California to the Pacific Ocean (Little Sur River, San Francisco, Gualala River, and Carsons, Humboldt County); to the Cascades in Oregon (Crater Lake, Salem, Beaverton, and Olney); Washington (Mount St. Helens, Mount Rainier, and Tacoma); British Columbia (Vancouver Island: Nootka Sound and Errington, the Queen Charlotte Islands, Hazelton, and Telegraph Creek); and Alaska (Forrester Island, Sitka, Kodiak Island, Frosty

Peak, Alaska Peninsula, and Lake Aleknagik). There is an isolated colony in the Sierra de la Laguna in the Cape district of Lower California.

Winter range.—The hermit thrush is found in winter **north** to Vancouver Island, British Columbia (Comox and Victoria); western Washington (Bellingham and Seattle); western Oregon (Portland, Corvallis, and Fort Klamath); eastern California (Grass Valley, Placerville, and Providence Mountains); extreme southern Nevada (opposite Fort Mojave); occasionally to extreme southern Utah (Zion National Park); central Arizona (Fort Verde and Salt River National Wildlife Refuge); southern New Mexico (near Salinas Peak); western Texas (Guadalupe Mountains) and southern and eastern Texas (San Antonio, Corsicana, and Gainesville); southeastern Oklahoma (Caddo); central Arkansas (Maumelle); southeastern Missouri (Cardwell and Tecumseh, occasionally); southern Kentucky (Bowling Green); southern West Virginia (Bluefield); Virginia (Lexington and Beulahville); eastern Maryland (Catonsville); southeastern Pennsylvania (Philadelphia); and central New Jersey (Princeton); occasionally north to Columbus, Ohio; Easton, Pa.; Orient, Long Island; Providence, R. I.; and the vicinity of Boston, Mass. **East** to central New Jersey (Princeton) and the Atlantic Coast States to southern Florida (Daytona Beach, Titusville, and Royal Palm Park). **South** to southern Florida (Royal Palm Park); the Gulf coast of the United States; Tamaulipas (Victoria); Mexico (Amecameca); Puebla (Mount Popocatapetl); and Guatemala (Coban, Tecpám, and Volcán de Fuego). **South** to Guatemala. **West** to Guatemala (Volcán de Fuego and Momostenango); Jalisco (Jonila); Sonora (Alamos); Lower California (Triunfo, San Ramón, and Todos Santos Island); California (San Diego, San Clemente Island, Santa Barbara, San Francisco, Eureka, and Crescent City); Oregon (Corvallis); Washington (Port Angeles); and southwestern British Columbia (Vancouver Island: Victoria and Comax).

The above ranges apply to the species as a whole, of which seven subspecies or geographic races are recognized. The Alaska hermit thrush (*H. g. guttata*) breeds in Alaska south to Cross Sound; the dwarf hermit thrush (*H. g. nanus*) breeds in the coastal region from Cross Sound, Alaska, south to southern British Columbia; the Monterey hermit thrush (*H. g. slevini*) breeds in the coastal belt of California from northern Trinity County to southern Monterey County; the Sierra hermit thrush (*H. g. sequoiensis*) breeds in the higher mountains from southern British Columbia to southern California; the Mono hermit thrush (*H. g. polionota*) breeds in the White Mountains, Mono and Inyo Counties, Calif., and the Charleston Mountains, Nev.; Audubon's hermit thrush (*H. g. auduboni*) breeds from southeastern British Columbia south through eastern Nevada to the mountains of

Arizona and to the eastern base of the Rocky Mountains in Montana, Wyoming, Colorado, and New Mexico; also in the Sierra de la Laguna, Cape district of Lower California; the eastern hermit thrush (*H. g. faxoni*) breeds from southwestern Mackenzie, central Alberta, Saskatchewan, Manitoba, and northern Minnesota eastward to the Atlantic Ocean. The winter ranges of the various races overlap.

Migration.—Some late dates of spring departure from the winter home are: Guatemala—Tecpam, April 17. Mexico, Guerrero—Omilteme, May 14. Texas—Somerset, May 6. Louisiana—New Orleans, May 15. Arkansas—Helena, May 10. Mississippi—Bay St. Louis, May 11. Tennessee—Nashville, April 25. Kentucky—Lexington, May 17. Alabama—Leighton, April 8. Florida—Pensacola, April 23. Georgia—Athens, April 22. South Carolina—Charleston, April 20. North Carolina—Charlotte, May 6. West Virginia—French Creek, May 6. District of Columbia—Washington, May 17. Pennsylvania—Wayne, May 11.

Some early dates of spring arrival are: West Virginia—Parkersburg, April 1. Pennsylvania—Carlisle, March 21. New York—Ballston Spa, March 31. Massachusetts—Amherst, April 9. Maine—Orono, April 16. Quebec—Montreal, April 14. Nova Scotia—Halifax, April 19. New Brunswick—Grand Manan, April 5. Newfoundland—St. Anthony, April 24. Ohio—Oberlin, March 28. Ontario—London, March 30. Indiana—Bloomington, March 31. Michigan—Ann Arbor, March 26. Illinois—Chicago, March 26. Iowa—Keokuk, March 30. Minnesota—Hutchinson, April 13. North Dakota—Fargo, April 1. Manitoba—Winnipeg, April 22. Saskatchewan—Regina, April 23. Colorado—Colorado Springs, April 21. Wyoming—Torrington, April 19. Montana—Great Falls, May 7. Alberta—Glenevis, April 28. Mackenzie—Simpson, May 6. Alaska—Ketchikan, April 16.

Some late dates of fall departure are: Alaska—Wrangell, October 12. Mackenzie—Simpson, September 6. Alberta—Glenevis, October 7. Idaho—Priest River, October 12. Montana—Fortine, September 20. Wyoming—Laramie, October 30. Colorado—Denver, October 20. Manitoba—Aweme, October 15. North Dakota—Argusville, October 21. Minnesota—Minneapolis, October 28. Iowa—Marshalltown, October 24. Wisconsin—Madison, October 24. Illinois—Waukegan, November 4. Michigan—Detroit, November 8. Ontario—Toronto, October 29. Ohio—Cleveland, November 3. Newfoundland—St. Anthony, October 24. New Brunswick—Scotch Lake, November 3. Quebec—Montreal, October 27. New Hampshire—Concord, November 5. Virginia—Lexington, October 15. West Virginia—French Creek, October 5. North Carolina—Raleigh, October 16. South Carolina—Spartanburg, October 10. Alabama—Leighton, October

16. Florida—New Smyrna, October 13. Mississippi—Ariel, October 14. Arkansas—Delight, October 14. Texas—Dallas, October 16. Mexico: Chihuahua—Guachochi, September 28. Guerrero—Taxco, October 16. Guatemala—Tecpam, November 4.

Banding records.—Banded birds have furnished some interesting records of migration and longevity. One banded at Demarest, N. J., on October 9, 1938, was found February 10, 1939 at Valdosta, Ga. One banded at Zion, Ill., on October 10, 1933, was caught about January 22, 1934, at De Queen, Ark.; another banded at the same place on April 30, 1939, was killed January 24, 1940, at Pasadena, Tex. One banded at Blue Island, Ill., October 1, 1934, was found about December 19, 1934, near Gould, Ark. One banded at Elmhurst, N. Y., October 24, 1935, was caught August 19, 1937, at Lakeport, N. H. One banded at Zion, Ill., October 4, 1932, was found about March 10, 1934, at Turkey, N. C. A bird banded at Almonesson, N. J., October 27, 1938, was captured November 19, 1938, at Hatley, Ga. This indicates a rate of migration of more than 30 miles a day.

A longevity record is furnished by a bird banded as an adult at Philadelphia, Pa., October 25, 1928, and found dead March 25, 1934, at Avera, Ga.

Casual records.—Three specimens of the hermit thrush have been collected in Europe: in Germany on December 22, 1825; on the island of Helgoland in the autumn of 1836; and one in Switzerland without date. In June 1845 a specimen was collected at Amaraglik, Godthaab District, Greenland. A specimen of the eastern race, was collected on Southampton Island, October 4, 1929, and one of the Alaska race near Barrow, Alaska, May 25, 1933.

Egg dates.—British Columbia: 10 records, May 11 to July 12; 6 records, June 4 to June 23.

California: 65 records, May 13 to July 22; 35 records, June 3 to June 21, indicating the height of the season.

Colorado: 22 records, May 14 to July 11; 12 records, June 12 to June 28.

Maine: 38 records, May 13 to August 15; 20 records, May 29 to June 12.

HYLOCICHLA GUTTATA NANUS (Audubon)

DWARF HERMIT THRUSH

HABITS

This is the small, dark race of the hermit thrush that breeds in the humid coast belt from Cross Sound, Alaska, southward to southern British Columbia. Ridgway (1907) describes it as "similar to *H. g. guttata* but coloration darker and browner, the color of the back, etc., more sepia brown, upper tail-coverts more russet, tail more chestnut, and spots on chest larger and darker."

Grinnell and Wythe (1927) state that the dwarf hermit thrush leaves the San Francisco Bay region in spring about the first of April, a late date being April 21 at Berkeley. And George Willett writes to me that "the species arrives in southeastern Alaska in late April and early May, and leaves mostly in September. It usually nests three weeks to a month earlier than *H. ustulata*."

Nesting.—Mr. Willett (MS.) reports two nests, each containing four eggs. One that he took at Ketchikan, Alaska, on June 8, 1924, was located three feet from the ground among the roots of a windfall in the woods; it contained four eggs, advanced in incubation, and was made of moss and lichens, lined with rootlets and leaves; the nest measured 100 by 60 millimeters in outside dimensions, and the inner cavity measured 72 by 37 millimeters.

He collected another nest at Petersburg, Alaska, on July 3, 1936; this nest was placed 3½ feet up in a young hemlock in the woods, and held four eggs, about half incubated; the nest was similar to the other in construction but of different dimensions; it measured 125 by 73 millimeters outside, and 62 by 35 millimeters inside. He remarks: "This latter is a very late nesting date for the region, young being usually hatched by the middle of June and full-grown about July 18."

S. J. Darcus (1930) mentions two nests found on Langara Island, in the Queen Charlotte group; they contained feathered young on June 10. "Both these nests were built on top of stumps, the one 8 feet from the ground, the other 6 feet."

Two nests were recorded by the 1907 Alexander Alaska Expedition, according to Dr. Joseph Grinnell (1909). At Idaho Inlet, on July 22, a pair had a nest that contained young nearly ready to fly; "the nest was built in a niche in a perpendicular moss-grown bank about four feet above the bottom." And on July 7, a nest was found at Glacier Bay that held four fresh eggs. "The nest was situated in a crotch formed by a small limb and the naked body of a ten-inch hemlock and was six and one-half feet above the ground. It was found by seeing the female fly from it, and was seemingly but a stray bunch of moss in which a cavity had been made by the bird."

Eggs.—Four eggs seems to be the usual number laid by the dwarf hermit thrush. These are like the eggs of the other hermit thrushes, usually ovate in shape and plain "Nile blue" in color, without markings. The measurements of 16 eggs average 22.3 by 16.6 millimeters; the eggs showing the four extremes measure **24.0** by 16.0, 23.9 by **17.3, 20.5** by 17.0, and 24.0 by **15.6** millimeters.

Behavior.—W. L. Dawson (Dawson and Bowles, 1909) gives a very good description of the characteristic behavior of hermit thrushes in general, which seems worth quoting:

As one passes thru the woods in middle April while the vine maples are still leafless, and the forest floor is not yet fully recovered from the brownness of the rainy season, a moving shape, a little browner still, but scarcely outlined in the uncertain light, starts up from the ground with a low *chuck*, and pauses for a moment on a mossy log. Before you have made out definite characters, the bird flits to a branch a little higher up and more removed, to stand motionless for a minute or so, or else to chuckle softly with each twinkle of the ready wings. By following quietly one may put the bird to a dozen short flights without once driving it out of range; and in so doing he may learn that the tail is abruptly rufous in contrast with the olive-brown of the back, and that the breast is more boldly and distinctly spotted than is the case with the Russet-backed Thrush.

Winter.—The dwarf hermit thrush spends the winter in California, Lower California, Arizona, and New Mexico. For the Fresno district of California, John G. Tyler (1913) writes: "From mid-October until March occasional examples of this thrush may be found in the willows along the ditches, where they seclude themselves for the most part in the gloomiest shady clumps of large trees. They are quite silent during the time they remain with us, and of such sluggish natures as to appear almost stupid at times. I have sometimes walked up to within five or six feet of one of these birds without causing it the least alarm. At a nearer approach it would leisurely hop to another branch, just out of arm's reach, where it would assume an air of indifference, and remain motionless for some time."

Grinnell and Wythe (1927) say that the dwarf hermit thrush is an "abundant winter visitant throughout practically the whole [San Francisco Bay] region. Arrives ordinarily about the middle of October; an early record is September 26, at San Geronimo. To be found in woods, in chaparral, in stream-side thickets, and in shrubbery of city gardens; in fact, it avoids only the most open ground of meadows, fields and hillsides."

HYLOCICHLA GUTTATA SLEVINI Grinnell

MONTEREY HERMIT THRUSH

HABITS

According to the 1931 Check-list, this small, gray hermit thrush "breeds in the Transition Zone of the coast belt in California from northern Trinity County to southern Monterey County. South in migration and in winter to Lower California, Arizona, and Sonora."

It is the smallest of all the hermit thrushes, and its general coloration is nearly as pale and ashy as in the Sierra hermit thrush.

Grinnell and Wythe (1927) record the Monterey hermit thrush as a "summer resident in small numbers in the most humid parts of the immediate coast district" in the San Francisco Bay region. It "adheres closely to the denser redwood growths on shaded slopes and in canyon bottoms."

Harry H. Sheldon (1908) says of a locality where he found it in Sonoma County: "In June of 1904 the writer made a collecting trip to the South Fork of the Gualala River, a small stream about forty feet in width slowly winding itself down a deep thickly wooded canyon. Its banks are bordered with a dense growth of huckleberry, and at their extreme edge the sweet azalea grows in myriads from a tangle of various ferns and lilies. In such places as this the Monterey Hermit Thrush (*Hylocichla guttata slevini*) makes his summer home."

Migration.—Harry S. Swarth (1904) records the Monterey hermit thrush as a migrant only in Arizona, and remarks:

At first it seems strange to find a bird belonging so decidedly to the Pacific Coast wandering as far as eastern Arizona, but when we consider that such species as the Hermit and Townsend Warblers, Cassin Vireo, and others, pass regularly through this region, it is evident that there is a regular line of migration from the Pacific Coast to the southeast, in spite of the formidable deserts that intervene, and might be expected to form an utterly impassable barrier.

I believe *slevini* to be a fairly common migrant in the Huachucas, though but few specimens were secured, for it is an extremely shy bird, and from the nature of the ground frequented, exceedingly difficult even to get sight of. *Auduboni* was found mostly in the pine woods, and *guttata* along the canyons, but *slevini* seemed to prefer the dense thickets covering the steep, dry, hillsides, an unpleasant place to travel in at any time, and almost hopeless ground in which to pursue a shy, secretive bird like the present species. The specimens secured were, a male shot on March 9, 1903, and two females taken on May 8th, and another on April 19, 1902.

Nesting.—In the locality mentioned above, Mr. Sheldon (1908) found several nests of the Monterey hermit thrush, with eggs or young or under construction. One was "in a clump of branches of an oak tree about eight feet from the ground above the stream"; the bird was working on this nest on May 27, but it was never completed. On May 30 he found another nest, "placed in the shoots of an alder on the bank of the river, and like our previous experience the bird saw us and the nest was abandoned." A nest previously located in process of construction was visited on June 3 and found to contain a complete set of three eggs. "This nest was placed in a bush of huckleberry on the edge of the stream three feet from the creek bed. It was composed of chips of dead wood, small branches of huckleberry, dead leaves and twigs, and held together with mosses and rootlets. The lining consisted of fine redwood bark, fibers, fine rootlets and the remains of dead leaves. * * * All nests found were placed from two to eight feet from the ground, their favorite nesting site being in patches of huckleberry and in all cases situated close to the stream."

Robert R. Talmadge gives me this description of two nests: "The first nest was composed mostly of green moss, with small rootlets, decomposed leaves, and small twigs. The lining was made of rootlets, decomposed leaves, and a little shredded redwood bark. The second nest was located while I was looking for a place to eat lunch. I had entered a small clearing and was approaching a small tan oak, when the bird flushed out into my face. This nest was quite different from the first, being composed mostly of small redwood twigs (with the needles still attached) and shredded redwood bark. There was a noticeable lack of moss, only a few bits being found. The lining was of rootlets, dried grasses and bark."

Eggs.—The Monterey hermit thrush lays three to five eggs, probably most often four. These are indistinguishable from the eggs of other hermit thrushes of similar size. The measurements of 30 eggs average 21.5 by 16.5 millimeters; the eggs showing the four extremes measure **23.2** by 16.9, 21.9 by **17.2**, **20.4** by 16.1, and 21.2 by **15.6** millimeters.

HYLOCICHLA GUTTATA SEQUOIENSIS (Belding)

SIERRA HERMIT THRUSH

HABITS

This is a gray hermit thrush, similar in coloration to both *auduboni* and *slevini*, but intermediate in size between the two.

Lyman Belding (1889a) described it as a distinct species under the name "big tree thrush," from specimens collected at Big Trees, Calaveras County, Calif. He says, in part: "In size between the Dwarf and Audubon's Thrushes. In color paler than either or any American thrush I have ever seen; both above and below considerably resembling *T. aliciae*, the spotting included, while its cheeks are still grayer than in *aliciae*. Tail and coverts about as light cinnamon as in *T. auduboni*."

It breeds in the various mountain ranges from southern British Columbia to southern California and migrates southeastward to Texas and northern Mexico. Samuel F. Rathbun tells me that it is the breeding form in the mountains of the Olympic Peninsula in western Washington, as well as in the Cascades; he found it in the former mountains in summer at 3,700 to 4,300 feet; in the vicinity of Seattle it occurs regularly, as a migrant only, in April and again from about the middle of August until late in fall.

J. Stuart Rowley writes to me that, in the high Sierra country of Mono County, Calif., he "found this bird to be a well distributed race from about 7,000 feet elevation upward to timberline."

Nesting.—Mr. Rowley says in his notes: "I have observed many occupied nests; and fresh eggs may be found on the same day as young birds on the wing throughout the latter part of June and into July. The average seems to be four eggs, but often only three are laid, and I have found several containing five eggs. The nesting sites chosen seem to be almost anywhere. I have found them in aspens, in lodgepole pine, in willow, and, along Mammoth Creek, in the *Artemisia tridentata* brush, much to my surprise. My notes show fresh eggs found on May 30 and fresh eggs observed on July 9."

Grinnell, Dixon, and Linsdale (1930) found several nests in the Lassen Peak region. The site of the first one, found May 28, 1927, "was in deep, dark fir woods with lodgepole pines and aspens close by. The nest was slightly over one and one-half meters above the ground, saddled at the intersection of a dead fir stem four centimeters in diameter slanting at a forty-five degree angle against a live young fir stem nine and one-half centimeters in diameter. The slanting stem and emanating dead twigs furnished most of the support." In some other cases, the nests were placed between small trees, usually an incense cedar and a young lodgepole pine, from 80 centimeters to a meter above the ground, supported by branches and twigs of the two little trees. A photograph of a nest in such a situation is shown.

Dr. Grinnell (1908) found many nests, both old and new, in the San Bernardino Mountains above an altitude of 6,300 feet. "They were all built in small firs or cedars usually growing in the shade of taller trees not far from the streams. The nests varied from eighteen inches to five feet in height above the ground, the average being about three feet." A typical nest "was three feet above the ground near the top of a diminutive fir tree growing a yard from the stream. * * * It was snugly ensconced against the main stem and was supported by horizontal branches. It was a compact structure deeply cup-shaped. The inside diameter was 2.40 and the depth 1.65 inches. Externally it measured 4 × 4.75 inches. It was composed largely of pine needles and weathered grass stems, and the cavity was lined with strips of cedar bark and fine dry rootlets."

Rollo H. Beck (1900) found an unusually high nest, which he recorded as a nest of Audubon's hermit thrush, but, as it was in the Sierra Nevadas, it undoubtedly belonged to a Sierra hermit thrush. He writes: "We were near the summit of the Sierras on the 6th of June, 1896, and while looking around in a grove of trees, I noticed a nest well out on a pine limb, thirty feet from the ground. On climbing the tree, the bird was seen upon the nest and flew off when closely approached. The nest is strongly built of twigs and bright yellow moss (*Evernia vulpina*), with a layer of fine dry leaves, within which is a heavy lining of fine grass stems. The nest contained four fresh eggs."

Taylor and Shaw (1927) show a photograph, taken by Mr. and Mrs. Finley, near the Third Crossing Bridge on the Washington Cascades, Paradise River. It was "in the branches of a scrub fir that hung down from the top of a rock wall a few feet above the rushing waters and not more than 20 feet from the railing of the bridge." The only nest found by Mr. Belding (1889a) "was in a hazel bush (*Corylus*) about three feet from the ground; was about five inches across the top and about half as deep; composed of small roots and lined with shreds of the bark of incense cedar (*Librocedrus*), with moss, lichens and dead leaves on the exterior."

Eggs.—The Sierra hermit thrush lays three to five eggs to a set, but most commonly four. These are similar to the eggs of other hermit thrushes of similar size. The measurements of 30 eggs average 21.8 by 16.4 millimeters; the eggs showing the four extremes measure **23.3** by 16.8, 22.8 by **17.3**, **20.3** by 16.3, and 23.0 by **15.5** millimeters.

Behavior.—Grinnell and Storer (1924) describe the behavior of hermit thrushes very well as follows:

> The demeanor of the hermit thrush is quiet and deliberate. When foraging on the ground it acts in much the same manner as a robin, hopping several times in quick succession and then halting upright and immobile for a few seconds to scan the immediate vicinity before going forward again. There is this important difference, however: The hermit thrush seldom forages out in the open, and if it does it never goes far away from cover, to which it can flee in case of need. When foraging on shaded ground strewn with dead leaves its characteristic performance is to seize a leaf in its bill and throw it to one side with a very quick movement of the head, following this with an intent gaze at the spot uncovered. A thrush will flick over leaf after leaf in this manner, every now and then finding some insect which is swallowed, as is a berry, at one gulp. Hermit thrushes thus make use of a source of food not sought after by other birds; fox sparrows may forage over the same ground, but they are after seeds, which they get at by scratching. The thrushes do not use their feet at all for uncovering food. The thrushes' legs are relatively long, so that the birds stand high, and have consequently an increased scope of vision.

Taylor and Shaw (1927) write: "Curiosity is a marked trait. Once while we visited our traps a thrush appeared within 15 feet. At short intervals it gave a whistled *twhit* or *whooit* call. Frequently, but not always, one, two, or three wing flirts were given at the same time as the call. The bird seemed torn between conflicting emotions, once or twice making as if to leave, but each time curiosity got the better of it and it remained. It cocked first one eye at the intruder, then the other. Once it scratched the corner of its mouth. It remained on the lower branches of a western hemlock usually 12 to 15 feet above the ground."

Hermit thrushes usually are seen in dense thickets in deep forests, or in the lower branches of the larger trees, but Mr. Belding (1889b) says that the Sierra hermit thrush sometimes "wanders at a considerable height through the foliage of the firs and other coniferous trees, when it is followed with much difficulty, even if its brilliant song is often heard. I shot the female type specimen while she was fluttering about seventy-five feet from the ground at the ends of fir twigs and catching insects in the manner of the warblers and tyrant flycatchers."

On the fall migration this, like other thrushes, often resorts to yards and gardens in towns and cities. Mr. Rathbun writes in his notes for August 30, 1913: "Another hermit thrush was seen this afternoon in the garden at the edge of the shrubbery. I was watering with the hose, and the bird would run out and dabble in the water. Seeing that it liked this, I created a little running stream, and the thrush took advantage of this in which to bathe. The bird acted very tame, and I played with it for fully ten minutes, driving it from place to place by means of the hose, and still it would not leave."

Voice.—The song of the Sierra hermit thrush is not inferior to the far-famed song of our eastern bird, which to my mind is the most uplifting of all bird songs; once heard in the picturesque surroundings of its mountain haunts, its charm can never be forgotten. Everyone who has heard it has praised it. When heard in contrast with the songs of other birds, any other song, however charming it may ordinarily be, seems like an intrusion on the soulful chant of this mountain minstrel.

Dawson (1923) writes of it: "Having nothing of the dash and abandon of Wren or Ouzel, least of all the sportive mockery of the Western Chat, it is the pure offering of a shriven soul, holding acceptable converse with high heaven. * * * Mounted on the chancel of some low-crowned fir tree, the bird looks calmly at the setting sun, and slowly phrases his worship in such dulcet tones, exalted, pure, serene, as must haunt the corridors of memory forever after."

HYLOCICHLA GUTTATA POLIONOTA Grinnell

MONO HERMIT THRUSH

HABITS

In the White Mountains of Mono and Inyo Counties in California, Dr. Joseph Grinnell (1918) discovered this decidedly local subspecies, which he found only in a limited range in these mountains between 8,000 and 10,000 feet altitude. He says of its characters:

Size large, between that of *H. g. sequoiensis* of the Sierra Nevada and of *H. g. auduboni* of the Rocky Mountains, nearest the former. Color of top of head and dorsum different from that in either of these races and, in fact, from that in any previously known race of Hermit Thrush. The tone of this coloration is the "olive-brown" of Ridgway (1912), and is close to that of the corresponding areas in the Olive-backed Thrush (*Hylocichla ustulata swainsoni*); it is if anything even more slaty. * * * The race *sequoiensis*, of the Sierra Nevada just across

Owens Valley to the west and in plain sight from the White Mountains, is ordinarily referred to as a pale-colored or even grayish-colored Hermit Thrush; but compared with *polionota*, the contrast in dorsal view is as of brown with slate-gray. The resemblance of *polionota* to the Olive-backed Thrush is striking. * * *

In an examination of hundreds of specimens of Hermit Thrushes from throughout the United States elsewhere than from the White Mountains, the writer has been unable to find one referable to the race *polionota*. It would seem that this subspecies, like some other migratory brids of the high mountains of the southwest, goes south in the fall to, and back again in the spring from, some far southern winter home without touching the lowlands within hundreds of miles of its restricted summer habitat.

M. G. Vaiden (1940) reports a specimen of this subspecies taken 5 miles south of Rosedale, Mississippi, on April 12, 1940; this specimen is now number 51587 in the collection of Dr. Louis B. Bishop, of Pasadena, Calif.

Dr. Jean M. Linsdale (1938) has extended the breeding range of the Mono hermit thrush into the Great Basin region, where he found it breeding commonly in the Toyabe Mountains in central Nevada, about 150 miles east of the California boundary. He says that it "seemed to be most numerous at about 8,000 feet, but nearly all the range where there were trees was occupied. The lines and groves of trees which grew close to streams were most certain to be occupied by hermit thrushes. Also they lived out over the ridges, on slopes covered with mountain mahogany, where the trees were close together, and where there was leaf litter on the ground. One factor of apparent importance in determining the presence of this thrush was the availability of shade. However, the shade was not dense in most of the territory occupied in this area."

Nesting.—Dr. Linsdale (1938) records some nine nests found by him in the Toyabe Mountains; one found on June 18 contained four small young, four on the 19th held four eggs each, two on the 21st three eggs each, and one found on June 24 contained three well-feathered young. These nests were all at elevations varying from 7,500 to 8,500 feet; four of them were in aspens, living or dead, three were in willows, and two were in sage; the heights above ground varied from 2 feet in a sage to 15 feet in a dead aspen, but only two were above 6 feet. Referring to the lowest nest, he says:

Another nest in a small grove of aspens at 8,000 feet was in the crotch of a sage bush, its rim only 2 feet above the ground. The rim, inside, measured 77 mm.; outside, 180 mm. Depth, outside, was 120 mm. The outer part of the nest was mainly the dead flowering stems of sage. The inner part was made of shreds of bark, rootlets, grass stems and black horsehair. Three eggs made up the set.

When I came within 3 feet, the brooding bird left and flew off silently. * * * With 1 or 2 exceptions when the bird may have been off the nest, all of the brooding birds showed a marked reluctance to leave. Most of them permitted approach close enough to touch them before starting. Then they usually dropped to a perch near the ground and moved away quietly.

A nest that was 3 feet from the ground in a 5-foot sagebush "was on a slope 15 feet from the margin of a grove of aspens which bordered a stream. The bush was on a northwest-facing slope where it was exposed to the sun for nearly the full day. Near it were grass, herbaceous plants, and *Symphoricarpus*."

There are three sets of eggs in the Doe collection, University of Florida, that were placed somewhat higher; one nest, containing four eggs, was located 12 feet from the ground in a small yellow pine; a set of three eggs came from a nest that was 20 feet up and 10 feet out on a limb of a lodgepole pine; and another set of five eggs was in a nest 30 feet from the ground in an aspen. These were all collected in the White Mountains.

Eggs.—The set for the Mono hermit thrush usually consists of three or four eggs, usually the latter number. These are very similar to the eggs of other hermit thrushes of similar size. The measurements of 30 eggs average 22.2 by 16.7 millimeters; the eggs showing the four extremes measure **24.1** by 15.3, 22.0 by **17.5**, **19.9** by 16.0, and 21.6 by **15.3** millimeters.

HYLOCICHLA GUTTATA AUDUBONI (Baird)

AUDUBON'S HERMIT THRUSH

HABITS

This mountain race is the most widely distributed of the western hermit thrushes, breeding from southeastern British Columbia and Montana, mainly in the Rocky Mountain region, south to Arizona and New Mexico, and in the Sierra de la Laguna in southern Lower California.

It was first recognized as distinct from the eastern hermit thrush by Baird (1864), who named it, based on a specimen from Fort Bridger, Wyo., of which he says: "The back is rather more olivaceous than in *pallasii*, the rump paler and less rufous, and the colors generally much as in *nanus*. * * * Whether the present bird be specifically distinct from *T. pallasii* or not, there is no doubt of its being a decidedly marked race, of larger size and grayer plumage above."

It is decidedly the largest of the hermit thrushes and is quite similar in coloration to the Sierra hermit thrush, the other mountain race.

It seems strange that this thrush should be found apparently breeding in the Sierra de la Laguna, so far removed from the remainder of the breeding range of this subspecies, with no other hermit thrush breeding in the gap, but there seems to be no doubt about it. Mr. Frazar collected six specimens in these mountains for Mr. Brewster

(1902), five of which were typical of this subspecies; and, as they were taken between May 11 and June 8, they were probably breeding there. "This Thrush, which has not been previously reported from any portion of Lower California, was found by Mr. Frazar only in the Sierra de la Laguna, where it inhabited deep, moist, shady cañons, and also, to some extent, dry pine woods. It was not numerous, but was seen almost daily during May, and up to the 9th of June when Mr. Frazar started for Triunfo. The males were in full song, and there can be little doubt that they and their mates were settled for the season and preparing to breed on this mountain."

Audubon's hermit thrush has well been called the Rocky Mountain hermit thrush, for everywhere its chosen summer home is at the higher altitudes in the mountains, in the deep recesses of the pine woods, in the open groves of aspens, or higher up in the dense forests of spruces and firs, even up to the tree limit. In Arizona and New Mexico it may be looked for in summer at between 7,000 and 12,000 feet elevation; and even as far north as Montana its range is between 4,000 and 6,000 feet altitude. Mrs. Bailey (1928) gives the following picture of its haunts in New Mexico:

At 11,000 feet, on Jack Creek below Pecos Baldy, we found them so surprisingly abundant in the dense spruce and fir forest that we named our camp Hylocichla Camp. From the woods above, below, and around us came their beautiful songs, the first heard in the morning and the last at night. At sunset, as we walked through the cool, still, spruce woods, its pale beards lit by the last slanting rays, involuntarily treading lightly to make no sound, from unseen choristers a serene uplifted chant arose, growing till it seemed to fill the remote aisles of the forest. Sometimes a silvery voice would come from the open edge of the dark forest, where the singer looked far down the mountainside and out over the wide mesa-clad plains—a wide view, the beauty and sweep of which seemed in rare harmony with his untroubled spirit.

Russell K. Grater tells me that this thrush is a fairly common summer resident in Zion National Park, Washington County, Utah, above 8,000 feet, nesting in June and July, in the fir belt.

Spring.—Audubon's hermit thrush probably breeds in some of the higher, spruce-clad mountains in Arizona, but Mr. Swarth (1904) met with it in the Huachucas only as a migrant between April 18 and May 19; the latter was in worn plumage and may have been a breeding bird. "I secured most of my specimens of *auduboni* in the highest parts of the range, feeding, not in the thick bushes and underbrush, as most of the thrushes do, but on the open ground under the big pines, scratching and working in the pine needles with which the ground was thickly covered. One or two specimens were secured in the canyons as low as 6,000 feet, but the great majority of the birds seen were along the divide of the mountain, from 8,500 feet upward."

Nesting.—I can find no record of Audubon's hermit thrush nesting on the ground, as is the common habit of the eastern hermit thrush. The nearest approach to a ground nest is that described in the data for a set in my collection, taken for Frederick M. Dille in Estes Park, Colo.; this was placed in what he called a "ground pine" and not 6 inches from the ground.

In New Mexico, according to Mrs. Bailey (1928), the nest is placed "in bushes or low trees usually in pine or spruce, but also in oak saplings, 3 to 10 feet from the ground; bulky, made almost wholly of bark and coarse grasses, outside covered with moss." Dr. Edgar A. Mearns (1890) tells of a nest that was built near his camp in the mountains of Arizona. "The nest was saddled on to the middle of the lowest limb of a large spruce, and the birds gathered material for its construction close about my tent with perfect freedom from shyness, accepting proffered bits of cotton for its completion."

William L. Sclater (1912) has this to say about nests in Colorado:

Gale's notes contain the record of a large number of nests found by him at various elevations, from about 6,000 to 11,000 feet; they were placed almost exclusively on spruce trees from about three to ten feet from the ground, generally in a spot near a mountain stream or close to a spring.

Nests were constructed of various materials, such as rotten wood, mosses, grasses and plant stems, and lined with rootlets, horsehair or fine grasses. All these materials have been found in the nests, though by no means in every nest. The construction varied considerably, but no clay or mud is used. The nests were very quickly completed; one begun on June 6th was finished on the 13th, and the first clutch of eggs laid by the 18th.

The nests of this thrush seem to be quite large and well built. D. D. Stone (1884) describes such a Colorado nest as follows: "Nest in small pine, five feet from ground, a few feet from edge of heavy timber. Parent glided off the nest and out of sight without a note. Nest, a slight base, and sides of twigs and coarse grass stems, within a compact wall ¾ inch thick, of green moss woven in with fine straw and rootlets. It is the most solid nest I ever saw, for one made without mud. Outer diameter 5¼ inches, height 3½ inches, inner diameter 2½ inches, depth 2 inches."

Eggs.—Except for an average difference in size, the eggs of Audubon's hermit thrush are similar to those of other races of the species. Four is probably the commonest number, but sets of three or five have been found. The color is light greenish blue, or "Nile blue."

The measurements of 40 eggs in the United States National Museum average 22.8 by 17.2 millimeters; the eggs showing the four extremes measure **24.6** by 15.3, 21.6 by **18.5, 21.1** by 18.1, and 22.9 by **15.2** millimeters.

Food.—The vegetable food consists largely of wild fruit or waste cultivated fruit. Among the wild berries eaten are pokeberry, serviceberry, holly, black alder, woodbine, elderberries, mistletoe berries, and the seeds of poison-oak. Animal food includes ants, caterpillars, beetles and other insects, and spiders. It evidently does no harm to cultivated crops.

Voice.—The exquisite song of this thrush is fully as beautiful and inspiring as that of its famous eastern relative, which, in my estimation, is one of our most beautiful bird songs. Mrs. Bailey (1902) expresses it very well as follows:

> As you travel through the spire-pointed fir forests of the western mountains, you know the thrush as a voice, a bell-like sublimated voice, which, like the tolling of the Angelus, arrests toil and earthly thought. Its phrases can be expressed in the words Mr. Burroughs has given to the eastern hermit, *"Oh, spheral, spheral! oh, holy, holy!"* and the first strain arouses emotions which the regularly falling cadences carry to a perfect close. The fine spirituality of the song, its serene uplifting quality, make it fittingly associated with nature's most exalted moods, and it is generally heard in the solemn stillness of sunrise, when the dark fir forest is tipped with gold, or in the hush of sunset, when the western sky is aglow and the deep voice rises from its chantry in slow, soul-stirring cadences, *high-up-high-up, look-up, look-up.*

Leon Kelso (1935), writing of it in Colorado, says: "They sing at all times of the day, but most often in the evening. June 17, 1933, they sang as late as 8:00 p. m. June 20, they sang as late as 8:20 p. m. One gave songs at intervals of 6–3–4–7–8–2–5 seconds. June 21, 7:00 a. m. the same bird sang at intervals averaging 5–6 seconds. At 4–4:30 p. m. it sang at 5–10 second intervals while the writer stood within ten feet of its perch. At 7:45 p. m. it sang at intervals of 4–5–5–5–4–6–5 seconds. All birds of this species ceased singing at 8:15 p. m. on this day, it then being quite dark."

Enemies.—Audubon's hermit thrush is listed by Dr. Friedmann (1929) as a rare victim of the Nevada cowbird; but he has only one definite record, that of a nest in the R. M. Barnes collection that held three eggs of the thrush and one of the cowbird.

Winter.—Audubon's hermit thrush goes farther south in winter than any of the other hermit thrushes. Dr. Helmuth O. Wagner (MS.) says of its winter haunts in Mexico: "In the winter time, from October 15 to April 4, you will find it in the forests around Mexico City and in the parks of the city. In the mountains I saw at all times only single birds, neither with other birds nor with those of its own species. They prefer to stay in the barancas and on the sides of the small brooks. The winter is the dry season here, and they are living only in the moister parts of the forests. In the city I saw them on the lawns which are watered each day. If the winter is very dry they travel to places where conditions are better; this winter, 1942–43, is extremely dry; I saw the last bird on November 13."

HYLOCICHLA GUTTATA FAXONI Bangs and Penard

EASTERN HERMIT THRUSH

CONTRIBUTED BY ALFRED OTTO GROSS

HABITS

The hermit thrush ranks high in the list of our favorite North American birds. The exquisite song of this modest bird of the northern woodlands has captivated the affections of a host of bird lovers. Those who have been privileged to hear its song possess delightful memories of associations with the hermit: perhaps a wooded border of some mirrored lake or some fern-carpeted woodland; or again they may have heard the fluted notes ringing across some brilliant sunset scene.

John Burroughs has beautifully expressed the inspiration, the elevating character of the emotions with which the hermit's song infuses us when he wrote the following lines in "Wake Robin": "Mounting toward the upland again, I pause reverently as the hush and stillness of twilight come upon the woods. It is the sweetest, ripest hour of the day. And as the hermit's evening hymn goes up from the deep solitude below me, I experience that serene exaltation of sentiment of which music, literature, and religion are but the faint types and symbols."

Unfortunately those who know the hermit only as a migrant are unfamiliar with this bird as the accomplished singer, for it passes on its migration without uttering more than a few uninteresting calls. Some of the earlier ornithologists were evidently unaware of its accomplishments. Wilson did not know of its song and Audubon as far as his personal acquaintance with the bird is concerned speaks only of its single plaintive note. One must meet the hermit in its nesting haunts of the northern woods to know this bird at its best.

O. Bangs and T. E. Penard (1921) found the two original names of the hermit thrush untenable. *Turdus solitarius* Wilson is preoccupied by *Turdus solitarius* Linnaeus. Wilson's description is of the hermit thrush, but the plate to which he referred represents *Hylocichla ustulata swainsonii* (Cabanis). *Turdus brunneus* Brewer is preoccupied by *Turdus brunneus* Boddaert=*Euphagus carolinus* (Müller). But from Brewer's article it is difficult to determine whether *Turdus brunneus* "Gmel." refers to *Hylocichla guttata pallasii* or *Hylocichla ustulata swainsonii*. Both names are thus of a composite nature, and the authors considered it best to propose an entirely independent name, *Hylocichla guttata faxoni* subsp. nov., for the eastern hermit thrush.

According to Ridgway (1907) the eastern hermit thrush is most like *Hylocichla guttata nana* of the six western subspecies, but the upperparts are of a lighter, more isabelline or cinnamomeous brown, spots on chest averaging larger, sides and flanks more buffy brown, and bill stouter. The various subspecies of the hermit thrush are of minor importance in a discussion of the habits and life history, and what is true of the eastern hermit thrush will in most instances also apply to the western forms.

Spring.—The migration of the hermit thrush through the United States is confused by the presence of wintering individuals. There are countless numbers of migration records early in March, but it is apparent that the peak of migration up the Mississippi Valley and into Canada is during April. An examination of some of the Canadian records of migration is of interest. According to J. H. Fleming (1907) the hermit is an abundant migrant at Toronto from April 13 to May 10. His earliest date of spring arrival is April 8. At Aweme, Manitoba, latitude 49° 42′ N., Norman Criddle (1922) in 19 years of observation found the average date of its first arrival to be May 2. His earliest record is April 19, 1917. Lynds Jones (1910) made extensive studies of bird migration on the sand spit of Cedar Point, Ohio. The comparative isolation of the spit from the mainland makes it the first step in the flight to Point Pelee on the Canadian shore of Lake Erie. In migration the birds are concentrated in this strip, which can be likened to the neck of a funnel. According to Jones the hermits are usually so numerous during migration that they spread well over the whole of the sand spit. The median date of spring arrival is April 2, the earliest March 21, 1903. The median date of spring departure is May 5, the latest May 20, 1907. It is seen that although a few individuals arrive at the spit in March, possibly individuals that wintered a relatively short distance south, the bulk do not arrive until April and do not leave until May.

The hermit thrush is the hardiest member of its group, for it is the first to arrive in spring and the latest to leave in autumn. Indeed, some individuals remain in certain sections of the southern limits of the breeding range, wherever there is an adequate food supply, to brave the cold winter. It dislikes snow, however, and usually manages to keep south of the line where snow remains on the ground for an extended period. It normally winters south of the 40th parallel to the Gulf States and west to central Texas. The migration starts in March, and by the middle of April it arrives in central New England, New York, southern Michigan, and Minnesota. During the first week of May it has reached the northern limits of its breeding range. It makes the journey by night and rests during the day. During the height of the migration large numbers are often seen in the parks and churchyards among the tall buildings and bustling life of our larger cities. After the ordeal of the nocturnal flight the birds are hungry, often exhausted, and at such times exhibit little fear and may be seen feeding about dooryards, allowing human observers to approach near to them. It they are caught in a snowstorm this behavior becomes even more pronounced. I have had individuals, benumbed by the cold, eat out of my hands, and one bird even allowed me to pick it up to be carried to the house to be warmed.

The hermits follow no special migration route in reaching their northern home except in the far West. Here they fly on a direct northwest route that takes them as far as the Mackenzie and Yukon Valleys. In fall the hermit starts southward in September, but it is well toward the end of October before the bulk of them have left their northern summer ranges. E. A. Preble tells me that he saw one early in December in Wilmington, Mass., about 1890.

Nesting.—The nest of the hermit thrush is a compact structure but often bulky in the amount of nesting materials used. The foundation and exterior of a typical nest are composed of twigs, strips of wood, bark fibers, dried grass, and ferns and ornamented on the outside by bits of green moss. The lining is made up of pine needles, delicate plant fibers, or fine rootlets. The interior dimensions of the nesting bowl are about 2¾ inches across by 2 inches deep.

The nest is generally built on the ground and in a natural depression of a knoll or hummock, often under a small fir or hemlock whose branches touch the ground, forming a kind of protective canopy over the nest. One nest found in northern Michigan was in a rather open space of woodland and was completely surrounded by blossoming bunchberries, and another nest was completely hidden from view by a luxuriant growth of ferns. I have found them along the edges of old wood roads and on the borders of pasturelands skirted by shrubbery and trees. In northern Maine the nests may be found in tussocks of the wet sphagnum bogs that are surrounded by growths of larches, spruce, and other coniferous trees. On Long Island the hermit frequents the hottest and driest barrens where the ground is carpeted with little else but bearberry and pine-barren sandwort. Near the site of the University of Michigan Biological Station, Douglas Lake, northern Michigan, the hermit is one of the commonest of the nesting birds. During July 1928 we found six nests in dry upland covered with a second growth that had sprung up after a severe fire that had raged through the section a few years before.

The hermit sometimes departs from its usual habit of nesting on the ground. Henry R. Carey (1925) reports finding a nest 5 feet up in a small hemlock, and Horace W. Wright (1920) found a hermit's nest resting firmly on several bean poles at a point 4 feet above the ground. A pair observed by John May was nesting in what appeared to be a typical robin's nest 2 feet up in a young hemlock. This nest had a foundation of coarse grasses and weeds, a middle layer of mud, and a lining of fine grasses. I found a hermit's nest on a barren shelf of rock of a perpendicular ledge adjoining a deserted feldspar quarry in Topsham, Maine. The shelf on which the nest was built was 15 inches wide and 3 feet long and about 7 feet above the ground. The nesting site though in an exposed situation was well shaded during most hours of the day by the dense foliage of several large hemlock trees. The nest was made of the usual nesting materials, but the twigs and leaves of the foundation were spread over an area of 12 to 15 inches. The nesting bowl of the deep cupped nest was well formed and firmly constructed.

Though we associate the hermits with lonely situations remote from the habitations of man, they have been known to nest about buildings. Miss Annie L. Warner, of Salem, Mass., wrote to Mr. Forbush (1929) that she found a hermit's nest with two well-grown fledglings about 7 feet from the ground, on a shelf under the eaves of a piazza of an occupied camp on Lake Winnipesaukee. Another hermit was reported nesting in a tin gutter under the eaves of the second story of a home at Holderness, N. H. (E. DeMeritte, 1920). Verna R. Johnston (1943) found a hermit thrush nesting on a rafter under a roof of a building at the University of Colorado Biological Station, at Boulder, Colo. The station is located at an elevation of 9,500 feet.

Eggs.—The eggs of the hermit thrush are ovate or elongate-ovate and a plain greenish blue in color. They are similar in appearance to the eggs of the Wilson's thrush but are of a much more delicate and lighter shade of blue. Occasionally the eggs are spotted. In correspondence received from Francis H. Allen he writes that one egg of a set of three found at Bridgewater, N. H., on August 1, 1883, has thinly scattered small brown spots. Another found near the same place on August 9 of the same year contained three eggs, one of which was spotted. Harry G. Parker (1887) writes that in two eggs in a set of three there were minute spots of black. An application of an acid wash failed to remove the spots. Others have reported similar markings on the eggs of the hermit thrush, but spotted eggs are by no means of common occurrence.

The number of eggs per complete set varies from three to six, but the vast majority of nests contain three or four eggs.

The measurements and weights in millimeters and grams of two typical sets of eggs are as follows:

NEST FOUND AT BRUNSWICK, MAINE, JULY 16, 1928

Long diameter	*Short diameter*	*Weight*
23. 2	18. 4	4. 05
23. 6	18. 6	4. 39
22. 8	18. 3	3. 95
21. 9	17. 1	3. 45

NEST FOUND AT BRUNSWICK, MAINE, AUGUST 5, 1930

Long diameter	*Short diameter*	*Weight*
24. 1	17. 2	3. 35
23. 5	17. 5	3. 34
24. 0	16. 5	3. 26

The measurements of 40 eggs in the United States National Museum average 22.1 by 16.8 millimeters; the eggs showing the four extremes measure **23.6** by 17.3, 22.6 by **18.0**, **20.1** by 17.5, and 20.8 by **15.8** millimeters.

Incubation.—The determination of the incubation period of 12 days is based on the study of two nests that were under continuous daily observation at Brunswick, Maine. On May 26, 1928, a nest of the hermit thrush located at the base of a small fir tree contained two eggs. On May 27 there were three, and on May 28 the set of four eggs was completed. Incubation, however, did not start until the following day, May 29. At 8 o'clock on the morning of June 10 there were three freshly hatched young, and the fourth hatched during the afternoon. A second nest was nearly completed but contained no eggs when it was found on June 10, 1940. This nest was in a natural depression located in a thick growth of blueberry vines. The first egg was laid during the morning of June 12. On June 16 there were four eggs and incubation started. The first egg laid hatched at 5 P. M. on June 28; the other three were hatched by dawn the next morning. The incubation period of the hermit thrush as reported by various observers varies from 10 to 13 days, but this wide range does not represent a real variation but probably is due to the failure to ascertain the exact time of the beginning of incubation as well as that of hatching.

In the nests I have had under observation only one of the pair, presumably the female, took part in the incubation of the eggs. The incubating bird was seen to leave the nest in search of food, but much of the food was delivered to her by her mate while she was attending her duties on the nest. During the 12 days of incubation the male spent much of his time singing and serving as guard against the intrusion by other birds or enemies that appeared in his territory. He often chose as a sentinel post the lower branch of a large pine tree that stood about 40 feet from the nesting site. At other times he perched at the tip of a small dead tree stub in an open situation where he was readily observed from the blind. Whenever there was an accidental noise or disturbance inside the blind to arouse his suspicions he would utter a *chuck, chuck* call accompanied by a characteristic sudden up-tilting and slow lowering of his tail.

The hermit is a wary bird, and during the first few days I spent in the blind the least provocation caused the bird to leave the nest with a whir of wings. But in the course of a few minutes after each such disturbance she flew back to a place within a few yards of the nest and from that point approached with caution, frequently stopping at some elevated knoll carefully to scrutinize the surroundings. She then crept close to the ground under cover of the vegetation, her progress being made known by the rustling of the leaves. She often took a circuitous route and when near sometimes circled about the nest on wing suspended in hummingbird fashion. Again after alighting she went along stealthily in a series of hops and when finally assured all was well went confidently to the nest. She adjusted the eggs with her bill and then settled on them, moving her body back and forth until the feathers of the breast were separated, permitting the eggs to come in direct contact with her warm body. This adjustment is repeated several times and not until it meets with her complete satisfaction does she settle down to the arduous task of incubation. The raised feathers of the neck and back then fall back to their normal position, the tips of the primaries are crossed over the rump, and the bill assumes an upward tilt. She is then motionless and her soft brown colors blend so into the lights and shadows of the surroundings that she is practically hidden from view. The eggs are turned at frequent intervals during the course of the day. From time to time the male, who seemed even more cautious than his mate, would timidly approach the nest, announcing his coming with a *wee* call. With a look of apparent admiration and devotion he delivered some choice insect or larvae in commendation of a task well done. Sometimes instead of bringing food he carried nesting material. This was graciously received by the female who merely cast it to one side of the nest. This behavior is of frequent occurrence among certain groups of birds such as the herons, gallinules, and hawks, but I have never before noted this behavior, a response to an emotional urge, exhibited in the Turdidae.

Once a red-eyed vireo unwittingly alighted in the small tree under which the nest was located. The male immediately uttered his war cry and dashed at the unwelcome intruder. He was joined by his mate and both birds chased the vireo into the dense woodland beyond. At another time a red squirrel making his way through the grass and vines passed within 2 feet of the nest, but strangely enough his appearance did not seem to excite the birds in the least. However, the female on the nest after having caught sight of the squirrel followed every move until he had passed and was well out of sight. On another day a hummingbird hovered for 30 seconds between the nest and the blind not more than 3 feet above the nest, but neither bird paid any attention to this unusual visitor. O. S. Pettingill, Jr. (1930), writes of a gartersnake that appeared among the leaves near a nest he was observing from a blind. The parent bird was much perturbed. She flew from the nest screaming alarm calls and hovering over the unwelcome guest in a defiant defensive attitude. It is apparent that through some previous experience or hereditary tendency the birds recognized the snake as an enemy to be challenged on sight.

Unless disturbed the female remained on the nest during the day, especially if it were cold and raining or when it was excessively hot. During the latter condition she would perch on the edge of the nest with her wings somewhat extended to keep the burning sun rays off the eggs or young. At one time when the temperature arose above 90° she kept her beak wide open and panted incessantly in order to retain her normal body temperature. Only in the early morning or late in the evening just before sunset did I see her leave the nest voluntarily. These trips were doubtless taken to supplement the food delivered to her by the male.

Young.—After the eggs are pipped hatching proceeds rapidly. The struggling embryo breaks the shell in two parts, the crack taking place near the greatest diameter. In the course of a few minutes the embryo is entirely free. Each time I saw this critical event happen the adult was away from the nest. On her return, after carefully inspecting the young, she picked up an empty half shell and carried it away returning immediately to remove the other part. The appearance of the young is an important event in the household and seems to excite the parents to greater activity. Both parents flit about nervously and seem most anxious to serve their offspring. They now exhibit less caution and more daring in approaching the nest. A small green larva was delivered and fed to the young in less than five minutes after it had come into the world.

At the time of hatching the young are nearly naked, being clothed by only a few scant tufts of dark grayish down on the crown and dorsal tracts of the body. Though the eyes are closed during the first two or three days, the young birds are most responsive to the approach of the adults at the very start. In fact, a mere touch of the rim of the nest is sufficient to initiate the feeding response—uplifting their heads, extending their gape, and displaying the bright colors lining the mouth. Both parents take an active part in the feeding of the young and at all times take meticulous care of the sanitation of the nest. The excrement is received in their beaks as soon as it is emitted. The young are carefully examined and even stimulated by a stroke of their bills after each feeding until the fecal sac appears. During the first few days it is eaten, but thereafter it may be carried away and dropped at some distance from the nest.

The food at first consists of small green larvae, but as the young become older, mature insects, small grasshoppers, moths, beetles, and spiders are added to the diet. While the nestlings are very small they are frequently unable to swallow the food brought to them. After a morsel is thrust into an open mouth or into different mouths without being swallowed, the adult will mince it in her bill, or if the larva is large and both parents are present each will grasp an end and pull it apart. Sometimes after repeated failures of the young to swallow the food it is eaten by the adults. At one nest there seemed to be a great deal of difference in the choice of the food delivered. One of the birds would invariably bring food of the proper size and tenderness, while the other would bring enormous larvae or winged insects such as large sphinx moths totally unsuited as food for the age of the young being fed. The latter may have been a young inexperienced parent with its first offspring. Perhaps it was the male! At least human fathers are not supposed to know much about proper infant feeding.

On the third day the eyelids of the young are parted, and from then on their reactions are more and more responses to sight rather than sound. On the fourth day the papillae of the remiges have pierced the skin, and by the fifth day the chief feather tracts are well defined. On the seventh day the tips of the primaries and secondaries are unsheathed, and those of the other tracts have tips which display the olive-brown and buffy colors. By the ninth day the young frequently preen their feathers, thus facilitating the unsheathing process, so that on the following day or two the full colors of the juvenal plumage are acquired. The young now exhibit evidence of fear when one approaches or when there is a disturbance near the nest. The tail feathers are well developed at this time but do not attain their full length until later when they are about six weeks old. When the young are 12 days old they are ready to leave the nest. If they are not frightened and not forced to leave the nest prematurely they are encouraged by the adults, who stay away from the nest and perch at some distance with an appetizing morsel in their beaks. The parents hop from perch to perch calling constantly until the hungry youngster responds. At such times I have seen one or more of the nestlings standing on the rim of the nest preening their feathers, flapping their wings, and going through all the gymnastics of a young osprey whose first venture away is by flight. Finally when the decision of the young hermit does come it leaps from the rim of the nest, flutters its wings, and then makes its way along the ground and through the vegetation in the direction of the coaxing adult. After a few yards of travel the youngster is rewarded with food. This performance of the adults is continued until all have left the nest in a similar fashion. If you attempt to follow the young they take a short flight at the same time, uttering a series of distress calls, which are a signal for the adults to come to their rescue. At such times the adults exhibit unusual bravery and may even dash at the human intruder in rage.

As is true with other ground-nesting birds a comparatively small percentage of young reach maturity. Miss Cordelia J. Stanwood (1910) states that out of 14 nests containing a total of 47 eggs only 19 fledglings left the nest. Others are lost after leaving the nest before they are able to fly well enough to perch well above the ground out of reach of terrestrial enemies.

The relative growth of the young during the 12 days spent in the nest can be shown by their daily weighings. The average daily weights of three young of an apparently typical family brood were 4.12, 4.93, 7.21, 10.12, 14.76, 16.98, 19.21, 22.35, 24.60, 25.13, 25.61, and 24.81 grams, respectively, for the 12 days. It will be seen that the weighings increase rapidly during the first week of nest life, but the proportionate increase diminishes as they grow older and was actually less on the twelfth than on the preceding day. The weight at the end exceeded six times the weight at the beginning of nest life.

The nesting season of the hermit thrush extends over a relatively long period from May to August, or about three months. O. W. Knight (1908) reports that he has found nests of the hermit thrush with full complements of eggs as early as May 1. Others have found nests with eggs during the first week of May. Miss C. J. Stanwood (1910) found a nest of the hermit thrush, containing three eggs, at Ellsworth, Maine, on August 22, 1909. The young left on September 8. Dr. C. Hart Merriam (1882) found a nest containing fresh eggs at Locust Grove, Lewis County, N. Y., on August 24, 1870. August nesting dates are by no means rare. This wide range in time of nesting dates, more than three months, is very suggestive that two or even three broods may be reared by a single pair of birds during one season. The incubation period is 12 days, and the time spent by the young in the nest is also only 12 days; hence a nest can be built and the young matured in the course of a month. The two distinct summer singing periods of the hermit are also suggestive of two nestings. It is well known that if the first nest proves a failure a second attempt will be made during the same season.

Plumages.—The plumages and molts of the hermit thrush have been described by Jonathan Dwight (1900) as follows:

Juvenal plumage acquired by a complete postnatal moult. Above, including sides of head, sepia or olive-brown, the rump russet, and everywhere spotted with large buffy white guttate spots bordered with black. The wings rather darker, the coverts and tertiaries with small terminal buffy spots. Tail burnt umber-brown. Below, white faintly tinged with buff, spotted with deep black, on the sides of neck, across the breast and on the flanks and crissum, the throat and breast, the fore part of the abdomen and flanks faintly barred. Bill and feet dull pinkish buff remaining pale when older. * * *

First winter plumage acquired by a partial postjuvenal moult, beginning late in August, which involves the body plumage, most of the lesser and median coverts, but not the rest of the wings nor the tail. Similar to previous plumage but without spotting above and the black spots below fewer. Above, including sides of head olive tinged mummy-brown, burnt-umber on rump and upper tail coverts. Below, white tinged faintly with buff on throat and breast, with olive gray on the sides and spotted heavily on the throat and faintly on the breast with large deltoid black spots. Lores and submalar lines black; orbital ring pale buff. The buff spotted coverts retained distinguish young from adults. * * *

First nuptial plumage acquired by wear, the upper surface becoming rather grayer and the buff below mostly lost.

Adult winter plumage acquired by a complete post-nuptial moult in August and September. Averages darker and lacks the tell-tale coverts and tertials of the first winter dress. Young and old become indistinguishable.

Adult nuptial plumage acquired by wear as in the young bird, from which it is usually distinguishable by the wing coverts.

The plumages and molts are alike in the two sexes.

Dr. Alexander Wetmore (1936) has determined the number of contour feathers in a number of passeriform and related birds. His counts of the contour feathers of three hermit thrushes is as follows:

Date	*Sex*	*Number of feathers*	*Weight of bird*	*Weight of feathers*
			Grams	*Grams*
October 21, 1933	Male	1,884	31.2	2.4
Do	Female	1,873	32.7	2.4
Do	?	1,828	31.9	2.3

Apparently albinism is not of frequent occurrence in the hermit thrush as only two cases have come to my attention. John H. Sage (1886) reports a partial albino hermit thrush taken at Portland, Conn., on October 27, 1885, which had the top of the head and the back light gray. Below it was white, the spots on the breast fairly distinct. The primaries and secondaries were a fawn color. A pure albino hermit thrush was shot at Stamford, Conn., by W. H. Sanford.

Food.—F. E. L. Beal (1915b) examined the stomach contents of 551 hermit thrushes, which were collected in 29 States, the District of Columbia, and Canada. These specimens represent every month of the year, though all the birds taken in winter were collected from the Southern States, the District of Columbia, and California.

Animal matter, consisting chiefly of insects and a few spiders, comprises 64.51 percent of the total amount of the food eaten by the hermit thrush. The insects consisted of beetles 15.3 percent, ants 12.46 percent, bees and wasps 5.41 percent, caterpillars 9.54 percent, Hemiptera (bugs) 3.63 percent, Diptera (flies) 3.02 percent, grasshoppers and crickets 6.32 percent, miscellaneous insects 0.27 percent, and spiders and myriapods 7.47 percent. Miscellaneous animals such as sowbugs, snails, and angleworms make up the balance of the animal food of 1.26 percent. Of the insects listed above less than 3 percent can be considered useful; the remainder according to Professor Beal are chiefly harmful to man's interest.

The vegetable diet of the hermit thrush (35.49 percent) consists largely of fruit, but little of this can be classed as cultivated. Beal found that 5.45 percent of the food eaten during September did consist of cultivated fruits but in most months the quantity was small, and in March, April, and May was completely wanting. The total amount of cultivated fruit eaten during the entire year was only 1.20 percent. Of the wild fruits (26.19 percent) 46 species were identified. A few seeds, ground-up vegetable matter not identified, and rubbish made up the remainder of the vegetable food, or 9.10 percent of the total.

S. A. Forbes (1880) in the examination of 150 thrushes obtained in Illinois found them destructive to useful predaceous beetles. The worst of the group in this respect was the hermit thrush, which maintained a high ratio of these beetles throughout the fruit season when the total insect food fell away rapidly. It is important in considering the insect food of any species to take into account the beneficial as well as harmful insects.

While in its winter haunts of the Southern States the hermit thrush feeds largely on wild fruits and berries such as dogwood, pokeberries, serviceberries, holly berries, blueberries, mistletoe, frost grapes, elderberries, spiceberries, mulberries, blackberries, and seeds of the greenbrier, Virginia-creeper, and sumac including the poison-ivy and poison-oak.

The hermit keeps close to wooded retreats, and hence the products of the farmer are seldom molested. The majority of the insects on which it feeds are injuries to trees and hence it can be considered a valuable tenant of the forest.

Lewis O. Shelley (1930) reports that in southern New Hampshire after a snowstorm on April 12, 1929, many species of birds including the hermit thrush made efforts to find earthworms and insects near his home. The thrushes became so tame that they readily took earthworms from his fingers. In notes received from F. H. Kennard he states that a hermit thrush was found in the middle of a meadow, warm but dead. There was a large earthworm protruding from its mouth which had choked it to death.

Coit M. Coker (1931) reports an interesting experience with a nesting pair of hermit thrushes in the Allegheny foothills of western New York. He states that in fully one-quarter of the trips made to the nest with food both adults brought small salamanders of two species, the Allegany and red-backed salamanders. During the hotter parts of the day fewer salamanders were brought, and this Mr. Coker attributed to the fact that the heat had driven the salamanders deeper under cover. Others have reported salamanders comprising a part of the food delivered to the young. While observing a nest of hermit thrushes at Brunswick, Maine, I observed one of the adults deliver a salamander about 2 inches in length to the bird at the nest. In this case the salamander was not fed to the young, then five days old, but she ate it herself. A similar case was observed at a nest at Douglas Lake in northern Michigan, indicating that salamanders are by no means a local menu.

During the summer of 1941 I had an opportunity to observe the food brought to the young throughout their life in the nest at Brunswick, Maine. The food was invariably held in the beaks of the adults so that it could be easily seen and often identified from the blind placed within 5 feet of the nest. The food the first three days consisted of small green larvae. During the first day the larvae were minced in the beak of the adult before they were delivered, and at other times the larvae if large would be divided in two by each of the pair of birds grasping an end of the worm and pulling until it parted. On several occasions larvae too large for the young to negotiate after they were thrust into the extended mouths were swallowed by the parent. After the third day winged insects, spiders, and ants were added to the diet. On the seventh and eighth days large moths, grasshoppers, and beetles were fed to the young without any mincing or tearing apart. During the many hours spent in the blind I did not see fruit, berries, or vegetable matter delivered to the young, but Henry R. Carey (1925) reports that fruit including blueberries and wild cherries was delivered to the young of a nest observed in the Pocono Mountains of Pennsylvania.

Daniel E. Owen (1897) kept a young hermit thrush in captivity from the time it left the nest on June 26 until July 31. During this period of about five weeks he made interesting observations on its behavior and especially of its food habits. Mr. Owen substituted its usual food with raw beef cut into bits about one centimeter long by half a centimeter wide. To facilitate swallowing the pieces of meat were dipped in water. On June 28 between 8 A. M. and 7 P. M. it was fed eight times and swallowed 27 bits of meat. After July 4 he weighed the bird's food as well as the bird itself. The bird's average weight during five days was 27.7 grams and the average weight of the meat eaten daily was 13.56 grams, indicating that it ate about 50 percent of its weight in meat. He experimented with earthworms and found that the thrush ate 19 worms between 8:30 A. M. and 1 P. M. He noted that worms from a dung heap were frequently rejected, whereas worms taken from cool black garden mold were eaten with a relish. The thrush ate 9 grams of worms an hour, so at this rate it would not take more than a few hours for it to eat its own weight in worms. Experiments were made to determine the time required for the food to pass through the alimentary tract by the use of blueberries, which dyed the bird's excretions. Only half an hour was required, which explains the enormous capacity the birds have for food.

Behavior.—H. R. Ivor (1941 and 1943) has observed the peculiar behavior of "anting" in many species of birds, including the hermit thrush. In "anting" the birds seize the ants and place them in their feathers, usually under the primaries of the wings. They may also crush the ants with their bills and rub the juices on the feathers, or the birds may dust themselves in anthills. Various theories have been advanced to explain this behavior: The ants are placed among the feathers to drive out ectoparasites; the bird anoints its feathers with the formic-acid secretions of the ant to repel ectoparasites; the bird eats the ants for the formic acid, which may be beneficial as a medication to increase muscular energy or to expel endoparasites; the bird places the ants in the feathers to have a reserve food supply during migration. These and other suggestions have been made. Further observations and study of this behavior will be required to enable us to interpret the true biological significance of "anting."

Voice.—As a boy living in central Illinois I knew the hermit merely as the thrush with the reddish-brown tail, and in those days I never heard its exquisite song as it passed through that part of the State on its way to and from the nesting grounds. It uttered nothing more than a protesting *quoit* or *chuck* when we intruded upon its transitory haunts in the few scattered wooded areas of that prairie section. It was not until I came to Maine to live in the midst of its breeding area that I fully appreciated this aristocrat of the bird world. In Maine this gifted songster is at its best soon after its arrival during the last week of April. At this season any visit during the early morning or evening hours to a particular evergreen forest traversed by a cool meandering trout stream is certain to be rewarded by the superb performances of this prima-donna songster. Indeed, the hermit has been given the tribute of being the most gifted songster in North America, and its song has often served as the inspiration for poetic writers.

M. Chamberlain (1882) described his impressions of the song of the hermit thrush as he heard it near St. John, New Brunswick, Canada, as follows:

> The music of the Hermit never startles you; it is in such perfect harmony with the surroundings it is often passed by unnoticed, but it steals upon the sense of an appreciative listener like the quiet beauty of a sunset. Very few persons have heard him at his best. To accomplish this you must steal up close to his forest sanctuary when the day is done, and listen to the vesper hymn that flows so gently out upon the hushed air of the gathering twilight. You must be very close to the singer or you will lose the sweetest and most tender pathetic passages, so low are they rendered—in the merest whispers. I cannot, however, agree with Mr. Burroughs that he is more of an evening than a morning songster, for I have often observed that the birds in any given locality will sing more frequently and for a longer period in the morning than in the evening. I prefer to hear him in the evening, for there is a difference; the song in the morning is more sprightly—a musician would say "has greater brilliancy of expression"—and lacks the extreme tenderness of the evening song, yet both have the same notes and the same "hymn-like serenity." The birds frequently render their matinal hymns in concert and the dwellers in a grove will burst out together in one full chorus, forming a grander *Te Deum*—more thrilling—than is voiced by surpliced choir within cathedral walls. On one occasion an Indian hunter after listening to one of these choruses for a time said to me, "That makes me feel queer." It was no slight influence moved this red-skinned stoic of the forest to such a speech.

Aretas A. Saunders, who has made intensive studies of many bird songs, has written his interpretations and analysis of the hermit's song, in personal correspondence as follows: "The song of the hermit thrush is a long-continued one, made up of rather long phrases of 5 to 12 notes each, with rather long pauses. All the notes are sweet, clear, and musical, like the tone of a bell, purer than the notes of the wood thrush, but perhaps less rich in quality. The notes in each phrase are not all connected. The first note is longest and lowest in pitch, and the final notes are likely to be grouped in twos or threes, the pitch of each group usually descending. Each phrase is similar to the others in form but on a different pitch, as if the bird sang the same theme over and over, each time in a different key. If one listens carefully for each note, however, two different phrases are rarely exact duplicates in form, but slightly varied, a likeness to certain symphonies of some of the great composers.

"In records of 38 different individual birds the pitch ranges from F″ to D#′‴, one tone less than two octaves. The average individual has a range of about an octave. In looking over my records it is quite apparent that birds in the Adirondacks have a greater range of pitch than those in Allegany State Park in southwestern New

York. Most of my records come from those two localities. The Adirondack birds average two tones over an octave, and the Allegany Park ones nearly three tones less than an octave. The average bird, from my records, has six different phrases, but some have as few as three, several have nine, while one bird in Allegany Park had 14.

"Nearly every individual has one or two phrases pitched considerably higher than the rest, and these very high phrases sound weak and of poorer quality than the others. This may be due, however, not to actual poorer quality but to man's inability to perceive the overtones of the very high notes. It is the overtones that cause the poor or rich quality or timbre of a sound."

Albert R. Brand (1938) writes that in his study of bird recordings on film it is revealed that certain high notes are inaudible to the human ear, and his field observations of certain birds including the hermit thrush seem to confirm the suspicion. He states that he has observed birds singing nearby through field glasses, he has seen their bills open as if emitting notes, yet he heard no sound. Mr. Brand (1938) reports the lowest note recorded for the hermit thrush was 1,475 vibrations per second, the highest 4,375 vibrations, with an approximate mean of 3,000 vibrations per second.

Henry Oldys (1913), after describing the usual song of the hermit thrush and comparing it with the human voice, analyzes an unusual song in developing a theory of the independent evolution of bird song. Mr. Oldys heard the unusual hermit thrush song at Pompanoosuc, Vt., in which he noted a very perceptible normal order in the basal notes and their independent phrases, and that order made a harmonic progression such as completely satisfies the requirements of human music. He concludes "that the evolution of bird music independently parallels the evolution of human music and that, therefore, such evolution in each case is not fortuitous, but tends inevitably toward a fixed ideal."

At Lost River, N. H., located in the midst of the White Mountains, both the hermit and olive-backed thrushes nest, and there I had an unusual opportunity to compare the songs and notes of these two songsters. The musical ability of the hermit is more varied than that of the oliveback. Its usual song dies out without the rising inflection of the latter and there is a pause after the first syllable, while in the oliveback's song there is no pause and the second syllable is strongly accented, the whole song being quickly delivered.

The alarm notes of the two thrushes are also quite different. The oliveback thrush when disturbed utters a metallic note, short and sharp, often ending in a querulous call. The alarm note of the hermit has a catbird quality about it, lower pitched and less metallic than that of the oliveback. The hermit has a nasal note of complaint uttered in two syllables, a chuck like that of the blackbird, and a lisp not unlike that of a cedar waxwing. The oliveback utters a similar nasal note, but it is more liquid in quality and the *cluck* of the hermit may be compared to the *puk* or *pink* of the oliveback. The lisp is peculiar to the hermit, while there is a queer multiple note of soliloquy peculiar to the oliveback.

Norman McClintock (1910) has recorded the various notes uttered by the hermit thrushes about a nest which he observed at close range from a blind. He frequently heard a note resembling *quirk* or *quoit*, which was uttered when the birds were slightly suspicious or when they mildly protested against the presence of an intruder. A second note was a high-pitched, thin, and wiry call resembling a cedar waxwing's note but pitched several tones higher. This note was used in warning the young of approaching danger. "To the little birds this call meant 'freeze'." McClintock continues:

A third note, which this pair of Hermits used signified extreme distress. This note sounded to me much like the note of a hoarse Canary. I can best describe it by the word *boyb*, spoken slowly and with a rising inflection. The note also reminded me of a mew of a kitten. *Boyb* was uttered by the thrushes with the mandibles well open, whereas the Cedar-bird call was made with the mandibles almost closed.

Besides the three notes described, there was a much used conversational note that evidently contained no implication of suspicion or trouble and was a strong contrast with the several notes already described. It was an exceedingly soft and sweet little note that could be heard but a few feet, and which I can best describe by *wee*. *Wee* was used by the parents to each other and to the young. It seemed, however, to be mostly employed to herald to the young the parents' approach with food. At a distance of six or eight feet from the nest a single *wee* from a parent would announce to the young the former's proximity. As the parent hopped closer, the *wees* were rapidly repeated, *wee-wee-wee-wee*, and the nearer the parent came to the nest, the softer the *wees* were uttered, until they were faint whispers. To these *wees* the young responded, during their first days, by erecting their heads and opening wide their mouths; but later, when they became more mature, they would rise to their feet upon hearing the first *wee* and energetically beg for food. * * *

The fifth, and only remaining note, was one I heard but twice and both times it came from the male. It was an indescribable explosive twitter of ecstasy made with fluttering wings. I first heard it on August 3, immediately after the male had been singing for four minutes. On another day, it was uttered in the presence of the female, who was close by and towards whom it was directed.

According to Miss Cordelia Stanwood (1910) the fledglings give a clear sweet whistle, *p-e-e-p*, a soft, husky, breathing sound, *phee-phee*, and occasionally *pit-pit-pit!*, an almost inaudible ventriloquial call.

The adults are also capable of a certain amount of ventriloquial power. Quite often when closely observing a hermit thrush sing while I was concealed in the blind only a few feet distant, the voice seemed to come from an individual located far away from the scene. The song probably could not be heard by an observer stationed 50 yards away. At such times the throat of the bird vibrates or pulsates but the mandibles are tightly closed, thus subduing the loudness and carrying power of the varied notes.

Horace W. Wright (1912) has made a study of the times of the awakening and the evensong of the hermit thrush. Out of a list of 57 species recorded the hermit took sixteenth place in order of the first voices heard early in the morning. Of 18 records of the hermit thrush the first song was 63 minutes and the latest initial song 45 minutes before sunrise. Of 55 species of birds studied for the latest evening song the hermit stands in fifty-first place. Out of 20 records the average number of minutes after sunset was 33, the latest final song 40 minutes after sunset, and the earliest final song 25 minutes after sunset.

The song season of the hermit thrush, unlike that of many of our song birds, is not limited to the time of the breeding season, but it is also in full song in its winter haunts in the south. Aretas A. Saunders (1929b) writes that he has heard the hermit commonly and in full voice in the pine forests of central Alabama. Otto Widmann (1907) in writing of the winter resident hermit thrushes of the peninsula of Missouri states: "He greets it with his most tender strains on his return in the fall, and sings aloud before he leaves it for the north." Many other observers have had similar experiences, of hearing the full song of the hermit in their winter haunts although this bird does not sing during its migration journey.

W. DeW. Miller (1911) observed 12 hermit thrushes which wintered in a grove of red cedars, in a sheltered valley near Plainfield, N. J., where there was an abundance of food in the berries of the flowering dogwood. He writes:

> I * * * heard three distinct call-notes from these birds, one, of course, the familiar low blackbird-like *chuck*. The two other notes do not seem to be commonly known, at least to those familiar with the bird only as a migrant. The first is a simple, high-pitched whistle, rarely loud; the second, a curious, somewhat nasal cry recalling the unmusical note of the veery.
>
> The Hermit Thrush seldom sings while with us in the spring, and the song is so low as to be inaudible if one is more than a few yards from the singer. On March 19, I was agreeably surprised to hear four or five of these thrushes singing through most of the afternoon, though it was raining at the time. The song of only one bird, however, was of sufficient volume to be heard at any distance.

Enemies.—The hermit thrush is subject to the usual enemies such as snakes, foxes, weasels, and skunks that molest ground-nesting birds. The domestic cat, which is so destructive to birds that nest about or near human dwellings, is less of a factor in the life of a bird that usually nests in remote situations seldom visited by cats. The stomach examinations of hawks and owls reveal that the hermit sometimes falls a victim to these predators.

Few birds are more valiant in resisting attacks on their brood and capable of creating a greater hubbub over the presence of marauders. Such an occasion invariably attracts other birds of the vicinity, adding a veritable chorus of protests. The parent birds, very tense, with crests erect, swoop and dash fearlessly, sometimes venturing so close that their wings strike the intruder. More often than not they are successful in driving the enemy to cover.

As is true with many birds, the hermit thrush is subject to infestation by a number of external parasites. Under ordinary conditions none of them prove fatal to the bird, but some of them may be very annoying to the host when they become abundant. Harold S. Peters (1936) has reported three species of lice, *Degeeriella eustigma* (Kellogg), *Myrsidea incerta* (Kellogg), and *Philopterus subflavescens* (Geoffroy); two species of bird flies, *Ornithoica confluenta* Say and *Ornithomyia anchineuria* Speiser; two ticks, *Amblyomma tuberculatum* Marx and *Haemaphysalis leporis-palustris* Packard; and the mite *Trombicula whartoni* Ewing. In addition to the above Peters (1933) also reported the tick *Ixodes brunneus* Koch as a parasite of the hermit thrush.

According to Herbert Friedmann (1929) the hermit thrush is a very uncommon victim of the cowbird, and at the time of writing he knew of only six definite records. Contrary to Friedmann's statement, I have found it to be a frequent victim. I have seen fewer than 15 nests of the hermit thrush, yet four of these were parasitized by the cowbird. On May 31, 1920, I found a nest near Brunswick, Maine, that contained three eggs of the hermit thrush and two eggs of the cowbird. This nest was destroyed by some predator a few days later. On July 8, 1928, a nest containing three eggs of the hermit thrush and one egg of the cowbird was found at Douglas Lake, northern Michigan. In this nest the cowbird hatched first and continued to thrive until it left the nest. Two of the eggs of the hermit thrush hatched, the third was sterile. Both of the two hermits were also successfully reared. On July 6, 1928, also at Douglas Lake, Mich., I found a nest containing two eggs of the hermit thrush and one egg of the cowbird. On July 12 the cowbird egg hatched, followed a day later by the hatching of one of the thrush eggs; the other thrush egg was sterile. The thrush and cowbird competed for food, but the cowbird proved more aggressive and maintained its lead in size throughout the nesting period. On July 19, when the cowbird was seven days old and the hermit six days old, there was a marked difference in weight and size and relative development of the feathers. By July 25 both the cowbird and thrush had left the nest. The adult hermit was seen feeding the cowbird nearby, but the young hermit was not seen, although I have no reason to believe that it perished.

A fourth nest of the hermit thrush parasitized by the cowbird was found at Topsham, Maine, on June 6, 1941. This nest contained two eggs of the hermit thrush and two eggs of the cowbird. No opportunity was presented to visit this nest a second time.

In addition to enemies and parasites the hermit thrush is subject to many hazards especially during the migration season. Along the coast of Maine the lighthouses exact their toll of these birds especially during the heavy fogs and storms which often prevail during that season of the year. Mention has already been made elsewhere of the destruction to the earlier migrants, which succumb to sudden excessive cold waves and late snowstorms, especially when the available food is covered by a deep fall of snow.

W. E. Saunders (1907) records a fall migration disaster in western Ontario as follows:

The early days of October, 1906, were warm and damp, but on the 6th came a north wind which carried the night temperature down to nearly freezing. Near there it stayed with little variation until the 10th, * * * the north wind brought snow through the western part of Ontario. At London there was only 2 or 3 inches, which vanished early next day; and the thermometer fell to only 32 degrees on the night of the 10th, and to 28 on the 11th, but ten miles west, there was 5 inches of snow at 5 P. M. Oct. 10, and towards Lake Huron, at the southeast corner, between Goderich and Sarnia, the snow attained a depth of nearly a foot and a half, and the temperature dropped considerably lower than at London. On that night, apparently, there must have been a heavy migration of birds across Lake Huron, and the cold and snow combined overcame many of them, so that they fell in the lake and were drowned.

Along the shore of the lake near Port Franks there were an estimated 5,000 dead birds to the mile along the beach. On the beach south of Grand Bend, Mr. Saunders counted 1,845 dead birds, including 20 hermit thrushes, in a relatively short time. On the beach at Sable River he found the dead birds even more numerous than at Grand Bend, the site of the above census. Mr. Saunders states that the bulk of the hermit thrushes had already passed by on October 6, yet they were well represented among the dead birds found.

Winter.—More than in the case of the other thrushes a considerable number of individual hermits spend the winter months in regions well north of the well-established winter range of the species. In correspondence from E. M. S. Dale he states that it is not unusual for the hermit thrush to winter in Ontario. He mentions one individual in particular that lived through the winter of 1941–42 at London, Ontario. It fed on currants supplemented by gleanings of suet and other food from the food shelf of the chickadees and downy woodpeckers. Its favorite spot was a corner of the front veranda, sheltered by a discarded Christmas tree, in the lee of which the currants were placed. It disappeared when the spring brought others of its kind on their journey north.

There are numerous winter records of the hermit thrush in southern New England, where it has been reported in favorable situations throughout Massachusetts and Connecticut even at times when it was very cold and the ground covered with snow. It has also been reported from various sections of New York and New Jersey during the winter months. South of this area it is a regular winter resident.

In South Carolina, according to A. T. Wayne (1910), the hermit arrives by October 23 and remains until the second week of April. The birds are not abundant until the middle of November, when they are apparently settled for the winter. Contrary to reports of the wintering hermits in New England, Mr. Wayne states that the species cannot endure a sudden change of weather, especially if very low temperatures prevail for even a few days, as great numbers perished on February 13 and 14, 1899. During January and February 1895, hundreds succumbed to cold weather, although the food supply was plentiful. Mr. Wayne states that the hermit is the least shy of the thrushes and can be readily approached within a few feet, especially during cold weather. According to W. P. Wharton (1941) the hermit has a strong tendency to return to the same winter quarters. Of 81 of the birds he banded at Summerville, S. C., he had 10 returns, or 12.34 percent.

In Florida, according to A. H. Howell (1932), the hermit is a common winter resident in the northern and central parts of the State but rare in the southern part. He states further: "During the winter season, these Thrushes inhabit thick hummocks and the borders of wooded swamps. [In many parts of the south because of this characteristic habitat the hermit is known as the 'swamp sparrow' or 'swamp robin.'] While not particularly shy, the birds are so quiet and retiring in disposition that they attract little attention as they feed on or near the ground."

In the Middle West there are winter records of the hermit thrush for Ohio, Indiana, Illinois, and Missouri, the majority from the southern sections of these States. In the winter of 1906–07 I found the hermit to be a very abundant bird in the wooded river bottoms of the Mississippi and Ohio Rivers in southern Illinois. Otto Widmann (1907), in writing of the wintering birds of the Peninsula of Missouri, states: "It is seldom heard to sing in transit, but may be heard in its winter home, where it frequents the same swampy ground as the winter wren adjoining the drier haunts of the fox, white-throated and other sparrows." The hermit is very abundant in the Mississippi River Valley south of Missouri, but the majority pass southward in October and northward in April.

HYLOCICHLA USTULATA USTULATA (Nuttall)

RUSSET-BACKED THRUSH

HABITS

The russet-backed thrush, the western race of a widely distributed species, lives in summer in a comparatively narrow range from Juneau, Alaska, to San Diego County, Calif., west of the Cascades and the Sierra Nevada. Throughout most of this range it is an abundant summer resident and a greatly admired songster. The 1931 Check-list gives it as a transient in Lower California, but M. Abbott Frazar collected for William Brewster (1902) four males in the Sierra de la Laguna, on May 4, 7, and 16, and a female at Triunfo on June 13, all of which they both thought were settled for the season and were breeding, or about to breed. There is a wide gap between southern Lower California and San Diego County, where no thrushes of this species are known to breed.

In the willow thickets in the lowlands of Los Angeles County, and especially along densely shaded streams, we found the russet-backed thrush to be a common breeding bird. Grinnell and Wythe (1927) record it as abundant in summer in the San Francisco Bay region, "nesting in lowland orchards, along willow-bordered streams, and in forested canyons, wherever in the whole region such occur." In the Yosemite region, Grinnell and Storer (1924) found this thrush breeding only among the foothills and in the valleys, in the streamside lowlands. Grinnell, Dixon, and Linsdale (1930), referring to the Lassen Peak region, write: "Russet-backed thrushes that lived in the section in summer were restricted closely to clumps of shrubby vegetation, chiefly willow and white alder growing in moist places—either bordering streams or on wet ground, as in the lower mountain meadows."

These haunts, west of the Sierra Nevada, are in marked contrast to those of the closely related olive-backed thrushes in the fir forests east of that range, and very different from the summer haunts of our eastern birds in the coniferous forests of northern New England and eastern Canada.

Spring.—The russet-backed thrush seems to be a rather late migrant in spring, advancing rather slowly. Grinnell and Wythe (1927) say that it "arrives later than most of the summer visitants in the San Francisco Bay region, about the last of April; an early date is April 15, at Berkeley." It seems to be much later farther north. We did not record it at all during the first two weeks in May in western Washington, though we made careful field notes every day; it is an abundant summer resident there, and Mr. S. F. Rathbun's notes indicate that it usually arrives around the middle of May, his earliest date being May 6. George Willett tells me that it arrives in southern Alaska mostly late in May, his earliest date being May 16.

Nesting.—The only nest of the russet-backed thrush that I have seen was evidently fairly typical of this subspecies. It was found on May 30, 1914, in Los Angeles County, and contained three fresh eggs; it was placed 8 feet from the ground in a crotch of a slender willow in a grove of willows in a damp spot; there was a foundation of dead leaves, twigs, and rubbish, mixed with mud, on which was built a superstructure of twigs, leaves, and plant fibers; it was lined with fine rootlets, fine fibers, and skeleton leaves.

Six other California sets in my collections were taken from nests in willows, blackberry tangles, or other low bushes; the lowest nest was only 2 feet from the ground. J. Stuart Rowley tells me that, in the vicinity of Los Angeles County, these thrushes seem to prefer the tangles of wild blackberries, or the willow thickets in the lowlands, as nesting sites.

Mr. Rathbun says, in his notes from western Washington, that this thrush nests from June 10 up to the middle of July, but the great majority of the nests will be found during the latter half of June. He has "found them generally well within the forest, and a favorite location is among the low growth along the forest's edge, particularly if in the proximity of the water." The nest "is generally built quite close to the ground in a variety of places, sometimes in the salal shrubs or on the top of fallen masses of the dead brackens, and many times in some low-hanging, rather dense branch of a bush. On occasions I have found nests at some considerable height above the ground near the extremity of a limb of some small tree, but these locations are out of the ordinary. The nests are almost always attractive in appearance and well made. They consist outwardly of an abundance of green mosses and dead leaves, and often with these are strips of thin, flat inner bark; within this at times is some dry, rotten vegetable matter quite smoothly moulded; and the nest is lined with dry grasses."

He describes an especially beautiful nest as follows: "It was built quite near the ground on a branch of a huckleberry bush. There was first used in its construction a few of the dead stalks of the bracken, into which was woven quite a mass of dead moss taken from the trunks of fallen trees, this forming a substantial wall for the nest. The outward covering was entirely of bright green, living strips of moss, so beautifully placed that nothing but this color showed, and this moss so arranged that the long strips hung downward, giving the nest a draped effect. In its color scheme the nest blended perfectly with its location, for the under part was the shade of the twigs on which it was placed, a neutral reddish-brown hue; whereas the color of the moss placed exteriorly blended with that of the leaves of the huckleberry. To carry out the illusion further there were interwoven in the green moss across the more exposed side of the nest, three of the brown bracken stalks, these giving the effect of being huckleberry branches and, also, affording additional support."

W. L. Dawson (1909) says of Washington nests: "In distance from the ground, nests varied from six inches to forty feet, altho a four or five foot elevation was about the average." Again he says "sometimes 30–60 feet high in trees."

Eggs.—The russet-backed thrush lays three to five eggs to a set, the commonest number being three or four; some say that three is the commonest number and some say four. The eggs are practically indistinguishable from those of the olive-backed thrush, which the reader will find described under that subspecies. The measurements of 50 eggs in the United States National Museum average 23.2 by 17.2 millimeters; the eggs showing the four extremes measure **25.4** by 17.8, 24.4 by **18.0, 20.8** by 16.8, and 21.3 by **15.2** millimeters.

Young.—The period of incubation is said to be 14 days, and the young birds are said to remain in the nest for about the same length of time, if not disturbed. I have no information as to whether both sexes incubate or not, and nothing seems to have been published on the care and development of the young except in Professor Beal's (1907 and 1915b) papers on food. "Two nests were carefully and regularly watched, and from these it was determined that the parent birds fed each nestling 48 times in 14 hours of daylight. This means 144 feedings as a day's work for the parents for a brood of three nestlings, and that each stomach was filled to its full capacity several times daily, an illustration that the digestion and assimilation of birds, especially the young, are constant and very rapid."

He examined the stomach contents of 25 nestlings, eight broods of young birds sacrificed in the cause of science, and found that their food was 92.60 percent animal and only 6.8 percent vegetable matter, a considerably higher percentage of insects than was found in the stomachs of adults. Caterpillars formed the largest item, nearly 27 percent; beetles amounted to 22 percent, including 7.7 percent of hard-shelled Carabidae; Hemiptera, including stink bugs, leafhoppers, treehoppers, shield bugs, and cicadas, made up 13.8 percent of the food; ants and a few other Hymenoptera amounted to 12 percent; only three stomachs contained remains of grasshoppers. "The vegetable food consisted of fruit (6.8 percent), mainly blackberries or raspberries, found in 11 stomachs, and twinberries in 1, and two or three other items, including a seed of filaree and some rubbish."

Plumages.—Mrs. Irene G. Wheelock (1904) says of a brood of three young russet-backed thrushes that were evidently about a week old: "They were sparsely covered with brownish gray down, and pin-feathers were just showing along the feather tracts."

Subsequent molts and plumages are apparently similar to those of the olive-backed thrush, which are more fully described under that subspecies.

Food.—Professor Beal (1907) examined 157 stomachs of the russet-backed thrush, taken mainly in the San Francisco region from April to November. The examination showed 52 percent of animal and 48 percent of vegetable food. Ants formed the largest item in the insect food, amounting to 16 percent; "Hymenoptera, other than ants (mostly wasps), bugs, flies, and grasshoppers, with some spiders, amount altogether to 12 percent of the year's food," but grasshoppers were found in only four stomachs. Beetles, only 3 percent of which were useful species, constituted 11 percent of the year's food; and caterpillars amounted to somewhat more than 8 percent.

The vegetable food consists mainly of fruit and only a trace of weed seeds. "It is probable that the greatest harm done by this bird is to the cherry crop, though undoubtedly it eats the later fruits to some extent. In May and June the fruit eaten reaches 41 and 38 percent, respectively, and this probably represents the greatest injury which the bird does, as most of the fruit was the pulp and skins of cherries." Other vegetable food included seeds of blackberries, raspberries, both wild and cultivated, and seeds of the twinberry, elderberry, coffee-berry, peppertree, and poison-oak. Fruit eaten in September amounted to 80 percent, the largest amount for any month.

Ian McT. Cowan (1942) records the russet-backed thrush among the species that he has seen feeding on the termite *Zootermopsis angusticollis* in southwestern British Columbia, where "the extensive areas of deforested land, strewn with decaying logs and stumps, provides ideal habitat for termites."

Behavior.—The russet-backed thrush is a close sitter when incubating and will allow a close approach to the nest; it then slips quietly off the nest and disappears in the shrubbery, keeping out of sight and uttering its peculiar notes of protest but offering no attempt at active defense. At other times it is a shy retiring bird, much oftener heard than seen; one hears its beautiful song or its characteristic alarm notes and attempts to follow it, but it fades away under the cover of the thick foliage, and one hears its notes again from some more distant point. It feeds mainly on the ground, as the other thrushes do, running over the ground and drawing itself up to its full height when it halts, as the robin does on the lawn. It seems less nervous than the hermit thrush, and is seldom seen to raise and lower its tail when excited.

Stanley G. Jewett (1928) tells an interesting story of a russet-backed thrush that assisted in feeding a brood of young robins in their nest. He and his sister watched the robin's nest for a period of four hours one afternoon, during which time the thrush made at least 12 visits to the nest and fed the young robins. "There were two robins and two thrushes near the nest during the entire afternoon. Although a systematic hunt was made for the thrushes' nest, it was not found."

Voice.—One who is familiar with the song of the olive-backed thrush in the East should easily recognize the the song of the russetback; the songs are very similar with the same rising pitch. Grinnell and Storer (1924) write:

By early June, and sometimes sooner, the Russet-backed Thrushes in Yosemite Valley are in full song and may be heard during the day as well as in the morning and evening hours. The song is set in character and each individual thrush begins his song on about the same key—not changing from song to song as does the Hermit. The first syllables of any individual's song are always on the same pitch, and full, clear, and deep; the remainder are more wiry, ascending, and sometimes the last one goes up so high in pitch as to become almost a squeal: *wheer, wheer, wheer, whee-ia, whee-ia, whee-ia*, or *quer, quer, quer, quee-ia, quee-ia, quee-ia*. The call note oftenest heard is a soft liquid whistle, *what* or *whoit*, sounding much like the drip of water into a barrel. An imitation of this note by the observer will often bring a thrush into close range. Now and then a thrush will give an abrupt burred cry, *chee-ur-r;* and again there may be a single whistle, louder and higher than the usual call. The song season lasts until early July, after which the birds become quiet. By the end of the month not even the call note is to be heard.

According to Mr. Rathbun, this thrush does not sing its full song immediately on its arrival in western Washington, for in his notes for June 5 he wrote: "Although in a good locality, where this bird was common, its song was seldom heard, and even then was not sung in full. The song of the russet-backed thrush is a succession of round, smooth notes, flowing easily in an ascending scale, the latter part having a reedy sound. It spirals upward, as it were, and will be heard at its best in the early morning and in the dusk of evening on a quiet day. Its song practically ceases by the end of July, except on rare occasions; then a period of silence seems to ensue; and about the first of September, one will again hear the call notes, and shortly thereafter the migration southward begins."

Enemies.—Dr. Friedmann (1929) records only one case in which this thrush has been imposed upon by the dwarf cowbird; there were two eggs of the thrush and one of the cowbird in the nest.

Field marks.—The russet-backed thrush is not likely to be confused with any other bird on the Pacific slope except with one of the hermit thrushes, but the uniform russet-brown of its back is easily distinguished from the contrasted brown back and rufous tail of the hermits;

furthermore, the haunts of the two, during the breeding season at least, are quite different.

Fall.—The russet-backed thrush spends a rather short season on its breeding grounds. Mr. Willett tells me that it is not seen in southern Alaska after late August; and Mr. Rathbun's notes indicate that it disappears from western Washington during the first week in September. Most of these thrushes have probably left California before the end of the latter month, though Professor Beal (1907) mentions specimens taken there in October and even November.

Winter.—The 1931 Check-list states that the russet-backed thrush "winters from Vera Cruz, Guatemala, and Costa Rica to eastern Ecuador and British Guiana," but it apparently does not go so far south. Ludlow Griscom (1932) says: "The intensive collecting of the last twenty years has failed to produce a single record of *ustulata* anywhere in the New World south of Costa Rica. This fact was correctly pointed out by Salvin in 1879, and ignored by everyone since."

In their report on the birds of El Salvador, Dickey and van Rossem (1938) record it as a "common winter visitant throughout the higher parts of the Arid Lower Tropical Zone and in lesser numbers in the lowlands. Extreme dates of arrival and departure are October 14 and March 15." They say further:

The russet-backed thrush differs radically from the two olive-backs in that it appears as a common winter visitant, while the olive-backs are, in the main, migrants. The distribution of the present race is apparently general over the whole of the hill country lying within the Arid Lower Tropical Zone. In the lower country it is decidedly rare, for only two or three birds were noted in the coyol-palm growth at Puerto del Triunfo in January, 1927. It is in the multitude of berry- and fruit-bearing shade trees growing above the coffee that the russet-backs are commonest. On Mt. Cacaguatique in November and December, 1925, they literally swarmed in suitable localities; some trees had constantly arriving and departing streams of these birds, with perhaps twenty-five or more in a tree at once.

Alexander F. Skutch has sent me the following interesting notes on the winter haunts and behavior of this thrush in Central America:

"On the Pacific slope of both Guatemala and Costa Rica—and probably also in the intervening countries—the russet-backed thrush winters in considerable numbers. It is particularly abundant in the zone of heavy, humid forests between 2,000 and 4,000 feet above sea level. Arriving in Guatemala early in October, and later in the month in Costa Rica, it is at first shy, retiring, and little in evidence, although occasionally a liquid *quit* reveals its presence in the dense undergrowth. After the beginning of the new year, it becomes an increasingly prominent member of the avifauna. Thus on the Finca Moca, a huge coffee plantation at the southern base of the Volcán Atitlán in western Guatemala, I found the bird abundant late in January in a tract of low, dense thicket dominated by scattered great trees that survived the destruction of the original forest. Here the thrush's liquid monosyllables were heard on every hand during the morning hours. It was surrounded by many other winter residents, including worm-eating, hooded, Kentucky, Wilson's, and MacGillivray's warblers.

"In the basin of El General in southern Costa Rica, it winters chiefly in the undergrowth of the high forest and, like its resident neighbors of the same habitat, never forms true flocks. During most of its sojourn here it is so retiring that, but for an infrequent call note, it might easily be overlooked, despite its abundance. Sometimes it ventures forth into the adjacent clearings, where the forest has been felled and burned to plant maize, then the land temporarily abandoned after the harvest. Such clearings are usually filled with a luxuriant growth of that widespread 'fireweed' of tropical America, the *jaboncillo* (*Phytolacca rivinoides*), a kind of pokeberry, which during the early months of the year bears a profusion of deep purple berries that attract a multitude of birds, from big toucans and finches to little tanagers, manakins, and honeycreepers.

"The russet-backed thrush sometimes joins the mixed flocks of small birds that follow the foraging swarms of army ants. Like most if not all of these birds, it does not devour the ants themselves but snatches up the insects and other small creatures driven from where they have lurked beneath the fallen leaves, in rotting stumps and crevices in the bark of trees. The thrush hovers about the outskirts of the swarm; and I have not seen it dash into the midst of the fray to seize a fugitive, in the manner of the tropical birds more adept at this kind of hunting. What strange company for a bird hatched among northern spruce and fir trees! Who that knows the russet-backed thrush only amid the severe simplicity of a northern coniferous forest could imagine it in the infinitely varied tropical silva, burdened with huge woody vines and a hundred kinds of epiphytes, where it consorts on intimate terms with such birds as manakins, woodhewers, antbirds, and ant-tanagers? Truly these migratory birds lead double lives.

"Late in February or early in March the russet-backed thrushes begin to sing in an undertone. Soon their slender liquid spirals of song arise in the forest on every side, not merely in isolated instances, but again and again and again. Their soft *quit* is also heard oftener now. In April they sometimes become so numerous that they have appeared to me the most abundant bird in the region, save only the migratory swallows passing northward in their myriads. Since the resident birds of the understory of the tropical forest are in the main parsimonious of song, at times early in April the russet-backed thrushes produce more music than any other species of the woodland. I am not familiar with this thrush on its nesting ground; but it seems that, although it sings so much here, it is always in an undertone, as while passing northward through the middle tier of the United States later in the spring. While earlier in the year they have been confined to the forest and its edges, these birds are now often seen in the bushy clearings, in hedgerows, and sometimes even in cultivated fields. At times in April, riding along the grassy cartroads of El General, I have heard the song and the fluid *quit* arising from every little patch of woodland that skirted the road. At this season one individual frequently pursues another; apparently their instinct to defend a territory begins to awaken while they are still so far from their summer home.

"While usually these so numerous birds are rather evenly distributed, with little tendency to flock, at sunrise on April 14, 1940, while watching great clouds of northward-bound swallows pass over the open crest of a hill, I saw about eight russet-backed thrushes together at the edge of a small tract of woodland, beside the open pasture. They created the impression of a migrating flock just come down to rest and forage.

"Late in April or the first days of May, while still to all appearances as numerous as ever, the russet-backed thrushes sing much less than at a slightly earlier date. I can explain this falling off of song, noted in several years, only by the hypothesis that the males depart first, leaving a preponderance of females that do not sing. With most other migrant song birds, the males that linger latest are the most songful at the time of their departure.

"Although the russet-backed thrush arrives late in El General, it also lingers late. My earliest date for its arrival is October 18, 1936. My last spring records are: May 10, 1936; May 4, 1937; May 11, 1939; April 21, 1940; and April 30, 1942. I have never seen the species anywhere on the Caribbean slope of Central America except late in spring, when migration was in progress. My actual records are: Barro Colorado Island, Canal Zone, April 4, 5, and 6, 1935; Vara Blanca, Costa Rica, 5,500 feet, April 3 to May 7, 1938; and Pejivalle, Costa Rica, 2,000 feet, April 15, 1941. During the year I spent on the Sierra de Tecpan in west-central Guatemala, studying the bird life between 7,000 and 10,000 feet, I recorded this species only on April 12 and October 19 and 20, 1933. I have not attempted to distinguish the races of *Hylocichla ustulata* in the field; but according to the analysis of Griscom (Bird Life in Guatemala, 1932) the birds wintering on the Pacific slope would all belong to the race *ustulata,* while the transients recorded on the Caribbean slope might be either this race or *swainsoni.*"

DISTRIBUTION

Range.—The range of the russet and olive-backed thrushes is from northern North America to central South America.

Breeding Range.—The breeding range is **north** to central Alaska (Tanana, Fort Yukon, and the Porcupine River); northern Yukon (La Pierre House); the Mackenzie River Valley (McPherson, Simpson, and Resolution); northern Alberta (McMurray); northern Saskatchewan (Black Bear Island and Reindeer River near its mouth); northern Manitoba (Norway House, Oxford House, Jack River, and York Factory); northern Ontario (Moose Factory); central Quebec (Rupert House); and southern Labrador (Upper Hamilton River and Paradise River). **East** to Labrador (Paradise River, Petty Harbor, and Caplin Bay); Newfoundland (Nicholsville); Magdalen Islands, Quebec; and Nova Scotia (Wolfville and Halifax). **South** to Nova Scotia (Halifax); southern Maine (Machias and Ellsworth); New Hampshire (Ossipee and Mount Monadnock); New York (Slide Mountain, Catskills); Pennsylvania (Laanna and the Pocono Mountains); the mountains of West Virginia (Shavers Mountain and Cheat Mountains); North Carolina (one record Mount Mitchell); northern Michigan (Kalkasta County, Douglas Lake, the Beaver Islands, and Newberry); northern Wisconsin, rarely (New London and Orienta); northern Minnesota (Cook County, Walker, and Fosston); southern Manitoba (Hillside Beach, Lake Winnipeg, and Aweme); southern Saskatchewan (Indian Head and the Cypress Hills); southern Alberta (Red Deer River region); south in the Rocky Mountains through Montana (Fortine and Belton); Wyoming (Yellowstone Park and Laramie); Colorado (Estes Park, Pikes Peak, and Grove Creek, Mesa County); northern Utah (Kansas, Parleys Park, and Salt Lake County); Nevada (Franklin Lake and Truckee Valley); and southwestern California (San Bernardino, Pasadena, and Santa Barbara). **West** to the Pacific coast of California (Santa Barbara, Santa Clara, Alameda, and Eureka); Oregon (Marshfield, Saline, and Portland); Washington (Cape Disappointment, Clallam Bay, and Blaine); British Columbia (Courtenay, Vancouver Island and Massett, Queen Charlotte Islands); and Alaska (Sitka, Lake Clark, McKinley Fork of the Kuskokwim and Tanana).

The breeding range as outlined applies to the entire species, of which three subspecies or geographic races are recognized. The russet-backed thrush, the typical race (*H. u. ustulata*), breeds from southeastern Alaska, the coastal region of British Columbia west of the Cascades, to southern California; the western olive-backed thrush (*H. u. almae*) breeds in the Rocky Mountain region west to eastern Nevada; the olive-backed thrush (*H. u. swainsoni*) breeds from central Alaska, northwestern Mackenzie, and northern British Columbia eastward to the Atlantic.

Winter range.—The winter range of the russet-backed thrush group is as yet imperfectly defined owing to the scarcity of carefully identified winter specimens. The russet-backed thrush winters north to central Mexico (Tres Marías, Minca near Mexico City, and Misantla, Veracruz); and **south** to Guatemala (Coban and Los Amates). Some race winters from El Salvador (Colima, Mount Cacaquatique, and Puerto del Triunfo) to southern Costa Rica (Basin of El General and Boruca). These have been recorded as *swainsoni* but it seems possible that some may be the more recently recognized *almae.* Specimens from Costa Rica in spring have been identified as this race.

The olive-backed thrush (*H. u. swainsoni*) winters in South America from the Santa Marta region of Colombia **east** to northwestern Brazil (Cocuy and Marabitanas); central Bolivia (Buenavista and Entre Ríos); and northwestern Argentina (Lules, Tucumán). **South** to northwestern Argentina. **West** to Peru (La Merced and Chinchao); western Ecuador (Zamora, Cuenca, and Quito); and western Colombia (San Antonio, Novita, and Santa Marta). Since there is as yet no record from Venezuela, it seems probable that a specimen taken at Roraima, British Guiana, on December 6 was a delayed migrant.

Migration.—In the migration, no attempt has been made to distinguish between the races. Some late dates of spring departure are: Ecuador—San José de Sumaco, April 18. Colombia—Valparaiso, March 24. Panama—Perine, April 22. Costa Rica—Volcán de Irazú, April 15; Guatemala—Patulul, April 3. British Honduras—Augustine, April 26; Mexico—Presidio, Veracruz, May 6. Florida—Tortugas, May 22. District of Columbia—Washington, June 2. Pennsylvania—Jeffersonville, June 9. Louisiana—Shreveport, May 13. Iowa—Sioux City, June 3. Texas—San Antonio, May 18. Kansas—Fort Riley, May 24. North Dakota—Charlson, June 5. Arizona—Tucson, May 30.

Some early dates of spring arrival are: Georgia—Macon, April 19. North Carolina—Chapel Hill, April 22. District of Columbia—Washington, April 18. Pennsylvania—Pittsburgh, April 24. New York—Rochester, April 22. Massachusetts—Harvard, May 5. Maine—Ellsworth, May 4. Nova Scotia—Wolfville, May 8. New Brunswick—Scotch Lake, May 17. Quebec—Quebec, May 23. Louisiana—New Orleans, April 2. Tennessee—Nashville, April 18. Kentucky—Guthrie, April 3. Ohio—Columbus, April 22. Ontario—Toronto, April 24. Missouri—Columbia, April 16. Iowa—Iowa City, April 17. Wisconsin—Madison, April 25. Minnesota—Minneapolis, April 30. Texas—Austin, March 2. Oklahoma—Norman, April 21. Nebraska—Lincoln, April 28. North Dakota—

Fargo, April 25. Manitoba—Aweme, May 10. Saskatchewan—Indian Head, May 5. Colorado—Denver, April 24. Wyoming—Laramie, May 5. Idaho—Lemhi, May 10. Montana—Miles City, May 11. Alberta—Edmonton, May 2. Mackenzie—Simpson, May 24. Arizona—Paradise, April 20. California—Witch Creek, March 9. Oregon—Sutherlin, April 13. Washington—Destruction Island, April 25. British Columbia, Queen Charlotte Islands, April 25. Yukon—Dawson, May 24. Alaska—Mount McKinley, May 12.

Some late dates of fall departure are: Alaska—Sergief Island, September 4. British Columbia—Courtenay, Vancouver Island, September 21. Washington—Yakima, October 17. California—Big Creek, October 30. Mackenzie—Simpson, September 10. Alberta—Athabaska Landing, September 14. Montana—Fortine, September 23. Wyoming—Laramie, October 20. Colorado—Fort Morgan, October 5. Manitoba—Shoal Lake, September 30. North Dakota—Fargo, October 7. Nebraska—Valentine, October 16. Minnesota—St. Paul, October 5. Iowa—McGregor, October 27. Wisconsin—North Freedom, October 15. Illinois—Lake Forest, October 17. Kentucky—Bowling Green, October 27. Ontario—Ottawa, October 21. Michigan—Ann Arbor, October 30. Indiana—Indianapolis, October 19. Mississippi—Bay St. Louis, October 31. Quebec—Montreal, October 1. New Brunswick—St. John, October 5. Nova Scotia—Pictou, October 11. Maine—Portland, October 20. New York—New York, November 1. Pennsylvania—Berwyn, November 6. District of Columbia—Washington, November 1. Cuba—Viñales, November 28.

Some early dates of fall arrival are: North Dakota—Argusville, August 25. Texas—Castle Mountains, September 11. Mississippi—Ariel, September 11. District of Columbia—Washington, September 1. North Carolina—Weaverville, September 2. South Carolina—Summerton, September 13. Florida—Pensacola, September 27. Mexico—Escuinapa, Sinaloa, September 13. Nicaragua—Bluefields, October 3. Costa Rica—La Hondura, September 27. Panama—Changuinola, October 15. Colombia—Chicoral, October 11. Ecuador—Zamora, October 23. Peru—Chinchao, October 26. Bolivia—Entre Ríos, October 17.

Banding records.—An olive-backed thrush (probably the form now recognized as Alma's) was banded at Madison, Minn., on May 25, 1937, and found dead about June 17, 1937, at Vulcan, Alberta. A bird banded at Fort Smith, Ark., on April 16, 1939, was found dead on May 2, 1940, at Fairport, N. Y.

Casual records.—There are several records of the occurrence of the olive-backed thrush in Bermuda in September, October, February, and April. The accidental occurrence of this species at Puerto Bertoni, Paraguay, is recorded without date. A specimen of *H. u. ustulata* was collected on January 22, 1894, at Los Gatos, Calif.

Egg dates: Alaska: 11 records, June 1 to July 2.

California: 124 records, April 15 to July 15; 66 records, May 19 to June 9, indicating the height of the season.

Maine: 51 records, May 29 to July 24; 26 records, June 7 to June 14.

New Brunswick: 43 records, June 6 to June 26; 27 records, June 12 to June 18.

HYLOCICHLA USTULATA SWAINSONI (Tschudi)

OLIVE-BACKED THRUSH

HABITS

The wide breeding range of the two eastern forms of this species includes most of the forested regions of Alaska, north and east of the Pacific coastal ranges, practically all the Canadian Zone in Canada and Newfoundland, and extends southward in the United States to northern California, east of the Cascades and the Sierra Nevada, Nevada, Utah, Colorado, northern Michigan, northern New England, and in the mountains to Pennsylvania and West Virginia.

This range includes that of the race *almae,* which did not appear in the 1931 Check-list. Dr. H. C. Oberholser (1898), in naming and describing Alma's thrush, says: "The present race differs from the eastern *Hylocichla ustulata swainsonii* in the more grayish, less olivaceous color of the upper surface, this being usually more noticeable on the rump and upper tail-coverts. The sides and flanks also average more grayish. No apparent difference in size exists. No comparison with *H. ustulata* proper is necessary, for *Hylocichla u. almae,* although geographically intermediate, is even less closely allied to *ustulata* than is *swaìnsonii.*"

H. S. Swarth (1922) made some interesting observations on the distribution of the thrushes of this species in the Stikine River region of northern British Columbia; he writes:

At Great Glacier, August 11, a young bird was collected, not yet able to fly, that is clearly referable to *swainsoni.* This last record is of considerable interest as it carries the breeding range of *swainsoni* westward in this region to a point about thirty miles from the coast, the habitat of *Hylocichla u. ustulata.* Although the habitats of the two subspecies thus approach so closely, there is no evidence of intergradation of characters between them. In the Stikine River series of *swainsoni* there is not one specimen of an equivocal character. On the contrary, these birds, like those from the Yukon region, show an extreme of grayness, compared with typical *swainsoni* from eastern North America, that carries them farther from *ustulata* in appearance than are specimens from the Atlantic coast.

In northern New England and in eastern Canada the summer haunts of the olive-backed thrush are in the spruce and fir forests of the Canadian zone, where it is one of the commonest and most characteristic of the birds. It is less common in the mature, dense coniferous forests than it is in the bordering growth of smaller trees, where young balsam firs are growing up, with a mixture of birches and a few other deciduous trees, but the presence of at least a fair proportion of firs or spruces seems essential. A preference is shown for the lower and damper sections of the forest, especially in the vicinity of woodland streams, but the birds are often found breeding in the dry upland coniferous woods. In the more mountainous regions of northern New England, the olive-backed thrush breeds at lower levels than Bicknell's thrush. Wendell Taber writes to me: "In my experience, there has always been a very sharp line of demarcation in habitat, altitudinally, between this species and Bicknell's thrush." But "the last oliveback will often be a few yards up grade from the first Bicknell's. I heard two olivebacks at an altitude of 3,800 feet or higher on Whiteface Mountain, Sandwich Range, N. H. On Mount Katahdin, Maine, my highest oliveback was at about 2,600 feet, with a Bicknell's also present."

Edward H. Forbush (1929) draws the following attractive picture of the haunts of this thrush in the Berkshire hills:

> Among the hills of western Massachusetts there remain isolated remnants of the spruce growth that clothed them in days of yore. There today in the murmuring forest, tall, straight columnar trees still stand, their serried ranks extending far up the mountain sides. As they fall in death, succumbing to age or the ax of the woodsman, the sun streaming in between the remaining trunks stimulates the seeds buried by birds and squirrels in the soft mold of the forest floor and starts a dense miniature forest of beautiful little spruces. In time these cover the ground to replace the ancient wood and hide the great, moss-covered, decaying trunks on the ground. Here and there young trees of moosewood and black birch are growing, and little brooks fringed by overshadowing ferns prattle noisily down over their beds of age-old moss-grown rocks. Here the winds whisper the secrets of the forest and here the Hermit Thrush with time and eternity all his own, sings his unhurried, ethereal lay. Jays call mournfully from the distant tree-tops, and at the foot of the slope we hear the strange chant of the Olive-backed Thrush.

Prof. Maurice Brooks writes to me: "Apparently olivebacks were found originally throughout the West Virginia spruce belt; when most of the original timber was removed the birds have seemingly been able to adjust themselves quite well to the brushy second growth areas, particularly if there was some spruce regeneration. They are not always restricted to spruce, however; I have found them a few times in hemlock, and in Canaan Valley they may occur where the forest is largely deciduous, although, so far as I have seen, there must always be some spruce, balsam, or hemlock present. One of our favorite spots is the fire lookout tower on Gaudineer Knob of the Cheat Mountains. The tower stands at 4,445 feet above sea level; the top of the mountain is covered with a wonderful growth of red spruce in the seedling and sapling stages."

In northern Michigan, the breeding haunts of this thrush are somewhat similar to those in the east. Dr. Max Minor Peet (1908), referring to Isle Royale, says: "The damp places bordering streams were a favorite resort, the birds being usually found on the lower border of the balsam and spruce or among the decay-leaves and rubbish at their bases. Owing to the dense shade the lowest branches usually died and dropped off, so for a height of three to five feet it was relatively open. It was this rather open, yet heavily shaded condition which seemed to be best suited to these thrushes during the breeding season. They were also found in the dense alder thickets and resorted to the border of the woods and the roadside during migration."

In the Stikine River region of northern British Columbia, Mr. Swarth (1922) found the olive-backed thrush breeding in quite different surroundings. He says: "This is a bird of the poplar woods and willow thickets of the lowlands, primarily, but we found it also in small numbers well up the mountain sides. On July 17 Dixon saw several at the upper edge of the spruce timber (about 4,000 feet) on the mountains." But all the nests they found were at the lower levels. Farther south, in the Rocky Mountains, these thrushes breed at higher elevations, 6,000 to 9,000 feet in Colorado.

Spring.—In Massachusetts, the transient olive-backed thrushes pass through during May or the first week in June. They come with the warblers and other late migrants, and, like the warblers, they are often seen in the tree tops, feeding on insects in the opening foliage. They are not wholly confined to the woods at this season, but are often seen in orchards, gardens, or parks, wherever there are trees or shrubbery.

The migration in Ohio is only slightly earlier, but the flight is sometimes very heavy. Milton B. Trautman (1940) says that, in the vicinity of Buckeye Lake, Ohio, "during the spring and fall periods of maximum abundance between 40 and 300 individuals could be daily recorded. Principally because of its large numbers this was the most conspicuous of the thrushes, particularly in spring when the males sang persistently. The Olive-backed Thrush inhabited the shrub layer of swampy remnant forests. It was present in small numbers about brushy fields, weedy and brushy fence rows, edges of brushy swamps, cattail marshes, and shrubbery near farm houses and cottages."

There is a heavy migration, both spring and fall, across Lake Erie, between Cedar Point, Ohio, and Point Pelee, Ontario, as described in some detail by Dr. Lynds Jones (1910) and by Taverner and Swales (1908).

What few banding records are available seem to indicate that substantially the same route is followed on both migrations while the birds are passing through some parts of the United States. Mrs. Arch Cochran (1935) reports that she banded two olive-backed thrushes at her banding station in Nashville, Tenn., on May 15, 1932. On September 18, 1932, both returned and were taken in the same trap. On May 30, 1933, one of them returned to the same location; and it was captured again on September 24, 1933, in the same trap. At least a part of the same route was followed on each of four migrations, an interesting record. (See remarks under "Fall.")

Nesting.—Practically all the nests of the olive-backed thrush found by my companions and me in northern New England were in spruces or balsam firs; of those recorded in our notes, 11 were in spruces and 10 in balsams; they were almost always in small trees where the forest growth was more or less dense; the height from the ground varied from 2 to 20 feet, but only two were above 7 feet; the nests were usually bulky and well made and were generally built on two or more horizontal branches and close to the trunk, though occasionally one was about two-thirds of the way out on a branch. Dr. Brewer (Baird, Brewer, and Ridgway, 1874) gives a very good description of the nest, as follows: "The nests average about 4 inches in diameter and 2 in height, the cavity being 3 inches wide by about 1½ deep. They are more elaborately and neatly constructed than those of any other of our thrushes, except perhaps of *T. ustulatus.* Conspicuous among the materials are the *Hypnum* mosses, which by their dark fibrous masses give a very distinctive character to these nests, and distinguish them from all except those of the *T. ustulatus*, which they resemble. Besides these materials are found fine sedges, leaves, stems of equisetaceous plants, red glossy vegetable fibres, the flowering stems of the *Cladonia* mosses, lichens, fine strips of bark, etc."

Six nests before me measure externally 4 to 4.5 inches in diameter and 3 to 3.5 in height; the inner diameter varies from 2.25 to 2.5 and the depth of the cavity from 1.5 to 1.75 inches. They vary considerably in general appearance and in composition. One very compact, dark-colored nest is made up largely of fine spruce or tamarack twigs, carried up the sides and into the rim, mixed with grasses, strips of weed stems, and black and green lichens; it is lined with dead leaves. Another is made almost wholly of grasses and weed stems, with only a few very fine twigs in the base; this is lined with skeletonized leaves and lichens. Three others are more or less intermediate in construction between the above two. A fifth is made largely of strips of the inner bark of the cedar, mixed with black rootlets and much rotten wood and very little black lichen; it is lined with finer strips of the above material. The sixth consists almost entirely of mosses and lichens of different colors, reinforced with fine twigs. The materials mentioned above by Dr. Brewer and below by Miss Stanwood appear in many of the nests, and several of them are profusely decorated around the base with loose strips of the outer bark of the yellow birch.

Miss Cordelia J. Stanwood, of Ellsworth, Maine, who has sent me some voluminous notes on the olive-backed thrush, describes one of her nests as follows: "Foundation, swamp grass, spruce twigs, bracken stipes, black and green usnea moss; cup of peat, taken up with roots so that it resembles a mud cup; lined with usnea moss, black and green, and some of the black, thread-like parts of the roots of decayed cinnamon ferns."

She says that it takes the thrushes about four days, on the average, to build the nest. Most of her nests were in spruces or firs, but one "nest rested in the crotch of a gray birch, formed by the bole of the sapling and a rudimentary branch, three feet above the ground. It was well surrounded and concealed by the branches of young firs."

An editorial note in the Journal of the Maine Ornithological Society, volume 5, page 27, mentions a nest found at Pittsfield that was in a tall cedar about 30 feet from the ground, apparently a record height. William Brewster (1938) records two nests, found at Umbagog Lake, Maine, that were out of the ordinary: "The first was built precisely like the nest of a Wood Thrush, on a prong of a dead birch some four feet above the ground. The position of the second was unique—in a hollow scooped in the earth that adhered to the roots of a fallen tree, and perfectly concealed by a portion of the bank which projected above it. The situation of this nest was in every way similar to that usually chosen by the Water Thrush."

J. R. Whitaker sent me a set of eggs from Grand Lake, Newfoundland, that was taken from a nest in an old birch stump, 4 feet above ground. Robie W. Tufts has sent me his data for five Nova Scotia nests, all of which were in spruces or firs, from 4 to 8 feet up.

Western nests are more often placed in deciduous trees and bushes. A. Dawes DuBois reports in his notes a Wisconsin nest that was "about 7 feet above ground, in the top of a maple sapling in a woods-clearing grown up with underbrush, in low ground." Of his two Montana (Flathead County) nests, one was "about 6½ feet up in a larch sapling, at a low place near a small spring," and the other was "about 7 feet from the ground, supported on a dead tree-limb leaning against a small maple." P. M. Silloway (1901) says that, in Montana, the olive-backed thrushes' nests have been found in fir trees, in low willow sprouts, and once in a syringa bush.

There is a set in my collection from Estes Park, Colo., that came from a nest that was placed near the base of a small bunch of willows on a meadow near a stream. And the nests found by Mr. Swarth (1922) in the Stikine River region, British Columbia, were all at the lower levels in willows or alders. Dr. Paul Harrington writes to me that, in central Ontario, he has found the nest in balsam, hemlock, tamarack, spruce (black and white), cedar, top of a stump, and rarely in a deciduous tree. One was 30 feet up in a birch.

Eggs.—The set of eggs laid by the olive-backed thrush varies from three to five; sets of five are apparently rare, and sets of three seem to be about as common as those of four. The eggs are usually ovate in shape, with occasional variation toward elongated-ovate or rounded-ovate, and they have very little gloss. The ground color varies from "Nile blue" to "pale Nile blue" or even lighter. They are generally more or less evenly marked with spots, small blotches, or fine dots of light browns, "hazel" to "cinnamon," or paler yellowish brown. Some are heavily marked about the larger end, and some are very sparingly marked with very pale brown; very rarely an egg appears to be nearly immaculate. The measurements of 40 eggs in the United States National Museum average 22.4 by 16.5 millimeters; the eggs showing the four extremes measure **26.7** by 18.3, 25.7 by **18.6**, **20.3** by 16.3, and 22.9 by **14.7** millimeters.

Young.—Miss Stanwood made some extensive and intimate studies of the nest life of olive-backed thrushes in several different nests and says in her notes: "I found the eggs hatching in 1908 on the twelfth and thirteenth days, and in 1913 on the tenth, eleventh, and twelfth days from the beginning of the incubation period. The young mature sufficiently to leave the nest in from 10 to 12 days. The mother olive-backed thrush appeared both to incubate the eggs and to brood the young. For the first few days after the babes were hatched she left the nest only for very brief periods; during this time she brooded the young and moved back onto the rim of the nest every few minutes to feed the youngsters digested food and cleanse the nest. She fed each nestling, in the beginning, every time she cared for them; standing astride the little ones, she dealt with one sturdy individual at a time; she pecked him and touched him with her beak until he held his head up and begged lustily for food. She also burrowed under the young to remove all parasites from the nest and from the sensitive bodies of the bantlings. Frequently she ministered to the babes as often as once in four minutes. When brooding she changed her position on the nest frequently, to accommodate herself to the wriggling young, shielding them with her body from the hot sun and the cold rain. Both the male and the female fed the bantlings fresh insect food, as well as digested food, and removed all waste matter from the nursery. At first tender insects were doled out to them.

"On the third morning I peeped through a crevice in the blind eight hours, during which period the female fed the young digested food 17 times, and both birds made 33 visits with fresh insect food. Spruce bud moths were fed on 15 occasions. From 2 to 12 moths were given at a repast, but mostly they were given by the beakful. Caterpillars were dealt out 13 times. The quantity varied from one to a billful, and the mouth of the olive-backed thrush is capacious. Usually the birds brought a goodly number. All four birds were fed on all these visits."

Her list of food given to the young includes "all colored inchworms, cutworms, earthworms, yellow-green, blue-green, and gray-green caterpillars, all tones of tan and brown caterpillars, rosy maple-moth caterpillars, rosy maple moths, June bugs, click beetles, flying ants, craneflies, geometrid and spruce bud moths, orange-colored worms, grasshoppers, larvae of cherry hawkmoth, and fruit."

The young thrushes increase in size and weight very rapidly. Miss Stanwood measured and weighed the young in four nests every day during their nest lives. One of these, during 11 days, grew in length from $1\frac{11}{16}$ to $4\frac{11}{16}$ inches; and the same bird increased in weight from 60 grains to $445\frac{3}{4}$ grains. For further details, the reader is referred to her published paper (1913). The feather tracts were in evidence by the end of the first day; on the second day the eyes were beginning to open; all the feather tracts were well indicated on the third day; by the end of the fifth day the bird was well covered with quills and pinfeathers; feathers began to appear on the seventh day; and by the end of the tenth day the young bird was well feathered and practically free of quill casings. With the gain in size and weight came an increase in activity and the development of the sense of fear.

Both parents assisted in cleaning the nest. At first the excretal sacks were eaten; the birds began carrying away some of the sacks on the ninth day, but some of the excreta were eaten up to the end of the nest life of the young. In a later note she says that the female begins to incubate *sometimes* after the second egg is laid, but *always* after the third egg is deposited.

Plumages.—The natal down in the young olive-backed thrush is very dark brown, Miss Stanwood says "so dark brown that it looked black."

The young thrush, in fresh juvenal plumage, is described by Dr. Jonathan Dwight, Jr. (1900), as follows: "Above, olive-brown, wings and tail darker, the feathers of the pileum, back, lesser, median and sometimes part of greater coverts and the rump with linear shaft streaks or terminal spots of buff. Below, strongly washed with buff on throat, breast and sides, and heavily spotted with black on the breast and sides of throat, the fore parts and sides of whiter abdomen indistinctly barred. Sides of head buff, spotted with black; orbital ring distinct, pale ochraceous buff; submalar stripes black."

The first winter plumage is acquired by a partial postjuvenal molt, beginning about the middle of August and involving all the body plumage and the lesser wing coverts, but not the rest of the wings nor the tail. This is essentially like the adult winter plumage, but the upperparts are somewhat less richly colored and the spotted median wing coverts are retained until the following summer.

At next postnuptial molt, which is complete in August and September, adults and young become indistinguishable. Both young and old birds in fall and early winter are richly colored, above deep olive-buff in young birds and rich brownish olive in adults; and the sides of the head, throat, and upper breast are washed with rich ochraceous-buff. There is apparently no spring molt, but wear and fading have produced a somewhat grayer back, and the buffy tints are much paler in spring birds.

Food.—Professor Beal (1915b) examined the contents of the stomachs of 403 olive-backed thrushes, from widely scattered localities in the United States and Canada, and distributed over nine months from March to November. The food consisted of 63.52 percent animal and 36.48 percent vegetable matter. He writes:

Beetles of all kinds amount to 16.29 percent. Of these 3.14 percent are of the useful Carabidae. The others belong to harmful or neutral families. Weevils or snout beetles (Rhynchophora) amount to 5.29 percent, a high percentage for such insects. One Colorado potato beetle (*Leptinotarsa decemlineata*) was found in a stomach taken on Long Island. Hymenoptera collectively aggregate 21.50 percent. Of these, 15.20 percent are ants—a favorite food of *Hylocichla*. The remainder (6.30 percent) were wild bees and wasps. No honeybees were found. Caterpillars, which rank next in importance in the food of the olive-back, form a good percentage of the food of every month represented and aggregate 10.30 percent for the season.

Grasshoppers are not an important element in the food of thrushes, as they chiefly inhabit open areas, while *Hylocichla* prefers thick damp cover, where grasshoppers are not found. An inspection of the record shows that most of the orthopterous food taken by the olive-back consists of crickets, whose habits are widely different from those of grasshoppers, and which are found under stones, old logs, or dead herbage. The greatest quantity is taken in August and September. The average for the season is 2.42 percent.

Diptera (flies) reach the rather surprisingly large figure of 6.23 percent. These insects are usually not eaten to any great extent except by flycatchers and swallows, which take their food upon the wing. The flies eaten by the olive-back are mostly crane flies (Tipulidae) or March flies (*Bibio*), both in the adult and larval state. Crane flies are slow of wing and frequent shady places. The larvae of both groups are developed in moist ground, and often in colonies of several hundred. With these habits it is not surprising that they fall an easy prey to the thrushes.

Hemiptera (bugs), a small but rather constant element of the food, were found in the stomachs collected every month, and in July reached 11.11 per cent. They were of the families of stinkbugs (Pentatomidae), shield bugs (Scutelleridae), tree hoppers (Membracidae), leaf hoppers (Jassidae), and cicadas. Some scales were found in one stomach. The total for the season is 3.76 per cent. A few insects not included in any of the foregoing categories make up 0.48 per cent of the food. Spiders, with a few millipeds, amount to 2.22 per cent, the lowest figure for this item of any bird of the genus *Hylocichla*. Snails, sowbugs, angleworms, etc. (0.32 per cent), complete the animal food.

The vegetable food consists mainly of small, soft-skinned fruit, mostly wild varieties. "Wild fruit (19.73 per cent) is eaten regularly and in a goodly quantity in every month after April. Weed seeds and a few miscellaneous items of vegetable food (4.04 per cent) close the account." He lists 50 varieties of fruits, of which only the following were found in ten or more stomachs: Blackberries or raspberries in 67, wild black cherries in 15, domestic cherries in 29, woodbine in 10, and elderberries in 15.

Miss Stanwood (MS.) emphasizes the spruce bud moth (see Forbush, 1929, p. 400) among the food of the young and mentions other moths, all of which are probably eaten by the adults also. The olive-backed thrush evidently procures much of its food among the foliage of the trees, as well as on the ground, and some of it in the air. Ora W. Knight (1908) says: "I have known them frequently to catch moths, flies and mosquitoes while on the wing, and locally applied names of northern Maine are Mosquito Thrush or Flycatching Thrush."

E. W. Jameson, Jr., has sent me the following note from Buffalo, N. Y.: "On September 20, 1942, at 10:30 A. M., I came upon a flock of 15 or 20 olive-backed thrushes feeding in the top of a 30-foot wild black cherry (*Prunus serotina*). The birds were in constant motion, flying back and forth between nearby black oaks and butternuts. When a bird had picked off a cherry, it manipulated it in its bill, so as to strip it of the fleshy part which it ate; and then it let the stone fall to the ground. In this way the thrushes were making a great supply of food available to a host of small mammals that lived below. A chipmunk trapped nearby had three fresh stones in its pouches, and pine mice, deer mice, and short-tailed shrews as well, probably utilized this source of supply."

Behavior. The behavior of the olive-backed thrush is similar to that of its western relative, though it is more given to frequenting the tree tops, especially on migrations. It is a close sitter on its nest but otherwise shy and retiring. Amos W. Butler (1898) gives his impression of the migrants, as he saw them in Indiana: "When surprised they fly upon the lower branches of a tree or bush, usually getting behind a limb or a tree trunk out of view, sometimes simply turning the back to the intruder and then sitting motionless. Often when frightened from this perch they will fly wildly away with a flight almost as erratic as that of Wilson's Snipe." Sometimes the bird will sit thus motionless for some time, but usually not for long, and one must seize the first opportunity to note the buff eye ring, for he is not likely to get another chance. If taken from the nest at the right age, a young thrush can become a tame and interesting pet; Miss Stanwood adopted one that was ten days old and was quite successful with it; she gives a brief account of it in her published paper (1913).

Voice.—The olive-backed thrush is a fine singer, but most observers agree that it is inferior as a vocalist to either the wood thrush or the hermit; some even place the veery as superior to it; all three of these thrushes have certain qualities in their songs that appeal more strongly to the listener than does the oliveback's song.

Aretas A. Saunders has sent me the following excellent description of the oliveback's performance: "The olive-backed thrush, like other thrushes, has a long-continued song, but the phrases of which it is composed sound so much alike that the average listener is likely to consider that the bird is singing the same song over and over. Each phrase consists of 5 to 15 notes, generally all of them of equal length and usually all connected. The phrase progresses upward in pitch, the lowest note being the first or second one in the phrase and the highest the last or next to the last. In a typical phrase the first note is lowest, while the second, fourth, sixth, etc., are each progressively higher than any previous note. But the notes between these are lower than the note that preceded them. It is as though the song progressed upward by going over a series of higher and higher mounds, each mound followed by a valley.

"The quality is sweet and musical but less rich than that of the wood thrush and not so pure and clear as that of the hermit. There is a somewhat windy quality about it, as though the bird were saying *whao-whayo-whiyo-wheya-wheeya*. Each phrase is slightly different in arrangement or pitch from the others, but the difference is less apparent than in other thrushes. Often a bird interpolates its call note *whit* between the phrases, or at other times a high piping note, much like the call of the spring peeper (*Hyla crucifer*).

"I have records from twenty-seven birds. The pitch ranges from E'' to F'''', half a tone over two octaves, but the majority of birds sing between A'' and C''''. Individual birds have from three to nine different phrases, the majority about six.

"It is not uncommon for the olive-backed thrushes to sing on the migration. I usually hear more or less of the song nearly every year in Connecticut; in fact I have missed it only 7 years in 25. The first birds to be seen are generally not singing, and the song is not heard till the species has been present for several dates. The average date for the first bird seen, in my records, is May 13, but the average date for the first song is May 19. Generally the last birds to be seen, late in May, are still in song.

"On the breeding grounds the birds sing till late in July or early in August; the average date of the last song in 15 years of observation is July 30, the earliest July 12, 1929, and the latest August 8, 1928. At Flathead Lake, Mont., where the oliveback is the most abundant breeding bird, 12 different individuals could be heard in the evening from the Biological Station. They sing later in the evening than any other diurnal bird. In Allegany State Park, N. Y., where four species of thrushes can be heard in the evening chorus, the last one to sing, as darkness comes on, is the olive-backed thrush. In the Adirondacks I met a party of people who mistook its late singing for that of the whip-poorwill."

As an aid in remembering the song, A. D. DuBois (MS.) thinks of it as saying "*whip-poor-will-a-will-e-zee-zee-zee*, going up high and fine at the close. Sometimes there is an extra *a-will*." Albert R. Brand (1938), who has recorded on films the virbration frequencies in the songs of so many birds, found for the olive-backed thrush an approximate mean of 2,925, the highest note 3,850 and the lowest 2,000 vibrations per second; the high note is considerably lower than that of either the hermit or the veery, but the low note of the hermit is much lower in pitch than either of the others. Comparison with the high frequencies of two of the shrillest singers is very striking. The corresponding figures for the grasshopper sparrow are 8,600 mean, 9,500 high, and 7,675 low; and for the blackpoll warbler the mean is 8,900, the high 10,225, and the low 8,050 vibrations per second.

Stewart Edward White (1893) studied the song of one of these thrushes that sang regularly near his cottage in Michigan; he figured that this bird sang 4,360 songs per day, and writes: "He sings on an average nine and a half times a minute with extreme regularity. During the song periods of morning and evening his constancy of purpose is remarkable; except to seize a passing insect, he never breaks the regular recurrence of his song. From a series of records it is found that he begins on an average about 3.15 A. M., and sings steadily (of course by that I mean ten times a minute, not constantly) until about 9.00 A. M.; he is nearly silent until noon, after which he sings occasionally for a minute or so. About 4.30 he begins again, and only ceases to retire for the night about 7.30 P. M."

Only a few observers have written about the night singing, but Dr. Wilfred H. Osgood (1909) has given us a pleasing account of hearing such a concert in Alaska:

Although rarely heard in the daytime except in cloudy weather, they sang almost continuously through the night. One of the greatest delights of summer camping in the Yukon Valley is to lie in one's blankets at night listening to the ringing chorus of these thrushes. If the camp be fortunately chosen, one hears not one or two but a score of songsters. First we may have a bird less than 20 feet from the tent whose every utterance is audible, varying from tones of the greatest depth and richness to exquisite inarticulate gurglings and confidential whisperings. Then a few rods farther away may be several others alternating with one another in a long-continued obligato, while still farther back in some small ravine are those whose songs are borne on the air with a slight reverberation, giving added charm. While we lie in delicious enjoyment of these nearer songs, a general sense of music pervades the air to the farthest echoes. Perhaps there is a momentary lull, a sudden silence crowded with expectation. Then from a deep canyon beyond the wooded ridge behind us comes a far-away note, faint but full of character, and though little more than an echo, still with a tone that thrills. In the same way other notes, or a whole chorus, faint but sweet, are borne from the distant thickets across the river.

Field marks.—The four common thrushes that migrate through the United States can be readily recognized by the color pattern of the upperparts. The wood thrush is more rufous on the head and more olivaceous toward the tail; the hermit thrush is just the reverse, more olivaceous toward the head and more rufous toward the tail, especially on the tail; the olive-backed and russet-backed thrushes are uniformly brownish olive above; and the upperparts of the veery are uniformly more rufous. Furthermore, the wood thrush is more conspicuously spotted with larger and blacker spots on the breast; and the veery is very faintly spotted there with obscure spots. The gray-cheeked and Bicknell's thrushes, which are rarer, cannot be easily distinguished from the oliveback by the color of the upperparts; it is necessary to see clearly the buff cheeks and the conspicuous buff eye ring of the oliveback to recognize it.

Enemies.—The nests of the olive-backed thrush are so conspicuously placed and usually so accessible that they are very vulnerable to the attacks of crows, jays, squirrels, other wild predators, and stray cats. Mr. DuBois says in his notes: "At nest 5, small pieces of broken egg shell in the nest and a larger piece on the ground indicated destruction by some animal. Nest 6 contained two fresh eggs, one of which had a large hole in the side and had been partly eaten. A chipmunk was chattering and running in the brush not far away. The thrush, nearby, had been calling its alarm note."

Dr. Friedmann (1929) records it as "an uncommon victim" of the cowbird, five cases having come to his notice.

Taverner and Swales (1908) say that this and the gray-cheeked thrushes, on migration at Point Pelee, "suffer greatly during the Sharp-shinned Hawk flights as mentioned before. During the periods of this Hawk's abundance little scattered piles of thrush feathers can be found every here and there through the underbrush."

Dr. Charles W. Townsend (1923) heard the distress cries of a pair of olive-backed thrushes and on investigation "found two of these birds flying about a Red Squirrel who sat erect on a fallen tree, holding in his fore-paws a partly eaten Thrush in the spotted juvenal plumage. The squirrel's face was smeared with blood and it was altogether a most lamentable spectacle."

Harold S. Peters (1933) lists as external parasites of this thrush a louse, *Myrsidea incerta* (Kellogg), and a tick, *Haemaphysalis leporis-palustris* Packard.

Many thrushes, as well as other birds that migrate at night, meet their death by flying against lighthouses.

Fall.—Mrs. Cochran's (1935) banding records, referred to in the paragraph on "Spring," suggest that the spring and fall migrations of individuals follow the same route. This is probably not true of the species everywhere. In the Connecticut Valley, Mass., it seems to be very irregular in its appearance from one year to another, and more common in the spring than in the fall; Bagg and Eliot (1937) "expect to see twenty to thirty individuals each May and only five or six each autumn." Actual count was kept for the first time in 1936 by Mr. Eliot; in May he saw over 50 olivebacks in Northampton alone; in the fall seven. In 1937 he saw about 30 in the period May 9–24.

Miss Stanwood tells me that "the olive-backed thrush remains until the wild fruits grow scarce, near the middle or last of September," in Maine. It passes through Massachusetts mainly in September, but some are seen in October; while with us it is associated with the gray-cheeked thrush, from which it can be distinguished with difficulty, and it may be found almost anywhere in open woodland glades, in the shrubbery along streams or along country roads, and often in the shrubbery of gardens or parks, wherever it can find a little cover.

E. S. Dingle tells me that the earliest date of arrival in South Carolina was September 13, 1916. Paul Griswold Howes (1914) has published an interesting paper on the fall migration of the olive-backed thrush, as observed at Stamford, Conn., from which I quote: "The night voices fill the September air; weird, almost awesome are these whistles of the migrating thrushes, guided by some unknown power through thousands of miles of space to their winter home in the tropics. It is thrilling indeed when one hears the sound high in the air and far in the distance. Gradually it comes closer as the bird flies steadily southward. As it passes, unseen, directly over head, again the cry floats down to earth and a fainter answering call in the north, tells one of a companion or perhaps a mate. Thus the voices echo back and forth across the sky from evening 'till early morn, when the birds drop down from the high road of travel to feed and rest in the friendly woods and thickets." He recognizes the notes of the olive-backed thrush on migration as different from any other bird with which he is familiar; "a singular mellow and almost plaint-whistle, sweet-toned and far reaching, seems best to describe the calls of this thrush. * * * As I have seen birds come down to rest as early as 5 A. M., and on the other hand, leave for the night's journey at 6 P. M., it is not unlikely that this species travels the air for eleven hours at a time. If they cover two hundred miles in a night, which is not a maximum figure by any means, their rate of speed while in the air would be slightly over eighteen miles per hour, which, as near as I could judge, was about the speed that the individuals traveling in the early morning were making." All the birds observed over his marked square were flying almost directly southwest, which would be part of an almost straight line from Newfoundland to New Orleans.

The fall migration is very heavy at times at Point Pelee, Ontario, on the islands in Lake Erie and on Cedar Point, Ohio. At the former point, according to Taverner and Swales (1908), "in September, 1905, the first arrived the 6th, becoming very common the 8th. It disappeared that night, but gradually increased again to the 13th, when it fairly swarmed all over the place, then slowly decreased in numbers to the end of our stay, the 16th." And Dr. Jones (1910) says: "On the day of my arrival on Pelee island, August 29, there were none found, nor any the next day in spite of a careful search, but with the first faint dawn of the 31st the peculiar notes of this bird were heard, and the full light revealed hundreds of them in the bushes and everywhere in the woods. They remained thus numerous until my departure the evening of the next day."

Mrs. Nice (1931) calls the olive-backed thrush a regular spring transient throughout Oklahoma, but remarks: "The lack of fall records is curious; for seven reasons [seasons?] I followed the fall migration with the keenest interest, but never once saw a thrush of any species." This is additional evidence that this thrush does *not* always follow the same route on both migrations. Perhaps the main fall flight may cross the Gulf of Mexico to southern Central America or to South America. Ludlow Griscom (1932) says that, in Guatemala, "October and November are the main months of the fall migration. November 15 is the latest date for western Panama, and there is one December record for Costa Rica." According to Dickey and van Rossem (1938), it occurs in El Salvador only as a spring migrant, or a "rare midwinter visitant."

Winter.—The 1931 Check-list says that the olive-backed thrush "winters from southern Mexico to Peru, Bolivia, Brazil, and Argentina." But Mr. Griscom (1932) says: "The Olive-backed Thrush winters over much of South America east of the Andes, and is a common spring and fall transient in Central America, unrecorded from the Pacific slope north of Costa Rica. * * * Its supposed wintering in Central America is based on assumption, and there is not a single critically determined specimen on record in this country [Guatemala], collected in January or February."

HYLOCICHLA USTULATA ALMAE Oberholser

WESTERN OLIVE-BACKED THRUSH

HABITS

Although the western olive-backed thrush was described many years ago, it was not until within comparatively recent years that it was admitted to our Check-list in the nineteenth supplement (Auk, vol. 61, p. 457).

In naming this race, Dr. Harry C. Oberholser (1898) gives a very full and detailed description of it and then remarks: "The present race differs from the eastern *Hylocichla ustulata swainsonii* in the more grayish, less olivaceous color of the upper surface, this being most noticeable on the rump and upper tail-coverts. The sides and flanks also average more grayish. No apparent difference in size exists. No comparison with *H. ustulata* proper is necessary, for *Hylocichla u. almae*, although geographically intermediate, is even less closely allied to *ustulata* than is *swainsonii*. * * *

"Alma's Thrush is a common bird in eastern Nevada, where it inhabits the growth of trees and bushes that fringe the mountain streams. In the Monitor and East Humboldt Mountains, it is apparently the most numerous species of the family."

We have no reason to think that the habits of this thrush differ materially from those of neighboring races that live in similar environments.

HYLOCICHLA MINIMA MINIMA (Lafresnaye)

GRAY-CHEEKED THRUSH

HABITS

The gray-cheeked thrush spends the summer farther north than any of our eastern thrushes, up to the tree limit and even beyond that point nearly or quite to the Arctic coast of the far northwest. Dr. E. W. Nelson (1883) describes its summer haunts in northern Alaska, as follows:

> In middle latitudes where our acquaintance is made with this bird we associate it with damp woodlands and sheltered glens, and it would seem almost incongruous to one familiar with it in such surroundings to look for it as an inhabitant of the barren stretches of arctic lands where for many miles not a tree raises its shaft

Such is its northern home, however, and throughout the entire arctic region north of Hudson's Bay to Bering Strait and across into Kamtchatka the bird is found in a greater portion of this range as an extremely abundant species. Wherever clumps of dwarf willows or alders have gained a foothold along the sterile slopes and hillsides in the north, a pair or more of these wanderers may be looked for. Along the entire Bering Sea coast of Alaska, and north around the shores of Kotzebue Sound, it is numerous among the many alder bushes found on these shores.

It was formerly supposed to be only a straggler across the Strait into northeastern Siberia, but, according to Thayer and Bangs (1914), it "is one of the American species that have extended their breeding range across Bering Strait, and it now breeds west at least to the Kolyma River region."

Dr. Joseph Grinnell (1900) says that "in the Kowak Valley it was to be heard from every willow bed and tract of spruces." From there eastward, along the northern fringe of stunted spruces, willows, and alders, the range extends to the delta of the Mackenzie, the Anderson River region, northern Labrador, and even to parts of Newfoundland.

Spring.—The gray-cheeked thrush is the champion migrant among our small thrushes, making the longest migration and, in some parts of its journey, the most rapid advance. Frederick C. Lincoln (1939) says: "An excellent example of rapid migration is that of the Gray-cheeked Thrush. This bird winters in Colombia, Ecuador, Peru, Venezuela and British Guiana and does not start its northward journey until many other species are well on their way. It does not appear in the United States until the last of April—April 25 near the mouth of the Mississippi and April 30 in northern Florida. A month later, or by the last week in May, it is to be seen in northwestern Alaska, the 4,000-mile trip from Louisiana having been made at an average speed of 130 miles a day." By reference to his map, it appears that the progress is much more rapid in the western part of the route, especially on the latter half of the journey, than it is up the Atlantic Coast States; by May 20 it has reached New England, having covered but little over 10° of latitude; while, during the same time, it has advanced over 30° of latitude, well beyond the 60° latitude in Mackenzie; and, during the next five days, it has advanced nearly 10° more to the northern limit of its range in Alaska. These dates and distances are probably only approximate averages.

During the spring migration the gray-cheeked thrush spreads out over all of the eastern United States, from the Atlantic coast to the Mississippi Valley, and only sparingly a little farther west. Dr. Nelson (1883) writes of the migration:

It passes by the groves and farms of the Northern States just as the buds are swelling and the warm, misty rains of spring are quickening into life the sleeping seeds and rootlets; filled with buoyant exultation it pauses now and then to pour forth those strange but pleasing cadences which once heard in their full sweetness will never be forgotten. But it has no time to tarry, and ere long it is already far on its way to the north. The strange, wild song which arose but a short time since in pleasant woodland spots and quiet nooks in southern groves is now heard by wandering Indians who seek their summer fishing-grounds by the banks of northern streams. Yet a little later and it troops in abundance near to the shores of the Arctic, where the Mackenzie and other rivers pour their spring floods into the icy sea. Down the Yukon these birds pass, using the densely bush-grown bank of the river as their highway, raising now and then their song which finds here fittest surroundings. Reaching the mouth of the Yukon, many wander along the coast of Bering Sea to the north, and some are said to cross the straits.

On migration gray-cheeked thrushes may be seen almost anywhere that they can find sufficient cover, in woodlands, or along the edges, in thickets along streams, in roadside shrubbery, in thick growths of young evergreens, and even in village yards or gardens or in city parks. They come along with the host of later migrants and are often in company with olivebacks, or in the east with the smaller Bicknell's; they can hardly be distinguished from the latter in life and must be seen under favorable circumstances to tell them from olivebacks. In the great wave of migrating birds that swept through the narrow timber belt along Maple Creek, southwestern Saskatchewan, on June 8, 1906, thrushes were very numerous. Only two were collected, one of which proved to be *aliciae* and the other a small specimen of a female *bicknelli* identified by my companion, Dr. Louis B. Bishop.

Lucien M. Turner, in his unpublished Labrador notes, says: "At the head of Hamilton Inlet, this thrush occurs in abundance, arriving about the 25th of May and remaining until the middle of September, breeding there plentifully even in the undergrowth surrounding the houses."

Nesting.—Breeding in a region where trees are stunted or replaced by low bushes, the gray-cheeked thrush builds its nest not far from the ground or even on it among low-growing shoots. A nest found by Johan Koren at Nijni Kolymsk, in northeastern Siberia, on June 15, 1912, was "placed on the ground among the stems of young alders," according to Thayer and Bangs (1914); it contained five fresh eggs. Dr. Grinnell (1900) says that, in the Kotzebue Sound region of northern Alaska, "the nests of this species were quite variously situated, according to environment. In willow and alder beds I found them within a foot of the ground built on the slanting or horizontal trunks. While in the spruce woods they were found as high as twenty feet, though commonly about six feet above the ground. A typical nest is of fine shriveled grass blades, incorporated when damp, and mixed with a small amount of mud. The lining is of fine dry grasses. When this structure dries it is remarkably compact and firm, in fact almost indestructible by the elements, for the woods were full of old nests some of which must have survived many seasons."

Dr. Brewer (Baird, Brewer, and Ridgway, 1874) gives a very good description of the nests, as follows: "The nests measure about 4 inches in diameter and 2¾ in height. The cavity is 2 inches deep, and its diameter 2½ inches. They are usually compact for the nest of a thrush, and are composed chiefly of an elaborate interweaving of fine sedges, leaves, stems of the more delicate *Equisetaceae*, dry grasses, strips of fine bark, and decayed leaves, the whole intermingled with the panicled inflorescence of grasses. There is little or no lining other than these materials * * *. The *Hypnum* mosses, so marked a feature in the nests of *T. swainsoni*, as also in those of *T. ustulatus*, are wholly wanting in those of *T. aliciae*."

Dr. Nelson (1887), however, found a nest at St. Michael that was "composed mainly of these mosses mixed with a small amount of coarse grass."

There is a set of eggs in my collection, taken by Rev. W. W. Perrett at Makkovik, Labrador, on June 27, 1899, that came from a nest made of small twigs, moss, and dry grass, lined with fine grass and a few dry leaves and rootlets; the nest was placed about 3 feet from the ground in a small spruce. Another nest in his collection was similarly constructed but placed on the ground under a small juniper. Another set in my collection was taken by J. R. Whitaker at Grand Lake, Newfoundland, on June 26, 1919; the nest was 18 inches from the ground on a prostrate tree. Dr. George J. Wallace (1939) also says that this thrush builds "a mossy nest." Herbert Brandt (1943) found the gray-cheeked thrush breeding in the Askinuk Mountains, near Hooper Bay, Alaska, "in the alder growth below an altitude of 500 feet," and says that "the outside of some nests is so profusely decorated with moss and lichens as to cover completely the grassy wall." The nests were in upright crotches, entirely without concealment, from 2 to 4.50 feet above the ground.

Eggs.—The gray-cheeked thrush lays ordinarily four eggs, sometimes only three and frequently five; Dr. Wallace (1939) says "from four to six."

The eggs are usually ovate in shape, but some are more elongated and some more rounded. They have very little or no gloss. The ground color is a light, greenish blue, from "Nile blue" to "pale Nile blue" or even paler. They are usually more sparingly and more faintly marked than the eggs of the olive-backed thrush, but with similar shades of brown, "hazel" to "cinnamon" or paler. Some eggs have fairly large blotches about the larger end, and some are uniformly and evenly sprinkled with minute dots. In some eggs the markings are so small and faint that the egg appears almost immaculate. The measurements of 50 eggs in the United States National Museum average 23.0 by 17.0 millimeters; the eggs showing the four extremes measure **24.9** by 16.0, 23.9 by **19.8, 20.6** by 16.8, and 22.4 by **15.8** millimeters.

Plumages.—The natal down of the gray-cheeked thrush does not seem to have been described. Dr. Dwight (1900) describes the juvenal plumage of an Alaskan specimen, as follows: "Above, greenish olive-brown, wings and tail darker, the pileum, back, wing coverts (except primary and greater) and rump with buffy white linear shaft streaks. Below, white, very faintly tinged with pale buff on the breast and sides, the breast and throat spotted with black, tending to barring on forepart of abdomen and flanks. Sides of head pale buff, black spotted; submalar streaks black; distinct orbital ring rich buff. * * * This dress is grayer and with less buff than the corresponding plumage of *T. u. swainsonii.*"

The first winter plumage is acquired by a partial postjuvenal molt, which involves the body plumage and the lesser wing coverts, but not the greater wing coverts, or the rest of the wings, or the tail. Dr. Dwight (1900) describes this plumage as "above, similar to corresponding plumage of *T. u. swainsonii*, the olive-brown usually darker with less yellowish tinge, especially on the head. Below, with no buff except a faint wash on the jugulum; the sides of the head and breast are therefore much grayer and the orbital ring distinctly white. The buffy edgings or terminal spots of the retained juvenal wing coverts are usually distinctive of the first winter dress." The postjuvenal molt of young birds and the postnuptial complete molt of adults are both, apparently, accomplished in August before migration. There is evidently no molt in spring, but wear and fading produce a slightly grayer effect. The sexes are alike in all plumages.

Dr. Wallace (1939) writes: "Breeding birds of the Newfoundland type vary from Brownish Olive, which is the migration color of *aliciae*, to Olive-Brown or Sepia. Thus they differ from summer specimens of the continental form by the almost complete lack of grayish suffusion in the dorsal plumage, the gray being replaced by brownish hues. Below, they are similar to the northern gray-cheeks, but with more buffy appearance on the throat and breast."

Food.—Professor Beal (1915b) examined the stomachs of 111 gray-cheeked and Bicknell's thrushes, combining the two subspecies. The analysis showed 74.86 percent of animal matter and 25.14 percent of vegetable food. Beetles were the largest item in the former, 33.32 percent; but only 2.83 percent were the useful Carabidae; the remainder belonged to harmful families, such as the Scarabaeidae, Elateridae, and the weevils. Ants amounted to 16.34 percent, and other Hymenoptera, as wasps and bees, were eaten to the extent of

5.60 percent; no honey bees were found. Caterpillars were third in importance, 8.81 percent. Grasshoppers were not a favorite food, amounting only to 1.72 percent. Other insects, including the remains of a seventeen-year locust in one stomach, amounted to 2.89 percent, spiders 5.77 percent, and a few other animals, such as crawfishes, sowbugs, and angleworms, to 0.41 percent.

Among the vegetable food, a few seeds of blackberries or raspberries were found, but they may have come from wild plants and amounted to only 0.15 percent. "Wild fruits of 18 different species (23.98 percent) make up nearly one-fourth of the whole food." The only fruits found in more than two stomachs were: Wild black cherries in 5, wild grapes in 5, flowering dogwood in 5, and elderberries in 3 stomachs.

Francis Zirrer, of Hayward, Wis., sends me the following note: "In spring this bird, and many others, subsist largely on larvae of various Diptera, inhabiting the rich woodland soil in such masses, that from the bottom of a small puddle of snow water a few yards square a good handful might be obtained, drowned and dead. During the fall migration the bird feeds often on the hairy caterpillars (Arctiidae) which at that time crawl everywhere. However, it does take considerable time and much pounding before the caterpillar is disposed of. But most of the fall feeding is done on the berries of *Aralia racemosa*. As shy as this thrush usually is, it will come directly under the windows of a lonely woodland cabin, if the plant grows there and the fruit is plentiful, which is usually the case, as this plant with its later period of blooming does not, as a rule, suffer from late frosts. This is the best opportunity for the dweller to watch this and the next thrush at the closest possible range through the window pane, often only 2 or 3 feet away. And, since the ripening of these berries on the truly enormous clusters proceeds slowly and over a period of several weeks, the birds will stay as long as there are any berries left, unless the weather turns exceptionally cold and nasty."

Behavior.—Like other thrushes of this group, the gray-cheeked is extremely shy, both on migrations and on its breeding grounds. We may hear it singing on the top of some bush or small tree, but, as we approach, it dashes down into the underbrush and disappears. It is even more shy than the oliveback. It is most often seen on the ground during migrations, hopping about in a characteristic thrushlike attitude, erect on its long legs and searching for food, most of which is found on the ground among the fallen leaves. It is seldom seen in the high treetops, as the oliveback so often is.

Wendell Taber writes to me that on May 17, 1942, in a cemetery at Nahant, Mass., "one of these birds seemed to have a definite preference for the top of a wooden fence. Although a number of us drove him away, he kept returning. Later, after we had walked through the cemetery and returned, the bird was again on top. A most peculiar action was the position of the bird; wings were extended from the body with primaries turned down and apparently touching the fence—in short a three-point landing. The bird did not appear to be wet or show any difficulty in flying."

Dr. Dayton Stoner (1932) "watched one of these birds for some time as it took its morning bath in one of the pools of shallow water in a bog. The ablutions were continued for several minutes and were accompanied by a great deal of fluttering and splashing of water. Then followed a most careful and meticulous preening and oiling of the contour feathers; particular attention being given the tail and large wing feathers, the vanes of which were run carefully through the mandibles so that the disengaged hooklets—if any—would be re-engaged. The entire procedure lasted between ten and fifteen minutes."

A rather peculiar action was noted by Cyril G. Harrold on Nunivak Island, Alaska, mentioned by H. S. Swarth (1934): "These birds have a habit of making for the boulder-strewn shore when alarmed and hiding under the large rocks, where it is very difficult to locate them."

Mr. Turner (MS.) writes: "I have observed this species on both the Pacific and the Atlantic Arctic regions, and cannot consider it as a shy bird, that is, difficult to approach. It is of a retiring nature, and in certain positions very apt to be overlooked, being oftener seen as it flits from one thicket to another than otherwise."

Voice.—The song of the gray-cheeked thrush is evidently inferior to those of the hermit and wood thrushes, and some think that the veery and even the oliveback are superior vocalists. This is largely due to the fact that the graycheek does not sing its best song while on migration, though it is often heard. Few of us have been privileged to hear it on its northern breeding grounds, where it is heard at its best and where the full, rich, sweet song is doubly appreciated in the barren surroundings of its summer home. Here it mounts some low tree or bush and pours out a most delightful melody. It ceases to sing, however, when the care of young nestlings absorbs its attention. Eugene P. Bicknell (1884) writes:

As a result of my experience with these birds, I have little hesitation in characterizing the song of the Gray-cheeked Thrush as weaker than that of the Olive-backed, entirely dissimilar in tone, and with a somewhat different disposition of the notes. Instead of musically outbursting, it is singularly subdued, and has a far-away and rather ventriloquial sound. It seems more the expression of some distant emotion revived in memory than of a suddenly felt present emotion which the song of the Olive-backed Thrush suggests. * * * The song of the Gray-cheeked Thrush commences low and reaches its loudest, and I think its highest, part a little beyond half its continuance. It is throughout much fainter and of less forcible delivery than the song of the Olive-backed species.

John A. Gillespie (1927) gives a very good description of the song: "Briefly described, the song in question commenced with a slurring 'wee-oh,' strongly suggesting the beginning of a common variation of the White-eyed Vireo's song. This was followed by two, and sometimes three, high pitched, staccato notes resembling 'chee-chee,' intermingled with almost inaudible cymbal-like tones. From notes taken at the time, the whole song might be represented as 'Wee-oh, chee-chee-wee-oh, wee-oh,' the latter half suggesting the Goldfinch in tone and execution."

To A. Dawes DuBois (MS.), the song of a bird that he heard in Illinois sounded like "*We-tichi-wheée, whitcheé-u*. The *we*, the first syllable of the *tichi* and most of the *wheée* are of approximately the same pitch, but the second syllable of the *tichi* is lower. The last portion of the *wheée* is tremulous and slurs downward, after which the *whitcheé-u* begins on the lower note, but is slurred quickly upward, and then gradually downward on the tremulous *cheé-u* ending."

The gray-cheeked thrush has a harsh scolding note and some short call notes like *what, chuck, pheu*, or *feé-a*, which probably express different emotions.

Lucien M. Turner (MS.) has this to say about the midnight song of the gray-cheeked thrush on its breeding grounds: "At the mouth of Whale River, Ungava, I was lying in my sleeping bag, preparing for a few hours' rest on the open ground, with naught but sky above me. The time was but a few minutes after midnight, and it was so still that the only sounds to be heard were the contending currents of the river, but a few feet distant, and the distress cries of a pair of semipalmated plovers whose nest was nearby. A drowsiness soon possessed me, but hearing a strange, clear song of a bird, which made the stillness tinkle with its music, I could not sleep so long as it continued. The next morning, I saw a pair of Alice's thrushes and knew they had sung their midnight song to the rising sun, for at this date, June 27, there is no darkness; the sun passes below the horizon, but leaves the daylight behind."

Field marks.—As its name implies, the best field mark is the gray cheeks. At short range and in good light, this character will distinguish this thrush from the oliveback, whose cheeks and eye ring are conspicuously buffy, though this character is not so well marked in the young bird in the fall. The plain, olive-brown back will distinguish this species from the wood thrush, veery, or hermit, all of which show some rufous on part or all of the upperparts. From

Bicknell's thrush the gray-cheeked can be distinguished only by size, a poor field mark.

Enemies.—All thrushes have plenty of enemies among the numerous predators, but few have the kind that persecute the gray-cheeked thrush on its breeding grounds. Dr. Nelson (1887) writes: "As soon as the breeding season is over they become less retiring and frequent the vicinity of villages and more open spots, where many are killed by the native boys, armed with their bows and arrows. Their skins are removed and hung in rows or bunches to dry in the smoky huts and are preserved as trophies of the young hunters' prowess. In the winter festivals, when the older hunters bring out the trophies of their skill, the boys proudly display the skins of these thrushes and hang them alongside."

Dr. Max Minor Peet (1908) mentions one that was killed at Isle Royale, Mich., by flying against a lighted window at night during a storm.

Only one louse, *Myrsidea incerta* (Kellogg), is reported by Harold S. Peters (1936) as an external parasite of this thrush.

Fall.—Gray-cheeked thrushes leave their northernmost breeding grounds by the last of August, or early in September, and travel southward over much the same routes that they traversed in spring, which covers practically all of Canada and the United States east of the Rocky Mountains. Dr. Wallace (1939) remarks that "apparently not even the Alaskan and Siberian inhabitants migrate along the Pacific coast. Habit is deeply intrenched in such migratory species; otherwise the Siberian representatives might better winter in southern Asia than to retrace their flight across the intercontinental peninsula, and migrate the whole length of North America to winter in South America."

Ludlow Griscom (1932) records it as "migrating along the outer islands of the east coast of Central America, very rare or casual on the mainland (4 records only)." But Dr. Wallace (1939) says: "Of the ten or more Central American records for this form only three appear to be from these outer islands, and two of these are of doubtful identity. Apparently they migrate over all the available routes to South America, by way of the West Indian chain of islands, along the Florida peninsula and across Cuba, by the outer islands along the eastern Central American coast, and particularly over Central America as a whole."

We seldom think of thrushes as migrating in flocks, but, at Isle Royale, Mich., Dr. Peet (1908) found these thrushes very abundant in September: "Large flocks were seen every day throughout the remainder of our stay [Sept. 12 to 22], the border of clearings and the roadways being the places where they were the most abundant."

They pass through Massachusetts in September and October, when they frequent mainly the berry-bearing thickets, feeding on the fruit of cornels and the berries of the deadly nightshade, barberry, spicebush, woodbine, wild grapes, the seeds of poison-ivy, etc. But we often see them in the shrubbery along the roadsides and even in our yards and gardens, feeding among the fallen leaves.

Winter.—The gray-cheeked thrush spends the winter in northern South America, mainly in Colombia and Venezuela, but also in British Guiana, Ecuador, and Peru. We do not know much about its winter haunts and habits. Dr. Wallace (1939) says that "they reach their winter quarters by late October, but apparently are not found in the valleys much later than that, seeming to move up into the mountains."

DISTRIBUTION

Range.—Northeastern Asia and North America, to northwestern South America.

Breeding range.—The gray-cheeked thrush breeds **north** to northeastern Siberia (Nijni Kolymsk and Pitalkaj); northern Alaska (Cape Blossom and the Kobuk River, rarely to Point Barrow); northern Yukon (Old Crow River and Lapierre House); northern Mackenzie (Aklavik, Franklin Bay, MacTavish Bay on Great Bear Lake, and Artillery Lake); northern Manitoba (Churchill); northern Quebec (Chimo); and northern Labrador (Nain). **East** to Labrador (Nain, Davis Inlet, and Cape Charles); Newfoundland (St. Anthony, Fogo Island, and St. John's); Miquelon and St. Pierre Islands, and Nova Scotia (Seal Island). **South** to Nova Scotia (Seal Island), and the mountains of northeastern United States: Maine (Mount Katahdin and Mount Abraham); New Hampshire (White Mountains and Moosilauke); Massachusetts (Mount Greylock); and New York (Adirondack and Catskill Mountains); southeastern Quebec (Gaspé County and Romaine); Hudson Bay region (Great Whale River, Quebec, and York Factory, Manitoba); southern Mackenzie (Hill Island Lake), northern British Columbia (Fort Nelson River and Atlin); southern Yukon (Lake Marsh); southern Alaska (Copper River, Kodiak Island, Nushagak, and Hooper Bay); and northeastern Siberia (Cape Tschukotsk). **West** to northeastern Siberia (Cape Tschukotsk and the tributaries of the Kolyma River).

The breeding range as outlined includes the entire species of which two subspecies or geographic races are recognized. The typical race, the gray-cheeked thrush (*H. m. minima*), is the northern race breeding from Siberia across northern Canada, Labrador, and in Newfoundland; Bicknell's thrush (*H. m. bicknelli*) breeds in Gaspé County, Quebec, Nova Scotia, and the mountains of northeastern United States.

Winter range.—In winter Bicknell's thrush has been found only on the island of Hispaniola (Morne Malanga, Haiti; and Puerto Plata, Sánchez, Aguacate, and San Domingo, Dominican Republic).

The gray-cheeked thrush continues to South America where it occurs from the Santa Marta region of Colombia east through the Orinoco Valley of Venezuela (Mérida, Barinas, Caicara, and the Mount Auyan-tepic region) to British Guiana (Kamakusa, the Kamarang River, and Bartica). It appears to winter at least casually in the Cauca Valley of Colombia (Puerto Valdivia); on the east slope of the Andes in Ecuador (Hacienda Machay in the province of Ambato and the Río Suno); and in eastern Peru (Chamicuras).

Migration.—The gray-cheeked thrush has an interesting migration route. The records indicate that the birds breeding in Siberia migrate through the Mississippi Valley of the United States. West of the hundredth meridian the species is known only by a few (possibly casual) records in eastern Wyoming and Montana, and there are very few records from southern Alberta, and none from British Columbia except north of the Peace River. It is apparently unknown in Mexico except in the islands off the coast of Yucatán, and there is one specimen from Guatemala; and several records for Cuba, Costa Rica, and Panama.

As many of the dates of migration are based on sight records, no attempt has been made to separate the races.

Some late dates of spring departure are: Colombia—Santa Marta, May 3. Guatemala—Uaxactún, April 28. Mexico—Cozumel Island, April 30. Bahamas—Cay Sal, May 19. Florida—Tortugas, May 22. North Carolina—Asheville, May 21. District of Columbia—Washington, June 3. Texas—Cove, May 21. Louisiana—Shreveport, May 16. Missouri—St. Louis, June 4.

Some early dates of spring arrival are: Florida—Pensacola, April 21. South Carolina—Aiken, April 24. Virginia—Lynchburg, May 5. District of Columbia—Washington, May 6. Pennsylvania—Harrisburg, May 1. New York—Rochester, May 2. Massachusetts—Boston, May 14. New Hampshire—Hanover, May 17. Maine—Dover-Foxcroft, May 1. Quebec—Quebec, May 21. Louisiana—Baton Rouge, April 22. Mississippi—Biloxi, April 21. Tennessee—Memphis, April 23. Indiana—Indianapolis, April 26. Michigan—Ann Arbor, April 25. Ontario—Hamilton, April 27. Missouri—St. Louis, April 30. Wisconsin—Racine, May 2. Texas—Houston, April 23. Nebraska—Red Cloud, April 29. North Dakota—Fargo, May 4. Manitoba—Aweme, May 5; Churchill, June 6. Alberta—Edmonton, May 13. British Columbia—Fort St. John, May 13. Alaska—Kobuk River, near the mouth, May 24.

Bering Sea—St. Lawrence Island, May 26. Siberia—Nijni Kolymsk, June 8.

Some late dates of fall departure are: Bering Sea—St. Paul Island, September 9. Alaska—Nome, September 8. British Columbia—Atlin, September 1. Alberta—Athabaska Landing, September 12. Manitoba—Churchill, September 6. North Dakota—Fargo, October 4. Minnesota—Minneapolis, October 1. Iowa—Iowa City, October 16. Michigan—Ann Arbor, October 7. Ontario—Point Pelee, October 9. Ohio—Columbus, October 6. Mississippi—Ariel, October 9. Texas—Cove, October 18. Quebec—Hatley, September 24. Maine—Phillips, September 20. Massachusetts—Harvard, October 5. New York—Astoria, October 23. Pennsylvania—Philadelphia, October 16. District of Columbia—Washington, October 24. North Carolina—Weaverville, October 28. Georgia—Athens, October 18. Alabama—Greensboro, October 20. Cuba—Habana, October 21. British Honduras—Toledo, November 13. Costa Rica—Tambor, November 2. Panama—Cocoplum, November 12.

Some early dates of fall arrival are: Minnesota—St. Paul, August 31. Illinois—Chicago, August 26. Louisiana—Thibodaux, September 21. North Carolina—Statesville, September 4. Georgia—near Atlanta, September 9. Florida—Princeton, September 19. Cuba—Habana, October 16. Costa Rica—Río Sicsola, October 4.

Casual records.—A specimen was collected at Godthaab, Greenland, in June 1845; and another in south Greenland in August 1852, and one in postjuvenal plumage was collected on the south fork of Cave Creek, in the Chiricahua Mountains in Arizona, on September 11, 1932.

Egg dates.—Alaska: 17 records, June 6 to July 8; 9 records, June 15 to June 23, indicating the height of the season.

Labrador: 5 records, June 17 to June 28.

Manitoba: 26 records, June 11 to June 30; 13 records, June 19 to June 22.

New Hampshire: 5 records, June 17 to June 28.

HYLOCICHLA MINIMA BICKNELLI Ridgway

BICKNELL'S THRUSH

CONTRIBUTED BY GEORGE JOHN WALLACE

HABITS

In 1881 E. P. Bicknell surprised American ornithologists by the unexpected discovery of a new thrush, apparently breeding in the Catskills, close to the stamping grounds of many of the foremost birdmen of that time. Specimens of the new form were submitted to Ridgway, who described them as southern subspecific representatives of the better-known and more widely distributed gray-cheeked thrush. Subsequent explorations soon disclosed the presence of the thrush on the nearby mountains of New England, causing ornithologists of that period to express both surprise and chagrin that the existence of a bird regularly breeding in the Northeastern States should have gone so long undetected.

Aside from its very scattered and restricted distribution in the Canadian Maritime Provinces, Bicknell's thrush proved to be a strictly montane form, limited in the breeding season to high elevations in the mountains of New England and eastern New York. It belongs primarily in the ecological niche extending from the 3,000-foot level up to timberline, in an environment of dense, stunted evergreens, where the rocky fir-clad heights are almost perpetually bathed in mists and clouds.

Because of the Bicknell's occupation of such a limited zone in only the most isolated regions, and because of its habit of nesting in the most inaccessible parts of these remote areas, the bird has long remained in comparative obscurity, its taxonomy and nomenclature confused, its distribution incompletely or erroneously recorded, its life history and ecology but little known. Excursions into its breeding haunts revealed many interesting but rather fragmentary and sometimes conflicting bits of information that at best were an inadequate record of the haunts and habits of the thrush. In 1935, however, the present writer had an opportunity to follow the bird through one complete breeding cycle, thus making it possible to clear up some of the obscure features regarding this interesting thrush.

Spring.—Bicknell's thrushes appear in the United States early in May, at which time they reach the southern Atlantic States from their west Indian winter home (the first continental spring records substantiated by specimens are for Charleston, S. C., on May 2, and the north Floridian coast on May 3). From there they move northward along the coast, passing through Washington, D. C., about the middle of May and arriving in southern New England during the latter half of the month (18 of the New England specimens were taken between the 20th and 30th of May, 5 earlier than the 20th, and 1 as late as June 11). During this period also they gradually move up into the mountains, presumably as soon as the retreating snows permit. They have been found returning to Mount Moosilauke in New Hampshire between the 25th and 30th of May (G. M. Allen, 1902), have been reported for Mount Mansfield, Vt., on May 25 (Davenport, unpublished bird list), and for Mount Greylock, Mass., between May 28 and 30 (Maynard, 1910). In 1935, in a considerably retarded spring, only a few were found in the snow-bound summit of Mount Mansfield on May 27, but the birds moved in rapidly during the ensuing few days (coincident with a thaw) and seemed to have reached their peak of numbers by May 31.

Courtship.—The belated arrival of the birds on their nesting grounds and the shortened season at high altitudes necessarily curtail the time available for mating and courtship. Though apparently the birds are unmated at the time of their arrival, nesting activities start promptly, leaving little time for prenesting nuptials. Such activities, moreover, take place largely in the dim light of evening, which makes observations on this phase of their breeding cycle difficult. Some probable sexual flights were observed on Mount Mansfield. The male pursues the female in swift flight, his crest feathers erected, and bill gaping. He often bursts into passionate song as the two dodge swiftly through the thickets. Such flights, however, are not restricted solely to the mating period, for during incubation the male frequently appears at the nest and drives the female away on a chase through the trees. Such a sexual flight was once observed in July; and apparently the pair later reared a second brood.

The evening is the great playtime of these thrushes. As the light of day retreats, the twilight-loving birds emerge from their concealing shelters and indulge in their remarkable flight-singing ceremonies until darkness envelops the woods. Occasionally they may be seen alighting upon stumps or rocks, a perfect picture of alertness, fluttering their wings tremulously as they dance nervously on their perch. Though these flight-singing performances are not confined strictly to the courtship period, their obvious concentration during the early part of the breeding season suggests that their major role is in nuptial affairs.

Young birds exhibit sex reactions at an early age. A young bird kept in confinement would, when a little over a month old, suddenly mount an object—a hand, foot, pencil, or small object on the floor—press its body down close upon it, flutter its wings, and open its beak. This behavior continued at intervals through the fall and winter, and though it may have been a mere playful impulse, to all appearances it looked like a reaction toward coitus with the opposite sex.

Nesting.—In spite of the special interest that attached to the home life of Bicknell's thrush in the period following its belated discovery, the revelation of actual nesting sites and pertinent life-history data has been slow, the bird's habit of breeding in isolated and relatively inaccessible habitats, and of building well-hidden nests in these places, necessarily hindering the gathering of such information. Known nesting records, published or disclosed through correspondence, have been given in some detail by Wallace (1939) and need not be repeated here. In 1935, the writer had an opportunity to follow the nesting cycle through an entire season on Mount Mansfield. The following nesting data are based largely on the 13 occupied nests studied that summer, supplemented by less detailed observations in 1933 and 1936.

Nest-building on Mount Mansfield in 1935 started during the first days of June, in less than a week after the arrival of the birds. On June 4, and again on June 6, birds were found laying materials for the foundations of their nests. On June 5 a bird was found working on a nest that was already probably several days old. Later observations on egg-laying, incubation, and hatching pointed to early June as the period of most active nest-building, with later nests probably due to some interruption in the nesting cycle.

Bicknell's thrushes build their nests by laying twigs and mosses loosely across one to several horizontal branches, usually close to the point where the branches diverge from the trunk. After a considerable mass of material is assembled, they start building up the side walls with moss and supporting twigs, and shape the cavity by sitting in the nest and working with rotary motions of the body until the cavity conforms to the size of the bird. In consequence, a sitting bird, thus perfectly adjusted to its nest, forms an efficient watershed for the eggs and young during rainy weather. Sometimes there is a delay of a day or two before the final lining is added. Nest-building activities were spread over the first half of June, though no individual structure is known to have required that long for construction. At two nests where an attempt was made to determine the nest-building period more accurately, the birds deserted.

Fresh green moss and small twigs comprise the bulk of the typical Bicknell's nest. Usually moss predominates with just enough twigs to give strength and stiffness to the structure. Long green strands of an abundant pleurocarpous moss (*Calliergon schreberi*) are most frequently used, often mixed with, or more rarely, entirely replaced by, the erect lighter green tufts of sphagnum. Supporting twigs are mainly spruce and balsam, usually with the addition of some minor items such as nonconiferous twigs, flower stalks, the pinnae or stems of ferns, dry leaves, shreds of bark, rotten wood, or hair. In one case the exterior of a nest was beautifully ornamented with a network of gray-green lichens. Moss, so readily available on the mountain for building purposes, is freely used by both ground- and tree-nesting birds, but none use it so freely as Bicknell's thrushes. Nests of the olive-backed thrush in the same region characteristically lack the mossy character of the Bicknell's nest.

Moss and twigs constitute only the body of the nest. The interior is filled with partially decomposed organic debris, apparently dug up beneath evergreen trees and no doubt replacing the inorganic mud used by robins and wood thrushes. Into the well-rounded cavity is then placed a lining of fine black rootlets, to which may or may not be added some dry, well-bleached grasses, and a leaf or two, the latter often dumped carelessly into the finished nest, and then left for a day or two, as if to mask the newness of the structure.

The dimensions of 20 nests were as follows: Outside diameter, 4.00 to 5.50, averaging 4.50 by 5.00 inches; inside diameter, 2.25 to 3.40, averaging 2.47 by 2.79 inches; outside depth, 2.75 to 3.75, averaging 3.34 inches; and inside depth, 1.50 to 2.50, averaging 1.81 inches.

The projection of twigs beyond the body of the nest was not included in the measurements of the outside diameter. Six nests had the interior cavity perfectly symmetrical, but the remaining 14 showed some distortion of diameter, usually due to compression of the nest against the trunk of a tree.

Nests are typically situated in small or medium-sized evergreens, usually placed where two or more horizontal branches join the main stem. Often it is set close against the trunk, so that the impression of the tree is permanently registered in the wall of the nest; less frequently it is a few inches to a few feet away from the trunk; and in three instances (out of more that 30) nests were found at the extremity of long low branches in large evergreen trees.

Though all the nesting records in the literature refer to nests in evergreens, three of the 30 or more structures located in the Mansfield region were situated in birches. The birds' apparent preference for evergreens is probably not alone due to the dominance of conifers over deciduous growth, but is no doubt correlated with the fact that birches (the only deciduous trees that are common at this altitude) are not leaved out early in June when nest-building begins, and thus would afford adequate concealment only for belated nests.

In height above the ground the nests varied from 3 to 12 feet. The average of 16 nests was 7.3 feet. Nests in the deep woods are commonly higher than those out in the dwarfed groves near the summit, the former being 8 to 12 feet high, the latter more often 5 or 6 feet. Tufts (Thayer, 1907) reported a Seal Island (Nova Scotia) nests 25 feet high and two others at 15 feet. Bent (MS.), however, found Seal Island nests at 8 to 10 feet, which closely corresponds to the heights for Mount Mansfield nests in the deep woods.

With few exceptions the nests are concealed in well-protected situations, usually in or on the margin of a dense evergreen tangle, so that the camouflaging character of the green mossy nests, and the fact that the wary birds seldom offer give-away clues to its location, make nest-hunting a painstaking and often unrewarding procedure. The finished nest, in most cases, is an exceedingly artistic and substantial structure. Though a few nests fall in ruins at the close of the nesting season, others, owing to their solid structure and protected situation, may survive through one or more winters, their mossy walls still green, still structurally perfect except for the debris-filled interior.

Eggs.—Three or four bluish-green, lightly spotted eggs are laid. Seven nests out of 13 had three eggs, and six contained four eggs, averaging 3.46 eggs per nest for the 1935 season on Mount Mansfield. Possibly this average is too low, since nests with three young were assumed to have had three eggs; but in two cases one and two eggs were known to disappear from a nest without affecting the remaining eggs. The brown spotting on the eggs is a variable feature; the eggs of the grayer-backed birds (there are two color phase types of Bicknell's thrush on Mount Mansfield) are less spotted, sometimes nearly immaculate, the eggs of the browner-backed birds more heavily spotted. However, there was often variation among the eggs in the same nest, with two or three eggs nearly immaculate and the other more heavily spotted.

In dimensions the eggs of eight sets (29 eggs) varied from 21.0 to 23.0 millimeters in length, averaging 21.9, and were from 16.0 to 17.5 millimeters in width, averaging 16.6. The eggs of the olive-backed thrush in the same area are typically larger and more heavily spotted.

Egg-laying in 1935 began on June 9 and continued through most of June, with one clutch of July eggs; but egg deposition after mid-June was believed to be due to the failure of earlier nesting attempts. Apparently one egg a day is laid until the set is complete. In the one instance observed the egg was laid at noon. Judged from hatching dates and from one observation, incubation starts with the deposition of the third egg, regardless of whether there are three eggs or four in the nests.

Young.—In the few instances that served as a reliable check the incubation period proved to be 13 to 14 days, but unfortunately three of the six nests for which egg-laying dates were known were broken up before hatching, and in another only one egg hatched, which could have been the fourth (last) egg hatching on the thirteenth day or an earlier egg hatching on the fourteenth day. In one study nest three eggs hatched after 13 full days of incubation, and the fourth young emerged half a day later. The three eggs in another nest hatched during the thirteenth day of incubation.

Considerable variability in hatching procedures was noted. At one nest the female apparently did practically all the work of liberating a feeble and underdeveloped chick, while at another nest a chick nearly freed himself by kicking open a slightly cracked shell during the absence of the female.

The newly hatched chick is blind, relatively helpless, and entirely naked except for wisps of wet natal down on the cephalic and dorsal feather tracts. Weighing less than 2 grams at birth, the birds showed a gain of 1 to 2 grams daily up to about 22 grams on their tenth day, after which taking weights and measurements proved impractical, causing the young to leave the nest prematurely. Weights tapered off considerably after the eighth day, presumably owing to rapid feather development. At nest-leaving age the tarsus, bill, and toes are practically full grown, but the wing is only half the adult size and the tail about one-fifth grown. More complete details on daily growth and development have been given by Wallace (1939).

Considerable variation occurs in feeding schedules at different nests and at different times, but for the most part a more or less standard schedule of 10–15 minute feedings is maintained, usually with a notable concentration early in the morning and a definite slackening early in the afternoon. The female usually proves more efficient and punctual in feeding operations. The male is apt to be irregular, but he effectively concentrates his efforts on critical periods, such as during heavy or prolonged rains, or in the early morning, at which time the female usually remains at the nest to brood. At one nest watched during a 2-hour period of slow rain the male alone maintained the standard schedule of 10–15 minute feedings, while the female stayed on the nest. During periods of mild weather, however, the male may disappear for hours at a time. Food brought by the male while the female is brooding may be parceled out by him to the young, or it may be taken by the female to distribute among the nestlings. At one nest three adults were found feeding the young. At another nest the male deserted, leaving all family responsibilities to the female. She succeeded alone, probably owing to the fair weather prevailing at that particular period, but it might have been a different story if inclement weather had intervened.

The food brought to the young was not closely or technically analyzed but was known to include many lepidopterous larvae (particularly geometrids), ants, wasps, and other Hymenoptera, and young grasshoppers (especially at one nest near a grassy opening). Large blister beetles (*Epicauta cinerea*), available on *Amelanchier* blossoms near one nest, were not utilized, even when gathered by hand and placed on the nest. A severe outbreak of sawfly larvae (*Pristiphora geniculata*), which threatened complete defoliation to the mountain ash, occurred that summer (1935), and many of the readily available caterpillars were brought to one belated nest of young; but they obviously did not like the strongly flavored larvae and often spit them out, as did a captive nestling force-fed on the spicy caterpillars.

The stomach contents of four dead nestlings consisted of the chitinous remains of beetles (one a cerambycid), lepidopterous larvae, ants and other Hymenoptera, a grasshopper nymph (*Melanoplus*), and two balsam needles.

Sanitation at the nest is scrupulously provided for by the adults. Excreta are picked out of the nest as soon as they appear and immediately swallowed. After a few days the young are strong enough to back up to the edge of the nest and deposit the fecal sac on the brim, where it is promptly picked up by one of the parents. Defecation typically takes place right after each feeding, the parent often waiting for the feces and picking them off the elevated cloaca as soon as they appear. As the young grow older, and digestion is perfected, thus reducing the amount of nutritive material in the feces, there is an increasing tendency for the parents to carry the excreta away instead of eating them at the nest.

Toward the close of the nestling period the young become very active—stretching, preening themselves, jostling, and clambering over one another to attain the topmost position. For some time before actual departure they appear interested in the outside world, peering over the edge of the nest, hopping and walking upon the brim, and even snapping—usually ineffectively—at passing insects. Naturally birds that are startled into a premature departure, as typically happens at disturbed nests, are not so active as those that remain the full nestling period. Nine- and ten-day-old birds were relatively quiet prior to their precipitous flight from the nest, but the more normal 11- to 13-day-old birds were very lively during their last days at home.

Dispersal after nest-leaving is relatively rapid, so that the young usually cannot be located in the vicinity of the nest the following day. They apparently scatter in different directions but are followed and cared for by their parents for an undetermined length of time. Young birds in juvenal plumage are frequently encountered in midsummer, usually only one in a place and attended only by one parent. Later both young and adults become increasingly hard to find, as they enter a quiet period of molting and seclusion that is seldom broken by a fitful outburst of song.

Plumages.—At birth young thrushes of this form are naked except for wisps of dark gray or blackish natal down along the cephalic, dorsal, and humeral tracts. The other feather tracts (alar, femoral, crural, ventral, and caudal) are not in evidence at birth, even as dark dots beneath the skin, but can plainly be seen by the second and third day, when they contrast sharply with the reddish, blood-suffused apteria. By the third day the quills of the primaries and secondaries are beginning to push through the skin, and on the sixth and seventh days feather tips burst through the ends of the quills. Feather development then proceeds rapidly, and by the tenth day the young appear to be nearly fully feathered in the buffy-brown, much bespeckled plumage characteristic of young thrushes.

This juvenal plumage is replaced by the first winter plumage during the postjuvenal molt in August. All the contour feathers except the remiges and rectrices are shed. Birds still in juvenal plumage were occasionally seen early in August, but before the end of the month, and particularly early in September, both young and adults began to emerge from their late-summer seclusion in full fall and winter dress. The adults undergo a complete postnuptial molt (shedding the remiges and rectrices) in August; the molt of the young is partial.

The postjuvenal molt was observed fairly closely in a confined second-brood bird, but the time of molt was not normal, since the bird was taken from a much belated nest. The initial stages may have escaped detection in spite of close watch, since in preening he frequently pulled out a feather and ate it; but by August 31 feathers were dropping uneaten to the floor of his cage. Raggedness was first noticed about the various head regions: The posterior nasal area was soon practically featherless, dotted by the incoming quills of the new plumage. Feathers next disappeared from the gular region, and then over the entire head and cheeks, new feathers quickly replacing those lost. The back and breast feathers had an unkempt appearance and gradually dropped out, their disappearance and replacement by the brownish-olive (unspotted) feathers of the first winter plumage indicating the progress of the molt. In another week feather replacement was practically complete, with one buffy juvenal feather clinging to the head and another persisting on the shoulder.

Normally there is not supposed to be a spring or prenuptial molt, and the slightly grayer breeding plumage is derived from the more olive-colored winter plumage by wear. The confined bird referred to above, however, grew a new tail in midwinter and in spring molted some wing and body feathers. This was interpreted as an abnormal molt due to poor feather condition, but the abrupt change in plumage from winter to spring in some Alaskan graycheeks suggests that spring molts in nature are not unknown.

Food.—During the breeding season Bicknell's thrush is strongly addicted to an insectivorous diet, for the amount of animal matter in the stomachs examined was practically 100 percent. W. L. McAtee has furnished the following data from the records of the U. S. Fish and Wildlife Service: Prof. F. E. L. Beal examined the contents of five stomachs, three from Mount Mansfield, Vt., and two from Slide Mountain, N. Y., all taken on the breeding grounds from June 22 to July 2. The total contents of the five stomachs included the following percentages of the various items: Formicidae, 39.6 percent;

Chrysomelidae, 18.6 percent; Lepidoptera larvae, 8.6 percent; Elateridae, 7.6 percent; Cerambycidae, 7.0 percent; Tipulidae, 6.4 percent; Carabidae, 3.0 percent; Hymenoptera, except ants, 2.4 percent; *Homorus undulatus*, 2.0 percent; Staphylinidae, 1.4 percent; Ephemeridae, 1.4 percent; miscellaneous Coleoptera, 1.0 percent; weevils, 0.6 percent; Lampyridae, 0.2 percent; and Diptera, 0.2 percent.

Their proclivity for ants is clearly shown, for in four of the birds 42 to 55 percent of the food consisted of these insects; the food of the other bird, however, consisted of only 5 percent Formicidae and 90 percent Chrysomelidae, with the other 5 percent Elateridae.

The list of food items found in two stomachs, taken on Mount Mansfield in June and examined by Dr. Clarence Cottam, is given in too much detail to be included here (see Wallace, 1939). The items that occurred to the extent of 5 percent or more are as follows: Carabidae, 24 percent, in one stomach only; Rhyacophilidae and Trichoptera, 10 percent each; Calliphoridae and Anthomyiidae, 8 percent each; Neuroptera and Phalangidae, 7 percent each; Diptera and Syrphidae, 6 percent each; and Chrysomelidae and Empididae, 5 percent each. Formicidae amounted to only 1 percent and that in only one stomach.

Late in summer and in fall these thrushes become partly frugivorous, taking considerable quantities of berries. Brewster (1906) writes that in autumn they subsist largely on berries—cornels, deadly nightshade, barberry, spicebush, wild grape, woodbine, and poison-ivy. Burns (1919) reports collecting a specimen of *bicknelli* at Berwyn, Pa., on October 6, "while feeding on poke berries."

Nevertheless their predilection for animal food is probably never lost, even in autumn, and what fruit they take seems more or less incidental to their habitual animal diet. Experience with two caged birds, though not an accurate criterion for determining what thrushes eat in nature, at least shows certain food preferences. These birds ate considerable quantities of the berries indigenous to Mount Mansfield—blueberries, bunchberries, snowberries, red-berried alder, and twisted-stalk—but they never could be induced to subsist entirely, or even largely, upon them, even for a short time. With their cage well supplied with berries they would clamor for insects and eat them greedily when brought. During the fall and winter the surviving bird lived chiefly on animal food (fresh meat) varied now and then with fruit. This presumably indicates that berries and fruit, which really may be consumed in surprising quantities at times, play only a secondary role in the Bicknell's diet, serving merely to supplement items of an animal nature.

An unexpected taste, discovered as the result of caring for caged birds, is an apparently natural desire for something leafy, two young thrushes in captivity taking greedily to fresh lettuce. Stomach analyses did not disclose this dietary feature, as leafy items, if present at all, are quickly passed, thus escaping detection. Possibly leafy matter may play a more important part in a bird's natural diet than commonly supposed. One caged bird was kept in good condition on a seemingly well-balanced diet of fresh meat (to replace insects), lettuce, and fresh fruit. The lettuce seemed to provide the bulky matter necessary for proper elimination.

The foregoing evidence relating to food habits of these thrushes clearly indicates their tendency to prey heavily upon injurious insects, and, as Beal (1915b) appropriately remarks, "the vegetable food, drawn entirely from nature's storehouse, contains no product of human industry, either of grain or fruit. Whatever the sentimental reason for protecting this bird, the economic ones are equally valid."

Behavior.—So many of the specific or racial characteristics of these thrushes have been mentioned in the preceding text that it is necessary to add only a few summarizing remarks under this caption. Ordinarily in their summer haunts only fleeting glimpses are caught of a brownish-olive form slipping quickly from view into concealment. Sometimes they are flushed when foraging on the ground among the ferns or brush or briefly viewed darting across an opening among the evergreens; or, more often, seen perched at evening upon a balsam spire, calling and singing. Because of their reputed shyness and wariness, their habit of occupying the most inaccessible places, and their almost unbroken silence during much of the year, they have come to be known, perhaps undeservedly, as one of the rarest, shiest, and least known of American passerine birds. Citations from early observers for the most part only serve to heighten this impression of obscurity, but Langille (1884) and Bent (MS.), in writing of their Seal Island experiences, and Torrey (1892), referring to observations on Mount Mansfield, reported them relatively friendly and not particularly shy.

Much of the obscurity prevailing regarding this form may stem from the fact that observers are seldom on hand at the time when the bird can best be studied. Their period of greatest activity is in June, at the height of the black-fly season, when the mountains are relatively unpopular. In addition the thrushes are so quiet during midday, owing to extensive morning and evening activities, that short daytime excursions into their haunts are seldom conducive of good results. At the height of the breeding season, however, the writer found these birds to be delightful study subjects. Though by nature somewhat shy and wary, nesting birds, particularly the female, seldom took very serious alarm at intrusions, and blinds were erected in close vicinity to several nests without the adults appearing unduly concerned. One young bird kept in captivity for a year proved to be a delightful pet, cheerful, playful, full of curiosity, and, aside from his dislike of being handled, was relatively fearless and friendly.

Voice.—The songs of our wood-inhabiting thrushes appear to be divisible into two distinct types—the slower, leisurely outpourings of separated phrases of greatly variable pitch, such as are characteristic of the wood and hermit thrushes; and the more continuous, slurred, chiming notes of more or less unvarying pitch typified by the veery and the gray-cheeked group. The Bicknell's song, by common consent, is of the veery type, distinguished from it chiefly by the characteristic break occurring a little past the middle of the song. The oliveback's song is somewhat intermediate in character, with the continuous notes exceedingly variable in pitch.

From the qualitative standpoint the Bicknell's song has been considered somewhat inferior for a thrush, no doubt largely because the veery type of singing is less popular than the rich, carefully expressed melody of the wood and hermit thrushes. It may, in fact, be an unperfected—or a degenerate—veery song: shorter, wilder, and higher-pitched, with less of the chiming quality that is so appealing in a veery's singing. There is, nevertheless, a special fascination in the song of the Bicknell's thrush, a wild, ringing, ethereal quality that is in perfect keeping with the evergreen solitudes it inhabits.

As a result of the numerous excursions in quest of Bicknell's thrush, considerable literature has been built up about its song. This has produced many fine descriptions, and a few erroneous accounts, such as the one quoted in Forbush's monumental work on New England birds, which likened the song to that of an olive-backed thrush. The literature on the singing of this bird, with many selected citations, has been reviewed at some length by Wallace (1939) and need not be reiterated here. The following concise summary may serve the place of the fuller description:

The song is introduced by two or three, usually two, low clucking notes, *chook-chook*, audible only at close range. The preliminary grace notes are hurriedly followed by two to four, usually three, high-pitched, vibrant, ringing phrases that slur downward, similar to but less marked than in the so-called intertwining circles of the veery that so suggest chiming. Usually on the third, but sometimes on the second or fourth, of these phrases there is an emphatic break, which is accompanied by both rise in pitch and increased intensity. This break comes a little past the middle of the song and is the peculiar feature that distinguishes the Bicknell's song from that of the veery and the northern gray-cheeked thrush. This climax phrase, consisting of several merged notes, is held for an instant, then runs imperceptibly into the closing notes, which are unemphasized. The final phrase seems to fade away as if in the ensuing silence there might be additional whispered notes which the human ear is not attuned to catch. The refrain is often repeated over and over, the singer seeming to hurry through in order to begin again. Not infrequently the song is begun by uttering the opening measure, but dropped without completion, as a whitethroat breaks off whistling; and conversely, the final part is often given without the prelude, though in many cases this omission is apparent rather than actual, the prelude simply not being audible.

The complete song, then, may be (inadequately) represented as *chook-chook, wee-o, wee-o, wee-o-ti-t-ter-ee.* This can be approximated by whistling through closed teeth, but the syllables are hard to distinguish, the many intertwining reverberations merging the notes inseparably. Little variation in the character of the song was noted in the Mansfield region, but judged by the accounts of other observers there may be some regional differences in other habitats. And, needless to say, word descriptions are hopelessly inadequate for giving a faithful picture of the technical character of the song.

Perhaps the most remarkable feature of the Bicknell's singing is its elaborate flight song, which, strangely, seems to have escaped the notice of the many early observers. Every evening during the height of the breeding season these thrushes indulge in flight performances over their respective territories, singing fervently on the wing. The flight songs start just after sunset (at about 8:20 P. M., E. S. T., early in June), when the light is dim and the woods full of shadows, and last for only 10 to 15 minutes, when darkness puts an end to all singing. In character the flight song is practically identical with the perching song, except that it is usually more hurriedly and breathlessly delivered. There is often a concentration of singers in a small area, so that the notes of several singers may be mingled. Though the birds probably have individual territories, the boundaries seem not to be strictly observed during flight singing, and shadowy forms drift from place to place, perching momentarily on trees, rocks, or stumps, and then take wing again. Flight singing is spread to some extent over the whole season of song but reaches its maximum development early in June before incubation is begun. This suggests a primarily prenuptial function.

Bicknell's thrushes exhibit a fairly definite daily and seasonal cycle of song, the latter correlated with nesting activities. Singing begins in the early dawn (before 3:00 A. M. on Mount Mansfield in June) but is somewhat fitful at this time, presumably because the birds are preoccupied with feeding as soon as it is light enough; and by the time a good meal is secured it is too light for optimum song production. During the prenesting season early in June, songs continue throughout the day, especially in misty weather, but by mid-June there is a decided falling off in daytime singing. The early afternoon in particular seems to be a rest period for the thrushes, but as the sun approaches the western horizon singing is gradually renewed, with flight ceremonies at twilight as the culminating feature of the day.

Song is at its height before and during nest-building early in June. The obvious decrease in song past the middle of the month is coincident first with incubation and later into early July, with care of the young. Singing during late June is due at least in part to interrupted breeding schedules, with a renewal of singing by males of the unsuccessful nests. At none of the nests studied did the male sing much during incubation or feeding, although before the nest was built, or if it was broken up, the nest site was the chosen location for song. Singing is renewed early in July at the end of the nesting cycle, regardless of whether the young are safely fledged. Such a recurrence of singing suggests renesting, but in only one case was a second nest found. This short July revival of song then abruptly wanes into almost complete silence late in July and in August.

Another feature of special interest in relation to song is the occurrence of singing among the females. At most of the nests under close observation the sitting bird was known to sing on the nest occasionally during incubation, hatching, and brooding. In one case the incubating bird stood up in the nest and sang repeatedly in broad daylight while an egg was hatching, but otherwise the singing was confined largely to the morning or evening hours. Nests were watched at all hours of the day and during every stage of the nesting cycle, and in all cases incubation and brooding seemed to be entirely the work of one bird, the female, without any substitution on the part of the male. The male often visited the nest and occasionally inspected the eggs, if the female was away, but was never known to sit on the eggs or brood the young. The female song does not differ materially from that of the male, except that it seemed to be of an inferior quality, and was usually, though not without exception, hoarse, thin, and weak.

The most characteristic call note of Bicknell's thrush is a harsh, penetrating, slurred whistle, similar to but harsher and higher-pitched than the familiar *wheu* of the veery. At times it approaches the sharply whistled call of the red-winged blackbird. Most other call notes are modifications of the typical one, adapted to express a variety of moods: high and piercing when used as a scolding or alarm note, lowered to an inquisitive *pe-irt* when used to express curiosity, or warbled more or less musically when the adults are at the nest. The female in particular often chirps and warbles to herself while nest-building, incubating, and brooding. And the parents seem to have numerous exchange calls, especially when feeding young. They also have a low *chook-chook* note, invariably used as a barely audible prelude to their song, and also as a scolding note. A rolling, wren-like *crr-rr-rr* is also occasionally heard.

A few concluding remarks on the vocal achievements of a young bird kept in captivity for a year may possibly add pertinent data to the moot question concerning the acquisition and inheritance of song and call notes. A caged bird, reared in nearly complete isolation from others of his kind since his fourteenth day, gradually acquired all the call notes characteristic of his race, most of which he had never heard before he gave them. The song, however, actually started when the bird was only 15 days old was always more or less random and experimental. It resembled the song of the adult in tonal quality but was not broken up into the proper sequence of phrases until he was taken back to his native haunts the following summer and given an opportunity to hear the wild birds sing. Even then he often reverted to his off-tune winter song, which had apparently been acquired at least in part by attempted imitation of running water, steam rushing into a radiator, or radio tunes.

This evidence is in general keeping with the belief that call notes of birds are largely hereditary but that the song is due to both inheritance and learning by imitation.

Field marks.—Little has been said in preceding paragraphs regarding the means of separating Bicknell's thrush from genetically or phenotypically similar forms, which has commonly been considered a problem of some magnitude in both field and laboratory. Over most of its summer range, however, the problem of identification is reduced to a minimum by the absence of all similar forms except the oliveback, which in the New England mountains is the only form likely to be confused with Bicknell's thrush. On Mount Mansfield the altitudinal range of these two birds overlaps slightly at the 3,000-foot level.

In good view the buffier-toned appearance of the oliveback's head and breast regions is a good field mark, contrasting plainly with the gray cheek of the Bicknell's thrush, and the light eye ring of the oliveback, when visible, is diagnostic. The oliveback's most characteristic call note is a weak, high-pitched *pip*, quite different from the Bicknell's sharp, harsh call; and the ascending spirals of the oliveback's song do not closely resemble the Bicknell's ringing, more or less even-pitched refrain. With a little experience, moreover, other minor differences, hard to describe, but quickly recognizable, are noticeable in their habits and movements. All of these criteria, except song in the fall, can be applied to migrating birds as well as to those in their summer home.

But the separation of the smaller Bicknell's thrush from the larger graycheek is a more serious problem and has rightly caused some New England field ornithologists, concerned with the now much-emphasized bird-listing practices, to question the reliability of sight records where the two birds occur together. Aside from probable differences in song (which could sometimes be used in spring), a slight disparity in measurements is the only criterion that can be used to separate these size races in either field or laboratory. Moreover, a female of the larger (gray-cheeked) race may in some cases be smaller than males of the smaller (Bicknell's) race, and there are, of course, no visible sexual differences in the birds. (Wing measurements in the larger form range from 95 to 108 millimeters, in the smaller form they range from 81 to 97, though this slight overlapping is apparently ignored in determination of skins where the sex is known.)

Obviously, then, only the extremes of these size races can be recognized by this means in the field. It has been suggested that the intermediate-sized oliveback makes a good standard for comparison; that is, in the rare situation where an oliveback chances to be present, a bird of the gray-cheeked group larger than the oliveback would be the larger northern form, and one smaller than the oliveback would be the "bicknellian" form. These comparisons need to be made with discretion, however, as a maximum-sized oliveback is larger than a minimum-sized northern graycheek, and, conversely, a minimum-sized oliveback would be smaller than a maximum-sized Bicknell. Only in dealing with average specimens of all three forms, then, would this interspecific size comparison be useful.

This leaves only vocal dissimilarities to supplement the usually inconspicuous disparity in size. To my ear the call notes of the two races offer no detectable differences. According to the few observers who are familiar with the songs of both races on their respective breeding grounds, however, there is a noticeable difference in song, the graycheek apparently lacking the break and rising inflection that is characteristic of the Bicknell's song. The limited singing that occurs in migration, however (rarely in fall, more frequently in spring), is often not full and clear, but merely snatches or fragments of the real song, which my ear usually is not able to distinguish with certainty as to race.

Most field observers are concerned only with separating the abovementioned size races, but perhaps it should be added, since many have tried to use color in separating the two forms, that use of color is more apt to confuse than clarify the matter. There are two color phases of Bicknell's thrush in New England—a brownish-backed and a grayer-backed form identical in size; and two similar color phases of the larger forms, the browner phase in the latter case largely restricted to Newfoundland and the gray-phase bird more typical of continental regions. Wallace (1939) has given a full description of these phases and their distribution.

From this it would appear that an observer reasonably familiar with the Hylocichlae can, in most cases, readily separate members of the gray-cheeked species from olive-backed thrushes but that only the extremes of the two size races can be safely determined in the field. Color differences are in some cases discernible in the field but are of no help in determining the size races, since both color types occur in both size groups.

Enemies.—Though the restricted distribution and relative rarity of Bicknell's thrush may be in large part due to their highly specialized habitat requirements, it may also in some way be correlated with their low breeding rate and the associated mortality factors that prevent further spreading. Tufts (Macoun, 1909) remarked that the Seal Island birds had several destructive enemies to overcome during the nesting season—the abundant crows and ravens, which constantly circled over the island, and the feral cats, which had been liberated and were breeding wild in considerable numbers. On Mount Mansfield there were no crows or ravens, and the thrushes encountered only natural predators, yet their nesting success during the summer of 1935 was low. Nine nests out of the 13 that were carefully watched came to complete grief, and, of the surviving four, two were only partially successful. Nesting success on this basis was only 25 percent; or, if calculated from the total number of eggs (45), reproductive efficiency was 24.4 percent, and the total loss of eggs and young 75.6 percent. In other words, 13 pairs of birds raised 11 young (0.85 per pair) to the nest-leaving age, a rate at which the adults would not be replacing themselves in two seasons. Considering postnest juvenile mortality, adult losses, and unmated birds, it seems doubtful if these figures can represent more than a stable, if not indeed, a decreasing population, though of course the results of a single summer cannot be considered typical of all years.

The main loss in 1935 was the unexplained disappearance of the eggs and young of 6 nests out of 13 (19 eggs or young out of 45 eggs, or 42.2 percent). This was believed to be largely if not entirely due to red squirrels, which were abundant and active arboreal hunters in the nesting groves. Weasels were present in the area, but they are mainly terrestrial hunters, like the more numerous chipmunks. Bobcats were seen only once, though subsequent winter-trapping returns, turned in to the State for bounties, indicated their presence had been generally overlooked.

Thus, more than half the known loss (42.2 percent out of 75.6

percent) was attributable to red squirrels or unassigned predators (one loss—not included in the above percentages—was fairly definitely assignable to a blue jay that chanced to pay one of its rare visits to the mountain from lower elevations), and in addition two infertile eggs and two deserted eggs suffered the same fate. Moreover, of the five empty nests of the year that were found in July, at least three gave evidence of having been plundered.

One nest came to grief through some disease, which may possibly have been *Protocalliphora* parasitism. When half grown the young became too weak and sickly to eat and slowly waned to the point of death in spite of the assiduous attention of the female parent. On the basis of 45 eggs this gives a 7 percent loss due to this disease, but if based on the number of young raised to the susceptible age (14) the loss would be 21.4 percent. Moreover, a nest observed in 1933 apparently suffered the same fate, the undeserted young dying in the nest when about half grown, and still another nest was found with the dried skeletons of half-grown young, a tragedy that may possibly have stemmed from this undetermined disease. In several of the study nests in 1935 the young, on about the sixth day, showed symptoms of sickness but recovered.

Other losses were due to infertile eggs or defective embryos (about 11 percent) and to deserted eggs (about 9 percent). Neither of these percentages is very accurate, however, as the former is based on the total egg count, 15 of which disappeared before hatching, and the adults of the deserted eggs renested, producing a second set of eggs, which was then lost to predators.

These heavy nesting losses appear to be typical of other local birds, in spite of which the blackpoll warblers, whitethroats, and juncos managed to maintain dense population, and the Bicknell's and olive-backed thrushes are certainly as abundant in their respective habitats as are their generic relatives, the hermits and veeries, at lower levels. It is entirely possible that the Bicknell's thrushes on Mount Mansfield are close to their optimum numbers and that the various limiting factors here noted merely tend to keep them at this favorable level. This introduces the age-old and probably unanswerable problem of whether their rarity and limited distribution is due to low productivity or whether slow reproduction is an adjustment to their enforced occupation of a restricted ecological niche.

Fall and winter.—The characteristic late-summer seclusion referred to above comes to a termination early in September with the manifestation of premigratory movements. The birds again become active in the woods, feeding among the groves, calling loudly and frequently, and sometimes bursting into short snatches of song. On September 10, my last day on Mount Mansfield in 1935, the thrushes were still abundant and active, their numbers apparently suffering little if any diminution due to migration. G. M. Allen (1902) reported them still on their breeding ground in Carter Notch, N. H., on September 15, 1900. Sweet (1906) heard them calling on Mount Abraham, Maine, on September 20. Migration records from southern New England likewise indicate that the thrushes remain in their mountain home until late in September. The earliest Massachusetts migration record is September 18, and the vast majority of the records fall on the final days of the month or early in October. From southern New England the thrushes proceed leisurely along the Atlantic coast, seldom occurring far inland, frequenting tangled beaches, fence rows, and woodlands and supplementing their insectivorous diet with generous quantities of wild fruit. Before the close of October, however, they leave the North American coast, probably anywhere between the Carolinas and Florida, and seek their only known winter home in the West Indies, where, high up in the mountains, they pass their days—no one knows how.

HYLOCICHLA FUSCESCENS FUSCESCENS (Stephens)

VEERY

HABITS

CONTRIBUTED BY WINSOR MARRETT TYLER

The veery makes its summer home in the half light of shady woods where a substratum of undergrowth deepens the shadows near the ground. Years ago in Lexington, Mass., there was a wooded swamp of perhaps 20 acres in which the veeries found ideal surroundings for nesting; many pairs came every spring and spent the summer there. Tall white pines, elms, and red maples grew thick in the moist soil, almost crowded together, and below them were smaller trees, great beds of cinnamon fern, patches of jewelweed, Canada lilies, and tangles of raspberry vines, and lower still, here and there, clumps of *Clintonia*—a rare plant thereabout. The wood was back from the road, always silent and undisturbed; it had stood unchanged for half a century, and the veeries had made this quiet spot their home for years; all through the summer the air trembled with their music. But now, many of the trees have been felled, letting in the sunlight, and few veeries nest there today.

Spring.—In the spring the veery takes a long journey from its winter quarters in South America—Colombia, Brazil, and British Guiana—by way of Central America to its breeding grounds in the United States and southeastern Canada. The species reaches New England early in May, a season when many of our summer resident birds have just arrived from the south. The Baltimore orioles are trumpeting in our orchards; the rose-breasted grosbeaks are whistling their sweet tune from our roadside trees; all the vireos are here, the red-eyed, the yellow-throated, the warbling, singing on their nesting grounds; the resident warblers are everywhere, squeaking out their tiny songs.

In contrast to these noisy travelers, the veery comes silently back to his summer home. We do not know that he is here until we go into his quiet, moist woodlands and hunt about for him. Here we find him easily, on the ground sometimes, not yet hidden by dense summer foliage, sometimes up in the trees where the little leaves, still unexpanded, do not conceal him. He is reserved and silent thus early in the season; we do not hear a note of his heavenly music, and we shall not for a week or two.

Horace W. Wright (1909) points out that veeries do sing occasionally during their spring migration. Speaking of veeries on the Boston Public Garden, he says: "The song has been heard, especially upon a damp day, audible above the din of the city."

William Brewster (1937) comments on the long interval of silence between the veery's arrival and its first song. Under date of May 19, 1899, he says:

> Wilson's Thrushes began singing to-day. * * * At evening there was general and protracted singing all around the Hill and in the blueberry swamp behind it, at least five or six birds taking part. All of them seemed to be in excellent form. Why is it that this species remains silent so long after its arrival? I saw the first this season on May 2, and by the 10th they were abundant. Living, as I do now, in the very midst of their favorite haunts, I should have known it had there been any singing before to-day. They have called a little at morning and evening and uttered the bleating notes but not once have I heard the song before this morning. Seventeen days is a longer period of silence than usual, however.

Courtship.—I find in my notes only one slight reference to the courtship of the veery, on May 18, 1914, and the two birds whose actions are described may possibly have been rival males, for the sexes are indistinguishable. The species arrived in Lexington, Mass., that year on May 7th, and I did not hear a bird sing until the 22d. The note reads: "Although the veeries in the swamp this morning were not singing, two birds were going through evidently a sort of courtship maneuver. They were perched low in a shrub, less than a yard apart, holding themselves motionless with the head drawn somewhat backward, the bill pointing upward at an angle of about 45° and turned slightly to one side. One of the birds slowly raised its tail to a marked angle with its body and moved it upward and downward very deliberately in the customary manner of the hermit thrush. The feathers of this bird's rump were elevated slightly and so separated that they appeared of a darker color than the back. The other bird stood perfectly quiet for a while. Then both birds sidled along the branch a little way, one moving toward the other, the other drawing away, both moving stiffly and very slowly."

Nesting.—The veery's nest is usually on the ground or very near it. In the moist woods where veeries build, if the ground is wet as it often is, the bird lays a thick foundation of dead leaves to protect the cup from moisture and thus often constructs a nest large for the size of the bird. When building, the veery seems preoccupied and pays little heed to the presence of observers. I found a veery early one May morning carrying nesting material—dead, very wet leaves—to a nest, a big loose heap of leaves, perhaps two-thirds completed. It was supported by three shoots of a small cherry shrub, at the side of a narrow, unfrequented path in the woods. The top of the nest was 10 inches, the base 5 inches, above the ground. The bird carried the leaves to the nest and arranged them in it while four of us stood a few yards away, and did not leave until we had stepped to within 5 feet of her. Later she sat on the nest while we walked past, so near the nest that our coats brushed against the branches that supported it. Several observers have reported similar tameness.

Elliott Coues (1874) gives an excellent description of a typical veery's nest which he found in Dakota, "placed," he says—

on a little heap of decaying leaves caught at the foot of a bush; resting on these, it was settled firmly in the crotch formed by several stems diverging at once from the root. The base of the nest was quite damp, but the floor was sufficiently thick to keep the interior dry. The nest was built of various slender weed-stems, grass-stalks, and fibrous strips of bark, compactly woven and mixed with dried leaves; the latter formed the lining of the base inside. The cavity is rather small, considering the bulkiness of the whole nest, measuring only about two inches and a half across by less than two in depth. The whole is as large as an infant's head, and of irregular contour, fitting the crotch in which it was placed, and bearing deep impress of the ascending stems of the bush.

T. Gilbert Pearson (1916) also gives a good description of a normal nest. He says:

The nest rested among the top limbs of a little brush-pile, and was just two feet above the ground. Some young shoots had grown up through the brush and their leaves partly covered the nest from view. It had an extreme breadth of ten inches and was five inches high. In its construction two small weed-stalks and eleven slender twigs were used. The nest was made mainly of sixty-eight large leaves, besides a mass of decayed leaf-fragments. Inside this bed was the inner nest, two and a half inches wide, composed of strips of soft bark. Assembling this latter material I found that when compressed with the hands it was about the size of a baseball.

Henry Mousley (1916) points out a distinction between the nests of the veery and those of the hermit thrush as he found them in Province of Quebec, Canada, "in that the lining has always consisted of dry leaves and rootlets, as against grasses and rootlets in those of the latter [hermit], which are also placed in drier situations."

The following records indicate the variation in the sites, composition, and surroundings of veery's nests: A. C. Bent (MS.) reports from Squam Lake, N. H., "three nests, all on or near the ground in heavy mixed woods. One was on a hummock, partially concealed by low plants, one in a low mountain-laurel bush about a foot up in plain sight between small maple saplings, and one a few inches above ground in a shady but open space in the woods well concealed between two very small laurels." F. W. Braund (MS.) sends Mr. Bent the following notes on veeries' nests: One, in Ontario, Canada, "in the fork of fallen limbs 2½ feet from the ground in thick undergrowth of dense woodland, made mostly of dead, moist leaves and lined with a few leaf skeletons"; another "on the ground in bushy, burnt-over woodland"; and a third at Andover, Ohio, "on the ground in second growth woodland of moist marshland, at the base of a small shrub near the water's edge, made of leaves, rootlets and grape vine bark." A. Dawes DuBois (MS.) sends these notes on the situations of veeries' nests found in New York State: "One foot above ground, supported by a large grape vine, at base of tree, well hidden in clump of weeds, beside a foot path; on the ground, in rather exposed situation by a small clump of blooming bluegrass, in a little opening in small woods, chiefly saplings; on ground, in weeds, in strip of low woods near a road; about 18 inches from the ground, in a small clump of briers, on steep bank of river, of usual materials, with some dried shreds of cattails in its exterior; by a stump at top of slope, in very exposed situation at side of lake-shore road, whole brown leaves placed in front in such a way that it became quite safe from observation of passers-by; on the ground, in woods, near a summer cottage."

Ora Willis Knight (1908), writing of his observations in Maine, states: "The nests are never situated on the ground as far as my experience goes, though one which I have found in the thick branches of a shrubby hemlock bush was nearly resting on the moss of a hummock in the thick woods. * * * The nests are often placed on top of the stub of a small tree around which the sprouts have started at heights of not over six feet from the ground. Other nests are placed in small evergreen bushes, in low alders and on dead stumps, in general two or three feet from the ground."

William Brewster (1906) shows that veeries sometimes depart radically from their usual breeding habits. He says, speaking of birds in eastern Massachusetts: "I have known veeries to breed in perfectly dry oak and pine woods, on the sides and summits of hills, in Lincoln and Concord, and in an orchard on the crest of a ridge in Cambridge, not far from the Pine Swamp. Mr. H. A. Purdie and I once found a nest built on the horizontal branch of an apple tree fully ten feet above the ground."

Dr. Thomas M. Brewer (1878) mentions a nest "built upon a horizontal limb of a tree, fifteen feet from the ground," and cites a case in which "Mr. George O. Welch several years since found a nest of this Thrush in Lynn at a height of twenty-five feet above the ground."

Edward H. Forbush (1929) reports the experience of "Mrs. Richard B. Harding, who spent a large part of a summer watching thirty nests of this species in New Hampshire and who has kindly given me her notes, is positive that the bird usually rears two broods there. The building of the nests required from six to ten days depending on the weather. By using a blind she was able to watch a nest from the time the young were hatched until they left it. Mrs. Harding says that both parents joined in guarding and defending the young. The male was most aggressive in driving other birds away from the nesting area and attacked red squirrels and chipmunks which trespassed upon his precincts, flying at them with great fury."

Eggs.—[AUTHOR'S NOTE: The set of eggs of the veery usually consists of four, but sometimes only three and occasionally five. These closely resemble small eggs of the wood thrush, or more closely those of the hermit thrushes, in shape and color, though the more elongated shapes seem to be less frequent in those that I have examined. They are only slightly glossy. The color usually varies from "Nile blue" to "pale Nile blue," with the usual variations seen in robins' eggs. They are almost always unmarked, but spotted eggs have been reported occasionally. Verdi Burtch tells me that out of 28 nests that he has seen, only two held spotted eggs. The spots are in various shades of brown. Paul F. Eckstorm mentions in his notes an egg that was "heavily spotted with pale brown in rather large spots." The measurements of 50 eggs in the United States National Museum average 22.4 by 16.7 millimeters; the eggs showing the four extremes measure **25.7** by 15.8, 23.4 by **18.0**, and **20.6** by **15.8** millimeters.]

Young.—Edward H. Forbush (1929), on the authority of Mrs. Richard B. Harding, gives the incubation period of the veery as 10 to 12 days, and, continuing the report of her observations, says: "The young were not fed by regurgitation on the first day, but with small hairless caterpillars together with soft white grubs and other small insects, all of which had been thoroughly bruised between the mandibles of the parent bird. This diet was continued for about four days. On the fifth day dragon-flies and slugs were added and a day or two later black swallow-tailed butterflies were added. The capture of many dragon-flies and butterflies indicates that this thrush is a skilful flycatcher as such insects are swift, erratic fliers."

Mrs. Harding (MS.) sends Mr. Bent the following pretty account of a young veery issuing from an egg-shell: "A pair of veeries had built their nest in a patch of laurel near the lake. Knowing that the clutch of eggs was ready to hatch, I examined the nest and found two tiny birds and one egg just hatching. Lifting it carefully out of the nest, I held it in my hand and watched the tip of the young bird's bill pierce the shell. Very methodically it drilled a series of holes around the large end of the egg, resting between efforts, then expanding its body until the shell began to crack open. Slowly the crack widened until the two halves separated, and the nestling freed itself from half the shell. The other piece remained on the top of its head like a blue hat for several seconds."

Frank L. Burns (1921) gives the period of nestling life as 10 days.

Plumages.—[Author's note: Dr. Dwight (1900) describes the juvenal plumage of the Wilson's thrush, a much better name than veery, as follows: "Above, including sides of head, deep raw umber-brown with dusky edgings and large guttate spots of tawny olive. Wings and tail tawny olive brown the greater coverts and tertiaries edged with tawny olive and darker tipped. Below, white, strongly tinged on jugulum, less strongly on the chin, breast, sides and crissum with tawny olive, heavily spotted or barred on the jugulum, faintly on the breast and anterior parts and sides of the abdomen with clove-brown, the feathers also barred with a subterminal tawny band. Submalar stripes dusky."

A partial postjuvenal molt, involving the contour plumage and the lesser wing coverts, but not the rest of the wings or the tail, begins about the middle of July and is finished in August. This produces a first winter plumage, which is like that of the adult at that season, except that the young bird retains until the next summer molt the juvenal greater wing coverts and tertiaries, which are edged with tawny-olive and are darker-tipped, thus distinguishing it from the adult all winter and spring. There is, apparently, no spring molt, the nuptial plumage being acquired by wear and a little fading of the buff shades and the spots on the breast. A complete molt occurs during the following July and August, when old and young birds become indistinguishable. Adults have but the one postnuptial molt in July and August. The sexes are alike in all plumages.]

Food.—Waldo L. McAtee (1926) gives the following summary of the veery's food:

This bird scarcely enters the orchard and garden, hence the fruit which it consumes (and that practically the whole of the vegetable food) is wild. It composes about four-tenths of the subsistence, the preferred kinds being juneberries, strawberries, blackberries, wild cherries, sumac and dogwood fruits, blueberries, wild grapes, and elderberries.

Beetles, ants and other hymenoptera, caterpillars, grasshoppers, and spiders are the principal constituents of the six-tenths of its food which the Veery derives from the animal kingdom. A few sowbugs and snails also are eaten. Click beetles (the parents of wireworms), round-headed and flat-headed wood borers, leaf chafers, junebugs, leaf beetles, the strawberry crown girdler, the plum curculio, clover root borers, bark beetles, plant bugs, and sawfly larvae are especially injurious insects devoured by the Veery.

The bird seems to do little or no harm, and feeds on various destructive insects, so deserves protection for its usefulness, as well as it does in an eminent degree for being an adornment to the forest, both in appearance and in song.

Edward H. Forbush (1907) speaks of the food thus:

The Veery feeds very largely on insects. Those which frequent the ground and the lower parts of trees are commonly sought. Ants, ground beetles, curculios, and grasshoppers are favorites. It goes to the fields sometimes at early morning, probably in search of beetles, cutworms, and earthworms. It has been seen, now and then, to eat the hairy caterpillars of the gipsy moth. It feeds considerably in the trees, and so takes many caterpillars; but is not usually seen much in gardens or orchards, except such as are situated near woods. In summer and fall it eats wild fruit, but seldom troubles cultivated varieties. Taken all in all, it is a harmless and most useful species.

We see the veery most commonly when it is feeding, down on the forest floor, hopping along—the hop almost a spring, a characteristic of the Hylocichlae—turning over the dead leaves and decaying vegetation, and snapping up the bits of food it finds there. It reminds us, as it feeds on the ground, somewhat of a fox sparrow, except that it does not jump up and throw the leaves backward with the strong scratching motion of the sparrow but thrusts about with its bill to expose its food.

Francis H. Allen (MS.), while watching a wood thrush and a veery feeding near at hand, noted that "while the wood thrush hopped along in the manner of a robin, more or less, the veery was continually flitting from a perch in a bush or tree (2 to 4 feet from the ground) down to the ground, where he picked up an insect or something of the kind, and then again to another perch."

Behavior.—The veery belongs to a group of small American thrushes much admired for the exceptional beauty of their song, their trim elegance of figure, and the quiet, reserved dignity of their manner, which has won for them the epithet of the aristocrats of North American birds. They make up the genus *Hylocichla*, composed of five species and several geographical races, which inhabit a large area in the United States and Canada. Members of the genus are found as breeding birds from the Austral to the Hudsonian Zones, inclusive, being arranged roughly in latitudinal belts that slightly overlap, and are made irregular by the effects of altitude. The veery breeds mainly in the Transition Zone but extends northward to lowlands in the Canadian region; it has the most southern range except the wood thrush whose breeding range extends far into the Southern States.

Frank W. Braund, in a letter to Mr. Bent, speaking of a spot in Ontario, Canada, where both the veery and the hermit thrush were breeding, brings out the fact that each species was restricted to its own habitat, although the two species were so near each other that they could sometimes be heard singing at the same time. He says: "The terrain was composed of rocky ridges, dry and lichen-covered, with moist, fern- and bracken-carpeted basins between them. The habitats of the veery and hermit were well defined: on each ridge I could hear a male hermit, singing, whereas in the moist bottoms, I heard the song of the veery. Standing on a ridge, I could hear alternately a hermit, a veery, a hermit, etc., until the notes faded into the distance."

Prof. Maurice Brooks (MS.) speaks of the veery's habitat thus: "The veery is a middle-of-the-mountain bird in the central Appalachian region. It is never found as a breeding bird in the warmer valleys, nor found abundantly at the spruce-clad heights. Where extensive hemlock stands occur, at whatever elevations, veeries are apt to be found. Along Cheat River, near Morgantown, W. Va., I have found them nesting at elevations around 900 feet. I had always thought of veeries as creatures of the wild, but during several seasons a group of nature students has camped near Lake Terra Alta, Preston County, W. Va., in a region where veeries are abundant. Every year our experience has been the same; during the first few days the birds are silent and not easily observed during the day, but as they become used to our presence they come into camp more familiarly, feeding around the tables even while we are eating. I have never seen another thrush of this genus so tame."

When we meet the veery in the forest we seldom get a clear view of it for very long a time; generally we have only a glimpse before it retires behind thick foliage or disappears among the crowded branches of a tree. Sometimes, however, we catch sight of a veery as it moves along a woodland path before us. Here, leaning forward with its head stretched out, it progresses by a quick series of long, springing froglike jumps—not scudding along like a robin—rising at each jump well above the ground—perhaps to leap over the dead leaves that usually litter the veery's pathway. When the bird pauses, it straightens up, standing almost upright, so that if it is facing us the white breast gleams out of the shadows, and holds the head high and often turned a little to one side, staring about with its large, soft eyes. It flips its wings as it goes on and sometimes leaps high in the air, giving a light beat with its wings, half flying as it leaps. All its motions are quick, deft, admirably coordinated, and as it stands motionless again, "in russet mantle clad," poised on its long, slender, wire-slim legs, it makes a delicate picture against the dark of the forest, like a lovely woodland nymph.

Voice.—The song of the veery is one of the strangest sounds in nature. The rendering "*whree-u, whree-u, whree-u, whree-u,*" as written by J. H. Langille (1884), suggests the form of the song well, a series of four or five downward-inflected phrases with a smooth transition in pitch, the final note prolonged and rolling, and each phrase a little lower than the one before it.

Perhaps because of its strangeness, the song is regarded variously by different observers; some considering it inferior to the other thrushes' songs; one, Ora Willis Knight (1908), passing it off as negligible, "if you can call it singing," he says; others perceiving in it qualities that raise it high above the song of any other bird.

We cannot think of it as a song in the sense of its being an expression of joy; it seems to express a calmer, deeper, holier emotion, like a hymn or a prayer. Bradford Torrey (1885) expands this thought in a delightful little essay on the veery's song:

To this same hemlock grove I was in the habit, in those days of going now and then to listen to the evening hymn of the veery, or Wilson thrush. Here, if nowhere else, might be heard music fit to be called sacred. Nor did it seem a disadvantage, but rather the contrary, when, as sometimes happened, I was compelled to take my seat in the edge of the wood, and wait quietly, in the gathering darkness, for vespers to begin. The veery's mood is not so lofty as the hermit's, nor is his music to be compared for brilliancy and fullness with that of the wood thrush; but more than any other bird-song known to me, the veery's has, if I may say so, the accent of sanctity. Nothing is here of self-consciousness; nothing of earthly pride or passion. * * *

And yet, for all the unstudied ease and simplicity of the veery's strain, he is a great master of *technique*. In his own artless way he does what I have never heard any other bird attempt: he gives to his melody all the force of harmony. How this unique and curious effect, this vocal double-stopping, as a violinist might term it, is produced, is not certainly known; but it would seem that it must be by an *arpeggio*, struck with such consummate quickness and precision that the ear is unable to follow it, and is conscious of nothing but the resultant chord. At any rate, the thing itself is indisputable, and has often been commented on.

Moreover, this is only half the veery's technical proficiency. Once in a while, at least, he will favor you with a delightful feat of ventriloquism; beginning to sing in single voice, as usual, and anon, without any noticeable increase in the loudness of the tones, diffusing the music throughout the wood, as if there were a bird in every tree, all singing together in the strictest time. I am not sure that all members of the species possess this power, and I have never seen the performance alluded to in print; but I have heard it when the illusion was complete, and the effect most beautiful.

Music so devout and unostentatious as the veery's does not appeal to the hurried or the preoccupied. If you would enjoy it you must bring an ear to hear. I have sometimes pleased myself with imagining a resemblance between it and the poetry of George Herbert,—both uncared for by the world, but both, on that very account, prized all the more dearly by the few in every generation whose spirits are in tune with theirs.

Frank Bolles (1891) also recognized the artistry of the veery's singing. He says: "The song of the veery has in it the tinkling of bells, the jangle of the tamborine. It recalls to me the gypsy chorus in the 'Bohemian Girl,' and when I hear it as evening draws on, I can picture light feet tripping over the damp grass, and in the shadows made by moving of branches and ferns I can see dark forms moving back and forth in the windings of the dance."

Henry Oldys (1916), writing of "Rhythmical Singing of Veeries," comments on the chordlike effect occurring in the bird's song, saying: "In the notations of Veeries' songs made by other musicians the closing notes have sometimes been represented as chords; but I believe that what these listeners heard were not actual chords, but broken chords, the separate notes of which were uttered so rapidly as to cause them to seem to blend in complete harmony." Sydney E. Ingraham (1938) published a photographic graph of a veery's song which is in accord with Oldys's opinion and speaks of the song as "like five-finger exercises on a harp." Thus Torrey's surmise, made over a half a century ago, is confirmed.

Aretas A. Saunders's (MS.) records show more or less variation in the song. "Some birds," he says, "sing the first slurs slowly and then the last two or three very rapidly; others have a simple, unslurred note at the beginning, or interpolated somewhere in the middle; some prolong the last note; some have slurs that rise and fall, like *rayeeoh*. The pitch of veery songs, according to my records, varies from D'''' to $B\flat''$, two tones more than an octave, which is much less than the range of other thrushes."

The commonest call note of the veery is a smoothly whistled *hee-oo* or *wheew*. This note may be used as an alarm note, but when much distressed, over danger to its young, for instance, the bird gives a note like *whuck*, low and guttural, very much like the quack of a catbird, or a long, loud, quavering *ka-a-a-a-a*, suggesting a red-eyed vireo's snarl. Dr. C. W. Townsend (1905) says that they "hiss like a Robin."

William Brewster (1936) speaks of a habit of the veery when singing: "I do not remember to have noted before that the Wilson's Thrush, like so many other birds, has favorite singing perches to which it resorts day after day. This, at least, is true of a bird which is breeding somewhere near the east end of Ball's Hill and which sings every evening in the large red oak on the edge of Holden's Meadow, sitting invariably not only on the same branch but actually on *the same twig* and always facing towards the northwest."

L. Nelson Nichols (MS.) adds: "The wheeling, weaving song is certainly at its best on a warm spring evening. If the moon comes up after sunset, the aroused veery is worth hearing."

Francis H. Allen (MS.) wrote in his notes for July 4, 1924: "One of several veeries singing had at least four different harsh, rasping notes, which it uttered between each two songs. Not all the notes were always present, but at least one of them always, I think. One was a high-pitched, disyllabic note, somewhat sibilant, which seemed more like a part of the song than the others. It generally immediately preceded the song proper. The bird sang constantly for some time till all the veeries stopped singing about 7:48, and the harsh notes were as regular as the song proper. The bird mewed occasionally, too. The strange notes suggested a catbird. One note was short and rather high-pitched, with a rising inflection. The other monosyllabic notes were lower-pitched and somewhat more prolonged."

Again he wrote, on May 21, 1935: "Early this morning I heard a veery singing over and over again, with hardly a second (I should say) between songs and keeping it up for some time—it may have been five minutes."

Field marks.—As indicated by the veery's old name, the tawny thrush, its back is a yellowish brown. This color and the almost unmarked breast serve to distinguish the veery from the other thrushes, which, although nearly alike in size and outline, have olive backs and are more heavily streaked or spotted beneath. Also the hermit has a reddish-brown tail, and the wood thrush has a reddish crown.

Enemies.—Veeries, like most ground-nesting birds, are exposed to attack by predatory mammals, such as red squirrels and chipmunks, noted under "Nesting."

William Brewster (1936) gives a long account of an attack of a snake on a veery's nest, and Harold S. Peters (1933) reports the finding of two species of flies and later (1936) one louse and another species of fly in the plumage of the veery.

Friedmann (1929), speaking of the veery vs. the cowbird, says: "A common victim; more so in some places than in others. At Ithaca, out of some thirty nests found, seven were parasitized. About twenty other records have come to my notice, ranging from Montreal, Maine, Connecticut, and New York, west to Illinois and Michigan. * * * This bird makes no attempt to get rid of the parasitic eggs foisted upon it and usually incubates and rears the young interlopers."

Fall.—The veery's song period comes to an end, here in eastern Massachusetts, within a few days of the middle of July, and after this date we see little more of the birds. In August, even when we walk through the veeries' favorite woodlands where they have been in evidence all summer, we rarely hear or catch sight of the birds, although some doubtless remain for a time, silent and inconspicuous, on their nesting grounds before starting on their migration. Taverner and Swales (1908) report from Point Pelee that in 1907 they "saw them almost daily from August 24 to September 2, after which none were noted, though we remained until the 6th."

Many of us who have long been attentive to the call notes of nocturnal migrating birds have heard year after year from the sky at night, during the latter half of August and early September, a clear, softly modulated, mellow whistle. On nights favorable for migration we might hear the notes every few seconds as the birds passed overhead—from high in the sky on starlight evenings, from nearer the ground, not far above the tree tops, on misty nights. These whistles correspond with no bird notes we ever heard in the daytime, and for years they remained a puzzle to ornithologists. Finally, in 1907, William Brewster, who had wondered since his boyhood what migrating bird might be the author of these notes, discovered that it was the veery.

Some years later (Winsor M. Tyler, 1916), when writing on the call notes of nocturnal migrating birds, with Mr. Brewster's permission, I told the bare facts of his discovery thus: "For a long time this call note remained a mystery to Mr. Faxon and Mr. Brewster, until finally Mr. Brewster, by a most fortunate chance, solved the problem. He was lying at dawn in his cabin on the shore of the Concord river, when he heard, far in the distance, the familiar whistle of the unknown migrant. The bird, still calling, flew nearer and nearer until it alighted in the shrubbery close by the cabin. Here it continued to call, but gradually changed the character of the note until, little by little, it grew to resemble, and finally became the familiar call of the Veery." Since Brewster's death much material from his journals has been published (Brewster, 1938), including a long account of the veery's nocturnal whistle, to which account the reader is referred.

DISTRIBUTION

Range.—Southern Canada to southern Brazil.

Breeding range.—The species breeds **north** to southern British Columbia (head of Crooked River; has been recorded north to Davids Lake, Crooked River; McBride, and Tete Jaune Cache); central Alberta (Peace River Landing, Edmonton, and Camrose); southern Saskatchewan (Carlton House, Prince Albert, and Hudson Bay Junction); southern Manitoba (Lake St. Martin and Shoal Lake); southern Ontario (Kenora, Port Arthur, Lake Nipissing, and Ottawa); southern Quebec (Quebec, Baie St. Paul, Kamouraska, Anticosti

Island, and Little Mecatine); and Newfoundland (Lewis Hills, South Brook, and St. John's). **East** to Newfoundland (St. John's); Nova Scotia (Jestico Island and Halifax); New Brunswick (Scotch Lake); Maine (Machias and Ellsworth), and the Atlantic Coast States south to northern New Jersey (Plainfield and Bernardsville); Pennsylvania (Scranton, Cresson, and rarely near Philadelphia); and south in the mountains to northern Georgia (Brasstown Bald); there are at least two records of breeding at Washington, D. C. **South** to extreme northern Georgia (Brasstown Bald); southeastern Kentucky (Black Mountain); northern Ohio (Youngstown, Oberlin, Sandusky, and Toledo); northern Indiana (Sedan); northern Illinois (Lacon, casually, and Chicago); northern Iowa at least formerly (McGregor and Spirit Lake); North Dakota, uncommon (Fort Rice and Oakdale); southeastern Wyoming (Torrington and Cheyenne); central southern Colorado (Fort Garland); northern Utah (Provo River and Salt Lake County); extreme northern Nevada (Mountain City); and northeastern Oregon (Enterprise, Fossil, and Prineville). **West** to central northern Oregon (Prineville and Rock Creek); central Washington (Yakima, Naches, and Coulee Dam); and south-central British Columbia (Kamloops, Bonaparte, Lac la Hache, and the head of Crooked River).

Winter range.—Very little is known of the winter home of this species, but it seems to be principally in southern Brazil. It has been recorded in winter as far south and west as Chapada and São Vicente in Matto Grosso and as far east as Santarém on the Amazon. Several specimens have been taken at Chapada, one of which has been considered identical with specimens from Newfoundland. The species is reported to be a "winter visitor" to British Guiana but without definite localities or dates. All records to date from Colombia are in October and therefore cannot be considered winter.

The recording of this species at Fort Brown, Tex., on January 1, 1877, by Dr. J. C. Merrill was evidently an error, as his published report gives no indication that the record was based on a specimen.

The ranges as traced are for the whole species of which three subspecies or geographic races are recognized. The typical race, the veery or Wilson's thrush (*H. f. fuscescens*) breeds from southern Ontario and Quebec southward and west to Indiana; the Newfoundland veery (*H. f. fuliginosa*) breeds in Newfoundland; the willow thrush (*H. f. salicicola*) breeds from Manitoba and Wisconsin westward.

Migration.—Some late dates of spring departure are: British Guiana—Camakusa, April 12. Cuba, Habana, May 4. Florida—Tortugas, May 22. Georgia—Athens, May 10. North Carolina—Chapel Hill, May 23. Louisiana—New Orleans, May 22. Mississippi—Biloxi, May 10. Arkansas—Helena, May 19. Kentucky—Lexington, May 24. Texas—San Antonio, May 18.

Some early dates of spring arrival are: Florida—Pensacola, April 15. Georgia—Round Oak, April 7. North Carolina—Raleigh, April 27. Virginia—Blacksburg, April 30. District of Columbia—Washington, April 9. Pennsylvania—Waynesburg, April 17. New York—Collins, April 16. Massachusetts—Amherst, April 30. Vermont—St. Johnsbury, May 1. Quebec—Montreal, May 3. New Brunswick—Picton, May 18. Louisiana—Madisonville, April 8. Mississippi—Tishomingo, April 20. Arkansas—Helena, April 18. Tennessee—Memphis, April 16. Illinois—Olney, April 20. Ohio—Oberlin, April 25. Michigan—Detroit, April 22. Ontario—Toronto, April 30. Iowa—Hillsboro, April 22. Wisconsin—Racine, April 28. Minnesota—Minneapolis, May 3. Manitoba—Winnipeg, May 2. Texas—Houston, March 27. Kansas—Independence, April 29. South Dakota—Yankton, May 2. North Dakota—Grand Forks, May 10. Saskatchewan—Indian Head, May 12. Wyoming—Torrington, May 6. Montana—Corvallis, May 8. Alberta—Edmonton, May 11.

Some late dates of fall departure are: Saskatchewan—Last Mountain Lake, September 1. Montana—Fortine, September 17. Wyoming—Green River, September 8. Manitoba—Aweme, September 12. South Dakota—Aberdeen, September 18. Minnesota—Minneapolis, September 21. Iowa—Davenport, September 23. Ontario—Ottawa, September 17. Michigan—Sault Ste. Marie, September 30. Indiana—Lafayette, October 2. Ohio—Cleveland, October 6. Kentucky—Lexington, September 27. Mississippi—Bay St. Louis, October 11. Louisiana—New Orleans, October 24. Newfoundland—Humber River, September 14. Quebec—Montreal, September 28. New Brunswick—Scotch Lake, September 22. Nova Scotia—Pictou, September 20. New Hampshire—Hanover, October 22. Massachusetts—Waltham, October 5. New York—Canandaigua, October 13. Pennsylvania—Doylestown, October 11. District of Columbia—Washington, October 1. North Carolina—Weaverville, October 10. Georgia—Augusta, October 1. Alabama—Autaugaville, October 22. Florida—Pensacola, October 18.

Some early dates of fall arrival are: Kentucky—Lexington, September 3. Mississippi—Bay St. Louis, September 7. Louisiana—New Orleans, September 12. Texas—Mission, September 20. North Carolina—Raleigh, August 28. Georgia—Athens, August 30. Florida—Tallahassee, September 11. Cuba—Habana, September 24. Colombia—Bonda, October 5.

Casual records.—The species is a rare or accidental migrant in Cuba. In New Mexico one was noted in the Taos Mountains July 17, 1904, and a pair seen August 5, 1910, at El Rito de los Frijoles that was reported to have bred there that season. Two specimens have been taken in Arizona; one at Tucson in May 1882, and the other at Fort Verde, May 13, 1887. Two specimens have been reported from Europe; one of unknown date in Pomerania, Germany, and one on Helgoland about 1833.

Egg dates.—Massachusetts: 32 records, May 22 to June 19; 26 records, May 28 to June 6, indicating the height of the season.

New York: 72 records, May 5 to June 30; 46 records, May 27 to June 10.

Ontario: 10 records, May 31 to June 26; 6 records, June 9 to June 15.

Washington: 9 records, June 6 to June 23.

HYLOCICHLA FUSCESCENS SALICICOLA Ridgway

WILLOW THRUSH

HABITS

This western subspecies of our familiar veery is described by Ridgway (1907) as similar to it, "but coloration duller, the brown of the upper parts less tawny (varying from deep isabella color to nearly broccoli brown), and brown streaks on upper chest and sides of lower throat averaging slightly darker." In general appearance it more closely resembles the olive-backed thrush than the veery, but it can be easily recognized by the absence of the buffy eye ring, which is so conspicuous in the oliveback.

Its breeding range covers southern Canada from British Columbia to Manitoba, and extends south in the western United States to central Oregon, Nevada, Utah, northern New Mexico, and central Iowa. It apparently migrates southward mainly between the Rocky Mountain region and the Mississippi Valley, straggling farther eastward, and spends the winter in South America.

Both its scientific and its common names were given to it because of its evident preference for willow thickets along the streams during the breeding season. In southwestern Saskatchewan, in 1905 and 1906, we often heard its veerylike song in the narrow timber belts along the creeks, where it was evidently common but seldom seen in the dense, shady thickets; we succeeded in collecting only two and found no nests. Laurence B. Potter, who lives in that region, says in his notes: "The willow thrush is certainly, in my experience, the shiest and most elusive bird, woodland or elsewhere. Unlike the well-behaved child, it is more often heard than seen, and to get a good look at this thrush by the ordinary methods is almost impossible. Certain observers have remarked a notable decrease in numbers of the willow thrush in recent years, and that is my own experience."

In Minnesota, according to Dr. Thomas S. Roberts (1932), "it is an inhabitant of low, damp woodlands, preferring especially tamarack swamps and thickets of poplar, willow, and alder, bordering streams and lakes. With the destruction of such conditions in southern Minnesota, the Willow Thrush, as a summer resident, has there greatly decreased in numbers in recent years; but in the north it is still one of the commonest of birds and its ringing song may be heard until mid-July, coming from all suitable places."

Nesting.—Dr. Roberts (1932) says that, in Minnesota, the nests of the willow thrush are placed "on the ground, sunk in the top of a mossy hummock; or near the ground, supported among alders or willows; or on tangled vines or fallen branches; often on the top of a stump among thick 'shoots.' Built of grasses, bark-fibers, small twigs, and moss, lined with fine grasses, rootlets, etc. No mud is used."

In Colorado, Denis Gale took only one nest of this thrush; "it was placed about three feet above the ground, in a low evergreen bush in a shady and wooded cañon" (Sclater, 1912). Dawson and Bowles (1909) report a nest taken near Spokane, Wash., that "was placed in the crotch of an alder at a height of two feet."

Eggs.—The willow thrush lays three to five eggs to a set, usually four. These are practically indistinguishable from those of the eastern veery. At least three sets of eggs have been reported in which some or all of them were spotted with fine dots or minute specks of dark or light brown or olive, usually sparingly and more or less obscurely. The measurements of 38 eggs average 22.9 by 16.9 millimeters; the eggs showing the four extremes measure **25.0** by 16.8, 23.4 by **18.0**, **20.8** by 16.8, and 22.9 by **15.8** millimeters.

Food.—The food of this thrush is doubtless quite similar to that of the closely related veery. Professor Beal (1915b) reported on the food of the two subspecies together. Mr. Potter writes to me from Saskatchewan: "In the spring of 1907, when range cattle lay dead in hundreds all over the western plains after a long hard winter, I obtained my best close-up views of willow thrushes which used to feed on maggots in a certain dead cow lying close to some willow bush. After a full meal, a thrush would be too glutted to do more than flop back to cover, if I got too close."

The general behavior of the willow thrush does not differ much from that of its eastern relative. Its voice is similar and it sings from within the thickets in which it lives, rather than from some loftier perch.

Dr. Friedmann (1929) mentions three cases in which this thrush was imposed upon by the Nevada cowbird; and, again (1934), he tells of two nests that contained five eggs each of the eastern cowbird! In one of these last two nests there were two eggs of the thrush and in the other only one.

HYLOCICHLA FUSCESCENS FULIGINOSA Howe

NEWFOUNDLAND VEERY

HABITS

Forty-five years elapsed after Reginald Heber Howe, Jr. (1900), described and named the Newfoundland veery before the A. O. U. committee added this subspecies to our Check-list in their twentieth supplement. Comparing it with the other two recognized races, Mr. Howe gives it the following subspecific characters: "Size slightly larger. Upper parts, especially on the head, distinctly *brownish*, much darker and not of the tawny shade of typical *fuscescens*, and *lacking* the greenish tinge of *salicicola*. Throat, lores, and upper breast suffused with *buff*, though perhaps less so than in *fuscescens* (in *salicicola* buff is practically absent), the upper breast and usually also the throat spotted *heavily* with *broad* arrow-shaped brown markings suggesting very strongly the throat and breast of *H. u. swainsonii*. The breast markings of both *fuscescens* and *salicicola* are narrow and more penciled and lighter in shade. Bill darker and heavier."

He gives its range as "Newfoundland (also possibly Anticosti and Labrador)."

Although we have no information on the habits of the Newfoundland veery, it seems fair to assume that they are not especially different from those of its western relative, for I found it fairly common along the Fox Island River and along other willow-bordered streams in western Newfoundland.

SIALIA SIALIS SIALIS (Linnaeus)

EASTERN BLUEBIRD

HABITS

The bluebird is well named, for he wears a coat of the purest, richest, and most gorgeous blue on back, wings, and tail; no North American bird better deserves the name, for no other flashes before our admiring eyes so much brilliant blue. It has been said that he carries on his back the blue of heaven and the rich brown of the freshly turned earth on his breast; but who has ever seen the bluest sky as blue as the bluebird's back? The early settlers in Plymouth Colony welcomed this friendly, cheerful songster, which reminded them of their beloved English robin redbreast, and they named it the "blue robin," an appropriate name still used among some children.

And, as our Pilgrim fathers welcomed it over 300 years ago, so do we today greet with joy the coming of this lovely, gentle bird each spring. Dull indeed would be the man that did not feel the thrill awakened by the first glimpse of brilliant color in the orchard and the cheery warbling notes borne to our ears on the first gentle breath of spring!

Before the English sparrows came, to crowd the bluebirds out, the latter came freely to nest in the boxes that we put up for them, or to occupy the natural cavities in the apple trees near our houses, even in the towns and villages. And the coming of the starling has driven them still farther away from our homes. So, now we must look for them in the open country, in the rural apple orchards, along the country roadsides, in open groves, and in burned-over or cut-over woodlands where there are plenty of dead trees and stumps with suitable hollows for nesting. They can be encouraged to remain, however, in any open region by putting up plenty of nesting boxes.

Spring.—The bluebird is a hardy bird; it does not go so far south in winter as most birds do, and it seeks the first favorable opportunity to return to its summer haunts. A few individuals may spend a mild winter in southern New England, but, as a rule, we may not expect to see the first arrivals here earlier than the first warm days in February; these are probably birds that have wintered not much farther south; and they may not stay long, as winter lingers in the lap of spring, late snowstorms and cold snaps may return and the venturesome birds are forced to retreat. But when the bluebirds come to stay, then we know that spring is really here. They are close rivals with the early robins and red-winged blackbirds, as harbingers of spring. W. E. Clyde Todd (1940) has expressed it very well, as follows:

> Of all our birds, this soft-voiced harbinger of spring is one of the most eagerly awaited. When winter begins to yield at last to the warming touch of the returning sun; when several days of clearing skies and southerly breezes have loosened the ice-fettered streams, drawn the frost from the ground, and given a balmy tang to the air; and when all nature seems in an expectant mood, vibrant with a new hope and a new promise—the Bluebird returns. * * * Its soft, pleasing warble, like the gentle murmur of a flowing brook in soothing cadence, awakens a sense of well-being and content in each responsive listener.

Bluebirds are seen more or less in winter over so much of their breeding range that the spring migration is not easily traced. Probably there is a gradual northward trend throughout all the winter range, with periodical retreats and advances influenced by weather changes. On Mount Mitchell, in western North Carolina, Thomas D. Burleigh (1941) seems to have noted a definite period of transition, for he found it "fairly plentiful" there "during the early spring months in the cut-over area (6,000 feet), occurring then in small scattered flocks. Extreme dates of occurrence are February 20 (1931) and

March 21 (1930). It may possibly breed sparingly at this altitude, although there are no actual records."

John Burroughs (1880) says:

In New York and New England the sap starts up in the sugar maple the very day the bluebird arrives, and sugar-making begins forthwith. The bird is generally a mere disembodied voice; a rumor in the air for two or three days before it takes visible shape before you. The males are the pioneers, and come several days in advance of the females. * * *

The bluebird enjoys the preeminence of being the first bit of color that cheers our northern landscape. The other birds that arrive about the same time—the sparrow, the robin, the phoebe-bird—are clad in neutral tints, gray, brown, or russet; but the bluebird brings one of the primary hues and the divinest of them all.

Many a disaster may overtake these hardy pioneers on their northward journey from the genial southland; perhaps they are more brave than hardy, for they suffer much, and many perish from the effects of sleet and snowstorms, and from freezing temperatures. Bagg and Eliot (1937) quote the following story from a Springfield, Mass., paper: "On March 28 a pair of Bluebirds came to the feeding station of Charles J. Anderson, 24 Eddywood Ave., Springfield, and after eating began to flutter and peck at the window. It was cold outside, so after talking to them through the glass, Mrs. Anderson let them in. The male was hardy, but the female manifestly required warmth. She was given warm milk to drink, and warbled her thanks. For three days, while the cold spell lasted, she returned periodically to get warm inside the room." They say that "Mr. Cross of Huntington has a photograph of twenty-two Bluebirds together which, caught in a heavy spring snowstorm, lived upon sumac berries and between feedings snuggled together, all fluffed up, on a small dead branch in the shelter of a building."

And Edward H. Forbush (1929) says that "in western Massachusetts and in Vermont during late spring storms many bluebirds have died huddled together in hollow trees, where they sought refuge from fury of the gale. During a storm a lady in Stowe, Vermont, heard a Bluebird calling in her living room and found two in the stove. They had sought shelter in the chimney and had come down the stovepipe."

Courtship.—The love-making of the bluebird is as beautiful as the bird itself, and normally as gentle, unless interrupted by some jealous rival who would steal his bride; then gentleness gives place to active combat. The male usually arrives a few days ahead of the female, selects what he considers to be a suitable summer home, and carols his sweetest, most seductive notes day after day until she appears in answer to his call. Then he flutters before her, displaying the charms of his widespread tail and half-opened wings, warbling in delicious, soft undertones, to win her favor. At first she seems indifferent to the gorgeous blue of his overcoat or the warm reddish brown of his ardent breast. He perches beside her, caresses her in the tenderest and most loving fashion, and sings to her in most endearing terms. Perhaps he may bring to her some delicious morsel and place it gently in her mouth, as an offering. Probably he has already chosen the cavity or box that he thinks will suit her; he leads her to it, looks in, and tries to persuade her to accept it, but much persistent wooing is needed before the nuptial pact is sealed. In the meantime a rival male may appear upon the scene and a rough and tumble fight ensue, the males clinching in the air and falling to the ground together, a confusing mass of blue and brown feathers struggling in the grass; but no very serious harm seems to have been done, as they separate and use their most persuasive charms to attract the object of their rivalry. At times, a second female may join in the contest and start a lively fight with *her* rival for the mate *she* wants. John Burroughs (1894) gives an interesting account of such a four-cornered contest, too long to be quoted here, in which the female of an apparently mated pair seemed to waver in her affections between her supposed mate and the new rival; and the latter seemed to have left the female of his first choice to win the bride of the other. However, after a much prolonged contest, the matter seemed to be satisfactorily settled, for two pairs of bluebirds finally flew off in different directions and started up housekeeping without further trouble.

But bluebirds are not always constant in their nuptial ties, even when they have raised a brood together successfully. Mrs. Nice (1930a) cites a case in which a male had a different mate for the second brood but returned to the first mate for the third brood, all in the same year. Seth H. Low (1934) has indicated, by banding at a station on Cape Cod, Mass., that bluebirds select different mates in successive seasons; he says: "In 1932 eight pairs of adults were banded at the Station. From two pairs neither bird returned. One adult from each of five pairs was captured nesting with a new mate. As it cannot be proved that each of the former mates were alive, it cannot be concluded that these birds were inconstant. Both adults did return from the eighth pair, but each took a new mate. No conclusions on mating constancy can be drawn from this one case."

T. E. Musselman (1935) writes: "During the first nesting period in 1935, I banded eighteen mothers. During the second nesting I found that none of these birds were in my nests, which leads me to believe that the mother bluebirds probably travel a number of miles between the first and second nesting and probably fly in small irregular bands with the broods of young birds. The second nesting is carried on by stray mothers which have formerly nested elsewhere."

If a male bluebird loses his mate, he quickly secures another. Dr. T. Gilbert Pearson (1917) tells of one that had three mates in a single season. The first two females were killed by a cat, but the third raised a brood, for "on a sunny hillside in the garden the cat was buried."

Nesting.—In the early days of my egg collecting, from 1880 to 1900, we always looked for bluebirds' nests in natural cavities in apple trees in old orchards, and fully 80 percent of our nests were found in such situations, though we found some in natural cavities in other trees and in old woodpecker holes. Nesting boxes were not so plentiful in those days as they are today. Two changes have taken place during the present century that have greatly modified the nesting habits of these birds. The old, decrepit apple trees have been pruned of their dead branches, the cavities have been filled, or the old trees have been removed entirely, thus destroying many favorite nesting sites for bluebirds, tree swallows, and some other birds. The old orchards have been replaced by new, young orchards, in which the trees are regularly pruned and sprayed, which is better for the apple crop but not so good for the birds. Furthermore, there has been an immense increase in the number of bird boxes put up by appreciative bird-lovers and by agriculturists who are now well aware of the economic value of the birds. The result has been that the bluebirds were not slow in adapting themselves to these two changes and in adopting these better types of nesting sites. So that, at least in settled communities, a great majority of the bluebirds now nest in the boxes.

To get the best results the boxes should be set on poles at no great height above ground, preferably between 8 and 12 feet, and in the open; to keep out starlings, the entrance hole should not be over 1½ inches in diameter; even then, there will be competition from tree swallows or house wrens, but the bluebirds are usually more than a match for these two.

Several large nesting projects have been reported where numerous boxes have been erected to encourage the birds to breed. One of these, part of which I have seen, centers around the great bird-banding station of Dr. Oliver L. Austin, at North Eastham, Mass. There are over 500 boxes in this project, chiefly around the main station, but also scattered at various distances away, from 2 to 9 miles north and south along the outer arm of Cape Cod. Most of the boxes that I have seen are erected on slender poles, within reach of a man standing on the ground, along lines of fences and around the edges of fields, bogs, marshes, and ponds. Most of them have been occupied by tree swallows, but many by bluebirds. It was here that Mr. Low made the studies of these two birds referred to under the two species.

For seven or eight years Dr. T. E. Musselman (1939) has been building bluebird boxes in quantity and erecting them on fence posts along the hard roads leading into Quincy, Ill. "The idea appealed to the popular fancy immediately," and he has received much help from school students of conservation and others. It took about 50 boxes to cover 38 miles of one road, and he placed 150 boxes along another 68 miles of road. He says: "All of these boxes are standardized, have removable tops, and by the time the entire project is complete will include nearly one thousand Bluebird boxes. Magazines and newspapers have printed copies of my plans and because of such publicity I feel that in many sections of the country, similar projects will be carried on." In a previous paper (1935) he says: "In no case did two birds nest closer than a quarter of a mile." His nests were placed from 3 feet to 10 feet above the ground, apparently mostly nearer 3 feet than 10, "and on posts away from human habitation. If the box is placed on the pasture side of a post away from the wires, cows use the box to scratch their backs, so I try to attach them to the wire-side of the post. This protects them from cattle and likewise makes it impossible for cats to molest them." He gives further useful instructions for making the boxes, to which the reader is referred.

Mrs. Amelia R. Laskey has sent me some elaborate notes on another interesting and successful project, of which she says: "Nest boxes for eastern bluebirds have been placed in Percy Warner Park and the adjoining Edwin Warner Park to increase the numbers of this species around Nashville, Tennessee. Starting in 1936 with 26 boxes, others have been gradually added so that 63 have been available the past three years." In one of her published papers (1939), she says that Percy Warner Park "consists of 2141 acres * * * much of it wooded hills, with many miles of winding automobile roads, bridle paths, and hiking trails, interspersed with picnic grounds, shelter houses, and homes of park employees. On the outer boundaries are numerous meadows, bordered on one or two sides with narrow thickets of trees and undergrowth. These meadows provide excellent sites for the Bluebird nest-boxes that have been placed there. * * * Of the 37 nest-boxes available in 1938, 36 were used, at least once by Bluebirds, with a total of 104 sets or 460 eggs laid, an average of 4.42 per nest."

A. Dawes DuBois has sent me his data for 15 nests, observed in Illinois, Minnesota, and New York. Five of these were in bird boxes, three in holes in fence posts, two in hollows in apple trees, two in other tree cavities, two in old woodpecker holes, and one was in a telephone pole.

M. G. Vaiden writes to me from Mississippi: "This bird is a fairly common nesting bird in the hill section of our State, especially from the central hills to the northward until reaching the Tennessee line. They select any suitable site where they think it possible to hide a nest, as a gate post and natural cavities in trees, and I found a nest in a drain pipe, where they were stacked for use and some 6 feet high. The bluebirds selected a pipe near the top of the pile."

Bluebirds have been known to nest in a number of other unusual places, such as empty tin cans or jars, in open hollows in the rotten tops of posts or stumps, and more than once in cliff swallows' nests, even in active colonies. Dr. Charles W. Richmond sent me, long ago, a clipping (Putnam and Wheatland, 1866) which reads as follows: "At the depot, the signal master called the attention of a number of the members to a pair of Blue Birds which had built a nest in one of the signal balls, from which a piece of the canvas had been torn. These birds, after raising one brood of young, had made another nest, by the side of the first, in which they had laid the eggs for a second brood. The signal ball, in which the nests were made, was lowered and hoisted about fifty times a day. The birds flying out as soon as the ball commenced its descent, and, alighting upon the fence nearby, would wait patiently for it to be hoisted again, when they would at once return to their nest."

Another railroad nesting site is mentioned by Charles R. Stockard (1905); it "was the hollow iron coupling of a flat car which stood for many weeks on a side track. The old style link and pin couple had a long hollow neck and back; in this neck a Bluebird had built its nest and deposited a set of five eggs."

A. L. Pickens writes to me that he "once found a bluebird's nest in a cavity in a steep earthen bank, some such a place as is usually frequented by the rough-winged swallow." Dr. Thomas S. Roberts (1932) tells the following interesting story of some very persistent bluebirds:

Many years ago there stood on the campus of the State University at Minneapolis two cannons, which were used every morning in artillery drill, and from which blank charges were frequently fired. A pair of Bluebirds selected one of these guns as a nesting-site. The nest was accordingly built but of course was removed next morning. This went on for several successive days, the nest built one day being destroyed the following morning. At length one morning the cadet whose duty it was to charge the gun failed to observe whether or not the nest was there and rammed down the cartridge with a will. When he tried to fire the gun, of course it would not go off; so the load was drawn and an examination disclosed the nest and the female bird jammed into a scarcely recognizable mass against the breech. Promptly the male secured another mate and the following morning the usual nest was in the gun. This continued for a day or two, when the cannon was stored for the season in a shed near by and a cavity in an adjoining tree was chosen for the nest, where peace reigned.

At least two combination nests have been reported. B. S. Bowdish (1890) mentions a bluebird's nest in the top of an old stump that held four eggs; under this in the same cavity was a nest of eight young mice. "The mice had access to their nest through a small hole in the bottom of the stump, and nothing separated them from the eggs but the material of the two nests." And Mr. Todd (1940) quotes an anecdote by J. Warren Jacobs concerning a bluebird appropriating the finished nest of a Carolina chickadee: "The nest was in two parts; one constructed by the Chickadee, and the other, which was the top story, was made by the Bluebird. The first story contained two [eggs] of the Chickadee, and in the next were five eggs of each species." I once found a flicker's egg in a bluebird's nest, together with five eggs of the bluebird; and in the same orchard there was a flicker's egg in a tree swallow's nest, with five deserted eggs of the swallow.

The nests of the bluebird are poorly and loosely built structures; this is probably all that is necessary in the snug cavities in which the nests are usually made, where a firmly built nest is not required. The nests are often made entirely of dried grass and weed stems, carelessly arranged; sometimes a few fine twigs are added; the lining may consist merely of finer grasses, or sometimes a little hair or a few feathers are added. The possible nesting sites are often pointed out by the male after he has attracted the female to his breeding territory, but she evidently makes the final choice. Both sexes help in building the nest, though most of the actual work on it is done by the female.

Wendell P. Smith (1937) made the following observations at his banding station at Wells River, Vt.:

> Nest-building did not proceed with uniform speed, especially in the case of an early beginning. There seemed to be some correlation with temperature, as cessation of activity coincided with lower temperature and resumption of construction began with the coming of warmer weather. The time required for a nest's completion differed in consequence. The shortest period recorded was four days, and the longest twelve days.
>
> Material was secured within a radius of seventy-five feet of the nest, and much of it within less than half that distance. In one case dried grass was used, while in the other, dead pine needles were obtained from the ground near by. Observations showed that the female performed nearly all the work of collecting. Between the completion of the nest and the laying of the first egg some time intervened, usually two or three days.

Dr. W. T. Harper (1926) has published some detailed observations on the building of a second nest by a pair of bluebirds. He concludes with the following summary:

> The most interesting points disclosed by these observations seem to be the following: First, the site for a second nest seems to have been selected while the first brood was still in the nest, and the male took the initiative in the selection. Second, the male laid the first foundation of the second nest, but the female did practically all the work while the male acted as watchman or boss. Third, work was faster at the beginning of the building operations and, as finishing touches had to be added, the work became constantly slower. Fourth, parts of four days were required to build the nest, most of the work being done between 6.30 and 10 A. M. Fifth, at least two hundred and eighty-nine trips with nesting

material were made by the female, the last fifteen of which were from a distance with material of fine texture, while the others were from less than 50 yards, with one or more pieces of dead grass. Sixth, the old birds, with young of both broods, returned to the vicinity of the two nests after an absence of about a month, and the old birds evidenced great interest in the second nesting-site and showed some jealousy when the young approached it too closely.

Ora W. Knight (1908) says: "Nest building is participated in by both parents, and I have known of a nest containing the full complement of eggs just seven days after the birds began building, indicating that the nest was completed in three days and an egg laid daily thereafter."

Henry Mousley (1916) "once witnessed a pair of these birds drive out a Hairy Woodpecker from a half completed nesting hole it had made, and after gaining possession of it they immediately set to work building a nest which was completed and four eggs laid in the remarkably short space of six days."

Alexander Sprunt, Jr., has sent me the following account of an unusual nesting site, as observed by Prof. Franklin Sherman, of Clemson College, S. C. Professor Sherman writes: "The nest is saddled on a horizontal limb of an oak, at about 12 to 15 feet above the ground, and about 15 feet out from the trunk of the tree, which is in the front lawn of the college hotel building, almost overhanging a much-frequented street or road. One or two small twigs give support to the nest, but it is not in any fork of the main limb—it is saddled on the limb itself, which is about 1¼ inches in diameter at the nest. During my stay of about 20 minutes the adult female made two visits to the nest and fed the clamoring young."

Eggs.—The bluebird may lay anywhere from three to seven eggs to a set; as small a set as three is unusual, five is a much commoner number than four, six eggs are often found, but sets of seven are rare. The eggs are ovate or short-ovate and are somewhat glossy. They are normally very pale blue or bluish white and always, as far as I know, are unmarked. Numerous sets of pure-white eggs have been reported; Dr. Musselman (1935) says: "In 1935 I was able to reach definite knowledge of the percentage of white eggs laid by Bluebirds. Of the 730 eggs recorded, 40 were albinistic in nature, or a total of 5.48 per cent. Fifty per cent of these white eggs hatched and the young were banded, and I am hoping that some of the young birds may return to this vicinity next year which will allow me to determine whether the trait of laying albinistic eggs is inherited." His hope was realized, for in his later paper (1939), he states: "This year [apparently 1938] I had the return of the first young female bird which had developed from a white egg laid in one of my boxes. * * * Imagine my delight in recording six albinistic eggs laid by this second generation bird. Of course, this one case is not sufficient to justify the conclusion that all female Bluebirds which hatch from albinistic eggs will in turn lay white eggs."

Mrs. Laskey, at Nashville, Tenn., has thrown considerably more light on this question of inheritance; I quote from her manuscript notes, as follows: "A number of individuals have laid white eggs, but there has been no evidence as yet to show this to be an inherited trait in this group. No. 36–146599, hatched April 1937 from an albino egg, was found in 1939 laying blue eggs. No. 38–121000, banded as an adult on April 6, 1939, was then incubating six white eggs. In 1940, one of those hatched from this set, N 6, laid five blue eggs in the adjoining meadow. The following year N 6 had moved on to the next meadow, laying six blue eggs in the second nesting period. From this hatch, N 22 was found in 1942 as she incubated six blue eggs. Thus, daughter and granddaughter of the white-egg-laying female were laying normally colored eggs.

"Five birds, known to have been hatched from blue eggs, laid white eggs (N 1, N 11, N 13, N 18, N 21). Only one, N 11, was found in two seasons. In April 1940, at 253 days of age, she began her first set of five in the box where she had been hatched from a set of four blue eggs. For the season she laid 5–5–5–4 white eggs, with only the third successful. She deserted her first two sets soon after completion and the young of the fourth set when they were five days old. She reappeared in the box in March 1941, laying five albino eggs, one blue-tinged. Four young were raised; one egg was sterile. On May 6 she began her second set of five white eggs but disappeared at the time this set was hatching.

"In 1942 there were more white eggs laid than in any previous season. They consisted of three sets of four, eight sets of five, two sets of six, and one set of seven. This total of 71 white eggs was 9.1 percent of the 774 laid this season. Incidentally, sets of seven bluebird eggs are rare; the 1938 and the 1942 sets are the only records in the Nashville area."

The bluebird is a persistent layer; if a set of eggs is taken, another will be laid within a very short time, as the two following accounts will show. Guy H. Briggs (1902) reports taking five sets of white eggs from one pair of birds during one season in a Maine orchard; the sets were all of five eggs, which he described as smooth and glossy, like woodpeckers' eggs. The sets were taken on May 1, May 27, June 13, June 24, and July 6, the nests being taken with the sets. Two of the sets were in the same cavity in an apple tree, and two others were in the same nest box. Between the last two dates only 11 days were required to build the bulky nest and lay five eggs. The bird had about half incubated the first three sets, but the last two sets were perfectly fresh. Thus, in about 76 days the birds had built five bulky nests and laid 25 eggs.

Arthur T. Wayne (1910) had an experience that almost equaled the above record. At Mount Pleasant, S. C., he took three sets of white eggs from a single pair of birds in one season, on March 30, April 12, and May 6; this bird laid another set late in May, and these were allowed to hatch. The interval between March 30 and April 12 was a short time in which to build a nest and lay four eggs.

The measurements of 50 eggs in the United States National Museum average 20.7 by 16.3 millimeters; the eggs showing the four extremes measure **22.9** by 15.8, 20.3 by **17.8, 17.8** by 16.0, and 22.4 by **15.2** millimeters.

Incubation.—The period of incubation is generally conceded to be about 12 days, though in some cases it may be a few days longer. The young birds remain in the nest 15 to 18 days, according to various observers, but probably the former figure is near the average. Both of these periods are evidently more or less variable according to circumstances. Mr. Smith (1937) noted that incubation "usually began with the completion of the clutch, but one instance was recorded where it began with the laying of the fourth egg in a complement of six. Of nine successful incubations of the two broods, the period consumed 14 days in four instances, 15 days in three, and 13 and 16 days respectively, in two instances."

Incubation is performed mainly by the female, but the male assists in this duty to some extent. Mr. Smith (1937) says:

In one instance the male was seen to take his mate's place upon the eggs three times in the course of three hours. The male of No. 2 pair fed his mate at intervals and maintained the semblance of a watch during her absence for food. Often the male would fly to the box, or a near-by limb, uttering rapid call-notes, whereupon the female would fly out and away for feeding. The male did not always remain near until his mate's return, but not infrequently he left shortly after the departure of the female. Absences, from meager observation, varied both in frequency and in regard to length of time. One nest was under observation from 3.15 to 5.30 P. M., and schedule is as follows: Female left at 3.27, returned at 3.35; remaining on the nest until 3.52. At 3.57 the male entered the nest and incubated until 4.17, when he left. At 4.18 the female returned to stay until 4.25. The male returned to the nest at 4.26, staying until 4.33 and returning again three minutes later for another period on the nest, which lasted until 4.48. The female entered the box at 4.59, and remained until 5.16. After four minutes absence, she came back and was still on the nest at the close of the observation period, ten minutes later. The male of No. 1 pair was not seen to take any part in the duties of incubation, although considerable time was spent in observation, three and a half hours being spent at one sitting.

Young.—Young bluebirds are fed and cared for by both parents more or less equally, but with considerable variation between different males. For instance, Mr. Smith (1937) says:

The male of No. 1 pair was not seen to feed the young. * * * The male of No. 2 pair, on the other hand, was particularly active and during some of the observation periods fed the young more often than the female did. The brooding of this pair was carried on exclusively by the female so far as we could learn. [During one hour, from 2.09 to 3.09 P. M.] the male brought food at 2.34:30, 2.37, 2.45, and 2.58:30. Total feedings for the interval were nine, five by female and four by male, and brooding lasted twenty-seven and a half minutes divided into five separate periods. This may be compared with an hour's observation five days later, the period extending from 2.54 to 3.54 P. M. Nine feedings occurred within this interval also, but six were by male and three by female and the brooding occupied twenty-nine and a half minutes divided into two separate intervals. * * *

In general the period passed by the young in the nest was eighteen days, one exception occurred in the case of No. 2 pair in 1933, when the first brood of four left the nest after seventeen days.

A brood watched by Mr. Du Bois (MS.) were in the nest just 15 days. And Ora W. Knight (1908) says: "The parents take turns in incubating and the eggs hatch in twelve days, the young leaving in fifteen days after they are hatched. Both parents feed them and carefully take away in their bills all the excrement voided by the young." Mrs. Laskey (1939) states that "the only Nashville record of a brooding male Bluebird is that of Simpson in April and May of 1937 when one individual was captured twice in a mail box on a nest containing eggs."

Mr. Smith's (1937) studies of the development of young bluebirds show that on the first day they varied in length from 31 to 41 millimeters; and that at the time of leaving, the 17th or 18th day, they measured 125 to 130 millimeters in length. "The eyes usually began to open on the 4th day, but in one instance this was delayed until the seventh day. Completion of the process required from three to five days. Tail-feathers appeared on the 8th day. Primaries became noticeable on the 4th day."

Mrs. Laskey (MS.) gives the following information on the success of hatching and rearing of the young, based on her study for seven years: "A careful analysis of the nesting data, accumulated through regular visits to the boxes, indicates that only 1,569 eggs of the 3,512 laid have been successful to the point of survival of the young to the age of 16–17 days when they normally fly from the nest. This is 44.67 percent of the total number laid and corresponds to percentages for birds building open nests. It is markedly lower than for hole-nesting species." Mr. Low's (1934) record for efficiency was decidedly better, varying from 62.7 to 87.5 percent.

Mr. Du Bois (MS.) gives the following account of young bluebirds leaving their nest in a fence post: "On June 19 the young were leaving their nest; only two remained within. I spent most of the afternoon trying for more photographs. After a long wait the male flew to a trolley bracket some 60 or 70 feet from the nest and sat there, and on the trolley wire, singing to the nestlings to come out. He kept this up for a long time. Occasionally a youngster would look out of the hole. They were hungry; they called to their parents in the musical young bluebird voice. But all afternoon the parents refrained from going to the nest to feed them. They merely came occasionally to try to coax the young ones out, by flying past, or by singing to them from some little distance. Finally, one of the youngsters—the one that had sat, two or three times, in the entrance way to look around—scrambled out on to the side of the leaning post, climbed partway around it, and flew across the car track to find a landing place on a horizontal guy cable, against a tree. Both parents fed it immediately; soon they returned and fed it again." He caught the young bird and returned it to the nest, but it came out again within a few seconds, flew over the pasture, and alighted on the ground. During the afternoon the parents had been busy feeding the other young that had left the nest earlier and were in the trees. The last youngster was still in the nest when he departed.

Bluebirds almost always raise at least two broods in a season, or at least attempt to do so; in many cases three broods are raised. As soon as the birds of the first brood are on the wing, the male takes charge of them, feeds them and teaches them to feed themselves. And the female immediately gets busy with her second nesting, either with the same mate or with another; as mentioned above, only a few days are needed to build the second nest, or lay the eggs in the same old nest, which has been renovated, if necessary. By the time the second brood is hatched the young of the first brood are well grown, are still in the general vicinity of the nest, and are able to assist in the feeding of the second brood of young, as has been frequently observed. After all the broods are fully grown, the family group keeps more or less together in the general vicinity of the nesting site until the time comes to wander about in fall, preparatory to migration.

Many yearling birds return the following spring to nest in the general vicinity of their birthplace. Mrs. Lasky (MS.) says: "Forty-two females, banded as nestlings, have returned to nest in the parks; also one banded elsewhere nested in the park, five miles from her birthplace. Numerous mated males, banded in the nest, are seen at the nests. The first eggs of 23 birds were laid at ages of 243 to 370 days, average 312. Egg-laying started on the average date of March 27 (1938 to 1942), nine days later than a group of 27 birds, two or more years old. Size of sets did not differ with age, five being the average.

Late and early hatched birds laid at approximately the same time the following spring."

Plumages.—Mr. Smith (1937) describes the natal down as "dark mouse gray." The young bird is in practically full juvenal plumage when it leaves the nest, except for the short tail. The two sexes are distinguishable in this plumage by minor differences. Dr. Dwight (1900) describes the juvenal plumage of the young male bluebird, as follows: "Above, slaty mouse-gray, the back lesser, median and a few inner greater coverts with white guttate spots bordered with sepia, the crown and rump much grayer and unspotted but sometimes with obscure transverse barring. Wings and tail dull azure-blue, the shafts and tips of remiges and rectrices dusky with faint whitish edgings; tertiaries and greater coverts edged with pale chestnut. Below, dull white, mottled on throat, breast and sides with sepia, the feathers centrally white bordered by the sepia and a rusty suffusion. Auriculars dusky mouse-gray mixed with white; lores grayish; conspicuous orbital ring pure white."

The young female is similar to the juvenal male, except that "the outer primary and outer rectrix have white outer webs, the blue is everywhere very much duller, and replaced with brown on the tertiaries and wing coverts, the edgings duller and the quills with duskier tips."

The first winter plumage is acquired by a partial postjuvenal molt, in August and September, the date depending somewhat on the date of hatching. This molt "involves the body plumage, wing coverts, tertiaries and tail, but not the rest of the remiges." This plumage is almost indistinguishable from the winter plumage of the adult male, though the colors are somewhat duller; Ridgway (1907) describes it very well, as follows: "Similar to the spring and summer plumage, but blue of upper parts slightly duller, more or less obscured on hindneck, back, and scapulars, by brownish tips to the feathers, and cinnamon-rufous of chest, etc., more purplish or vinaceous in hue."

Dr. Dwight (1900) says of the first winter female: "In first winter plumage the blue is obscure and confined to the wings, tail and rump, the back is dull grayish chestnut, grayer on the crown. The sides of the head are gray and white mixed, the orbital ring white. Below, the throat, breast and sides are reddish cinnamon, tingeing also the grayish white chin; abdomen and crissum dull white."

The adult and first nuptial plumages of both sexes are acquired by wear, which removes the edgings and brightens the whole plumage. The following postnuptial molt, beginning about the middle of August, is complete.

Food.—In its food habits, the bluebird is one of our most useful birds. It does practically no harm to human interests and it destroys large quantities of harmful insects. In his analysis of 855 stomachs, taken in every month in the year, Professor Beal (1915a) found that the food consisted of 68 percent animal and 32 percent vegetable matter. He says: "Orthoptera (grasshoppers, crickets, and katydids) furnish the largest item of animal food, amounting to a good percentage in every month, and in August and September aggregating 52.68 and 53.47 percent, respectively. The month of least consumption is January, when they amount to 5.98 percent, and the average for the whole year is 22.01 percent. * * * Beetles constitute the second largest item of animal food, and for the year average 20.92 percent of the diet. Of these, 9.61 percent are useful species, mostly predaceous ground beetles (Carabidae). Few birds exceed this record of destruction of useful beetles. * * * This destruction of useful beetles has been considered by some writers a blot upon the fair name of the bluebird." Various other beetles of a more or less harmful nature, such as May-beetles, dung beetles, weevils and others, are eaten in lesser amounts.

Ants amount to 3.48 percent, and other Hymenoptera (wasps and bees) to only 1.62 percent of the bluebird's food. Only one worker honey bee was found in one stomach. Hemiptera (bugs) average 2.75 percent for the year; stink bugs predominated, and remains of chinch bugs were found in one stomach. Lepidoptera, in the form of caterpillars and a few moths, form an important and regular article of food, averaging 10.48 percent for the year, the third largest item of animal food. Other insects, spiders, myriapods, sowbugs, snails, and angleworms, with a few bones of lizards and tree frogs, made up the remainder of the animal food.

Beal's analysis showed that "the vegetable portion of the eastern bluebird's food is largely fruit and mostly of wild species. Practically all of the domestic fruit taken was in June and July. Cherries and raspberries or blackberries were the only fruits really identified, though some pulp may have been of cultivated fruit. The most important vegetable food of the bluebird is wild fruit. The maximum quantity is eaten in December, when it amounts to 57.64 percent. January comes next, but after that month the amount decreases rather abruptly to zero in May. * * * The average for the year is 21.85 percent. At least 38 species of wild fruits were identified and probably more were present but not recognizable." Seeds are eaten sparingly, and grain was found in only two stomachs. Miscellaneous matter includes seeds of sumac, both the harmless and the poisonous kinds, poison-ivy and bayberry, amounting to 7.84 percent for the year. Beal includes long lists of insects and vegetable matter eaten.

Bluebirds obtain their food in the air, in the trees, and on the ground. In the air they are not so expert as the flycatchers and cannot catch the swifter insects, but they are often seen fluttering along near the ground after low-flying insects or darting out from a perch on some high tree to snap up passing insects, sometimes darting about with a hovering flight for a considerable distance from their perch. Francis H. Allen writes to me: "One September day I saw about a dozen of them feed thus for an hour or two, the air being full of dancing gnats." Once, he saw "a male feeding for a long time on the ground on a lawn, progressing in straight lines for considerable distances. He fed much as a robin does, but hopped instead of running and did not pull out worms."

In the trees bluebirds dart about among the foliage for flying insects, or pick caterpillars, katydids, and other insects from the leaves and twigs. Fruits and berries must be picked mainly from the trees and bushes.

But by far the greater part of their insect food, such as grasshoppers, crickets, beetles, etc., is found on or near the ground, and one often sees a bluebird sitting on some low perch, a fence post or wire, or some low tree, watching for its prey. Then it suddenly darts down, seizes something from the ground, and returns to its perch or another lookout point. Perhaps it may flutter down and, hovering just above the grass tops, seize a grasshopper and alight on the ground to eat it or return with it to its perch. Sometimes it stands on the ground and looks around, or actively searches for beetles or crickets; if its prey takes wing, the bird may flutter along after it and catch it in the air.

Behavior.—Bluebirds are generally regarded as gentle and lovable birds and rightly so, for such is their ordinary demeanor. If undisturbed they are friendly with their avian neighbors. But they can be aggressive, and even fierce in standing up for their rights against aggressors. In the competition for nesting sites they have often been known to compete successfully with English sparrows and tree swallows, attacking and driving them away when they attempted to usurp their nesting box. Other larger birds are often driven away from the vicinity of the bluebirds' nest; the male stands guard while the female is incubating, feeds her occasionally, and drives away unwelcome intruders, even human beings. Once, while I was introducing a young boy to the mysteries of bird study, we were vigorously attacked; one of our party had removed the female and was holding her in his hand; and while the boy was examining the nest the male flew at him so savagely that he lost his balance and fell flat on his back. Mr. DuBois (MS.) had a bluebird fly at his head in a very determined manner several times while he was examining a nest with young; it did not actually strike him but came very near it.

And Francis H. Allen tells in his notes of a similar experience; he writes: "The parents were very solicitous and very bold; whenever I approached the nest they swooped at me, making a 'clopping' noise with their bills and uttering a harsh chattering note. The male was the more active of the two in the demonstrations. I could hardly help dodging when he launched himself at my head."

William A. Taylor sends me the following account of a swallow-bluebird feud at the Moose Hill Sanctuary in Sharon, Mass.: "Each spring for years past these two species have fought for the possession of a particular nesting box just back of the house. As a rule, the bluebirds won out, but this year they were outnumbered, and the swallows held possession and the bluebirds were forced to take another box some 35 feet away. For a time peace seemed to prevail; but one morning, when the swallows had eggs and the young bluebirds were about to leave the nest, I became aware of a commotion about the swallows' box. As I watched, both male and female bluebirds emerged with swallow eggs, which they dropped to the ground. The swallows left the neighborhood but, much to my surprise, returned after four days and, finding the bluebird box vacant, laid a second clutch and brought forth their young on July 3. The bluebirds raised their second brood in the swallows' first box, thus resulting in a complete exchange of boxes."

Edward A. Preble (MS.) refers to a swallow-bluebird experience at his boyhood home in Wilmington, Mass. A nesting box was made with two apartments, side by side. Each spring its occupancy was a matter of sharp contention. But one spring the battle soon ended by a compromise. The two pairs proceeded to build in adjoining rooms, and both brought out their broods in relative peace.

The bluebird, like many other birds, has been seen shadow boxing or fighting his own image in a windowpane or other reflecting surface. John Burroughs (1894) gives an amusing account of such behavior. He tells a story related to him by a correspondent; a pair of bluebirds had a nest on the observer's porch and a pair of vireos had a nest with young in some lilac bushes but a few feet away; for several days the male bluebird was seen to feed the young vireos repeatedly, greatly disturbing the old vireos; his correspondent writes: "Sometimes the bluebird would visit his own nest several times before lending a hand to the vireos. Sometimes he resented the vireos' plaintive fault-finding and drove them away. I never saw the female bluebird near the vireos' nest."

With kind treatment and a little encouragement, bluebirds may become very tame, confiding, and friendly. C. F. Hodge (1904) tells an interesting story about how he trained a whole family of bluebirds, old and young, to become friendly with all the members of his family; he began coaxing them to his windowsill with mealworms, of which they seemed to be very fond, and finally had them feeding out of his hand.

The reader is referred to an interesting study of the territorial, nesting, and other behavior of the eastern bluebird by Ruth Harris Thomas (1946), which is published in too much detail to be included here.

Voice.—The bluebird is no great singer; he cannot begin to compete with the greater songsters of the famous thrush family; but his short contralto notes of greeting, as we hear them early in spring, are most welcome and pleasing to the ear, full of richness and sweetness, and even expressing affection. He really does not need to sing; his simplest notes are full of music and fully satisfy the hungry ears of the listener.

Aretas A. Saunders (MS.) has sent me the following full description of the song: "The song of the bluebird is soft, sweet, rather short, and warble-like. It consists of three to eight notes grouped in phrases of one to three notes each, with very short pauses between them. It is repeated every few seconds, and frequently two different songs are alternated. In the latter case it often happens that one song ends with a rising slur and the other with a descending one, so that it gives the effect of a question and an answer: *Ayo ala loee?* – – – – *alee ay lalo leeo!*

"The song is never so loud as those of other thrushes. It varies less in pitch and between individual birds. The range of pitch, from 24 records, is only 4½ tones, from F ''' to A ''. Many individuals vary only 2½ tones or 3 tones in the entire song. Though the song is comparatively simple, it is always pleasing, perhaps largely because the soft tone and lack of very high-pitched notes prevent any shrillness.

"Bluebirds sing from March to July or August. The song does not always begin when the first migrants arrive. In 8 out of 29 years of observation in Connecticut, bluebirds were singing when first noted in arrival. In other years several days elapsed before song began. The average arrival is March 10, but the average first song is March 18. The earliest date of beginning of song is March 3, 1923, and the latest April 2, 1940. Since the bluebird is never very common in the North and has periodical periods of scarcity, I often hear very little song in summer. In only eight years have I heard the song in July or August. In these years the average date of the last song is July 26, the earliest July 11, 1926, and the latest August 11, 1932.

"According to my observations, the male bluebird sings abundantly during courtship and nest-building, following the female about as she makes trips to and from the nest for nesting material. But as soon as incubation begins, the song ceases abruptly and is not renewed until the young of that brood have left the care of the parents and it is time to start a new nesting.

"The call notes of the bluebird are fully as musical as the song. These notes may be 2- or 3-syllabled, *oola, aloo, oolaloo,* or *aloola.* They may be heard frequently in the fall migration, as flocks of the birds fly over in October and November. The alarm note, given when the young are just out of the nest, is the only harsh sound I have heard from this bird; it sounds like *chat* or is often doubled to *chatat.*"

Mr. DuBois (MS.) writes the fall note as *juüit* or *Juliet,* which seems to be a good rendering of it. I have heard this plaintive fall note early in spring, before the real song season begins. To John Burroughs (1871) the bluebird seems to say "Bermuda! Bermuda! Bermuda!" The song has often been expressed in other syllables, such as *turwy, cherwee, cherey-lew,* or *tura-lee,* in soft, liquid, musical tones. W. E. Saunders (1887) once heard, and saw clearly, a bluebird imitating the *kay-kay* note of the blue jay; he "found that after the bluebird had warbled from four to seven times, the next warble would be prefaced with the Jay note."

The bluebird has about the lowest-pitched voice of any of the passerine birds; the crow's voice is decidedly lower, and that of the Baltimore oriole is slightly lower on the average but has a higher range. According to Albert R. Brand (1938) the bluebird's voice has an average mean frequency of 2,550, a maximum of 3,100, and a minimum for the lowest note of 2,200 vibrations per second.

Enemies.—Bluebirds seem to have no human enemies; everybody loves the gentle birds and appreciates that they are very useful and harmless tenants in our orchards and about our farms and gardens. But they have plenty of natural enemies to contend with. Cats readily climb to many of their most accessible nests and can reach in and pull out the young or the incubating parent; snakes climb into some cavities and destroy the eggs; red squirrels and blue jays invade the nests and eat the eggs or young; and house wrens often puncture the eggs, so as to appropriate the nest. Mrs. Laskey (1942) reports for that season: "A total of 174 sets, 774 eggs, were laid. From this large number only 261 young, 33.7 percent, left the nest box safely. Predation was heavy, 81 nests being entirely unsuccessful. Among these, 18 mother birds and 46 nestlings are known to have been destroyed by cats and 55 eggs failed to hatch through the loss of the incubating females. A boy robbed 11 nests of eggs; 42 were rifled of their contents by snakes. A 54 inch specimen collected in one of the boxes last year after eating the young was identified by Dr. Jesse M. Shaver, Peabody College, as a Southern Pilot snake (*Elaphe obsoleta obsoleta.*"

Dr. Herbert Friedmann (1929) says that the bluebird is "a very uncommon victim" of the cowbird, and cites about 15 records. Later (1934) he reports seven additional records and states: "Although the bluebird is still to be considered a rather infrequent victim of the cowbird, it is by far the most often parasitized of hole-nesting birds." Mr. DuBois writes to me that he found five bluebirds' eggs and two cowbirds' eggs in a box in his yard; all the eggs hatched, except one bluebird's egg, which was found on the ground, punctured; the other six eggs hatched, but the sun was very hot and most of the young perished from the heat; one cowbird and possibly one bluebird survived, though he could not find the latter.

Dr. Musselman (1942) once found in one of his boxes a filthy nest with four half-grown bluebirds cuddled in the bottom; and above them was a two-thirds-grown starling sitting complacently on the smaller birds; "the droppings of the larger bird had soiled and in one case almost covered the head of one of the tiny birds below; one eye was entirely covered and there was a stench which is unusual about such a nest." He destroyed the young starling, washed the young bluebirds, rebuilt a clean nest and returned the young bluebirds to it; the mother bluebird accepted the change and raised her young successfully. "In the many years that I have carried on my Bluebird experiment, I have never before found a Starling roosting in or employing one of my boxes for a nest site. In fact, only upon three or four occasions have I found Cowbird eggs in the normal nest. Only when somebody has removed the top of a box thus allowing an approach of the female Cowbird through the aperture above has there been molestation on the part of the Cowbirds."

Competition for nesting sites is one of the bluebird's greatest troubles. House wrens have always been aggressive competitors, but the bluebirds have generally been able to resist them and sometimes to evict them. Edward R. Ford has sent me the following note: "When young bluebirds left the 6 by 7 by 7 inch nesting box, June 20, I cleaned it out at once. By noon of the same day, house wrens took possession and began filling it with twigs. A few days later I noticed that bluebirds were still about the box, and when I looked into it on June 29 it held three bluebird's eggs. When the second brood had flown, August 2, an investigation showed that the bluebirds had assumed ownership before the wrens had completed the usual true nest in the twig mass and had made a scanty one of their own with a few dry grass stems."

When the English sparrows came the bluebirds had to face a determined competition; often the bluebirds were more than a match for the sparrows; but when the sparrows came in groups or droves they were too much for the bluebirds to resist; fortunately, the sparrow population is not so formidable as it once was, since its numbers have declined some. But the introduction of the starling gave the bluebirds another setback; these large, powerful birds can easily drive out the bluebirds and occupy any of the larger cavities; many old apple orchards that formerly housed bluebirds are now preempted by starlings. Bluebirds are safe from these intruders, however, in many of the properly constructed bird boxes; if the entrance hole is not over 1½ inches in diameter the starling cannot enter; but a 1¾-inch hole might allow the starling to use the box.

Bluebirds are generally able to contend with tree swallows, to drive them out or to defend their homes against them. A housing feud between these two species is mentioned above, under "Behavior." Flying squirrels, deer mice, and even bumble bees have been known to appropriate suitable cavities for bluebirds.

Harold S. Peters (1936) lists two species of lice, one fly, and two species of mites as external parasites of the eastern bluebird. Doubtless there are other forms of vermin that infest the nests.

I have left until the last the bluebirds' most formidable enemy, Jack Frost, the agency that has destroyed more of them than all other enemies put together; countless thousands have succumbed to extreme cold, snowstorms, and cold, ice-forming rainstorms. Bluebirds seem to be very vulnerable to these elements in winter and even in spring. The most notable of these catastrophes occurred during the winter of 1894–95, the season of the "big freeze" in the Southern States. Amos W. Butler (1898) describes the event as follows:

> The weather was warm until after Christmas. December 27 and 28 it became quite cold in this latitude [Indiana]. The Bluebirds were forced farther southward beyond the limits of the severe weather. There it remained warm until late in January. On the 24th of that month the temperature as far south as South Carolina remained near the zero mark. It turned warmer that night and the next day, January 25, the weather was bright and clear. The day following was Friday. It rained, then snowed; the wind came down from the northwest with great velocity and the temperature fell rapidly. Everything was ice-bound or snow-bound to the Gulf of Mexico. Then followed weeks of unusual severity. By the end of the severe weather in April, it is said, but few Robins or Bluebirds could be found. The destruction of bird life must have been enormous. The Bluebirds seem to have been almost exterminated. Few, indeed, returned to their breeding grounds in the north and from many localities none were reported in the spring of 1895.

Bluebirds began to increase slowly during the next few years, but it was five or ten years later before they seemed to have reached normal numbers. A lesser reduction in their numbers in the East occurred as a result of the very cold winter of 1911–12 in the Southeastern

States, but this was more local in its effect, and the birds soon recovered from it. Dr. Musselman (1939) writes:

In the seven years that I have been banding and studying Bluebirds through the use of bird-boxes, we have had three severe freezes in April after the majority of the Bluebirds had laid their full quota of eggs. Nearly always I found complements of frozen eggs deserted by the mother. Later, a second grass nest was built directly over the old eggs, then the new mother would begin her nesting activities. Seldom did the original mother return to her old nest. The unfortunate feature about such a catastrophe is not alone the destruction of fifteen hundred to two thousand eggs, but it is the fact that the nesting period is advanced about two weeks. This means that these Bluebird boxes which are very much in demand by several types of birds have eggs in them at the time the House Wren (*Troglodytes aedon*) returns. The number of pierced eggs has been correspondingly large on the years of such freeze. During normal years the baby Bluebirds are in the nest at the time of the wrens' return. Generally they are not molested. On normal years the nesting is so timed that when the first batch of young Bluebirds desert the nest, the House Wrens have already established themselves elsewhere. When the Bluebirds return later for the second nesting, there is little danger that piercing of the second complement of eggs will take place.

In addition to the frozen and punctured eggs, he found on several occasions the frozen bodies of the incubating birds where they had died on their nests; and once two birds were found frozen to death in a single box.

Field marks.—Bluebirds are so well known and so conspicuously colored that they are easily identified. Even the spotted young have bluish wings and tails.

Fall.—Dr. Winsor M. Tyler has sent me the following sketch: "Bluebirds are all along the roadsides this morning—a windless, warm, October day. They are gathered sociably in companies of half a dozen or more and keep near together like a big family, one bird following another when it flies. They are quietly musical as they flit about, giving the gentle *whit* call, the soft chatter, the velvety *turwy*, and sometimes a phrase of song. It is easy to imagine that the bluebird's song was evolved from a repetition of the *whit* note, perhaps by way of the *turwy;* a slight change in the tone of voice, making it mellower, louder, and sweeter, lengthening the notes a little, and there is the song.

"The birds perch on dead branches, wires, or fence rails, scanning the ground as from observation posts, sitting upright with the tail straight down; they explore holes in the apple trees, peering in, sometimes entering the cavities, calling to one another; they drop to the grass or to the hard, surfaced roadway where they catch up something with a deft peck. The bluebird's shadow at this season, the myrtle warblers, come down to the road, too, and act in the same way.

"In flight the bluebirds are very charming at this time of year; a leisurely flip of the wing carries them along silently with just enough momentum to keep them afloat in the air, and they often sail for a long way, drifting along with open wings. In contrast to the goldfinches and purple finches they fly only a short distance before alighting again. We shall see few more bluebirds before winter comes. This little company is already on its way south, yet they seem in no hurry to leave New England. How leisurely the bluebirds are as they flit about in fall!"

Only in the northern part of its summer range can the fall migration be satisfactorily traced, but there it is sometimes quite conspicuous. Robie W. Tufts writes to me that bluebirds are uncommon in Nova Scotia, but during October 1937 a flock containing "some hundreds" was observed in Annapolis County. "These were seen at the peak of their abundance for only a short time, but bluebirds were seen more or less constantly for a few days after the main flight had passed. Considering the relative scarcity of these birds in Nova Scotia, the origin of same is a mystery to me."

In Massachusetts we usually see them passing through in October and November. Out in the open country on clear days with a northwest wind, we often hear their sad farewell notes drifting down around us from all directions; and, looking up into the blue sky, we see large numbers flying over, high in the air, widely scattered or in small detached flocks, and all floating along in a generally southward direction; we know that they are leaving us, and we are sorry to see them go. They sometimes turn up in unexpected places; on November 1, 1915, a flock of eight appeared at our shooting club among the sand dunes of Monomoy Island; the next day they were joined by 10 more; these were two clear, warm days, but the following day it blew a gale from the northwest, with heavy clouds and some rain; the bluebirds had departed.

Edwin A. Mason writes to me from Groton, Mass., that on November 3, 1942, at 8:00 A. M., "it was raining, with a fairly strong wind blowing from the NNW. Birds from the tops of tall bare willows caught my ear. There, throughout the tips of the tree's branches, was a flock of bluebirds. They were moving occasionally from twig to twig, constantly talking back and forth. Very soon the major part of the flock took to the air. This made it possible to count them. The surprisingly large number of 28 were winging their way through the rain in a SSW. direction, with the wind quartering them somewhat, but still substantially on their tail. Evidently the flock had paused to rest and despite the rain considered it a good time to continue on its migration. Three birds hesitated to join the flock, one of them starting out after it only to return. These three probably tired birds remained, calling back and forth, as the main body of their erstwhile traveling companions went winging away southward through the dull leaden sky, their voices and shapes gradually diminishing as the vastness of the murky sky enveloped them."

The flock observed by Mr. Mason was not "surprisingly large," for the birds are often seen in larger flocks, sometimes as many as a hundred, though usually more or less scattered. Late in summer and early in fall mixed flocks of old and young desert their breeding resorts and wander about the open country and woodland, often associated with similar flocks of roving robins, all of which are much wilder and more restless than they are about our grounds in nesting time. Bagg and Eliot (1937) state that, in the Connecticut Valley in Massachusetts, "in October, transient Bluebirds are abundant, and natives come back as if to say good-bye to their homes, and sometimes carry nesting-material into their boxes, in that Indian Summer of the procreative instinct that many birds evince on warm October days."

Referring to the Buckeye Lake region in Ohio, Milton B. Trautman (1940) writes: "The first southbound migrants were noted during the first half of September, and until the end of the month a rather gradual, daily increase in numbers was observed. The migration reached its peak in October, when the bird was as numerous as in spring. In autumn its lisping note, uttered from overhead or from a fence post or tree, was one of the most pleasing and familiar of all fall bird calls. The Eastern Bluebird was very conspicuous during the calm, warm 'Indian summer' days of late October—such weather was called 'bluebird weather' by local sportsmen."

At Point Pelee, Ontario, the migration is often conspicuous; on October 29, 1905, according to Taverner and Swales (1908), bluebirds were there in numbers. "Here numbers were feeding on the bare sand with the Prairie Horned Larks. It was in the waste clearings beyond Gardner's place, however, that the greatest numbers were found. Here they were in flocks almost as dense as blackbirds. When flushed from the ground they generally flew to some of the numerous clumps of bushes growing here and there in the open and, when they lit and were viewed from a little distance, they were in sufficient numbers to give the whole bush a decidedly blueish cast."

Winter.—A few bluebirds spend the winter in southern New England, especially in mild seasons and more commonly near the seacoast, feeding on bayberries with the few wintering myrtle warblers or on the seeds of sumacs. They take shelter in the dense growths of red cedars, which protect them from the cold winds and furnish some berries for food. They roost in hollow trees or in bird boxes, sometimes several together. Mr. Forbush (1929) cites William C. Wheeler, of Waltham, Mass., as having twice seen one go to roost in an old robin's nest. Dr. Harold B. Wood writes to me that bluebirds were common all through the winter of 1913–14 at Slocum, R. I., which is five miles west of Narragansett Bay in the central part of the State.

Bluebirds sometimes winter in the more northern parts of the Mid-western States and even in southern Ontario. There are winter records for Point Pelee. And E. M. S. Dale says in his notes from London, Ontario: "Although the bluebird is one of our earliest spring migrants, it was not until December 27, 1937, that we found any here in winter. On that date we found four birds about a bit of marshy ground, where some springs had kept the snow melted and gave them a chance to obtain food. The ground was covered with snow; in fact, we were taking a hike on skis and snowshoes when we found them. The temperature had been down to 8° below zero a few nights before. They were still there on January 1 when we went out to begin our New Year's list."

From the Carolinas southward bluebirds are present all through the year, but they are probably not the same individuals, the local breeding birds having moved southward to be replaced by others driven down from the north. M. P. Skinner (1928) says: "This seems all the more probable because during cold spells I found Bluebirds gathered in large flocks of as many as seventy birds in most unusual places. They did not seem to be familiar with the country and its supplies of food and water. But with warmer weather these large flocks of strangers disappeared and the familiar birds were found again in the usual small groups."

In their winter resorts they are found in the more open woods, such as the flat pinewoods of Florida, seeking the denser growths only for shelter and spending most of their time for food in the more open places, such as cotton, corn, and sugarcane fields. In such places they are often associated with myrtle, pine, and the palm warblers.

A. L. Pickens tells me that "the sheltered nooks selected by individuals are interesting. A flock, I once observed, selected the cracks between the logs of a cabin in which cotton that had not been ginned was stored. Packed thus against the logs the cotton afforded a heat retainer, while the upper log gave shelter and the lower footing. One bird I saw took possession of an old summer-tanager nest for a winter dormitory."

M. G. Vaiden tells in his notes of a winter disaster not mentioned above: "For some reason, probably the terrific winter of 1906 when sleet was 4 to 6 inches deep over a great part of central Mississippi with a complete freeze-up of the ground for some 4 to 6 inches deep, when some trees were frozen and the trunks burst open, the bluebirds of this area, the normal breeding population, were frozen to death or died of hunger and thirst, and the nesting of the bluebird in the hill section certainly fell away considerably." He believes that the breeding birds of that area remain and mingle with the migrants from the north, rather than migrating farther south.

When all the vicissitudes with which bluebirds have to contend are considered, it is not strange that there seem to be no records of great longevity. Mrs. Laskey says in her notes: "So far none of my banded nestlings have been found after three years. The high rate of mortality through predation is doubtless the main factor in this prevailing short life span. The oldest bluebird of record in the Parks group is an adult female, banded in May 1938 and nesting there each year. Her latest capture was in April 1942, when she was at least four years old. Another female, banded at my home as an adult in April 1936, was retrapped each year until November 4, 1939, when she was at least in her fifth year of age."

DISTRIBUTION

Range.—North America east of the Rocky Mountains from southern Canada to El Salvador and Honduras.

Breeding range.—The eastern bluebird breeds north to southern Saskatchewan (Eastend, Lake Johnston, Indian Head, and Hudson Bay Junction); southern Manitoba (Aweme, Lake St. Martin, and Winnipeg); southern Ontario (Emo, Port Arthur, Rossport, Lake Abitibi, and Ottawa; casually to Moose Factory); and southern Quebec (Blue Sea Lake, Quebec, Kamouraska, Point de Monts, and Havre St. Pierre). **East** to eastern Quebec (Havre St. Pierre and Anticosti Island); Prince Edward Island (Alberton); Nova Scotia (Halifax), and the Atlantic Coast States to southern Florida (Jacksonville, Miami, and Royal Palm Park). **South** to southern Florida (Royal Palm Park and Deep Lake); the Gulf coast of the United States; Veracruz (Jalapa); Guatemala (Panajachel); Honduras (San Juancita), and El Salvador (Mount Cacaguatique). **West** to El Salvador (Mount Cacaguatique and La Reina), Guatemala (Antigua and Duenas); Oaxaca (Cerro San Felipe); Jalisco (La Laguna); Sinaloa (Plomosas); extreme southeastern Arizona (Sierra del Parajarita and Santa Rita Mountains); central Texas (Kerrville, San Angelo, and Wichita Falls); central Oklahoma (Wichita Mountains and Fort Reno); western Kansas (Garden); extreme eastern Colorado (Holly; occasionally Denver; has occurred at Pueblo and in Estes Park); southeastern Wyoming (Cheyenne and Laramie, possibly Newcastle); rarely to central Montana (Billings and Great Falls); and southwestern Saskatchewan (Eastend). The eastern bluebird is also resident in the Bermudas but is more numerous in winter.

Winter range.—In winter the bluebird withdraws from the northern part of its breeding range. The northern limit of wintering varies from year to year according to the severity of the weather. The winter range extends **north** to extreme southeastern Arizona (Fort Huachuca); northern Sonora (Bavispee River); northern Coahuila (Sabinas); central Texas (Fort Clark, San Angelo, and Wichita Falls); central Oklahoma (Wichita Mountains and Tulsa); eastern Kansas (Wichita and Topeka); eastern Nebraska (Fairbury and Omaha); western Iowa (Sioux City); casually at Yankton, S. Dak.; Minneapolis and Duluth, Minn.; Madison, Wis.; central Illinois (Knoxville and Rantoul, casually to Rockford); southern Michigan, casually (Kalamazoo, Ann Arbor, and Detroit); northern Ohio (Oberlin and Akron); southern West Virginia (Charleston and Bluefield); southern Virginia (Lynchburg); Maryland (Washington, D. C., and Cambridge); southeastern Pennsylvania (Philadelphia); southern New York (Shelter Island, Long Island, and Rhinebeck, casually); and casually to southern Massachusetts (Northampton, Taunton, and Cape Cod).

The eastern bluebird has extended its range westward within a generation or two. At Portage la Prairie, Manitoba, in 1884 it was referred to as a "recent arrival." In 1909 Macoun did not mention any occurrence of this species in Saskatchewan; in 1922 it was found breeding in the Cypress Hills of southwestern Saskatchewan.

The bluebird apparently is casual in winter in Cuba; a specimen was collected in April 1860 and a flock of seven seen near Habana on February 24, 1917.

The foregoing range applies to the whole species, which has been divided into several subspecies or geographic races. The eastern bluebird (*Sialia s. sialis*) occupies the range east of the Rocky Mountains except southern Florida and southern Texas; the Florida bluebird (*S. s. grata*) is found in the southern half of Florida, the Tamaulipas bluebird (*S. s. episcopus*) is found in northeastern Mexico and the lower Rio Grande Valley in Texas; the azure bluebird (*S. s. fulva*) occurs from southeastern Arizona south in the tableland of Mexico at least as far as Jalisco. Other races occur south of the United States.

Migration.—Some early dates of spring arrival are: New York—Syracuse, February 27. Vermont—Rutland, March 6. Maine—Waterville, March 12. Massachusetts—Wilmington, February 22. Quebec—Montreal, March 12. New Brunswick—Scotch Lake, March 19. Prince Edward Island—Alberton, March 25. Ontario—Toronto, March 18. Illinois—Chicago, February 28. Michigan—Grand Rapids, February 18. Minnesota—Minneapolis, March 6. South Dakota—Dell Rapids, March 21. North Dakota—Fargo,

March 22. Manitoba—Margaret, April 3. Wyoming—Laramie, April 24. Saskatchewan—McLean, April 4.

Some late dates of fall departure are: Saskatchewan—Eastend, October 12. Wyoming—Laramie, November 7. Manitoba—Aweme, October 26. North Dakota—Fargo, October 24. South Dakota—Yankton, November 27. Minnesota—Lanesboro, November 19. Wisconsin—New London, November 3. Michigan—Ann Arbor, November 27. Quebec—Montreal, November 17. New Hampshire Durham, October 30. Maine—Livermore Falls, November 6. Prince Edward Island—Alberton, November 2. New Brunswick—St. John, November 4. Nova Scotia—Halifax, October 2. New York—Rhinebeck, November 12.

Banding records.—A few recoveries of banded bluebirds indicate their migration or wanderings. One banded on Cape Cod, as a young bird, on June 23, was found the following December at Merry Hill, N. C. Another young bird banded at the same place on May 17 was caught on December 1 at Soperton, Ga., and one banded May 28 was caught November 7 at Durham, N. C. A young bird banded on July 26, 1932, at Manchester, N. H., was killed on February 14, 1933, at Sharpsburg, N. C. One banded at East Durham, N. Y., on July 15, 1931, was found dead in April 1932 at Neuse, N. C. A nestling banded at Ottawa, Ontario, on June 24, 1936, was killed on March 3, 1937, at Lake City, Fla. An adult banded at Ravenscliff, Ontario, on May 24, 1938, was found in March 1939 at Blairsville, Ga. An immature banded at Quincy, Ill., on May 8 was caught the following December at New Waverly, Tex.; another banded at the same place on June 7 was found the following January at Frost, La. An immature banded at Dallas City, Ill., on May 24, was found dead on December 26 at Sugarland, Tex. One banded at Knox City, Mo., on May 18, 1935, was shot on November 17, 1935, at Morrows, La. A nestling banded at Lisle, Ill., May 20, 1938, was caught on January 23, 1940, at Stockton, Ga. An immature banded at Gates Mills, Ohio, July 18, 1927, came down a chimney on June 2, 1928, at Milwaukee, Wis. An immature banded at Pomfret, Conn., July 29, 1931, was found dead on April 20, 1932, at Lunenburg, Mass.

Egg dates.—Florida: 18 records, March 17 to June 19; 10 records, April 10 to May 10, indicating the height of the season.

Illinois: 34 records, March 26 to June 20; 17 records, April 2 to April 29.

Massachusetts: 45 records, April 16 to July 17; 16 records, April 23 to May 8; 17 records, May 13 to June 10.

West Virginia: 38 records, April 3 to May 25; 24 records, April 3 to April 15.

SIALIA SIALIS FULVA Brewster

AZURE BLUEBIRD

HABITS

The history of this bluebird is very brief, for very little seems to be known about it, as it has rarely been seen north of the Mexican boundary.

Under the name of *Sialia sialis azurea* Baird, William Brewster (1885) writes:

Three Bluebirds obtained in the Santa Rita Mountains [Arizona] in June are doubtfully referable to this subspecies. One of the two males (No. 1855, F. S., June 18) has the blue above of that greenish shade said to be characteristic of *azurea*, but the other (No. 1856, F. S.), taken the same day, does not differ in this respect from *sialis*, the tint of the blue being precisely the same. Both are peculiar in having the under parts (excepting the usual dingy white space on the abdomen, crissum, and tail-coverts) nearly uniform pale brownish-orange, paler and yellower, in fact, than in the female of *sialis*, and with scarcely a tinge of the usual deep reddish-brown. This characteristic is not mentioned in descriptions of *azurea*, nor do I find it in any of the dozen or more Mexican and Guatemalan examples before me. The Santa Rita female (No. 1897, F. S., June 20), is still paler beneath, as well as browner above than the female of *sialis*. All these specimens differ further from *S. sialis* in having rather longer wings and tails, in this respect agreeing with *azurea*. In the event of their proving distinct from the latter, which seems probable, I propose for them the name *fulva*. Whether distinct or not, the bird is new to Arizona, no form of *Sialia sialis* having been previously reported from that Territory.

It would seem from the above description that these Arizona specimens are merely intermediates or hybrids (if we may use that term with subspecies) between these two races of *S. sialis*, as they show a mixture of the characters of both forms. Mr. Ridgway (1907) used the name *fulva* in the main text of his Bulletin 50, but he apparently changed his mind on it, for, in a footnote on page IX of the table of contents, he says: "This should be *Sialia sialis azurea* (Baird). (See Addenda, p. 887.)"

The 1931 Check-list says that the azure bluebird "breeds mainly in the Transition Zone from the mountains of southern Arizona south to Jalisco, Oaxaca, and Vera Cruz. Winters south to northern Guatemala."

Harry S. Swarth (1914) calls it "rare in summer in the high mountains of extreme southern Arizona." He cites the record of the Brewster specimens, mentions a specimen taken by Dr. A. K. Fisher at Fort Huachuca on April 30, 1892, and says: "The species is not of regular or of common occurrence in either of these mountain ranges, where *Sialia mexicana bairdi* is the common breeding bluebird; in fact the above records are the only ones known to me, though the region has been visited frequently by collectors."

Eggs.—There are not enough eggs of this subspecies available to make comparisons, but there seems to be no reason to expect them to vary to any extent from those of the eastern race. The measurements of seven eggs average 20.0 by 16.7 millimeters; the eggs showing the four extremes measure **20.4** by 16.6, 20.0 by **17.2, 19.7** by 16.8, and 20.0 by **15.9** millimeters.

SIALIA SIALIS EPISCOPUS Oberholser

TAMAULIPAS BLUEBIRD

Based on four specimens, sent to him by Dr. Louis B. Bishop, from northeastern Mexico, Dr. Harry C. Oberholser (1917) named the above subspecies in honor of Dr. Bishop and described it as "similar to *Sialia sialis fulva*, but blue of upper parts rather darker, and anterior lower parts very much darker." He gives a full description of the adult male, and says of its range: "State of Tamaulipas, Mexico, north to the Rio Grande Valley in central southern Texas."

We seem to have no information on its habits.

SIALIA SIALIS GRATA Bangs

FLORIDA BLUEBIRD

HABITS

Outram Bangs (1898) named this bird and gave it the following subspecific characters: "Size of *S. sialis sialis;* bill larger and stouter; tarsus and foot larger; color of upper parts clearer blue, less purple. In *Sialia sialis sialis* about smalt blue, and in *S. sialis grata* about French blue."

Arthur H. Howell (1932) says that this bluebird is resident "nearly throughout peninsular Florida, from about Lake County south to Royal Palm Hammock." He remarks further: "The Florida Bluebird lives chiefly in the open pine forests where there is an abundance of rotting stubs suitable for nesting sites. The birds are not at all shy, and their sweet, mellow whistles add charm to the desolate wastes that compose so much of central and southern Florida. The nests are in hollow stubs or fence posts, usually from 4 to 25 feet above the ground."

SIALIA MEXICANA BAIRDI Ridgway

CHESTNUT-BACKED BLUEBIRD

HABITS

This handsome race of the Mexican bluebird breeds farther east than but not so far north as the western bluebird, according to the 1931 Check-list, "mainly in the Transition Zone from Utah, Colorado, and central western Texas south to Durango and Zacatecas. Winters from southern Utah and southern Colorado south to Sonora and Zacatecas."

For a clear understanding of the characters separating the races of *Sialia mexicana*, as well as the individual variation within each of the subspecies, the reader is referred to an extensive paper on the subject by Robert Ridgway (1894). Later Ridgway (1907) described the chestnut-backed bluebird more concisely as—

similar to *S. m. occidentalis*, but adult male with whole back and scapulars uniform chestnut, producing a large and conspicuous dorsal patch; cinnamon-rufous of under parts more extended, always extending broadly across chest, sometimes covering whole breast; adult female with upper parts browner than in *S. m. occidentalis*, the back and scapulars hair brown to between sepia and prouts brown, usually in strong and abrupt contrast with the mouse gray or hair brown of pileum and hindneck; young much darker and browner than those of *S. m. occidentalis* or *S. m. anabelæ*, with under parts more heavily streaked or squamated and the streaked areas more or less strongly suffused with pale fulvous or rusty brownish. Decidedly larger than *S. m. occidentalis*, with smaller bill.

There is, of course, intergradation with the adjacent races and considerable individual variation in the distribution of the blue and chestnut areas.

Aiken and Warren (1914) write of its haunts and habits in El Paso County, Colo.:

While this species is common almost everywhere on migration, though probably never ranging quite as high as the next species, it breeds mainly in the yellow pine region between 7,000 and 8,000 feet, where it outnumbers the Mountain Bluebird. July 17, 1899, on the Divide north of Peyton, Aiken saw 20 Chestnut-backed to 5 of the Mountain Bluebirds, and it is probably more numerous on the Divide than anywhere else in the County. The two species are sometimes found in mixed flocks in the spring, especially when the weather is stormy. The appearance on the plains of this Bluebird during the spring migration is but for a short time, as it goes into the mountains and onto the Divide by the first of April, but the storms which usually come early in May drive the birds down in small flocks which remain until the weather clears and the snow melts. At these times the birds often become much emaciated and some die from starvation, being unable to obtain food while the snow is on the ground.

Harry S. Swarth (1904) says of its haunts in the Huachuca Mountains, Ariz.:

During February and the early part of March I found the Chestnut-backed Bluebirds quite numerous in the lower foothills, and on the plains immediately near the mountains, being entirely absent from the higher parts of the range, where the snow still lay deep on the ground; but about the middle of March they began to move upward, and by the first of April there were none to be seen except in the higher pine regions, their breeding grounds. Here they remained through the summer in the greatest abundance, none being seen below 8,000 feet, and being most numerous along the divide of the mountain. About the middle of August they began, to some extent, to move down to a lower altitude once more, for the evening of August 12th a small flock was seen flying overhead near the base of the mountains.

Russell K. Grater has sent me the following interesting note from Zion National Park, Utah: "This bird is a summer resident above 7,500 feet and is commonly seen in the open forest glades near meadows. Usually they range in groups of six or more before and after the nesting season. In winter they drift down from the highlands into the canyon bottoms but always prefer wide canyons with some open areas. Frequently they are found drifting through the piñon-pine–juniper forest, where they eat extensively of the fruit of the mistletoe. Some birds, when collected, had no other undigested food except the mistletoe. Nests have been found in hollow trees, usually about 7 to 15 feet from the ground. Eggs are laid in May and June and two broods are not infrequent. A nesting record of more than passing interest was recorded while I was living at Grand Canyon National Park. In May 1935 a hollow in an old piñon pine was utilized as a nesting site by both a chestnut-backed bluebird and a black-eared nuthatch. Both birds would occupy the hollow at the same time, sitting side by side. Each bird apparently fed only her own brood, and two families were raised by each set of parents during the summer. These birds were quite tame and were not at all disturbed when the hole into the hollow was enlarged sufficiently for a study of this unusual event. Careful records of the event were kept by more than one interested bird student at the park. This is the most unusual bird relationship thus far observed at Grand Canyon."

The chestnut-backed bluebird nests also in old woodpecker holes and in bird boxes, lays from four to six eggs, and in all other respects does not differ materially from the western bluebird, which is treated more fully on the following pages. Dr. Friedmann (1929) lists this bluebird as a very rare victim of the dwarf cowbird.

The measurements of 30 eggs average 21.5 by 15.9 millimeters; the eggs showing the four extremes measure **23.6** by 16.5, 23.4 by **16.8,** **19.7** by 15.1, and 20.0 by **15.0** millimeters.

DISTRIBUTION

Range.—Western North America from southwestern Canada to central Mexico.

Breeding range.—The bluebird of the *mexicana* group breeds **north** to southern British Columbia (Beaver Creek, Vancouver Island, Alta Lake, 150 Mile House, Edgewood, and Newgate). **East** to southeastern British Columbia (Newgate); western Montana (Fortine, Columbia Falls, Lolo, and Laurel); western Wyoming (Yellowstone Park and Pinedale); central Colorado (Estes Park, Golden, Colorado Springs, Wet Mountains, and Fort Garland); central New Mexico (Truchas

Peak, Ribera, Capitan Mountains, and Cloudcroft); western Texas (Guadalupe Mountains and Davis Mountains); southern Coahuila (Carneros); western Tamaulipas (Miquihuana); western Veracruz (Las Vigas and Cofre de Perote); and eastern Puebla (Tochimilco). **South** to Puebla (Tochimilco); Morelos (Huitzilac); and Michoacán (Mount Tancitaro). **West** to western Michoacán (Mount Tancitaro and Mount Patamban); southern Durango (El Salto); western Chihuahua (Colonia García); northern Sonora (San Luis Mountains); central northern Lower California (Sierra San Pedro Mártir, El Rayo, and 20 miles east of Ensenada); the Coast Range and sometimes the coast of California (San Diego, Los Angeles, Santa Barbara, Monterey, Nicasio, and Humboldt Bay); Oregon (Gold Beach, Elkton, Newport, and Olney); Washington (Aberdeen, the Olympic Mountains, and Mount Vernon); and southwestern British Columbia (Vancouver Island; Alberni and Beaver Creek).

Winter range.—Through much of its range the species is resident or only retires to lower altitudes in winter. It winters **north** to southern British Columbia (Comox, Chilliwack, and the Okanagan Valley); to the valleys of the Cascades in Washington (Bellingham and Dungeness Spit, occasional at Spokane); Oregon (Portland, Salem, and Corvallis); eastern California (Lassen Peak region); western and southern Nevada (Carson City and Searchlight); Arizona (south rim of Grand Canyon, Salt River, and Tucson); central New Mexico (Socorro, Albuquerque, and Las Vegas); and southwestern Texas (Frijoles, Marathon, and Fort Clark). In migration, possibly occasionally in winter, it has been found as far east as Kerrville, Tex. There is one record of occurrence in winter near Polson, Mont., near the south end of Flathead Lake.

The ranges as outlined are for the species as a whole, which has been divided into several subspecies or geographic races. The chestnut-backed bluebird (*S. m. bairdi*) breeds from Utah and Colorado south to Durango and Zacatecas; the western bluebird (*S. m. occidentalis*) breeds from southern British Columbia and western Montana south to southern California; the San Pedro bluebird (*S. m. anabelae*) breeds in the Sierra San Pedro Mártir and the Sierra Juárez of northern Lower California.

Migration.—Some early dates of spring arrival are: Colorado—Colorado Springs, March 4. Wyoming—Yellowstone Park, March 13. Montana—Fortine, March 9. Utah—Salt Lake City, March 6. Idaho—Rathdrum, March 3.

Some late dates of fall departure are: Idaho—Meridian, October 27. Utah—Kanab, October 21. Montana—Fortine, October 28. Wyoming—Yellowstone Park, September 8. Colorado—Beulah, October 24.

Egg dates.—California: 104 records, April 4 to June 29; 58 records, May 2 to May 31, indicating the height of the season.

Colorado: 9 records, May 6 to June 16.

Oregon: 29 records, April 18 to June 29; 15 records, April 18 to May 21.

Lower California: 6 records, May 15 to June 12.

SIALIA MEXICANA OCCIDENTALIS Townsend

WESTERN BLUEBIRD

HABITS

The 1931 Check-list states that the western bluebird breeds from southern British Columbia to southern California, but it may breed farther north. Theed Pearse, of Courtenay, tells me that the large numbers that pass through that portion of Vancouver Island indicate that the breeding range extends considerably north of that point. The range extends eastward to northern Idaho and western Montana.

Samuel F. Rathbun records it in his notes as a common species about Seattle, Wash., from early in spring to late in fall, and of frequent occurrence during the winter. It is found in logged-off sections, along highways, and about isolated farms and clearings, where there are a few tall dead trees.

In the Lassen Peak region of California, according to Grinnell, Dixon, and Linsdale (1930), "in summer western bluebirds were to be observed about clearings and in those places where the cover of trees was sparse. Individuals and pairs when not in flight were most often seen perched in the tops of dead or dead-topped trees."

Grinnell and Storer (1924) state that, in the Yosemite region, "in the spring and summer months the local Western Bluebird population is confined almost entirely to the blue oak belt of the western foothills and hence within the Upper Sonoran Zone. * * * In the fall months, however, Western Bluebirds appear at many *up*-mountain localities not previously tenanted by the species."

Howard L. Cogswell writes to me: "On May 6, 1936, I saw one pair accompanied by an immature on the oak-bordered Flintridge Golf Course, near Devil's Gate Dam, Pasadena; and on June 28, 1940, I saw a pair feeding half-grown young in the San Gabriel River Sanctuary, south of El Monte, in a typical Lower Sonoran riparian association, at an altitude of only 300 feet above sea level."

Courtship.—The only information I can find on the courtship of the western bluebird is contained in a short note from Theed Pearse. He watched a pair mating; the male, that had been sitting beside the female, mounted her; and then the female mounted the male, which then flew away. A similar performance was seen a few days later; the male seemed to flatten himself out for the female to mount.

Nesting.—The nesting habits of the western bluebird do not differ materially from those of its eastern cousin. The only two nests that I have seen were in nowise out of the ordinary; one, found on May 14, 1911, near Tacoma, Wash., was in a cavity 10 feet from the ground in a dead oak stub; it contained four eggs; the other, in Ventura County, Calif., on April 7, 1929, was in an old hole of a California woodpecker on the under side of a limb of a sycamore tree. The nests were very simply and carelessly built structures of dry grass and a few feathers.

Referring to the Yosemite region, Grinnell and Storer (1924) write:

Old woodpecker holes are occupied when available, but failing to find one of these the birds will use some naturally formed opening in a tree. The decay of stubs of medium-sized branches often results in the formation of cavities in the heart wood of an oak which are appropriate in form and size for use by the bluebirds. * * * The nest found near Lagrange was in a blue oak on a hill top. It was in a naturally rotted-out cavity at a height of 9 feet from the ground. Distant but 17 inches in the same stub was the nest of a Plain Titmouse. The bluebird's nest was 6½ inches below the rim of the opening and the sparse lining upon which the 4 eggs lay consisted chiefly of dry foxtail grass. Another nest seen at Smith Creek, east of Coulterville, was 14 feet above the ground in a black oak. A natural cavity about 11 inches deep by 5 inches in diameter had been filled for a depth of 4 to 5 inches with soft materials. Entrance was afforded to the nest on two sides; on the one side was a hole about 2½ inches in diameter, while there was a much larger opening on the other side, so that the nest was easily visible from without.

In the Point Lobos Reserve, Monterey County, according to Grinnell and Linsdale (1936), "all the nests were in cavities in pines or pine stumps at heights ranging from five to forty feet, averaging twenty-two feet. On the few occasions when material was seen being carried to the nest, the female was doing the work. Usually, however, the male was present and showed an interest in the procedure. Once a male at a nest spent more time there than did the female, going in and out and moving the materials. Several times a male was seen to feed his mate."

Western bluebirds are as easily encouraged to nest near our homes in bird boxes as are their eastern relatives; here they often meet serious competition with violet-green swallows or western house wrens, but the bluebirds are generally the masters of the situation; they can defend their homes against such intruders and have been known to drive out the swallows from their occupied nest. Where natural cavities, woodpecker holes, or bird boxes are not available the bluebirds will build their nests in any suitable cavity in a building, or even in a cliff swallow's nest. They are very persistent in their attempts to raise a family; if a set of eggs is taken the birds will lay a second set of eggs within a very short time and, if necessary, a third set.

Eggs.—The western bluebird has been known to lay three to eight eggs to a set; probably sets of three are incomplete; sets of four, five, and six seem to be almost equally common; I have heard of only one set of eight, in the collection of Sidney B. Peyton. The eggs are pale blue and practically indistinguishable from those of the eastern bluebird in every way. The measurements of 50 eggs in the United States National Museum average 20.8 by 16.3 millimeters; the eggs showing the four extremes measure **23.3** by 16.3, 20.1 by **17.5, 19.1** by 16.8, and 20.3 by **15.2** millimeters.

Young.—The period of incubation does not seem to have been definitely determined, nor does it seem to be known how long the young remain in the nest; probably both of these periods are similar to those of the eastern bluebird. Incubation of the eggs and brooding of the young seems to be done entirely by the female. Harriet Williams Myers (1912) watched a nest near Los Angeles for 1 hour and 35 minutes. "During this time the female left four times, staying away five minutes once and eight the other times. Her times for brooding were respectively twenty-two, eighteen, ten, and twenty-four minutes. Almost invariably during this and subsequent watchings the female did not leave the nest until the male came to it." Four days later, during an hour and a half, "the female left the nest four times as before. The longest interval of staying away was twenty-seven minutes; the shortest two minutes. The longest interval of brooding was sixteen minutes; the shortest thirteen." Eight days later when there were young in the nest, she watched it for an hour. "During the hour fifteen trips were made to the nest, the feeding being very equally divided. In fact, with two or three exceptions, the birds were both at the nest at once each of the fifteen times."

James Murdock has sent me a photograph of a western bluebird at its nest in a hole in a yellow-pine log. While he was photographing he noticed that two males and one female were feeding the young; he and his companions saw the two males enter the nest. He writes to me: "One at a time they flew directly in front of us and entered the nest, one coming out while the other went in. Both had food in their bills."

Kenneth Racey (1939) writes: "The bluebirds are most amusing in the way they keep house and care for the young and they go through the same performance each year. The first brood is brought off and then within a few days the old birds are busy laying and brooding again. When the second brood fledges, both families, usually eight young and the two adults, join in one flock and remain in the neighbourhood, visiting the garden every few days until it is time to leave for a warmer climate. The young of the first family are fed and cared for while the second clutch of eggs is being incubated."

Plumages.—The sequence of plumages and molts evidently parallels that of the eastern bluebird. The adults of the two species are more unlike than are the juvenals. In the western bluebird the sexes are distinguishable in the juvenal plumage. In the young male the head and neck are plain sooty gray, the interscapular region is from "Verona brown" to "olive-brown," conspicuously streaked with white, and the rump, upper tail coverts, and the lesser and median wing coverts are dull slate color; the breast and sides are "warm sepia," heavily streaked or spotted with white; the wings and tail are much like those of the adult female, but the blue is brighter and the tertials are margined with pale grayish brown. The young female is similar to the young male, but all the colors are paler and duller, and the breast is more heavily streaked with white. Ridgway's (1907) account indicates that only the young female has the interscapular region streaked with white, but all the young birds in the considerable series of both sexes that I have examined are so streaked, and Dawson's (1923) account agrees with my findings.

The postjuvenal molt begins early in July in some birds and not until the middle of August in others, this probably depending on the date of hatching. This produces the first winter plumage, in which young birds become very much like the winter adults of the respective sexes. Ridgway (1907) says of the young male at this season: "Similar to the adult male in winter plumage, but the blue lighter and less violaceous and (except on rump, upper tail-coverts, rectrices, and remiges) duller, the feathers of pileum and dorsal region more broadly tipped with grayish brown; chestnut of under parts rather lighter, the feathers with paler tips." The first winter female differs from the adult winter female in about the same way.

There is apparently no spring molt, but the grayish-brown edgings have worn away, giving a brighter appearance; first-year birds can then be distinguished from adults by the juvenal wings and tails, which have been retained through the winter.

One-year-old young birds and adults have a complete postnuptial molt in August and September, at which old and young become indistinguishable. Ridgway (1907) says that, after this molt, the blue of the upper parts of the male is "slightly obscured by narrow brownish tips to the feathers, and that of the chest and breast by pale grayish brown tips." These tips wear away during winter, producing the full spring brilliancy. Of the adult winter female, he says: "Similar to the spring and summer plumage, but brighter in color, the pileum and dorsal region decidedly bluish, and ruddy brown of under parts more chestnut."

Food.—Professor Beal (1915a), in his food studies, treats the western bluebird, the chestnut-backed bluebird, and the San Pedro bluebird all together, as subspecies of the Mexican bluebird. In the 217 stomachs examined, the food was found to consist of 81.94 percent animal and 18.06 percent vegetable matter. Among the animal food, grasshoppers formed the largest item, with an annual average of 21.29 percent. Caterpillars were a close second, averaging 20.25 percent. Useful beetles, mostly Carabidae with a few ladybirds, amounted to 8.56 percent; and other beetles, all more or less harmful, accounted for 15.44 percent of the food. Ants constituted 5.38 percent, and other Hymenoptera only 1.26 percent; no honey bees were found. Flies and a few other insects, spiders, myriapods, angleworms, snails, and sowbugs were eaten in very small quantities.

The vegetable food was made up mainly of fruit, mostly all wild species or waste cultivated fruits picked up late in the season. "Rubus fruits (blackberries or raspberries) were found in 4 stomachs, prunes in 1, cherries in 1, and figs in 3." Elderberries and mistletoe berries proved to be favorite foods; weed seeds were eaten sparingly, no grain of any kind was found, and a few other items, such as seeds of poison-oak and other *Rhus* seeds, made up the balance. Of the food of the nestlings, he says: "The real food consists of grasshoppers and crickets, 90 per cent, and beetles, 3 per cent, the remainder being made up of bugs, caterpillars, and spiders. * * * The remains of 11 grasshoppers were found in one stomach and 10 grasshoppers, a cricket, and a beetle in another."

Theed Pearse tells me that he has seen western bluebirds feeding on the berries of the Virginia-creeper, taking them from the vines; he has also seen them "hawking" insects in the air with a very pretty butterfly flight. They often dart out into the air from some high perch and catch the insects in flight, but more often they watch from some low perch and flutter down to catch their prey on the ground, or hover along over the tops of the herbage to catch the flying insects that they have disturbed. Frank A. Pitelka (1941) saw some of these birds soaring in a strong wind while feeding; he writes:

> The bluebirds would fly to a position in the up-draft some 6 or 8 feet above the ground, there hover for a second or two, and then soar for a few seconds. On a number of occasions, one or two of them remained in a soaring position without movement of wings for 6 to 8 seconds. The birds were foraging for insects, which they caught by dropping quickly from their position in the air. It appeared that the wind was blowing insects upward over the hill slightly above the grass. The bluebirds, hovering or soaring and looking down, watched for them; when prey was sighted, the bird turned about face and flew back to catch up with it. Such a behavior was observed on both the open slopes and

in the draws. As many as four birds were noted hovering and soaring at one time in a few yards of area.

Dr. Grinnell (1904) describes an interesting method of feeding, as follows: "In Palm Canyon great numbers were in evidence among the giant palms. A dozen or more would be seen clinging to each pendant cluster of dates obviously attracted by the fruity outside pulp. While thus feeding upon the fruit of the palms, the noise made by the seeds dropping into the dry brush at the bases of the lofty trees was so great as to give the impression, before the true cause was discovered, that some large animal was trampling through the undergrowth."

Grinnell and Storer (1924) give an account of how these bluebirds help to spread the seeds of the mistletoe. Only the soft pulp of the berry is digested, and the seed, which is coated with a film of mucilaginous material, passes through the bird's alimentary canal in condition to stick where it falls. Should the seed happen to fall on the right kind of host tree, it would, under proper conditions, germinate and start a new plant of this objectionable parasite. The birds swallow so many of these seeds that probably some new plants are started each year. They write of the feeding process: "The birds individually will seek perches about clumps of mistletoe, either on adjacent parts of the tree or on the twigs of the parasite itself. Berries will be picked off and swallowed in rapid succession. Each bird, as it gets its fill of berries, flies to some nearby perch and sits there quietly. The process of digestion is a rapid one, and before many minutes have elapsed enough of the berries will have gone from the bluebird's gullet into its stomach to permit of further feeding. Thus the day is spent, alternately in feeding and digesting."

Behavior.—The same authors have this to say on this subject:

In general demeanor the Western Bluebird is much like other members of the thrush family, being of deliberate or even phlegmatic temperament. When perched it sits quietly, not hopping about as do many small birds such as sparrows and warblers. It ordinarily seeks a perch which will command a wide field of view, as on some upper or outer branch of a deciduous tree. * * * Upon taking to flight bluebirds make off in the open, high in the air, uttering their soft call notes now and then as they fly. The high course of flight and the repeated flight calls are suggestive of the behavior of linnets under similar circumstances. Sometimes the flight is so far above the earth that the birds are quite beyond the range of vision of an observer stationed on the ground, only the mellow call notes giving indication of the passage of the birds overhead. When bluebirds are in flocks the formation is never compact or coherent; individuals move here and there among their companions and single birds or groups join and depart at intervals.

Voice.—Several observers have condemned the song of the western bluebird with faint praise. Dawson (1923) says: "The Eastern Bluebird warbles delightfully; therefore, the Western Bluebird *ought* to—but it doesn't. In an experience of some thirty-nine years, the author has never heard from the Western Bluebird's beak an utterance which deserves the name of song, or anything more musical than the threefold *miu.*" Ralph Hoffmann (1927) writes: "In the breeding season the male utters a low *chu, chu, chu,* apparently his only song, but before dawn a camper among the pines hears a chorus of the rich call notes repeated from all sides as if from birds flying about in the darkness." And Grinnell and Storer (1924) say that the "song is a very simple affair, just the common call notes uttered over and over again with monotonous persistence."

Winton Weydemeyer (1934), however, gives this bluebird credit for more musical ability, and contributes the following account of the singing of a bird near his house at Fortine, Mont.:

The first attempt at singing was noted at 4:40 A. M. (in full darkness) on April 19. For several minutes without pause one of the birds from a perch rendered an endless song consisting of the common call note, *few,* repeated over and over, regularly but with varying inflection. On succeeding mornings the notes gradually became more varied. The following description was jotted down on the morning of April 26: "Bluebird from perch began singing at 4:35 (quite dark), sang for about 40 minutes. Sang without pause for about fifteen minutes first; later snatches of song successively shorter, intervening pauses longer. Song a succession of call notes (3 different phrases); notes same as given separately in daytime, but connected in a series to form a typical 'song.' Song louder and more energetic than that of the Mountain Bluebird, just as the call notes are louder and more vigorous. Tempo much like that of Robin's song. *F-féw, f-féw, f-féw f-féw, eh-eh, féw, f-féw, eh-eh, féw, eh-eh, féw, f-féw* . . . The *eh-eh* is a common phrase given with the call note *few* (or *tew*) during the day. It resembles the short catch notes of Ruby-crowned Kinglet and Cassin Vireo."

Three days later, on April 29, I awoke in the darkness at 4:20 a. m. to find a bluebird already singing. I wrote down his song thus: "*Ic-ic té, téw, ic-ic, téw, ic-ic towée, towée* (often *two-lée, two-lée*—more musical), *ic-ic, téw, ic-ic twoée, towée* . . ." These songs, with minor variations, were given throughout the season. * * *

The singing of these birds resembled the usual song of the Western Robin even more closely than does the song of the Mountain Bluebird as observed in this locality. In the darkness I often found it difficult to tell whether a song was given by a Western Bluebird a few hundred feet away or by a Western Robin at a greater distance. To me the Western Bluebird's singing, from a musical standpoint, is less enjoyable than that of its quieter relative, the song of the Mountain Bluebird being softer, more subdued, and more pleasingly modulated.

During the early part of the season, in April, while the Western Bluebirds were pairing and selecting houses, the males during the day frequently gave a double note that was not heard later in the season. This was a musical *pa-wée,* much resembling a goldfinch's call. This was also coupled with the common call note to form a series of phrases which perhaps constituted a "mating song"; *Pa-wée, few few. Few few fa-wée. Fa-wée. Few few fa-wée. Pa-wée. Pa-wée, few, few* . . . Another phrase sometimes given at this season I noted as *etherick tóe,* the first double note resembling a common phrase of the Western Robin's song.

Field marks.—There are several partially or largely blue birds on the Pacific slope, but not one is so intensely blue as the male western bluebird; the rich blue of his head, wings, and tail and the deep chestnut on his breast are distinctive marks. The jays are all much larger, the blue grosbeak does not have the solidly chestnut breast and its stout bill is quite noticeable, and the lazuli bunting is a much lighter blue and has white wing bars. The colors of the female are similar to but much paler and duller than those of the male. The mountain bluebird, in its pale soft colors, is not likely to be confused with the western, as there is no chestnut in the plumage of either sex.

Fall.—Theed Pearse writes to me that, on Vancouver Island, some birds, possibly local breeding individuals, move southward very early, during the last of June or the first two weeks in July. Other migrating parties pass through in August and as late as October 20, consisting of bluebirds, Audubon's warblers, robins, and cedar waxwings. Mr. Rathbun tells me that the western bluebird is a very common migrant during the latter part of October in western Washington. Grinnell and Storer (1924) write:

The manner of association during the season of molt has not been observed, but by September flocks have been formed which include both adult and immature birds, and in this fashion they spend the winter. The flocks, in observed instances, included from 6 to 25 members. Sometimes other birds are associated. In Yosemite Valley we saw Western Bluebirds in company with Audubon Warblers on one or more occasions, and Mr. C. W. Michael (MS) reports Western and Mountain Bluebirds together there during November of 1920. Western Bluebirds and Robins are frequently seen together during the winter months though the two do not flock with each other in the usual sense of the word."

They say that there is an up-mountain movement of the bluebirds in fall in the Yosemite region and that they are seen at high altitudes all through the late fall and early winter. "The attraction for these birds at these higher altitudes is the abundant supply of food in the form of mistletoe berries. This food supply, rather than weather, short of extremely severe storms, seems to be the factor regulating the stay of the bluebirds in the mountains. That snow alone is no particular deterrent to the birds' stay is shown by our observations made on the stormy morning of December 10, 1914, at Mirror Lake, when bluebirds were flying about actively, now and again alighting on the snow-weighted mistletoe clumps. Masses of the snow would be dislodged and shower the observer beneath, but the birds themselves seemed in nowise discommoded."

Mr. Cogswell writes to me: "This species is varyingly numerous in lowland areas from late in summer to early in spring but is usually abundant in the foothill areas around Pasadena and in the more open mountain canyons below snow level. Several were seen in Bear

Valley, San Bernardino Mountains, on December 28, 1941, with a foot of snow on the ground. Together with their frequent companions, Audubon's warblers and house finches, they rove over the foothill mesas and into the outlying edges of town in small flocks (5 to 15), feeding on insects chiefly, but also at least occasionally on grapes and other berries."

Mr. Rathbun tells me that in mild winters some of these bluebirds remain about his grounds off and on all winter, investigating his bird boxes, and begin their nest building early in spring.

SIALIA MEXICANA ANABELAE Anthony

SAN PEDRO BLUEBIRD

HABITS

It seems to be generally accepted now that this subspecies is confined in the breeding season to the Sierra San Pedro Mártir and the Sierra Juárez in northern Lower California, though Ridgway (1907) mentions a number of bluebirds from upper California, as far north as Mount Lassen, that show more or less the characters of this form. For a critical study of the races of the Mexican bluebird, the reader is referred to an extensive paper on the subject by Ridgway (1894), in which some doubt was cast at first on the validity of this subspecies.

A. W. Anthony (1889) described this bird and named it for his wife, Anabel Anthony. He gave its subspecific characters as "differing from *S. mexicana* in slightly larger form, in the bay of the breast, which is divided by the blue of the throat, restricting it to patches on the sides of the breast, and in the almost entire absence of bay on the scapulæ." The female "differs from the females of *S. mexicana* in my collection in the more pronounced blue of the head and larger size." He includes in its range Mount Lassen, Calif., Puget Sound, Utah, and Nevada; this extension of range is probably based on certain specimens from these localities that show some intermediate characters, such as those mentioned below by Ridgway.

Mr. Ridgway (1907) described the San Pedro bluebirds as "similar to *S. m. mexicana*, but with bill larger and stouter; blue of upper parts averaging less violaceous (more ultramarine), back more often mixed with chestnut laterally; adult female with back and scapulars grayish brown, forming a definite dorsal patch, distinctly defined against the brownish gray or dull grayish blue of pileum and hindneck, and, with the grayish brown or brownish gray of throat, breast, etc., paler than in *S. m. mexicana*." In a footnote he shows the variation in the color pattern in 43 adult males from the San Pedro Mártir Mountains; 21 have no chestnut whatever on back or scapulars; 18 have the back chiefly blue; 4 have the back about equally blue and chestnut; 30 have the chestnut of the breast divided by the blue of the throat, a character of *S. m. mexicana;* 11 have the chestnut of the breast continuous anteriorly; and 2 do not belong to either category. This analysis shows that the characters of *anabelae* are not absolutely constant, even in its restricted range. In another footnote he says: "California specimens are not typical of this form, but are much nearer to it than to *S. m. occidentalis*, from which they differ in larger size, more restricted areas of chestnut (though this character varies greatly in both forms), and, on the average, decidedly richer or more violaceous hue of the blue."

Since the Sierra San Pedro Mártir has become the birthplace of so many local subspecies, it seems worth while to include here Mr. Anthony's (1889) description of the region.

About one hundred and fifty miles south of the United States boundary, and midway between the Pacific Ocean and the Gulf of California, lies a high range of mountains, which is marked upon the later maps of the peninsula as "San Pedro Martir." The region embraces a series of small ranges which rise from an elevated *mesa*, having a mean elevation of about 8,000 feet, and an extent of sixty by twenty miles. In these mountains are born the only streams that this part of the peninsula affords, and an abundance of pine timber is found throughout the region. Many of the ranges on the eastern side of the San Pedro Martir rise to an elevation of 11,000 feet, or even, in one or two places, to 12,500 (?) feet.

Arising as the region does from the dry, barren hills of the lower country to an elevation higher than any other on the peninsula or in Southern California, and presenting in its alpine vegetation and clear mountain streams features so different from the dry manzanita and sage-covered hills of the surrounding country, it is not unnatural to suppose that its animal life would be found to differ in some respects from that of the surrounding hills.

J. Stuart Rowley writes to me: "Near La Grulla in the Sierra San Pedro Mártir, of northern Lower California, I took a set of five slightly incubated eggs of this race of bluebird on June 12, 1933. So far as I can determine, the habits of this bluebird are no different from those of the western bluebird of the north." Charles E. Doe, who now has this set in his collection at the University of Florida, in Gainesville, tells me that the nest was in a pine stub 8 feet from the ground.

Eggs.—The eggs of the San Pedro bluebird are probably indistinguishable from those of the species elsewhere. The measurements of 26 eggs average 20.9 by 16.2 millimeters; the eggs showing the four extremes measure **23.1** by 16.5, 20.2 by **16.9**, **19.4** by 16.2, and 19.6 by **15.2** millimeters.

SIALIA CURRUCOIDES (Bechstein)

MOUNTAIN BLUEBIRD

HABITS

The mountain bluebird is not so gaudily or so richly colored as the western bluebird, but it is no less pleasing in its coat of exquisite turquoise-blue. As it flies from some low perch to hover like a big blue butterfly over an open field, it seems to carry on its wings the heavenly blue of the clearest sky, and one stands entranced with the purity of its beauty. As Mrs. Wheelock (1904) says: "No words can describe his brilliancy in the breeding season, as he flies through the sunny clearings of the higher Sierra Nevada, or sits like a bright blue flower against the dark green of the pines." The male certainly is a lovely bird, and the female is hardly less charming in her coat of soft, blended colors.

It occupies a wide breeding range throughout the western half of Canada and the United States, west of the Great Plains. It was formerly called the Arctic bluebird, a decided misnomer, for it is in no sense an Arctic bird, being found in the northern part of its range only in summer. Another old name, Rocky Mountain bluebird, was more appropriate, for it is one of the characteristic birds of the western mountains. Using the latter designation, H. W. Henshaw (1875), who recorded it as very common in Utah and Colorado, and in northern New Mexico and Arizona, says: "I have usually found it during the breeding season in the wild, elevated districts, from 7,000 feet upward, where it frequents the more open spaces, where aspen groves alternate with the remains of pine woods, the broken stubs of which, charred by fires which have swept through again and again, are seen on every side. * * * In the neighborhood of Santa Fé, they breed commonly, and here were noticed in the vicinity of houses, seeming in fact to be as familiar and as much at home as does our own bluebird in the East."

Robert Ridgway (1877) writes of it in the same general region: "Its favorite haunts are the higher portions of the desert ranges of the Great Basin, where there is little water, and no timber other than the usual scant groves of stunted cedars, piñon, or mountain mahogany."

Mrs. Bailey (1928) says that, in the Sangre de Cristo Mountains in New Mexico, "we found a nest in a grove of aspens on the edge of the open grassy mesa at 10,300 feet, we found families of old and young going about together at 11,000 feet. * * * On August 11, we were much pleased to find a flock of the Bluebirds, together with Red-shafted Flickers and Chipping Sparrows, at 12,300 feet, on a protected slope in the dwarf timberline trees on the south side of Truchas."

In the Yosemite region, according to Grinnell and Storer (1924), it is chiefly a bird of the Hudsonian Zone, the greater part of the population being found at altitudes above 8,000 feet, "and from there it ranges up to the highest meadows found in our maintains short of timber line." The highest points at which these bluebirds were seen were at about 10,500 feet, and the lowest were seen at 7,400 feet at Mono Meadow in June.

Russell K. Grater writes to me from Zion National Park, Utah, that "this species is resident during the summer months above 7,500 feet. During the winter months, these birds wander to lower elevations, but seldom enter the deep canyons, apparently preferring more open valley country."

Aretas A. Saunders (1921b) calls it a common summer resident throughout Montana and says that it "breeds in the Transition zone and less commonly in the Canadian. In the eastern part of the state breeds in the pine hills, farther west, in cottonwood groves, about ranch buildings, and in the more open types of coniferous forests in the foothills of the mountains. In the Canadian zone, it is sometimes found about the edges of mountain parks, but it is never as common at such elevations in Montana as it is in the Transition."

Spring.—At Chelan, Wash., during what they called a typical season, Dawson and Bowles (1909) noted that "the migrations opened with the appearance, on the 24th day of February, of seven males of most perfect beauty." These observers continue:

They deployed upon the townsite in search of insects, and uttered plaintive notes of Sialian quality, varied by dainty, thrush-like *tsooks* of alarm when too closely pressed. * * * On the 15th of March a flock of fifty Bluebirds, all males, were sighted flying in close order over the mountain-side, a vision of loveliness which was enhanced by the presence of a dozen or more Westerns. Several flocks were observed at this season in which the two species mingled freely. On the 27th of the same month the last great wave of migration was noted, and some two hundred birds, all 'Arctics' now, and at least a third of them females, quartered themselves upon us for a day,—with what delighted appreciation upon our part may best be imagined. The males are practically *all* azure; but the females have a much more modest garb of reddish gray, or stone-olive, which flashes into blue on wings and tail, only as the bird flits from post to post.

Norman Criddle (1927), of Treesbank, Manitoba, writes:

The male bluebirds always arrive a few days in advance of the females, but it is not long before the latter appear upon the scene and in an astonishingly short time pairs have taken possession of a nesting site and the females are taking nesting material into boxes. This haste in constructing a nest is difficult to appreciate because the birds do not, as a rule, actually start domestic duties for some time afterwards. * * *

The male bird is an extreme optimist and nearly any hole meets with his approval, but his mate is not so easily satisfied and many of his selections are discarded as worthless. It is interesting to watch this home seeking, to see the male put his head into a hole followed by the female; should she enter it, he flutters his wings in the height of enthusiasm, but should she turn away unsatisfied, as she does nine times out of ten, then he appears dejected for a few moments, but speedily recovering, endeavors to entice her into other holes the whereabouts of which he appears to have discovered beforehand.

* * * Old pairs probably remain united providing the male is able to overcome his rivals in battle but not, I suspect, otherwise. The males have been observed to fight vigorously and these combats have continued intermittently for weeks before one bird finally admitted defeat. The female is always a witness to these encounters, in fact she often follows the fighters from place to place, but I have not been able to discover that she takes any part in them and she apparently accepts the victor as a matter of course.

Nesting.—All the bluebirds appear to have very similar nesting habits, and the mountain bluebird is no exception to the rule; almost any cavity and almost any location seem to suit them. A. D. Henderson tells me that as far north as Belvedere, Alberta, the mountain bluebird is a common breeder, nesting in flicker holes in the woods, as well as in bird boxes around the buildings. A. Dawes DuBois (MS.) reports two Montana nests; one was in a hole in a burnt stub, 20 feet from the ground; apparently the other was in a hole in a bank of the Teton River, in the prairie region; he saw a pair investigating several holes in the bank and they were "especially interested in one hole, which looked like a kingfisher's nest tunnel; both of them went into it."

Mr. Grater tells me that in Zion National Park, Utah, "nests have been located in old trees or in old woodpecker holes. These nests are usually only a few feet from the ground." Dr. Jean M. Linsdale (1938) reports two very low Nevada nests; one was "in a hole close to the top of a piñon stump 3 feet high," and the other was "4 feet above the ground on the east side of an aspen trunk." Another was found "beneath the roof at the corner of a house. * * * The nest was composed of grasses and lined with a few chicken feathers; it held 6 eggs. Both adults were present, and they flew about excitedly or perched on a telephone pole 25 feet away." Still another "nest was in a cavity made by flickers in the side of a house."

Referring to the Yosemite region, Grinnell and Storer (1924) write:

At Mono Lake Post Office a pair of Mountain Bluebirds had appropriated to their uses a ledge in a woodshed, entrance to which was gained through a hole in the wall. Here at the height of 10 feet from the ground a loosely woven nest had been constructed. This nest was made of shreds of bark many of which showed evidence of having been freshly pulled from the trees for the purpose. There were included also numerous chicken feathers from the nearby farmyard. The dimensions outside were roughly 6 or 7 inches in diameter and 2½ inches in height. The depression for receiving the eggs was 3½ inches wide and 1½ inches deep. After one brood had been reared this nest was re-lined to receive a second set of eggs.

In addition to the more normal nesting sites in all kinds of cavities in various kinds of trees, in holes in banks, in old woodpecker holes, and in the more recently adopted bird boxes, mountain bluebirds' nests have been found in crevices in cliffs and among rocks, in old nests, and in almost any available cavity about human habitations and ranches. Miss Catherine A. Hurlbutt says in her notes: "I once observed these birds carrying nesting material into a cliff swallow's nest under the eaves of a barn, evidently appropriated from the rightful owners, as there were swallows in all the surrounding nests; also another in a chipmunk hole in the bank of a road cut about a couple of feet above the surface of the road. I am not sure that the young were successfully raised in either nest."

According to Mrs. Wheelock (1904) both male and female cooperate in building the nest. The materials used are quite varied; almost any available material seems to satisfy them. R. C. Tate (1926) lists the following material in Oklahoma nests: "Stems of wild oats, rosin weeds, goldenrod, sticktights and milkweeds, and rootlets of prickly pears, stinking sumac, scrub oak and ticke-grass." And the Macouns (1909) say that a nest found in a clay butte at Medicine Lodge, Saskatchewan, "was wholly composed of the outer bark of the old stems of *Bigelovia graveolens*, a composite plant that grew in profusion near the site of the nest. It contained seven light blue eggs. Another nest taken under the same conditions along Frenchman river, Sask., on June 21st, was built of the outer bark of sage brush (*Artemisia cana*)."

Eggs.—The mountain bluebird lays four to eight eggs to a set. Five and six are the commonest numbers, and sets of seven are not extremely rare. They are usually ovate and are somewhat glossy. The color is pale blue or bluish white, averaging paler than those of other bluebirds; very rarely they are pure white, much less often than those of the eastern bluebird. They are apparently always unmarked. The measurements of 50 eggs in the United States National Museum average 21.9 by 16.6 millimeters; the eggs showing the four extremes measure **24.9** by 16.8, 21.8 by **17.8**, and **19.8** by **15.2** millimeters.

Young.—Mrs. Wheelock (1904) writes:

Fourteen days are required for incubation, and in this the male often, but not always, shares. When not on the nest himself he brings food to his mate, calling to her in sweetest tones from the outside before entering the doorway. The newly hatched young are of the usual naked pinkish gray type, looking as like tiny newborn mice as birds. On the second day down begins to appear in thin hairs on head and back; on the fourth or fifth day the eyes show signs of opening; on the sixth day they open, and the down is well spread over the bodies.

Up to this time they have been fed by regurgitation, the adult swallowing each bit first to moisten or crush it; but from the fourth day on fresh food is given occasionally, and from the sixth or seventh day all the food given is in the fresh state, not regurgitated. Crickets, grasshoppers, beetles, butterflies, and worms are their menu, with a few berries. The young Bluebirds double in weight every twenty-four hours for the first week, and in twelve days are growing a respectable crop of feathers, though the bare skins is still distressingly visible. Their breasts gradually take on the soft, mottled light and dark, and the upper parts have a hint of blue among the grayish brown on the wings and tail. One would suppose that this blue on the upper parts would be too conspicuous, but when the youngsters leave the nest and perch on the soft gray of the dead trees, they become almost invisible in the strong sunlight.

On Mount Rainier, on July 18, Taylor and Shaw (1927) discovered a nest of the mountain bluebird 30 feet up in a dead stub, and watched the parents feed the young for over half an hour and summarized their observations as follows:

The young were fed 14 times in 34 minutes, the feedings averaging 2.4 minutes apart. The male fed twice in this time, the female 12 times. The rate of feeding established by this set of observations is approximately 22 times an hour. If this is maintained for five hours in the morning and another five hours in the afternoon, 220 feedings a day would be indicated. But on Mount Rainier at this time of year there were nearly 17 hours of daylight. If the birds averaged 22 feedings an hour for 17 hours, a total of 374 would be indicated. This figure may be closer to the truth than the former one. The male was a shy bird and usually paused on a short branch one to three or four minutes, afraid to go to the nest while the observer was about. The mother was far less cautious. She usually perched for a moment on a branch near the nest and then went directly to it, often entering the cavity and apparently covering the young for a moment.

Probably throughout most of its range the mountain bluebird rears two broods in a season, or tries to do so, perhaps sometimes three; but in the northern portion of its range and on the higher mountains, where the nesting dates seem to be later, it may have time for only one.

Plumages.—Ridgway (1907) describes the young male in juvenal plumage, as follows:

Pileum, hindneck, back, and scapulars light brownish gray or drab-gray the interscapular area usually more or less streaked with white; rump and upper tail-coverts light ash gray; remiges and rectrices as in adults, but with distinct terminal margins of white (duller on remiges), the tertials dusky gray with pale gray or dull whitish margins; middle wing-coverts brownish gray margined terminally with dull white or brownish white; greater-coverts dull blue, margined terminally and edged with pale gray or whitish; a conspicuous orbital ring of white; lores grayish white, suffused with dusky in front of eye, and margined above by dusky; auricular region brownish gray or pale brownish gray, indistinctly streaked with paler; throat and upper chest pale gray (passing into dull white on chin), indistinctly streaked with whitish; chest, sides, and flanks squamately streaked (broadly) with grayish brown or drab, the center of the feathers being white; rest of under parts white.

He says of the young female: "Similar to the young male, but blue of wings and tail much duller and (especially that of wings) greener; color of back, etc., browner."

The postjuvenal molt begins early in August, or earlier, the date varying somewhat according to the date of hatching; I have seen a young bird that had nearly completed the molt on August 20, but usually the first winter plumage is not complete until some time in September. This molt involves all the contour plumage and the lesser and median wing coverts but not the rest of the wings or the tail. In the first winter plumage young birds are practically indistinguishable from adults.

One-year-old birds and adults have a complete postnuptial molt in summer, beginning sometimes before the middle of July, and in some cases it is not completed until well into September. Ridgway (1907) describes the winter plumage of the adult male as "similar to the summer plumage, but blue of upper parts duller, that of pileum, hindneck, back, and scapulars more or less obscured by pale brownish gray margins or tips, the greater wing-coverts and tertials edged with whitish or pale grayish; blue of under parts washed, more or less strongly, with pale brownish gray or grayish brown, especially on chest and sides of breast."

Of the adult female in winter, he says: "Similar to the summer plumage but, coloration slightly deeper, especially the buffy grayish of under parts."

There is apparently no spring molt, the colors becoming brighter by the wearing away of the brownish gray edgings. The whole plumage becomes very much worn before the postnuptial molt.

Food.—Professor Beal (1915a) examined only 66 stomachs of the mountain bluebird and says: "The contents consisted of 91.62 percent animal matter to 8.38 percent vegetable. This is the highest percentage of animal matter of any member of the thrush family herein discussed and is equal to some of the flycatchers." The largest item in the animal food consists of beetles; taken collectively they amount to 30.13 percent, but, of these, 10.05 percent belong to three useful families, predaceous ground beetles, tiger beetles, and ladybirds. Weevils amount to 8.11 percent, "the highest record for any American thrush." Ants were eaten to the extent of 12.51 percent, a record "not exceeded by any other bluebirds or robins." Other Hymenoptera, bees and wasps, amount to 3.80 percent; Hemiptera, bugs, to 3.89 percent, consisting of small cicadas, stink bugs, negro bugs, assassin bugs, and jassids; and flies to only 0.92 percent. Lepidoptera, mostly caterpillars, are a regular article of food, amounting to 14.45 percent for the year. Orthoptera (grasshoppers, locusts, and crickets) are the largest item of food, averaging 23 percent for the year. "Very curiously January shows the greatest consumption, 70.33 percent; August, the normal grasshopper month, stands next with 53.86 percent."

Of the vegetable food, he says: "As with most of the other thrushes, the vegetable portion of the food of the mountain bluebird consists principally of small fruit. The currants and grapes found were in all probability domestic varieties, but as the grapes were from stomachs taken in December and January, and the currants from one taken in April, they can have but little economic significance." Other items listed are elderberries, sumac seeds, and unknown seeds. In their winter haunts these bluebirds feed largely on mistletoe and hackberry seeds, on the drupes of the Virginia-creeper and on cedar berries, as well as other wild fruits.

Dr. George F. Knowlton tells me that out of 172 stomachs of this bird collected in Utah 47 contained adults and 4 held nymphs of the beet leafhopper, *Eutettix tenellus*, a destructive pest.

Grinnell and Storer (1924) describe its feeding habits very well, as follows:

> In the nesting season and indeed through most of the year the Mountain Bluebird subsists upon insects. These are captured in two totally different ways, according to the habits of the insects sought. For beetles and others which fly through the air a bluebird will take position on a boulder in a meadow or on the low outswaying branch of some tree and dart after the insects which pass by. For insects which live on the ground, such as grasshoppers, the bird mounts 10 to 20 feet into the air over the grassland and then by fluttering its wings rapidly, hovers in one place for several seconds and intently scans the surface below, like a Sparrow Hawk when similarly engaged. If something is sighted the bird drops quickly to the ground and seizes it; otherwise the bluebird moves a short distance to a new location which is given similar scrutiny. It thus examines the ground in a manner recalling that employed by the robin though from an aerial location where its scope of view is much greater though less thorough.

This hovering habit has been noted by numerous observers and seems to be characteristic of the species; it seems to prefer this method of foraging to using posts and wires as lookout points, from which to dart down onto its prey, as the other bluebirds generally do.

Behavior.—The mountain bluebird is not a swift flier, probably not making more than 17 or 18 miles per hour on its ordinary short flights. John G. Tyler (1913) says that "a company of these bluebirds in flight may be identified at a distance by their peculiar manner of poising for a few seconds on rapidly beating wings, then flying ahead in undulating swoops. They are often seen in company with Linnets, the two species frequently perching for many minutes in neighborly manner on telephone wires. The bluebirds take wing one at a time and fly ahead at the approach of an intruder, the different units of a flock sometimes becoming quite widely scattered."

Claude T. Barnes writes to me: "It is interesting to see them alight upon snow-covered telephone wires. The snow bothers them, and the flock will not alight until one of their number has ventured upon the wire and shaken himself, thus jarring off the snow."

These, like other bluebirds, are ordinarily gentle birds, but they are able to defend their nests against aggressors and can become aggressive themselves in the competition for nesting sites, or even drive out the rightful occupants. Dr. Alden H. Miller (1935) tells how he saw them drive a pair of hairy woodpeckers from their partly finished hole and appropriate it for themselves:

During the first day, June 5, commotions were frequently noticed at the Woodpecker's tree. The trouble was instigated by a pair of Mountain Bluebirds (*Sialia currucoides*). Whenever the Woodpeckers alighted near the holes, both Bluebirds attacked by diving at them, uttering harsh notes and apparently snapping their bills. Such attacks often lasted five minutes. Evidently the Woodpeckers were too much disturbed by them, possibly also by us, and deserted. During the last two days at camp, no more fights were seen and the Bluebirds were carrying nest material to the tree. The Woodpeckers stayed in the grove, often close to camp, but did not go to the trees near the nest. Since the Bluebirds were just beginning to build, the Woodpeckers were clearly the first occupants and had been dispossessed. Irrespective of other factors which may have contributed to their departure, there was no doubt of the intention of the Bluebirds to displace them.

On June 6 the female Bluebird went to an unfinished Robin's nest just over the tent and settled in it, much to my surprise. She plucked material from the margin and flew to her own nest hole. The Robins added to their nest later that day. The Bluebird, symbol of happiness and gentleness, became to us a different character, whose actions, viewed anthropomorphically, were aggressive and piratical. Interspecific competition for nest material and nest site were enacted before us.

Voice.—The song of the mountain bluebird seems to have appealed quite differently to various observers. Dawson (1923) evidently did not hear it at its best, or did not appreciate it, for, after describing its call and alarm notes, he says: "Other songs the birds have none. * * * The entire song tradition, including the 'delightful warble' attributed to the bird by Townsend, appears to be quite without foundation." And that close observer of bird songs, Aretas A. Saunders, writes to me: "In all my experience with this species in Montana, I never heard it sing. On a number of occasions I camped near nesting bluebirds and heard morning awakening songs of various other birds, but nothing from the bluebird. It may be, as Weydemeyer suggests, that some individuals do not sing, or perhaps I never happened to be awake at the right place and time to hear it."

Ralph Hoffmann (1927) says: "The Mountain Bluebird at all times is singularly silent. An occasional low *terr* is its commonest note, uttered by a flock in flight. The only song which the writer has heard is the repetition of a few short notes, like the syllables *kĕ kŭ* or *kŭ, kŭ, kŭ.* When concerned about their young, the parents utter a vigorous *tschuk, tschuk.*"

On the other hand, some observers have been more favorably impressed. Francis H. Allen says in his notes: "On several occasions late in September and early in October of 1929, in Colorado, I heard the song of this species. It was a beautiful, clear, short warble, higher-pitched than that of *S. sialis* and hardly suggesting it. The call notes were sweet and soft, very much like those of the eastern bluebird, but not so clear or so loud."

Claude T. Barnes writes to me from Utah: "Sitting on a clothesline, a female bluebird resented my approach by a *click*, several times repeated; then at last it sang its very soft, charming song: *Trlll, trlll, trlll*, a mellow roll uttered without opening its bill. The number of *trlls* varied from one to half a dozen, but I am positive the bird never once opened its bill during the ten minutes that I watched it without taking the glasses from my eyes. This habit is, no doubt, somewhat responsible for the subdued nature of the lovely bird's song."

Mrs. Wheelock (1904) refers to the song as "a sweet clear 'trually, tru-al-ly,' like that of the Eastern species, and a mellow warble." And Winton Weydemeyer (1934a) writes:

On frequent occasions during the last seven summers I have forsaken the comfort of my bed to enjoy their subdued, gentle singing. For one must be an early riser indeed if he wishes to hear the Mountain Bluebird's song. Singing commences in full darkness, and continues for a few minutes to as much as an hour, ceasing soon after daylight. * * * Only once during the past seven years have I heard the bird's song at any other time of the day: At 9 o'clock on a dark, rainy morning in March of 1932, a Mountain Bluebird gave weakly a few snatches of its usual daybreak song.

In form, the song is almost a replica of the familiar caroling of the Western Robin; but it is given very softly, crooningly, with an unmistakable quality of the Bluebird's gentle call. Though I have not determined the distance at which the singing can be heard, I doubt if it is audible at seventy yards. The notes are repeated over and over, without a pause, for as much as thirty minutes at a time.

Because of the marked resemblance to the song of the Robin, * * * the song of the Mountain Bluebird appears to be a possible illustration of retrogression in the evolution of bird song. It seems probable that the song, at some time in the past, was louder and more varied, and was sung more commonly, than it is now; and that it is gradually being lost, even as the species is losing other thrush-like characters. If this be so, it is possible that some or all of the Mountain Bluebirds in some parts of their breeding range are already songless, as the testimony of many writers indicates.

Field marks.—The male mountain bluebird is unmistakable in his beautiful coat of plain cerulean or turquoise blue, with only a shading of dusky in his wing tips. The female is modestly clad in a pale brownish or buff gray, with a tinge of bluish on the upper surface. Both can be distinguished from the western bluebird by the entire absence of chestnut. The blue is much paler than in most other blue birds, except the lazuli bunting, which is a smaller bird and has white wing bars.

Enemies.—Other hole-nesting birds are competitors for nesting sites, notably flickers and English sparrows, and probably the ordinary predators take their toll. Mr. Criddle (1927) says of these two enemies: "Flickers are quite important factors in the survival of young bluebirds, even though they often provide the adults with nesting holes. Two instances came to my attention of the parents being driven from their nests by flickers, in both cases resulting in the deaths of the young by starvation. * * * Male mountain bluebirds are able to defend their nests against all intruders of their own size, this includes the house sparrow which has somewhat of a reputation for ousting other species. The sparrow, however, is no match for the bluebird in open fight and despite its persistency, it has never been observed to get possession of a nesting box occupied by the latter." He has seen the bluebirds drive away kingbirds, crows, and squirrels, but the flickers are generally too formidable. House wrens have nested near the bluebirds, but he has never seen the former do any damage to the latter. Among the enemies of the bluebirds he mentions Cooper's and sharp-shinned hawks and weasels but thinks the latter are not so destructive as the squirrels, which are better climbers. Cold rain and snowstorms early in spring or late in fall often prove fatal to young, or even adults.

Dr. Friedmann (1938) could find only one record of cowbird parasitism; a nest found by T. E. Randall in Alberta contained four eggs of the bluebird and one of the Nevada cowbird.

Fall.—Early in August, family parties begin to gather into small companies or larger flocks and gradually drift along southward from the more northern portions of their summer range, the migration continuing well into November. They do not ordinarily move in compact flocks. At this season one often sees detached companies of mountain, western, or chestnut-backed bluebirds wandering about the open country, sometimes loosely associated with sparrows, juncos, warblers, flickers, and other small birds, on their way south. But many bluebirds spend the winter in the greater part of their summer range, moving from the mountains down into the valleys.

Winter.—Ralph Hoffmann (1927) says: "In winter any extensive open country, such as the wheat fields of interior California, is visited by flocks of mountain bluebirds."

These bluebirds do not wholly desert the higher elevations even in winter; Russell K. Grater writes to me: "During late February 1942, I had occasion to be on ski patrol into Cedar Breaks National Monument. At that time the snow was several feet deep on the open. The altitude is around 10,300 feet. Here on the snow fields were many mountain bluebirds, darting hither and yon, apparently catching something on the snow. An investigation revealed that the snow was literally alive with hundreds of tiny winged insects and the birds were making the most of the abundance of food. These birds were observed commonly throughout the open country on this trip."

George F. Simmons (1925) says that, in the vicinity of Austin, Tex., the mountain bluebird "appears in these comparative lowlands usually during very cold weather or during the rare snowstorms and accompanying freezes, which remain for several days."

DISTRIBUTION

Range.—Western North America from Alaska and northern Canada south to northern Mexico.

Breeding range.—The mountain bluebird breeds **north** to east central Alaska (Fairbanks, Lake Mansfield, and Ketchumstock); southwestern Yukon (Dawson, probably; Selkirk, and Lake Lebarge); possibly southwestern Mackenzie, since it has been taken in summer at Fort Franklin, Great Bear Lake (type specimen), and at Hay River and Fort Resolution on Great Slave Lake; north central Alberta (Lesser Slave Lake and Boyle); southern Saskatchewan (Cypress Hills, Wood Mountain, and Yorkton); and southern Manitoba (Aweme, Lake St. Martin, probably, and Richer). **East** to southeastern Manitoba (Richer); western North Dakota (Fort Union, Arnegard, and Medora); western South Dakota (Short Pine Hills and the Black Hills); northwestern Nebraska (Pine Ridge and other points in Dawes and Cheyenne Counties); eastern Wyoming (Laramie Mountains, Cheyenne, and Sherman); central Colorado (Estes Park, Denver, Colorado Springs, Pueblo, and Fort Garland); extreme northwestern Oklahoma (Black Mesa region); and the mountains of New Mexico (Halls Peak, Pecos Baldy, Ribera, and the Sacramento Mountains). **South** to southern New Mexico (Sacramento Mountains and Silver City); central Arizona (White Mountains and Mogollon Mountains); and southern California (San Bernardino Mountains). **West** to central southern California (San Bernardino Mountains), the Sierra Nevada and Cascade ranges in California (Walker Pass, Sequoia Park, Fyffe, Mount Sanhedrin, and Mount Shasta); western Oregon (Swan Lake, Fort Klamath, Saddle Mountain, and Portland); western Washington (Mount St. Helens and Mount Rainier); western British Columbia (Horseshoe Lake, Oyster River, Vancouver Island, Smithers, Atlin, and Bennett); and eastern Alaska (Taku River, McCarthy, and Fairbanks).

Winter range.—The mountain bluebird winters **north** casually to central Washington (Yakima and Spokane), southwestern Idaho (Meridian); northern Utah (Salt Lake City and Provo), and central Colorado (Grand Junction and Denver). **East** to central Colorado (Denver, Colorado Springs, and Pueblo); eastern New Mexico (Las Vegas and the Guadalupe Mountains); western Chihuahua (Pacheco);

and eastern Sonora (Nacori and Alamos). Irregular or casual in winter east to eastern Kansas (Manhattan); Oklahoma (Caddo); and Texas (Bonham, Corsicana, San Antonio, and Corpus Christi); and Nuevo León (Monterrey). **South** to southern Sonora (Alamos). **West** to Sonora (Alamos and Sonoyta); northern Lower California (San Ramón, Rancho San Pablo, and Guadalupe Island, one record); the coast valleys of California north to San Francisco Bay (San Diego, Santa Barbara, occasional on Santa Cruz Island, Salinas, and Sonoma); east of the Coast Range in Oregon (Medford and Portland); and east of the Cascades in Washington (Yakima).

Since the first decade of this century the mountain bluebird seems to have increased and spread eastward, at least in the northern part of its range. At Eastend, Saskatchewan, it is reported to have increased noticeably between 1910 and 1922; at Aweme, Manitoba, it was "rare" in 1890, and "common" before 1928; at Lake St. Martin, Manitoba, in 1931 the Indians reported it to be a comparatively recent arrival. A young bird, thought to be this species, was shot but not recovered, June 10, 1931, at Churchill and an adult specimen was collected there on August 6, 1938.

Migration.—Some early dates of spring arrival are: Nebraska—Hastings, March 5. North Dakota—Charlson, March 19. Manitoba—Aweme, February 28. Saskatchewan—Indian Head, March 16. Colorado—Fort Collins, February 12. Wyoming—Wheatland, February 28. Montana—Helena, March 12. Alberta—Camrose, March 14. Utah—Corinne, March 8. Idaho—Meridian, February 17. British Columbia—West Summerland, February 14; Atlin, April 13. Yukon—Dawson, April 20. Alaska—Dyer, April 20.

Some late dates of fall departure are: Yukon—Carcross, September 24. British Columbia—Atlin, September 24; Chilliwack, November 6. Alberta—Glenevis, October 15. Idaho—Priest River, October 13. Montana—Fortine, November 10. Wyoming—Laramie, November 3. Utah—Ogden, November 4. Colorado—Walden, November 5. Saskatchewan—Eastend, October 14. Manitoba—Brandon, October 28. North Dakota—Argusville, October 11.

One very interesting banding record is available. A young bird banded at Camrose, Alberta, on June 9, 1939, was caught previous to November 30, 1939, at Wingate, Runnels County, Tex.

Casual records.—At Cape Etolin, Nunivak Island, Alaska, three specimens were collected on September 23 and 28, 1927. At Barrow, Alaska, two specimens were collected on June 5, 1930, and one on May 20, 1937. All five of these birds were females. There are two records for the species in Minnesota: a pair seen closely on April 5 and 7, 1935, near St. Cloud; and one to five seen from December 1942 to March 14, 1943, at Duluth.

Egg dates.—Alberta: 8 records, May 21 to June 19.

California: 34 records, April 5 to July 17; 17 records, June 9 to June 18, indicating the height of the season.

Colorado: 30 records, May 2 to June 12; 18 records, May 18 to June 4.

New Mexico: 6 records, May 5 to June 6.

OENANTHE OENANTHE OENANTHE (Linnaeus)

EUROPEAN WHEATEAR

CONTRIBUTED BY BERNARD WILLIAM TUCKER

HABITS

The European wheatear is a bird of extremely wide distribution, ranging across Europe and Asia and extending, in the words of the A. O. U. Check-list, "to northern and east-central Alaska and south to the mouth of the Yukon and the Pribilof Islands." With such a range it might be expected to occur in widely varied types of surroundings, and this is in fact the case. Always a bird of open country, it finds suitable breeding grounds from bare hillsides, downs, and rabbit warrens in England to the steppes and highlands of Asia and the Arctic barrens and hilltops of all three northern continents.

In Alaska it seems especially to frequent the mountains and barren hilltops. Dall (Dall and Bannister, 1869), who saw many at Nulato in May, was informed by natives that they were abundant on the dry, stony hilltops, where the reindeer congregate. In much the same way Nelson (1887) was informed at St. Michael, Norton Sound, that they were "common on the bare mountains in the interior, frequenting the summer range of the Reindeer." Osgood (1909) records that on August 7, the day after his arrival at the head of Seward Creek, east-central Alaska, "a party of Wheatears, two adults and several young, was found flitting about some rock piles near camp. Thereafter one bird or more was seen on every trip into the higher parts of the mountains." He observed that they frequented slides and heaps of small broken rock almost exclusively. Dixon (1938) gives particulars of its occurrence in the Mount McKinley National Park in the Alaska Range. A male, one of a pair that was evidently breeding, was collected on May 29, 1936, high up on a mountainside 1,000 feet above timberline. A young one barely able to fly was obtained on July 14, 1926, at Copper Mountain, where two families were subsequently seen, and in 1932, when they were numerous, nesting pairs were seen at Sable Pass, Savage Canyon, and near Double Mountain. In August of the same year several families were seen at Highway Pass, along the highroad which was then under construction.

Not much seems to be recorded about the habits of the wheatear in America, but in the extreme north of Europe I have seen wheatears in surroundings probably not dissimilar from their Alaskan haunts. On the high fells of the Arctic coast of Norway, dreary wastes of frost-shattered rocks and stones, a few wheatears share the desolation with snow buntings, but in Scandinavia at any rate they are found commonly on the lower ground and may be met with breeding in rocky places anywhere from sea level. Farther south in Lapland, in the conifer forest belt, wheatears are by no means confined to the high ground above timberline, or even mainly found there. They breed freely in forested country where there is a certain amount of open, rocky ground and are there quite arboreal in their habits, far more so than in England, as will be described under "Behavior." It seems, therefore, that in Alaska the wheatear is more restricted in its habitat than in comparable latitudes in Europe. The Alaskan birds average slightly larger than the European ones (Ridgway, 1907) but have not been considered distinct enough to separate.

Spring.—In temperate Europe the wheatear is one of the earliest migrants to arrive. Birds begin to reach the British Isles in the second week in March, the main arrival extending from the end of the third week to mid-April, though some passage continues till mid-May and even later (N. F. Ticehurst, 1938, vol. 2). An average arrival date in the English Midlands is March 23. Between the latitude of the Mediterranean and that of England the birds must travel rapidly on their way north, for the passage in the former region seems to begin little earlier than the date of the first arrivals in the British Isles. Meinertzhagen (1930) states that the spring passage in Egypt begins early in March (earliest March 3) and that between about March 18 and April 10 the birds are abundant. Alexander (1927) found that the passage near Rome extended from March 24 to May 2. The interval between the dates of arrival in Britain and in the far north of Europe is much greater than that between the former and the beginning of the spring passage in the Mediterranean countries. Thus Blair (1936) observed the first arrivals at Vadso on the northern shores of Norway on May 20. At Nijni Kolymsk on the Arctic coast of east Siberia the first arrival in 1912 was not seen till May 31 (Thayer and Bangs, 1914); the conditions here are more severe than in Arctic Norway. Data for Alaska are rather scanty, but arrival seems to be no later than in northern Scandinavia in spite of the much greater distance from the birds' winter quarters. Even as far north as Point Barrow, Murdoch (1885) observed the first arrival in 1882 on May 19, when the ground was still covered with snow except in a few places, but the birds remained only a few days, passing on to the northeast and not being seen again. Dall (Dall and Bannister, 1869) saw numbers at Nulato on May 23 and 24, and Nelson (1887) obtained specimens on May 26 and 28 but does not state definitely whether any had been observed previously. He stresses the "curiously irregular distribution of this bird in Alaska in different seasons," its appearances in numbers in particular localities in one year and complete absence in another. This is characteristic of many species of birds in Arctic and sub-Arctic countries.

Dall and Bannister (1869) speak of wheatears being seen on arrival in Alaska in "large flocks," but probably this should be understood as meaning no more than a large number in one place. In Europe, although considerable numbers may be met with on migration in places near the coast, they are generally scattered and show little evidence of real gregariousness. Moreau (1928) notes that in Egypt "in spring males preponderate in the earlier half of the migration, but there is no striking segregation of the sexes at any time," and this would seem to hold good as a general statement.

Courtship.—The usual display of the wheatear takes the form of hoppings and bowings round the female and a common posture is with the body somewhat tilted forward and the tail fanned so as to show off its conspicuous pattern of white and black. Bertram Lloyd (1933) has described a case in which a female was being courted by two males, one of which, when all three flew off together, continued his tail-fanning display in the air, fluttering along about 15 feet above the turf and singing spasmodically the while. The observer writes of this as an aerial display flight, but as there is no other record of similar behavior it seems rather doubtful whether it can be properly so called, if the term is understood to imply a regular feature of the bird's behavior; it seems not improbable that the carrying over of the tail-fanning of the more usual ground display into flight may have been something of an aberration under stress of excitement. Another performance of a paired male accompanying his mate as she prospected rabbit holes for a suitable nest site is recorded by Miss E. L. Turner (1911), who says: "He would rise into the air some distance then drop like a stone within a hairsbreadth of the ground, much after the manner of an ecstatic lapwing. When this display of daring ceased to impress, he would hop round her, suddenly turn his back and spread out his tail feathers so that the white parts were brought into view."

Occasionally under the influence of extreme excitement a male wheatear has been observed to behave in a remarkably frenzied and ecstatic manner, as has been vividly described by Bertram Lloyd, already quoted:

Their chosen ground was a rough shallow trench in the turf, some eight to ten yards long, with a number of rabbit-scrapes round about it. Actually this "trench" was merely a rough irregular patch about a foot wide and three inches deep, where the turf had been dug out. Twice as I watched, the male Wheatear rose hurriedly in the air * * * and sang a few hasty notes * * *; but mostly the pair were quietly feeding or perching upon clods, flirting their tails in their characteristic manner, and were by no means loquacious. Presently, however, when the female was at one end of the "trench" and the male some distant, near the other end, the latter turned smartly about and facing his mate, suddenly threw himself into the air for a few feet with wide-spread wings. Then dropping sharply to the ground he proceeded to perform a kind of almost rhythmic dance, violent sexual emotion thus freeing itself, I suppose; for he leapt from the centre of the shallow trench to the edge of its bank (perhaps six inches) then back to the centre a little further forward (eight or ten inches) then on to the bank again, and so on in rhythmic progression, with the utmost celerity. He thus seemed merely a little whirling mass of fluffed-up feathers and quick-darting legs, till reaching his mate, who was watching (and possibly admiring) him from her end of the trench he cast himself flat before her, lying taut, with head to earth and wings and tail widely outspread. In this wise he certainly displayed very effectively a great part of his plumage to his demure and sober-coloured mate, the sun glinting on the fresh blue-grey of his back, and the wide-spread wings throwing into clear relief the vividly contrasted whiteness of his rump and the fine markings and lines of his head and chin. I could see his form quivering with excitement while he lay thus prone for a few seconds. * * * Then, rising, he flew quietly off accompanied by his mate, and they resumed feeding a few yards away.

This observation was made on May 5, 1924, near Tring, England. That the utilization of a depression for the purpose of dancing to and fro across it was not merely an individual trick is shown by the curiously similar account given by the well-known observer Edmund Selous (1901). He describes in detail the behavior of two male wheatears on March 30, when a female was in the vicinity, and repeatedly observed a frenzied dancing display, much as Lloyd describes, the performer generally selecting some depression in the ground in order to dart to and fro across it in a state of the greatest excitement.

Finishing here, he runs a little way to another such depression, enters it, and coming out again, acts in precisely the same way, making the same little rapidly moving arch of two black up-and-down pointed wings, moving now this way, now that, now forwards, now backwards, from edge to edge of the trough, perching each time on each edge of it, but so quickly, it seems rather to be on the points of the wings than the feet that he comes down. Wings are all one sees; they whirl forwards and backwards, backwards and forwards, making a little arch or bridge, the highest point of which, in the centre—which is the point of the upper wing—is some two feet from the floor of the trough, whilst the point of the lower one almost touches it.

After a time the performance would cease, to be resumed by one or other after an interval during which both hopped about together on the turf. During the display the non-performing bird appeared to show little interest, but sometimes after it they would dart at one another as if to attack, only to separate when almost in the act of closing. Once a variant of the dance was observed: "He now hardly rises from the ground, over which he now seems more to spin in a strange sort of way than to fly—to buzz, as it were—in a confined area, and with a tendency to go round and round. Having done this a little, he runs quickly from the hollow, picks a few little bits of grass, returns with them into it, drops them there, comes out again, hops about as before, flies up into the air, descends again and dances about."

Presently the female reappeared, the rivals meantime hopping or flitting about, singing or maneuvering around one another in a rather indefinite fashion until at last a fierce fight took place. This, however, appeared to lead to no positive result, the combatants separating as if by mutual consent to resume their hopping about, singing, and displaying. After another fight and more of the same sort of behavior, one of the males flew to the female, "and these two are now in each other's company, singing, flying and twittering for some ten minutes. It would seem as though she had made her choice, and that this was submitted to by the rejected bird, but just before leaving at six o'clock all three are together again." Here, unfortunately, the account ends. At the end of March the birds would probably not long have arrived on their breeding grounds, and it seems clear that Selous was fortunate enough to witness most of the drama of pair formation, though it is a pity that he missed the last act. But in the case of Lloyd's observation, the birds were evidently paired, as would be expected from the date, and it is evident that the "dancing" display may be an outlet for sexual excitement at widely different stages of the breeding cycle.

Nesting.—The wheatear is territorial in its breeding habits. The nest is normally always under cover. In Britain, especially in the lowlands and on the less elevated hills and downlands, it is very commonly in a rabbit burrow. Otherwise the normal site is a cavity under or among rocks, or in heaps of stones, and doubtless this is the type of situation normally used in Alaska. In Europe it may also sometimes be placed in an old stone wall or drain pipe, and near human habitations old tins and all sorts of other artificial receptacles may be occupied. Occasionally a somewhat more open site may be used, and Boyd Alexander (1908) has recorded that on the Kent coast it is not unusual to find nests built in a mere depression on the bare beach, but in most districts such a situation would be considered highly abnormal.

The male may inspect nest sites before the arrival of a female, but later both sexes or the female, with the male accompanying her, have been observed doing so and it appears probable that the female exercises the final choice (Nethersole-Thompson, 1943). The nest is generally built by both sexes, though principally by the female. It is rather loosely built cup of grasses, bents and roots, and sometimes moss, lined with fine grass or rootlets, hair and feathers, and bits of wool or vegetable down.

Eggs.—The eggs of the wheatear are a uniform, delicate, pale blue, only occasionally showing some red-brown specklings. The Rev. F. C. R. Jourdain (1938, vol. 12) gives the measurements of 100 British eggs as: Average, 21.2 by 15.9 millimeters; maximum, 24.8 by 15.4 and 21.8 by 17 millimeters; minimum, 19 by 15.3 and 19.4 by 14.6 millimeters. I am not acquainted with any measurements of Alaskan eggs. The normal clutch in Europe is six, sometimes five and occasionally seven, but Jourdain mentions that complete sets of three or four may also sometimes be found and very rarely eight. The species is double-brooded on the south coast of England (Walpole-Bond, 1938) but is usually single-brooded, and it seems safe to state that it must be so in Alaska. In the British Isles eggs are found late in April or early in May, but chiefly in May. In Germany the period is given as the second half of May to June (Niethammer, 1937). At Vadsö, northern Norway, where the conditions no doubt more nearly approach those in Alaska, full clutches were found on June 23–25 (Blair, 1936).

Young.—Incubation is performed chiefly by the female, but Miss E. L. Turner (1911) and H. L. Saxby (1874) record the male also taking a share. Niethammer (1937), however, quotes a case of a captive pair in which the task was performed entirely by the female. Incubation begins on completion of the clutch (D. Nethersole-Thompson, 1943), and the period is given by Jourdain as about 14 days. Caroline and Desmond Nethersole-Thompson (1942) in a detailed survey of eggshell disposal by British birds state that the eggshells are removed from the nest, though small pieces may be found crumbled up in the lining. They have observed a female carrying a shell and have found many shells in nesting areas. The young are fed by both sexes, but according to Miss Turner, already quoted, if the birds are disturbed or suspicious one member of the pair, not necessarily always the same sex, may be much more easily put off from feeding than the other. She records that at several nests she was photographing the hens alone brought food to the young, the males diligently collecting it and supplying their mates, but refusing to face the camera. But she quotes a case of another nest at which the photographer counted 32 visits by the male and only one by the female. The feces of the young ones are carried away by the parents, but when old enough the nestlings deposit their droppings at the mouth of the nest hole. This is recorded on the authority of Dr. H. J. Moon in a paper by R. H. Blair and B. W. Tucker (1941) summarizing the results of an enquiry on nest sanitation and published information on this subject. Oscar and Magdalena Heinroth, the results of whose amazing achievement in rearing most of the birds of central Europe from the egg are embodied in their great work "Die Vögel Mitteleuropas" (1926), found that young wheatears leave the nest in about 15 days and can fly rather well at 19 days. Field observations on the fledgling period seem to be lacking, as are exact observations on the length of time that the young are tended by the parents after leaving the nest. Family parties may be seen when the young are well grown, but in England full-grown young birds may be seen alone in August. The Heinroths found that the young could feed themselves well and adequately at 26 days.

Plumages.—The plumages of the common wheatear are fully described by H. F. Witherby (1938, vol. 2) in the "Handbook of British Birds." The down of the nestling is dark gray, long, and fairly plentiful except on the femoral tract; the mouth is pale orange inside, with no spots; and the flanges externally are very pale yellow. The juvenal plumage (both sexes) is quite distinct from that of adults. The upperparts are grayish to buffish brown, each feather with a pale subterminal spot and narrow dark brown tip, producing a somewhat spotted general effect. The white upper tail coverts are usually very narrowly tipped with brownish, the throat creamy white and breast pale buffish, the feathers lightly tipped with brown, giving a rather mottled or obscurely barred appearance. The wing quills, greater coverts, and tail feathers are like those of the adult female. These feathers are retained when the rest of the juvenal plumage is exchanged in August for that of the first winter, in which both sexes resemble the adult female in winter except that occasionally some of the new median coverts have small white wedge-shaped spots at the tips as in the juvenal. The first nuptial plumage, assumed early in the year, is like that of the adults, except that in males the upperparts are a browner, less pure and clear gray, and the wing and tail feathers a browner black.

Food.—The wheatear, like its allies, is mainly insectivorous. The Rev. F. C. R. Jourdain (1938, vol. 2) gives the following summary: "Coleoptera (small Carabidæ (*Amara*), Staphylinidæ, Curculionidæ (*Otiorrhynchus*), and Elateridæ), Diptera (Muscidæ, Tipulidæ and larvæ), Hymenoptera (*Bombus* and ants), Lepidoptera (*Euchelia, Zygæna*, larvæ of *Arctia caia*, etc.), Orthoptera (grasshoppers), etc. Also small land Mollusca (*Helix, Clausilia*), centipedes, and spiders. In winter ants and beetles."

Though based mainly on British data, this may be taken as giving a good idea of the usual diet. Vegetable food does not seem to have been recorded in Europe, and so it is rather curious to find that the stomach of a single bird recorded by Cottam and Knappen (1939) taken at Hooniah Sound, Alaska, May 12, 1920, contained food almost entirely of plant origin. It included "thirty unidentified bulblets and fragments of many more, totalling 96%, and undetermined plant fiber, 3%. The animal matter consisted of one jumping spider (*Attidæ*), 1%; fragment of another spider, trace; and fragment of an undeterminable insect, trace. Gravel constituted 5% of the total content." It may be noted that Bean (1883) found wheatears at Cape Lisburne in August feeding on grass seeds and fruits of *Saxifraga*. The indigestible portions of the food, mostly the chitinous parts of insects, are disgorged as pellets, a habit probably found in all insectivorous passerines, though in many it has not been recorded.

Behavior.—The wheatear is a lively and sprightly bird, constantly active. As already mentioned, it is a characteristic species of open country, with a liking for rather stony and waste places. In describing its behavior it will be convenient to follow, but slightly to expand, the condensed and necessarily somewhat "telegraphic" account that the writer gave in the "Handbook of British Birds" (1938, vol. 2). It is essentially terrestrial in habits, moving over the ground in a quick succession of long hops, sometimes so rapidly that it seems to run, frequently halting on some little eminence or flitting a short distance from one such perch to another, or making little fluttering dashes into the air after insects. At rest the carriage is rather upright, but it is seldom long still, constantly bowing and bobbing and at the same time spreading the tail and moving it up and down. When perching off the ground it usually does so on fences, walls, rocks, or heaps of stones, sometimes on bushes, but in England not often on trees. Where there are scattered trees on its breeding ground it may sometimes be seen to perch on them, but it has been repeatedly observed that on migration the Greenland wheatear is much more disposed to perch in trees than birds of the present race. That an Arctic race breeding on treeless barrens should show a greater liking for trees than one accustomed to more temperate latitudes may seem odd, but is none the less a fact. In the days when it was usual to read human feelings and motives into the actions of animals it might have been suggested that the disposition to perch on trees was sufficiently explained by the very fact of a sojourn in lands where any such inclination could not be indulged. Nowadays such explanations will hardly satisfy. In actual fact there is evidence that the habit is a peculiarity of northern-breeding wheatears whether they nest in treeless regions or not. In Lapland, where birds of the present race breed regularly in more or less open and rocky places in forested country, they perch freely on the tops of the tall pine trees, and it has recently been recorded that among wheatears in at least some parts of Scotland the habit is far more prevalent than is the case in England. The behavior of the Alaskan birds where trees are available does not seem to have been noted. The flight of the wheatear when sustained is undulating like that of so many small birds, but more often than not one sees it fly for a short distance only, low over the ground with a rather flitting action, to settle not far away on some little hillock or low perch and repeat the performance if followed. In its pursuit of insects it will sometimes hover in the air for a moment, and occasionally where the air currents are favorable a more sustained hovering has been observed as the bird prospects the ground for food. In places where conditions permit, especially on migration, wheatears have an inclination to take shelter underneath rocks or stones, or in other holes and crannies. They will do this to avoid birds of prey and other causes of alarm and will roost in such places when they can.

Voice.—The note of the wheatear most commonly heard is the scolding or alarm note, a hard *chack-chack* like two pebbles struck together. This may be prefaced by a shriller note, *weet-chack-chack*, and this should perhaps be regarded as a more definite alarm note, regularly heard when the birds have young. The *weet* is also used alone. The food call of the young, both before and after fledging, is a tremulous, shivering wheeze or rattle, but well-grown young birds have also a very distinct, twanging *teek*, and a not dissimilar note may at times be heard from adults. The song, which may be heard in England from the time of arrival to about mid-July, is a short, pleasantly modulated warble rather incongruously mixed with creaky or rattling sounds often suggestive of a handful of little pebbles shaken together. Imitations of other birds may sometimes be incorporated, and these may on occasion be very good indeed, as has been especially stressed by Saxby (1874) writing of wheatears in Shetland. Probably it is largely an individual trait. The song is most characteristically delivered in a little fluttering song-flight, in which the bird rises to no great height and then glides down again with spread tail. But it is also regularly uttered from some little eminence on the ground, such as a clod or stone, from rocks, fences, stone walls, and the like, occasionally from bushes or even telegraph wires, and in northern Europe also from the tops of trees (see the preceding section), though over much of its range this would be considered most unusual. Occasionally the song may be heard at night. It is evident that on migration the wheatear begins to sing only when approaching its breeding grounds, for although Misses Baxter and Rintoul (1914), the well-known Scottish ornithologists, record that birds on the spring passage are frequently heard singing on the Isle of May in the Firth of Forth, song from migrants in the Mediterranean region has apparently never been recorded. Nor does the species normally sing in the winter quarters in Africa, as do some passerines, though Meiklejohn (1941) has recorded one heard singing in Tanganyika on December 25.

Nicholson (1936) has described a subdued subsong, heard from a young bird in August, a low inward warbling reminiscent of the song of the skylark (*Alauda arvensis*).

Field marks.—The wheatear has been compared to a bluebird as to shape, flight, and feeding habits, though it is actually somewhat smaller and its coloring is very different. The outstanding marks of the species are its white rump and white tail with broad black terminal band and black central feathers, combined with black or blackish wings. This pattern is as distinctive of the somewhat spotted young as of the adults. The adult male in breeding plumage is a clear gray above with a black mark through the eye bordered by white above and with whitish underparts, more or less tinged with sandy buff on the breast. In autumn and winter he is much browner and more like the female, which is brown above and buff below, with the blacks of the male replaced by dark brown and the dark facial mark often quite obscure. If birds of the present and Greenland races are seen together, at any rate in Britain, where it is no very exceptional experience since the latter is a regular passage bird, the Greenlander appears a noticeably larger and stouter bird and often seems to have a bolder, more upright carriage. It should be noted, however, that the Alaskan wheatears also tend to be rather large. In spring typical males of the two races are tolerably distinct in coloring and many Greenlanders can be recognized in the field with considerable confidence by observers with a good eye and the requisite experience, owing to the birds deeper buff underparts and rather more brown-tinged, less pure gray upperparts. But a good many individuals could not be distinquished in the field by coloring. In Britain it is also noticeable that the Greenland bird is quite decidedly more prone to perch in trees than is the common form, but as further explained under "Behavior" this point in itself must not be relied on too much. In any case information is lacking as to the habits of the Alaskan birds in this respect.

Enemies.—The wheatear is liable to the attacks of hawks in the same way as other small birds frequenting open ground, and its nest to the occasional depredations of ground vermin as are other species of similar nesting habits.

In the southeast of England, man would formerly have had to be reckoned as its chief enemy, for wheatears were at one time greatly esteemed by epicures and were trapped in fantastic numbers on the autumn migration by the shepherds of the South Downs of Sussex. The traps, or "coops," as they were called, were made by the simple device of cutting a small trench in the turf and laying across it, grassy side downward, the sod removed in making it. This formed a passage or cavity with a way in at either end in which a horsehair noose, or, according to other accounts, two such nooses, were fixed. Descriptions of the precise construction of these traps differ somewhat and some are perhaps inaccurate, though no doubt the details did vary. Two types for which there is firsthand authority are figured by Walpole-Bond (1938) in the latest treatment of the subject. In both of these the excavation was T-shaped, the crosspiece of the T formed by the removal of an oblong sod about two-thirds as wide as long (dimensions of about 8 by 11 inches have been mentioned), while the other limb of the T, across which the sod was placed, was longer and relatively narrow. Owing to the birds' predilection for running into shelters of this kind, and being deceived by the appearance of a clear way through, they were very easily taken in such snares.

The scale on which these traps were operated and the numbers of birds taken were astonishing. In Yarrell's "History of British Birds" (1874) it is stated:

One man and his lad can look after from five to seven hundred of them. They are opened every year about St. James's Day, July 25th, and are all in operation by August 1st. The birds arrive by hundreds, though not in flocks, in daily succession for the next six or seven weeks. The season for catching is concluded about the end of the third week in September, after which very few birds are observed to pass. Pennant, more than a century since, stated that the numbers snared about Eastbourne amounted annually to about 1,840 dozens, which were usually sold for sixpence the dozen, and Markwick, in 1798, recorded his having been told that, in two August days of 1792, his informant, a shepherd, had taken there twenty-seven dozens; but this is a small number compared with the almost incredible quantity sometimes taken, for another person told the same naturalist of a shepherd who once caught eighty-four dozens in one day. In Montagu's time (1802) the price had risen to a shilling a dozen, and it is now much higher, through the greater demand for and smaller supply of the birds.

In the course of the nineteenth century the numbers taken gradually declined and the practice has fortunately long died out, though Walpole-Bond (1938) mentions that up to as recently as 1902 wheatears were still sent in fairly large numbers to certain hotels at Brighton and perhaps Eastbourne as well. It appears, however, that these were chiefly "caught in clap-nets spread ostensibly for Starlings" and "that very few were captured by the old-fashioned method."

A list of ecto- and endoparasites recorded from the species is given by Niethammer (1937).

Fall and winter.—From what has been said in the preceding section it will be apparent that wheatears may occur in large numbers in coastal districts on the autumn migration. But it is rarely, if ever, that they are in anything that can properly be called a flock; when the numbers are large the birds are usually widely scattered, and although they may be said to exhibit social tendencies to the extent of indulging in a certain amount of chasing and play they are essentially individualistic, flushing singly or in little groups of two or three if disturbed and not flying in close or coordinated formations. This is true of the Mediterranean region as well as of more northerly localities in Europe, and it doubtless applies equally to America and Asia. In its African winter quarters it is also usually met with singly, as has been noted, among others, by Lynes (1925) and Moreau (1937), frequenting open country, especially where there is bare soil, anywhere from the low hot steppe to above timberline. In Tanganyika Moreau records its occurrence from sea level up to 10,000 feet.

With regard to the time of departure from the breeding grounds information is naturally much scantier for the high north than for the British Isles and temperate Europe. At Nijni Kolymsk, east Siberia, the species is recorded as late as September 21, 1911 (Thayer and Bangs, 1914). In Alaska birds are on the move in August and are to be seen at least till the end of the month and probably later. Nelson obtained specimens from August 20 to 25; family parties were still to be seen in the Alaska Range on August 24 (Dixon, 1938); Bishop (1900) shot one at the Aphoon mouth on August 27; and a young male is recorded from Kruzgamepa on August 30 (Thayer and Bangs, 1914). In the Pribilofs it has occurred on September 1 (Mailliard and Hanna, 1921). Singularly little is known about the movement of the Alaskan and east Siberian birds between their breeding grounds and the remote winter quarters in east Africa and southern Arabia. Hartert and Steinbacher (1938) state that on the Siberian east coast it is only known as a very scarce passage bird. "The autumn migration even in the case of the most easterly birds must thus be to the southwest, although isolated specimens are known from Japan, the lower Yangtsekiang, and the Philippines." In England the earliest departures from the south coast are early in August, and movement continues till the third week of October (Ticehurst). In central Europe the period of the passage is given as from the end of August or beginning of September to mid-October (Niethammer). In central and south Italy it is from about mid-August to the end of September (Alexander; B. W. Tucker). In Egypt the first birds arrive on the Mediterranean coast about August 15, and they are common by the 21st. Passage diminishes about October 21, and the latest record is November 17 (Meinertzhagen). In Kenya and Uganda considerable numbers arrive in mid-September, and these are augmented throughout October and November (van Someren). It will be understood that the European and African data quoted in this and the "Spring" section are far from exhaustive but are selected as representative.

DISTRIBUTION

Breeding range.—Breeds throughout the greater part of Europe, including the whole of the British Islands, northward to latitude 71° in Norway and 70° in Russia, also in north Sweden and Finland and in Novaya Zemlya, southward to north Spain, south France, Corsica, Italy, Sicily, Greece, and south Russia. In the Mediterranean region it breeds chiefly in mountainous country. Extends westward through Asia Minor, Syria, northern Iran, Aralo-Caspian region, and central Asia to Mongolia and Siberia. In America, in northern and east-central Alaska, south to the mouth of the Yukon and the Pribilof Islands. Attempts have been made to distinguish other races in Asia, but these do not appear to be well founded. For discussion, see Meinertzhagen (1922) and Hartert and Steinbacher (1938). The race *Oenanthe oenanthe nivea* (Weigold), of central and southern Spain, is considered valid by Witherby (1938). Other races are described from Crete and northwest Africa.

Winter range.—South Arabia and east Africa; also, but rarely, in southern and eastern Iran and Mesopotamia. The A. O. U. Check-list and Ridgway, no doubt following Seebohm (1881) in the "British Museum Catalogue," include India in the winter range, but there appears to be no good evidence that any winter regularly in India.

Spring migration.—The earliest date recorded on the American Continent appears to be Point Barrow, May 19, and the next (many birds) Nulato, May 23. Other particulars will be found under the section "Spring."

Fall migration.—Latest dates recorded in America appear to be: St. Paul Island, Pribilof Islands, August 29 and September 1; Kruzgamepa, August 30; but it probably occurs later, as on the Arctic coast of east Siberia it has been recorded as late as September 21. Other particulars will be found in the section "Fall and winter."

Casual records.—Jourdain (1938, vol. 2) notes occurrences in Waigatz, Kolguev, Yalmal, Spitsbergen, Bear Island, and the Canaries. Other casual records could no doubt be added for Asia and the western Pacific.

OENANTHE OENANTHE LEUCORHOA (Gmelin)

GREENLAND WHEATEAR

HABITS

The 1931 Check-list gives the following interesting distribution for the Greenland wheatear: "Breeds in the Arctic Zone from Ellesmere Island and Boothia Peninsula east to Greenland and Iceland and south to northern Quebec. Winters in West Africa, migrating through the British Isles and France. Casual in migration or winter in Keewatin, Ontario, New Brunswick, Quebec, New York, Pennsylvania, Louisiana, Bermuda, and Cuba."

The wheatears are characteristic Old World birds, having a wide Palearctic distribution, in which there are several Eurasian forms. The species has extended its range into the northeastern corner of North America in the form of the Greenland wheatear, and into the northwestern corner in the form of the typical wheatear of Europe and Asia. Neither form, though both are more or less established as breeding birds at the two extremities of Arctic North America, has established any regular migration route on this continent. The steps which led up to the discovery of the two migration routes and the separation of the Greenland subspecies have been fully explained in an interesting paper by Dr. Leonhard Stejneger (1901), to which the reader is referred. He gives the diagnostic characters of the Greenland bird as "larger than *Saxicola oenanthe*, the length of wing varying between 100 and 108 millimeters; color similar, but the rufous tints more bright on the average." The wing measurement of the typical subspecies seldom equals 100 millimeters and is usually much less.

He adds the following comments on the strange distribution and probable expansion of range of this interesting species into the Nearctic region:

> The Wheatear, the most widely distributed species of the genus *Saxicola*, thus extends its range across the entire palaearctic continent from the Atlantic to the Pacific Ocean. At both extremities of its home continent, however, it has expanded its range into the New World, and no one who follows on the map the route of the retreating winter migrants can for a moment be in doubt that these routes really represent the way by which the species originally invaded America. It would be difficult to find a more beautiful example to illustrate that now well-known law which was first formulated by Prof. Johan Axel Palmén, of Helsingfors. Moreover, no better example could be found for demonstrating the necessity of minute discrimination in ascertaining the characters by which these "migration route races," as Palmén calls them, are characterized.
>
> It seems that one more lesson can fairly be drawn from the differentiation of the Greenland race, viz, that the Greenland-Iceland-England route must be considerably older than the Alaska-Tchuktchi-Udski route, since it has resulted in the establishment of a separable race. A consideration of the further fact that no regular migration route could have been effected between Greenland, Iceland, and Great Britain during the present distribution of land and water in that part of the world also leads us back to a period when the stretches of ocean now separating those islands were more or less bridged over by land. For such a condition of affairs we shall have to look toward the beginning of the glacial period. At that time it must, therefore, be assumed that the Wheatear extended its range into Greenland. The advent of the typical form into Alaska, on the other hand, is probably one of very recent time, an assumption corroborated by the somewhat uncertain and erratic distribution of the species in that northwestern corner of our continent.

Migration.—The Greenland wheatear has the most remarkable migration route of any of our passerine birds, which evidently follows the ancestral route by which the species originally invaded Greenland and North America. From its winter range in western Africa it passes through England mainly in May, but often in April and sometimes as early as the first of April (Witherby, 1920); it has been known to arrive in Greenland as early as April 4–12 (Chamberlain, 1889); and it has been taken several times in Quebec in May and June, where it probably breeds occasionally as far south as the north shore of the Gulf of St. Lawrence. The wheatear is undoubtedly a very hardy bird, but it is remarkable that it can survive on that long, northern route, which has to be covered at a season when that inhospitable region is locked in ice and snow and when food must be very scarce.

The return trip in fall over the same route is made largely in September and October, though a late straggler was taken at Godbout, Quebec, on November 9, 1886 (Comeau, 1890), and the last one was seen that year in southern Greenland on October 5 (Chamberlain, 1889). At these dates winter must have held an icy grip on those northern countries.

Nesting.—The former scientific name, *Saxicola*, and the old common name, stonechat, well indicate the haunts and the nesting habits of the wheatear, for it seems to prefer to live on the barren, stony slopes, where there are loosely piled stones and boulders scattered about over the open spaces. Its nest is usually well hidden under stones, or in crevices among the rocks or in cliffs or walls. Referring to the nesting of the Greenland wheatear, Montague Chamberlain (1889) writes: "It builds in locations similar to the Snowflake, though it commonly selects a spot under 600 feet high. The situation of the nest is also similar, though the present species goes farther into the heaps of stones—sometimes as much as four feet or more. A favorite situation for the nest is the wall of a house or a stone fence."

John Ripley Forbes (1938) found a nest on Baffinland that was tucked in a crevice in a cliff 8 or 10 feet high. "It was constructed of dry grasses and beautifully lined with the white feathers of the ptarmigan. The entrance to the crevice was so small that it would not admit my hand through the entrance. The crevice ran some distance back into the rock and, during another visit, I found the young had left the nest on hearing my approach and had retreated into the rock, to return after I had left."

Eggs.—Six or seven eggs make up the usual set for the Greenland wheatear, but as many as eight and even nine have been reported; these are larger numbers than are usually laid by the European bird. The eggs are similar to those of the European wheatear, but are slightly longer on the average.

Witherby's Handbook (1920) gives the average of 40 eggs as 21.8 by 15.8 millimeters.

Food.—Clarence F. Smith has sent me the following information on the food of the Greenland wheatear, taken from a paper by T. G. Longstaff (1932): The stomachs of eight wheatears from western Greenland were examined; they held Coleoptera and their larvae in six stomachs, parasitic Hymenoptera in six, Diptera and their larvae in five, Lepidoptera adults in two and their larvae in two, and Heteroptera in two. Spiders, gastropod shells, shrimps, harvestmen, Trichoptera, and Neuroptera also were found. All stomachs contained some animal matter. Plant food was found in seven stomachs, including crowberry (*Empetrum nigrum*) in six and seeds of *Vaccinium uliginosum* and *Juniperus communis.*

Behavior.—All the habits of the species have been so fully and well described in B. W. Tucker's excellent account of the European wheatear that it hardly seems necessary to say anything more about them here. However, the following account by Dr. Harrison F. Lewis (1928) of the behavior of a Greenland wheatear that he saw at Natashquan, Quebec, is of interest:

> It had a way of standing quietly for some time in one place, puffed out to a sturdy, rotund figure, perched on slender feet, with the high, rounded dome of its gray head rising directly from the upper end of its body. When it was darting on foot after insects on the ground it depressed and extended its head, and appeared wide and flat. It pursued its prey with quick little darting runs, now here, now there, with head lowered. Sometimes it would flit about for greater or less distances rather restlessly, at other times it would stand entirely still, except for quiet but alert turning of the head, for minutes together.
>
> When the bird was restive because of my close approach and observation, it would stand erect and regard me attentively, and occasionally would give a quick little bob and jerk, as though feigning to spring into the air, yet not moving its toes all the while. Sometimes such a jerk was accompanied by a quick, nervous flirt of the wings, exposing for a flash the white about the upper part of the tail, and sometimes it was not so accompanied. Of course, when the bird flew the white of the upper part of the tail was very conspicuous.

CYANOSYLVIA SUECICA ROBUSTA (Buturlin)

RED-SPOTTED BLUETHROAT

CONTRIBUTED BY BERNARD WILLIAM TUCKER

HABITS

The bluethroat as a species extends right across Europe and Asia and just reaches the New World. A number of races have been separated, 15 being recognized by Hartert and Steinbacher (1938). In most of these the striking brilliant blue gorget has a chestnut-red spot in the center, but in the Middle European form, *C. s. cyanecula*, the spot is silky white. In several other races the spot is either white or else both types occur. The bluethroat that occurs in western Alaska is treated binomially in the A. O. U. Check-list, but it may be assumed to be the east Siberian race, *C. s. robusta* (Buturlin). This is a large, dark, richly colored race with a wing measurement, according to Hartert (1910), of 75–80 millimeters. Grinnell (1900) found this bird evidently breeding near Cape Blossom, Kotzebue Sound, and the A. O. U. list also records breeding at Meade Point on the authority of Bishop, *in litt.* No further particulars about the latter record are available to the writer, and as Grinnell's account appears to be the only one published it may be quoted in full. He writes:

I met with this species in the vicinity of Cape Blossom on July 3, '99. The locality was the side of a ravine between two hills of the first range, about a mile back of the Mission. This hillside was of a gentle slope, and was clothed with thick patches of dwarf willows one or two feet in height. I was tramping along the bed of the ravine when I heard a harsh note, entirely unfamiliar to me, from the brush a little to my right. I started up the slope so as to be in more open ground and get a better view, when I caught a moment's glimpse of the author of the strange note, as he flew hurriedly close along the ground to a distant bush. The note and bearing of the bird reminded me more of those of a wren, and not until I finally had the bird in hand did I have any idea of its identity. By hiding and making squeaking noises I succeeded in attracting the bird within range, and secured it. It had an insect in its bill, and so I judged there must have been a nest in the vicinity. But after waiting a long time I failed to see or hear any other Bluethroat, and as it was late in the day I started on my return to camp. I had proceeded about a quarter of a mile when I heard that faint harsh note, unmistakable after once learned, among the calls of Tree and Savannah Sparrows and Yellow Wagtails, on a similar hillside. I soon obtained a good view of this Bluethroat, and it, too, had an insect in its bill. It was less shy than the first one. I had no doubt of a nest this time, and selecting a point of observation behind a bush, waited and watched. At last I gave it up, intending to return the next day. But that proved to be my last day with the birds at Cape Blossom * * *. The single specimen obtained of *Cyanecula suecica* is an adult male in somewhat worn plumage. That this species was breeding at Cape Blossom, I have no doubt, and I can easily see how I could have previously entirely overlooked it, on account of its unfamiliar habits and notes.

Others have seen birds which may have been only vagrants. Thus, Nelson (1887) records a party of seven, of which one was obtained, met with by Dr. Adams at St. Michael on June 5, 1851. They were not seen again and the natives were said not to be familiar with them. Again, Bailey (1926) found birds fairly common on Wales Mountain on June 10, but none were noted after June 11, and Friedmann (1937) records one from St. Lawrence Island in August.

[AUTHOR'S NOTE: The latest information comes from Henry C. Kyllingstad, of Mountain Village, Alaska, who writes to me as follows: "The following may be an addition to the known range of this bird. On July 11, 1943, a small bird with a nervous manner flew into the Government schoolyard at Mountain. It had a loud clear call somewhat like that of the water thrush, but I did not hear it sing. I had only a brief look at it, but it was close, not more than 15 feet away, and I was able to see the brilliant blue patch on the throat and the characteristic shape of the tail, though the bird did not remain still. I had no idea what the bird could have been until some weeks later when I received a copy of E. W. Nelson's 'Report Upon Natural History Collections Made in Alaska,' when I recognized it from the plate on page 220. I am sure it could have been no other bird. On three subsequent occasions I saw birds with a similar manner and call within a short distance of Mountain Village."]

As so little is recorded about the habits of the bluethroat in Alaska we must rely upon observations on the well-known western form in northern Europe, where there is no reason to suppose that the behavior or habits differ in any essential way from those of the eastern race.

The bluethroat would better merit the name of "northern nightingale" than the red-winged thrush, to which this title has been applied, for not only is it actually a close relative of the nightingale, but it has a fine and varied song and can hold its own with the best songsters anywhere. As might be gathered from Grinnell's experience, it is a bird of the Arctic willow and birch scrub, and in Europe and Asia it is one of the common and characteristic passerines of the far north and is found breeding from sea level wherever suitable ground occurs. It is found principally in swampy localities, though probably not so much from any special attachment to wet ground as such as because it is here that it finds the sort of scrub vegetation that it delights in. Its rich and musical song, pleasantly contrasted with the cheerful but much simpler performance of the Lapland longspur, enlivens the lonely and monotonous—yet to the naturalist fascinating—tundra country beyond the forest limit, where it can still find sufficient cover for its liking in the water-logged hollows. In the forest belt it is found among the scrub of the moorland tracts, in open swampy places in the woods, and in the more luxuriant willow thickets along the rivers or where the forest has been cleared around farmsteads and habitations. In the most southern parts of its range, as in the case of many other Arctic forms, it is only at high altitudes that the red-spotted bluethroat finds congenial conditions, so closely is the association with an Arctic or sub-Arctic type of habitat ingrained in its makeup. Yet the structurally identical white-spotted bluethroat of temperate Europe is mainly a lowland bird.

Courtship.—The display of the bluethroat has been described by several observers. In the characteristic display posture, which may be associated with vehement singing, the head is stretched almost straight upward or even actually inclined backward, displaying the blue gorget and red or white spot to the full, while the tail is strongly cocked or jerked up and down and the wings drooped. This posture seems to have been first described by Ziemer (1887), who observed it in the male of a captive pair of the white-spotted form in an aviary. He writes (translation):

> While the female now sat quiet and apparently indifferent in the middle of the cage on a little eminence, the male for some time ran about restlessly and evidently excited, jerking his tail and calling from time to time. Then he began to sing, at first softly and intermittently, then gradually louder and more continuously, until finally he drooped the wing tips even more than usual, so that they almost trailed on the ground; then he fanned out his tail and cocked it up to beyond the vertical, laid the head so far back that it almost touched the tail, and then, singing with all his might, pirouetted round the female in this attitude, thus showing off his finery to the full and from time to time making bowing movements. Bustling about in this way he moved only his feet, while he remained in the same stiff posture, so that it looked as if he were being driven by clockwork.

It is necessary to exercise some caution in basing accounts of bird behavior on observations under the necessarily somewhat unnatural conditions of captivity, but in this case we have sufficient supporting evidence from field observation to justify the conclusion that the behavior described was essentially normal, and Ziemer's account is valuable because it enables us to fill in some details of a display that is difficult to observe adequately in nature because it is generally performed in cover and often, it appears, toward dusk. Otto Natorp (1928) records observing a pair of white-spotted bluethroats in the dusk, the male moving round the female with the head stretched upward and backward and the tail strongly cocked, just as Ziemer describes, and he states that he observed a similar display on another occasion. Again, Aplin (1903) describes a male of the red-spotted form singing ecstatically with head and neck stretched up, bill pointed nearly upward, tail flirted up and down or held at rather less than a right angle with the body and wings drooped, while the female was creeping and hopping about in the Arctic birch scrub close by. In all these cases the display as described seems to be fairly definitely a courtship performance, but it should be observed that in the related European robin a closely similar posture, displaying the red, instead of blue, breast, has been shown to be an aggressive or threat posture and not sexual at all. This naturally raises the question whether the same may not be really true of the bluethroat, but the recorded accounts do not favor this supposition and more observation is desirable. It would not be unprecedented for fundamentally the same posture to be used in different situations in two species.

Finally it must be mentioned that although the male sings freely from perches he also has a somewhat pipitlike song flight, which must be reckoned a form of display, in which he rises singing in the air and glides down again with spread wings and tail.

Nesting.—[AUTHOR'S NOTE: Only a few North American nests have been reported, all in the vicinity of the Meade River, about 30 miles inland from Barrow, Alaska. Three of these are in the Wilson C. Hanna collection, two being taken on June 18, 1932, and one on June 19, 1936, and all containing six eggs each. A set of eggs in the collection of Dr. Louis B. Bishop was taken there in July 1928. And there is a set in the Chicago Academy of Sciences, taken in the same locality on June 20, 1932.

There are two sets of eggs of this species in the Thayer collection in Cambridge. One was taken by Koren, near Nijni Kolymsk in northeastern Siberia on June 14, 1905, from a nest "in a hollow in the ground in a thicket of small willows and heathberry plants." The other came from Tornea, Lapland, evidently the western race, and was taken from a nest among the roots of a bush, on June 23, 1900. The base of the nest consists largely of string, cottony substances, leaves, and very fine twigs or rootlets; the main nest is made of strips of inner bark, interwoven with grasses, rootlets, and fine twigs, mixed with a little green moss and plant down; it is lined with cattle or reindeer hair. Externally it measures 3 inches in diameter and 2½ inches in height; the inner cup is about 2 inches wide by 1¼ deep.

Henry J. Pearson (1904) mentions two nests of this species found in Russian Lapland. One, found on June 24, "was placed on a bank facing south, and was made of moss outside, well lined with fine grass." The other, found June 29, "was in a clump of grass on the edge of the lake, surrounded in fact by water and only two inches above its surface. I expect the water had risen since the nest was made, as it was eight inches above the usual level."]

Eggs.—[AUTHOR'S NOTE: Six seems to be the usual number of eggs laid by the red-spotted bluethroat; all the Alaskan sets reported were of this number; Pearson (1904) reported a set of four and a set of seven; and others have said that seven eggs in a set are not unusual.

The two sets of six eggs each in the Thayer collection are ovate and somewhat glossy. The ground color varies from "tea green" to "deep lichen green," and the eggs are mostly very faintly sprinkled with the finest dots of brownish olive or dull, pale brown; in some eggs the markings are concentrated into a brownish cap at the large end; some eggs appear almost immaculate.

Mr. Hanna has sent me the following description of the three sets in his collection: "The eggs are ovate to short-ovate. The shell is close-grained and shows little or no gloss. The uniform ground color is grayish olive. One set is almost without superimposed markings, while another set has slight, fine markings of drab. The third set has more superimposed markings of drab or buffy brown, heavier on the large end, and in three of the eggs to form a small, faint wreath."

The measurements of 45 eggs average 19.1 by 14.5 millimeters; the eggs showing the four extremes measures **21.2** by 14.8, 19.0 by **15.3, 16.1** by 13.8, and 19.3 by **13.7** millimeters.]

Young.—Incubation, according to Jourdain, is apparently by the hen only, but information is scanty, and the period is about a fortnight. Both sexes feed the young. According to Chislett (1933) nest sanitation is the special duty of the male, but it seems rather unlikely that the female never takes part. Jourdain states that the young leave the nest at 14 days but are not then fully fledged. Only a single brood is reared in the short Arctic summer.

Plumages.—The plumages of the western form of red-spotted bluethroat are described in detail by Witherby (1938, vol. 2) and the eastern race, to which doubtless the American birds belong, is described by Hartert (1910) as a large, dark, strongly colored form with large red breast spot and with wing measurement of 75–80 millimeters.

Witherby had not examined a nestling of the red-spotted form, but describes one of the European white-spotted bluethroat (*C. s. cyanecula*), which can safely be assumed to be similar, as having dark slate-gray down, fairly long and plentiful, distributed on the outer and inner supraorbital, occipital, humeral, and spinal tracts. The mouth inside is orange, without tongue spots, and the flanges are whitish yellow externally. The juvenal plumage is very dark with light streaks. The upperparts are blackish brown, each feather with a median buff streak broadening at the tip; the throat, breast, and flanks are similar. The tail and wing quills are like those of the adult, and the wing coverts have buff tips. In the first winter plumage the male resembles the adult male in winter but has less blue on the throat, which is more or less whitish but may show some rufous in the center, while the chestnut breast band is paler than in the adult. The female is also like the adult, but the throat is whiter and lacks the dark spots that the adult females show. The primary coverts and outer greater coverts have buff tips in both sexes.

Food.—Bluethroats are primarily insectivorous. Jourdain (1938, vol. 2) notes: Diptera and their larvae (Culicidae, Tipulidae, small black flies, etc.), Coleoptera and larvae, and also some aquatic insects. Small snails are also recorded, as well as worms, and some seeds and berries are taken in autumn. Gurney, quoted by Jourdain, mentions *Acocephalus nervosus*, *Philaenus spumarius*, and a shell of *Littorina rudis* in the case of a migrant on the English east coast.

Behavior.—Bluethroats are, generally speaking, skulking birds, keeping to the ground or near it in the cover of vegetation, but will come out to feed in the open when undisturbed. With the exception of the singing male and to some extent of birds with young, bluethroats do not show themselves readily away from cover. They spend most of their time on the ground among rank vegetation or, in the case of migrants, at any rate in Europe, often among root crops in cultivated fields. In thick cover the bird creeps about in a mouse-like fashion, but if the observer remains concealed or keeps very quiet it may emerge into the open. Here the carriage is seen to be noticeably erect, and the bird moves over the ground with long hops or sometimes in little runs. The tail is usually cocked up in a perky fashion and is frequently flirted up and down and from side to side, being somewhat spread at the same time. It is also somewhat spread when the bird alights. When suspicious or slightly uneasy it has a nervous bobbing action, another of the characteristics emphasizing relationship with the European robin. If it is driven from cover, or for that matter when making a voluntary flight, it travels low and seldom for more than a short distance, quickly diving into cover again. In doing this it has been observed of migrants that the line of flight generally flattens out at the last moment instead of the bird dropping down into cover vertically as so many species do. On the breeding ground, in spite of its predilection for cover, it will at times—and not only when singing—perch quite freely in the open on some low bush or other perch, but any slight disturbance will quickly send it into shelter again.

Voice.—The note most commonly heard is a scolding, hard-sounding *tacc, tacc,* but it has also a more plaintive *hweet,* which seems to be more definitely an alarm note, and a rather soft, croaking *turrc, turrc.*

The delightful song of the bluethroat has charmed all naturalists who have heard it in its northern haunts and has earned for it among the Lapps of its native country a name meaning "a hundred tongues." It is a loud, sweet, and remarkably varied performance with, in parts especially, a distinct family resemblance to that of the nightingale, though never quite so rich and full. As is the case in the nightingale's song, each phrase is a repetition of the same note or simple combination of notes, and although some, including a striking metallic *ting, ting, ting,* which has been compared to a note struck on a metal triangle, seem to be common to all birds, there is a great deal of variation not only in the notes and phrases in the repertoire of any given bird but also between those of one bird and another. Not all the notes are musical; here and there more churring or other nonmusical sounds occur, but this does not detract from the beauty of the song as a whole, any more than does the same feature in the song of the nightingale.

In the unbroken daylight of the Arctic summer the bluethroat pours out his song at all hours. During spells of genial weather in these high latitudes, the night hours, with the sun still up and the strong light of noon no more than gently and restfully subdued, cast a strange, indefinable glamor and "other-worldliness" over the landscape, and it is at such times that the song is heard to perfection. But it may be heard too under very different conditions when the vagaries of a late spring in the Arctic lead the human observer to take a much less romantic view of his surroundings. To his own intrinsic merits as a songster the bluethroat adds that of being an excellent mimic, and Seebohm and Harvie-Brown (1876) have described how a whole variety of notes and calls of other species may be run together "in such a way as to form a perfect medley of bird-music, defying one who is not watching to say whether or not the whole bird-population of that part of the forest are equally engaged in the concert at the same time." The song is generally delivered from a more or less exposed perch, often on a low bush or tree, but not infrequently, in contrast to the bird's habit at other times, at a fair height on the top of a young conifer or birch tree or, where such things are found, even on telegraph wires. The song is also uttered in a special display flight, which has been referred to already, and Seebohm (1901) states that on first arrival it may warble in a very low undertone, evidently indicating what would nowadays probably be called a subsong.

Field marks.—For a European observer unfamiliar with the species one can compare the general form and build of the bluethroat to that of a rather slim robin, but for ornithologists in America, to whom "robin" means a quite different bird, no such ready comparison offers itself. Perhaps the actual form and outline of the bird may be described as rather like that of a small, long-legged bluebird, but the coloration is entirely different. The upperparts are dark brown. The handsome blue throat and breast are fully developed only in the adult male in the breeding season, and in a sense is a better field character, since it is present at all ages and in all seasons, is the rufous base to the dark brown tail. This is conspicuous when the bird spreads its tail or when it flies away from the observer, and in a migrant that flits up among bushes or, as may happen in Europe, in a field of roots or potatoes, and quickly drops into cover again, it may serve for identification when little else is seen. The bright blue bib of the breeding male shades into blackish below and below this again is bounded by a chestnut-red band of the same color as the spot in the middle of the bib. These bright colors are somewhat obscured in fall by pale tips to the feathers. The female has the bib whitish, defined by a dark breast band and stripes at the sides, with some dark spots, and may have some indication of a red center spot. There is a good deal of individual variation and the young birds in the fall are much like the adults, but distinguished as described under "Plumages." The underparts other than the bib are whitish.

Enemies.—No special data are available on this subject, but the secretive habits of the bluethroat probably protect it to a considerable extent from birds of prey.

Fall and winter.—As very little is recorded about the eastern form of bluethroat on its fall migration we must again rely largely on observations in the west, and although the precise conditions under which they were made are obviously not reproduced exactly in Asia, nevertheless they will serve to give some idea of the habits of bluethroats at this season. In Great Britain the species is hardly ever met with inland but is a regular passage migrant on parts of the east coast and still more upon certain islands. It also occurs in numbers in the German island of Helgoland. On migration it may be found skulking among *Suaeda* bushes or other scrub or rank vegetation on coastal sand dunes, or it may even frequent cliffs. Where there is some cultivation, as on Helgoland and on Fair Island, its British counterpart, it has a special liking for potato and root fields, where it gets most of its food hopping about under or close to the shelter of the protecting leaves and is difficult to flush. It is also much attracted by gardens, where these exist, and here if it is not disturbed it may become very confiding, for it is secretive and wary rather than shy. Gätke (1895), the famous ornithologist of Helgoland, writes: "If during one's garden occupations one pays no special attention to the bird, or pretends not to notice it, it will for hours long hop around near one, at twenty, fifteen, or even a less number of paces off. * * * If, however, it becomes aware of being watched, it vanishes swift as lightning, in long bounds, under some shrubs or among some bushes." Gätke adds, however, that on the fall migration it frequents the potato fields exclusively and never comes into the gardens, which form its chief resort in spring. But this, like some of his other statements, is somewhat too sweeping and dogmatic, for I have myself watched bluethroats in Helgoland gardens in fall.

In their winter quarters bluethroats are solitary in habits and retain that attachment to cover near water and in swampy places that is observable on the breeding ground. They are found in such places as the outskirts of reed beds and canebrakes, the borders of lakes or streams or of irrigation channels among crops, as well as, where they occur, in suitable hedgerows, gardens, and other cultivation. R. E. Moreau (1928) states that occasional snatches of song may be heard in winter quarters in January and February. Hugh Whistler (1928) writes:

From September until May the Bluethroat is a common species in India either as a passage migrant or a winter visitor, but its movements have not yet been properly worked out. It does not breed nearer than Ladakh. Although extremely common at certain times and places it escapes observation through its skulking habits. It is a bird of the ground and heavy cover, preferring dampish spots, such as reed-beds on the edge jheels, tamarisk thickets in river-beds, heavy standing crops and similar situations. In these it feeds on the ground, only occasionally ascending to the top of the bushes to look around or to sing a few bars of its beautiful song.

Ordinarily it is only seen when one walks through cover, as it dashes up at one's feet and flies a few yards before diving headlong again into obscurity where it runs rapidly along the ground in short bursts; at the end of each course of running the tail is elevated and slightly expanded; the dark brown tail with its bright chestnut base is very conspicuous in flight and readily leads to identification. The alarm note and ordinary call is a harsh *tack*, but on its breeding grounds this Bluethroat is a fine songster and mimic.

DISTRIBUTION

Breeding range.—The species *Cyanosylvia suecica* breeds in most of Europe, except Portugal, Italy, and the British Isles (though except in the north it is represented by the white-spotted form, *C. s. cyanecula*) and across Asia eastward to east Siberia and Manchuria, south to Armenia, Iran, Ladakh, and southern Mongolia. A number of races are recognized. The range of *C. s. robusta*, the form to which the Alaskan breeding birds in all probability belong, is given by Hartert and Steinbacher (1938) as covering the larger part of Siberia from the Taimyr and the lower Tunguska to the Tchuktschi Peninsula.

Winter range.—The winter range of the species comprises North Africa, Iraq, Iran, India, and southeastern Asia. Hartert (1910) gives that of *C. s. robusta* as Burma, Assam, Farther India, and China.

Spring migration.—The bluethroat is a somewhat late migrant, the European birds passing through Britain and Helgoland for the most part in May and reaching their breeding grounds late in May or even in June. In Chihli, northeast China, *C. s. robusta* also occurs on passage in May (La Touche). Thayer and Bangs (1914) record the first arrival of this race on its breeding grounds at Nijni Kolymsk on May 31. In Alaska, Nelson (1887) records a party at St. Michael on June 5.

Fall migration.—The passage of European bluethroats in Britain takes place from the end of August to the second week of October, but mainly in the latter half of September. The passage of *C. s. robusta* in northeast China also takes place in September and October (La Touche).

CALLIOPE CALLIOPE CAMTSCHATKENSIS (Gmelin)

GREATER KAMCHATKA NIGHTINGALE

HABITS

One of the pleasantest surprises and one of the most important results of our expedition to the Aleutians in 1911 was the addition of this beautiful and attractive bird to the North American list. It was a mere straggler, of course, for no other ornithologist had ever reported it as occurring at all regularly on these islands. My assistant, Mr. McKechnie, shot one of these birds on Kiska Island on June 17, and saw two others at the same time. They were near the beach about some old buildings, but were too wild for him to secure any more. Dr. Alexander Wetmore, another member of our party, saw one there two days later, but was unable to secure it.

Dr. Leonhard Stejneger (1885) records the capture of an immature specimen on Bering Island on January 29, 1883, where he says that it is only an accidental visitor. It is therefore a rarity even in the more western Commander Islands, and the capture of our specimen, now in the United States National Museum, extends its range several hundred miles eastward.

It is a common bird on Kamchatka and an exquisite songster, of which Dr. Stejneger (1885) writes: "Kamtschatka's Nightingale, one of the loveliest birds I ever saw or heard, breeds plentifully round Petropaulski, especially in the sunny alder-groves on the slopes above and behind the town. In the late spring of 1883 I shot the first male arrivals on the 22d of May. It was absolutely silent, creeping shyly among the lower branches of the bushes. During the following autumn I met several in the latter part of September. They were found especially in a narrow valley on the eastern side of the graveyard, the same place where Kittlitz, more than fifty years ago, had collected his specimens during the same season of the year. About the 1st of October all had left."

Austin H. Clark (1910) says that this "was the most abundant bird about Petropaulski and also the best songster. Its fine, clear song was the most characteristic bird note of the place, and was heard from sunrise to sunset. This species shows a preference for hillsides covered with scrubby growth, in which it is very adept at concealing itself. It is also common on the lowlands where any little clumps of bushes occur sufficient to afford it shelter. Most of its time is spent on or near the ground, but the song is usually delivered from the tops of the bushes or the lower limbs of small trees. If surprised in such a situation, the bird is very quick to take refuge in the thick underbrush."

Spring.—La Touche (1925) says that the "ruby-throat," as he calls it, "winters in Formosa and as far as India and the Philippine Islands. It nests in N. W. China, in Siberia, Kamtschatka, the Kurile Islands, and Yezzo, and it passes through South Japan in numbers, as I saw many migrants at sea off the South coast of Hondo during the month of May. * * * Although not a conspicuous bird in South China, it must be fairly common at times of passage. I once met with a number in a bean-field near Chinkiang, but that was the only time when I saw many together."

Tsen-Hwang Shaw (1936) says that it passes through Hopei Province in China in May, and that it is found in cultivated fields and reed beds. "Its food consists of ants, wasps, small beetles, and other insects."

The following two paragraphs, on nesting and eggs, are contributed by Bernard W. Tucker.

Nesting.—Dybowski, quoted by Taczanowski (1872), says: "It nests on the ground in very well concealed places, either in heaps of boughs and small bits of wood swept together by the floods, or else in thickets or dense grass, or under the shelter of hillocks. The nest is found only by accident; we found only a few, although the bird is so numerous. The nest is domed and has an opening at the side. It is constructed of dried marsh grass, and lined with a few bents. Although artistically built, the structure is weak, and it is difficult to remove it without destroying its original shape."

Eggs.—The same authority writes: "Late in June the female deposits five oval-shaped eggs; some, however, are rather elongated, others shorter and stouter; and they have a slight gloss. The ground color is greenish blue; and the entire surface is marked with very pale brick-red (almost imperceptible) spots, which are rather more thickly scattered round the larger end. They measure from 18.8 by 15.3 millimeters to 21.4 by 16 millimeters."

In his later, posthumously published "Faune Ornithologique de la Sibérie Orientale," Taczanowski (1891) quotes measurements of 16 eggs, which, excluding an obviously abnormal one measuring 25.5 by 16.6 millimeters, show an average of 21.11 by 15.5 millimeters; maximum, 22 by 16 millimeters; minima, 20.4 by 15.2 and 22 by 14.8 millimeters.

Plumages.—As this pretty little bird has not been described in North American manuals, it seems worth while to include Shaw's (1936) description of the adult plumages, as follows:

Adult male. Entire upper parts olive brown, feathers of forehead and crown faintly edged paler and centred darker; * * * tail brown edged with olive brown; a superciliary line from base of bill to the eye and a broad moustachial stripe white; lores and under the eye black; ear regions olive brown with whitish shaft-streaks; chin, throat, and fore neck glossy scarlet surrounded by a narrow black line; sometimes the scarlet feathers fringed with white; upper breast brownish gray, shading into buffish gray on lower breast and flanks, and into nearly white on the centre of abdomen; under tail coverts buff.

Adult female. Differs from the male in having no scarlet or black on the chin and throat; these parts being dull white; breast buffish brown; lores brownish black; cheeks pale brown; the superciliary line indistinct and the moustachial streak dull.

A bird collected by Dr. Stejneger (1885) on Bering Island on January 29, 1883, presumably a young male, "has the throat and chin white, with some mottlings of the lovely scarlet, which adorns these parts in the adult male."

La Touche (1920) says: "The female of this bird is generally described as having the throat white, but old females have sometimes a considerable amount of the ruby colour. Two of these birds taken at Shaweishan on the 8th of May and 27th of October have the throat as richly colored as young males, while two others taken in the same locality on the 1st of May and 29th of October have the edges of the feathers just tinted with red. The general plumage of these birds is that of the adult female."

Behavior.—Hamilton M. Laing (1925) writes:

This charming little thrush, with the ruby-jewelled throat, and the song that is said to be of the angels, was quite common and nesting in the woods surrounding Petropavlovsk. It was also found high on the hilltops where it could be called out of every alder tangle. Though in pose like most thrushes while hopping on the ground, when they perched they elevated the tail jauntily and took on a "perky" appearance. They were never seen taking elevated perches and were always found in the shrubbery. They were the first to answer the decoy call of the bird in distress and always followed the deceiver for a time to voice mild-mannered protest. When alarmed or curious the male often gave a little whistle, a note not heard from any American thrush.

Vaughan and Jones (1913) say that "the Siberian Ruby-throat is rather a rare winter visitor" in southeastern China, "but it has such very skulking habits that it is seldom seen. One was shot at Kong Mun from a boat in mistake for a rat, as it was running among some reeds close to the water's edge."

All writers seem to agree that the Kamchatka nightingale is one of the most popular cage birds among the Chinese, on account of its brilliant scarlet throat and its charming song. Capt. H. A. Walton (1903), writing of the birds of Peking, says: "Many Ruby-throats were caught during May. They seem to thrive well, for a time at least, on a mixture of finely chopped up raw meat and bean-flour paste; but a bird that has passed safely through the winter in captivity commands a good price."

Voice.—Although everyone praises the angelic song of the Kamchatka nightingale, no one seems to have given a very good description of it. But Mr. Laing (1925) says of it: "At the time of our arrival,

July 15, the song season was evidently over. Only once was a song suspected from this bird. On the morning of the 22nd about 4 miles from town, 'a single charming song was heard from the wooded hillside—a thrush for a certainty. It had the quality. It was inexplicably sweet—as fine as a Hermit Thrush—even finer, sweeter, and quite as light and sentimental.' Let us hope the Kamchatkan Nightingale was the author."

Mr. Tucker contributes the following items: "Impressions of the song vary somewhat. Dybowski says that the rubythroat is one of the pleasantest songsters in Dauria, but describes the song as soft, quiet and somewhat unvaried. Others are less sparing in their praise: Clark (1910), for example, speaks of its fine, clear song and (1945) calls it 'the finest and most persistent songster in Kamchatka'; David and Oustalet (1877) describe it as equally remarkable for the vivacity and grace of its movements as for the beauty of its song. Seebohm (1879), again, speaks of a wonderfully fine song, richer and more melodious than that of the bluethroat and scarcely inferior to that of the nightingale. In any case the song is so much esteemed by the Chinese that it is one of the most popular cage birds in northern China. In the wild state the song is usually delivered from the tops of the bushes or the lower limbs of small trees. If surprised in such a situation, the bird is very quick to take refuge in the underbrush (Clark, 1910). Laing (1925) observed that 'when alarmed or curious the male often gave a little whistle, a note not heard from any American thrush.' It has also a harsh scolding note *tic, tic, tic,* mentioned by Seebohm."

Field marks.—"The rubythroat is a small bird of the thrush type with uniform brown upperparts and, in the case of the male, a bright ruby-red throat and broad white superciliary and mustachial stripes. The female is less distinctive. She has the throat white, sharply defined from the buffish breast, the superciliary stripe buffish white, and no mustachial stripe" (Tucker, MS.).

Fall.—The southward migration of the Kamchatka nightingale seems to be mainly coastwise and even largely at sea. Mr. Clark (1910) says that the United States National Museum has specimens of this larger, northern form from the following localities: "Hakodate, Yezo (2); at sea off Kinkesan Light, Hondo; Yaeyama Island; Amoy, China; Malate, Philippines. * * * During the first two weeks of October, when we were about the southern Kurils and the eastern coast of Yezo and Hondo, these birds were frequent visitors to the ship. One was captured on October 10, several miles east of Kinkesan Light, on the coast of Hondo."

La Touche (1920), referring to northeastern Chihli, in northern China, says: "The Common Ruby-throat is scarce at Chinwangtao in spring, when it passes in May. During the autumn passage it is very abundant, and passes then from about the 10th of September to the end of that month."

Shaw (1936) reports it as passing Hopei Province in September and even in the first part of October.

Winter.—Referring to the Kamchatka nightingale in its winter home in the Philippines, John Whitehead (1899) writes:

This beautiful migrant from the north is common in the highlands of Luzon, being met with from the coast-line up to the summit of the highest mountains. It is shy and easily alarmed, passing most of its time in the thick tangled growth, where pursuit is almost impossible. I have seen this species on the slopes of Monte Dulungan, in Mindoro, and in Negros (within a few yards) in a native garden. At Cape Engaño, in the month of May (30th), a female of this species flew into my tent and settled for a moment on one of my collecting-boxes; the birds were then migrating north, and were common in some low plants amongst the seadrift. The natives call this bird "Kerin," a word which resembles its note, but it also has an alarm-cry, not unlike the croak of a frog. It is a frequenter in North Luzon of the overgrown banks of rocky streams, and is decidedly more active after sunset, flying about after dark, when its note "kerin" may be heard.

DISTRIBUTION

CONTRIBUTED BY BERNARD WILLIAM TUCKER

Breeding range.—*C. c. camtschatkensis* breeds in Kamschatka and the Kurile Islands. The typical race ranges from the Altai to the Ussuri region, southward to Manchuria and Transbaikalia, northward to about the Arctic Circle in Yakutsk and the Yenisei region, and sporadically in western Siberia and west to Perm; also north China (Kansu, northern Szechwan).

Winter range.—India, Tenasserim, South China, Hainan, Formosa, the Riu-Kiu Islands, and the Philippines. On passage in Japan.

Casual records.—Ufa and Orenburg Governments in Russia, Caucasus, south France, Italy.

MYADESTES TOWNSENDI (Audubon)

TOWNSEND'S SOLITAIRE

HABITS

Audubon (1840) named and figured this rather puzzling bird from a single female obtained by that pioneer naturalist J. K. Townsend near the Columbia River; this one specimen remained for a long time unique. It is now known to have a wide distribution in the mountain regions of the West, from central eastern Alaska and southwestern Mackenzie to southern California, Arizona, and New Mexico.

Its status has at last been fixed as a member of the thrush family, though at first glance it would hardly seem to belong there. It looks and acts much like a flycatcher, with its somber colors and flycatching habits. In flight the light patches in its wings and the white in the tail suggest the mockingbird. Its feeding habits remind one of the bluebirds. But its song is decidedly thrushlike, though not equal to the songs of the star performers in this gifted group, and its spotted young proclaim its close relationship to the thrushes.

During the breeding season the solitaire is a bird of the mountains, at various altitudes in different parts of its range. In New Mexico it breeds mostly above 8,000 feet and from there up to 12,000 feet, ranging up to timberline and above it, among the stunted spruces and dwarfed willows, in summer. In Colorado, its breeding range is not much lower, from 7,000 to 10,000 feet. In his notes from Zion National Park, Utah, Russell K. Grater says: "This bird is resident throughout the year in elevations from 7,500 to 10,500 feet and is commonly seen in the lower canyons in the winter. It appears to frequent the more deeply shaded, narrow canyons much more than the more open situations." Farther north it breeds at much lower altitudes.

Its favorite haunts in the mountains are the open forests of pines and firs on the gentle slopes, which it seems to prefer to the more densely wooded and more shady forests, though it is sometimes found there also. Steep, rocky, fir-covered slopes are often favored, within the Canadian Zone. And it sometimes finds a congenial summer home in the wider canyons, where the high rocky walls support a scattered growth of stunted cedars and offer suitable crevices for nesting.

Nesting.—Townsend's solitaire is a lowly nester. It usually places its nest on or near the ground, often sunken into it, but generally the nest is protected from above by some form of overhanging shelter, which also helps to conceal it. Many nests have been found partially concealed at the base of a fir or pine, where a small cavity had rotted out or been burned out by forest fires; many such cavities exist on the fir-clad slopes of the mountains. Another common nesting site is under the overhanging bank on the side of a narrow mountain trail, where the sitting bird may be flushed by a passer-by; in such a situation the nest may be sunken into a hollow in the earth and is often concealed under the overhanging roots. A cavity under a rock or a crevice among rocks is sometimes chosen, or a rotted out cavity in a dead stump, rarely as high as 10 feet above the ground, may be used. Among the tangled roots of a fallen tree the birds may find a suitable cavity, especially if a large stone has fallen out and left a tempting hollow.

Grinnell, Dixon, and Linsdale (1930) mention a nest, found close to Lake Helen in the Lassen Peak region in California, that "was in a nook (20 by 20 centimeters) formed by three rocks on a dry, rocky ridge. The cavity had a little dry moss in the back part of it and a spray of grass at the entrance. A few hemlocks stood above and below the site on the slope, but none was nearer than fifty meters. The nest, composed of sticks and twigs, was lined with needles from silver pine."

Mrs. Bailey (1928) says of a nest in New Mexico: "When climbing Pecos Baldy, on a flat-topped grassy ridge at 12,000 feet, where Pipits were nesting, and Horned Larks flying around with grown young, we flushed one of the Solitaires from an old charred log and to our surprise discovered its nest fitted into a burned hollow underneath, resting on the ground roofed over by the log. In this case the nest was made from material close at hand—grass and weed stems."

A most unusual nesting site is illustrated by C. Andresen (1942), who published a photograph of "a solitaire's nest built in an open cupboard of a table in a camp ground at Lake Almanor, Plumas County, California. On June 12, 1942, the nest had 3 eggs and one of the birds was incubating."

A very good description of a nest of Townsend's solitaire is published by A. W. Anthony (1903), furnished by J. W. Preston, to whom the nest was sent:

At the base of the nest is a quantity of disintegrated trash such as bits of bark, pieces of weed stalks and finely broken old grass stems and blades, with some dirt and dust which had evidently been scratched up from the bottom of the cavity. On this slight platform are dead sticks and twigs, from larch and pine, intermixed with much old faded grass, pine needles and leaves of fir, and with some bulbs and rootlets of different grass-like sedges. The materials have been drawn into the burrowed-out cavity in the bank, leaving two-thirds of the material outward from the true nest, which is of fine dry grass stems and blades finely shredded and formed into a neat, well-rounded rather shallow cup. I note a few sprays of the long, black moss so common among the fir trees of the mountains. The structure before me is oblong in outline, being ten inches long by five wide, and three and one-half inches deep. In the inner end is formed the neat, symmetrical nest, cunningly resting in so great an amount of superfluous matter. The inside measurements are one and one-half inches deep by two and nine-tenths across. The structure is of course, somewhat compressed in boxing.

Grinnell and Storer (1924) describe a typical nest, as follows: "It was in a cut bank, three feet above the road and two feet below the top of the bank, in a depression in the earth between rocks and at the base of a young fir tree the outstretching roots of which partially concealed the nest. As is usual with the solitaire, a straggling 'tail' or apron of material extended down the bank a foot or so from the nest proper. The constituent materials of the latter were slender dead fir twigs and old, brown needles of sugar and Jeffrey pines. Inside, the nest was about 3 inches (80 mm.) across and 2 inches (50 mm.) deep."

The nests are not all as large as the one so fully described above;

they vary greatly in size to fit the cavity occupied; the material used may amount to merely a few handfuls or less, and rarely none at all is used; the material is often only carelessly thrown into the cavity, so that it looks like a wind-blown mass of rubbish that had lodged in a depression; and the long tail or apron straggling out below adds to the delusion.

J. K. Jensen (1923) reports a nest in which no nesting material was used: "It was in a clay bank beside the road in the Santa Fe Canyon. The bird had evidently scratched the little pocket out in which the eggs were deposited. The four eggs were resting on the bare ground, and there was not even a suggestion of nest building."

Eggs.—Townsend's solitaire lays three to five eggs, most commonly four and only rarely five. They are usually ovate but sometimes short-ovate or elongate-ovate, and very rarely slightly pointed. They are only slightly glossy. The eggs are entirely different from the eggs of other North American thrushes and are often very beautiful. The ground color is usually dull white, but sometimes a very pale light blue or bluish white, or more rarely greenish white or yellowish white, and very rarely with a shade of pinkish white. They are more or less evenly covered with small spots or very small blotches or scrawls of various shades of brown, reddish brown, yellowish brown, or darker browns, together with underlying spots or blotches of "ecru-drab" or "lavender-gray." The markings are sometimes concentrated at the larger end, or consolidated into a ring of spots.

Two published descriptions are worth quoting. The eggs sent to J. W. Preston, of Baxter, Iowa, are described by Mr. Anthony (1903) as follows: "The ground color of the eggs is faint greenish-blue, blotched and marked with pale chestnut and lavender. Some of the spots are large, and a number of irregular markings resembling written characters appear, well scattered over the surface, but heavier about the larger end. Two of the eggs are less heavily marked, the specks and spots being smaller. These eggs appear somewhat elongate."

Dr. Joseph Grinnell (1908) says of the eggs taken in the San Bernardino Mountains:

> The four sets of eggs taken, conform to one general type of coloration, though there is some variation. All the eggs of each of the four sets are practically identical among themselves. Two extremes of coloration may be described. In one style the ground color is white, with the palest possible tint of grayish-blue. The markings are so profuse as nearly to obscure the ground, doing so completely about the larger ends. These markings vary from brick red, through an unbroken series of tints to very pale lavender; but a vinaceous tint prevails. The markings are in the nature of blotches and finer dots and points, often blurred together. In the other style of egg the ground is white with a decided pale blue tint, spattered with blotches and spots of lengthwise trend. These are thickest at the large end, bold and distinct, not running together, and are in color lavender, vinaceous, brick red and burnt sienna.

The measurements of 50 eggs average 23.5 by 17.2 millimeters; the eggs showing the four extremes measure **26.5** by 18.1, 22.9 by **18.3**, **20.8** by 17.3, and 22.8 by **16.2** millimeters.

Young.—No one seems to have worked out the incubation period for the Townsend's solitaire. Probably what nests have been found have been robbed by egg collectors. Nor does it seem to be known how long the young remain in the nest, nor at what rate they develop. Mrs. Wheelock's (1904) brief experience with a brood of young solitaires would seem to indicate that both parents assist in the care of the young and are very solicitous for their welfare. The late dates at which fresh eggs have been found suggest that two broods are often raised in a season, but the evidence is not conclusive.

Plumages.—Ridgway (1907) describes the striking juvenal plumage very well, as follows: "Pileum, hindneck, back, scapulars, rump, upper tail-coverts, and lesser and middle wing-coverts conspicuously spotted with buff, each feather having a single spot of this color, approximately rhomboid or cordate in shape, the feathers broadly margined with blackish, causing a somewhat squamate effect; under parts pale buff or grayish buff, the feathers margined with black or sooty." The wings, except the coverts, and the tail are as in the adult; the greater wing coverts, which are not renewed at the postjuvenal molt, are tipped with buff, which fades out to white during winter, and the tips are largely worn away before spring.

The postjuvenal molt begins early in August and is usually completed before the end of September. This involves all the contour plumages and the lesser and median wing coverts but not the rest of the wings nor the tail. It produces a first winter plumage, which is practically indistinguishable from that of the adult.

Adults have one complete postnuptial molt, beginning sometimes as early as the middle of July and continuing mainly through August. June and July birds are usually in much-worn plumage, and many are in fresh plumage again before the end of September. There is but little seasonal change in plumage; there is, apparently, no spring molt, but wear reduces the extent of the white on the greater coverts and the tertials, and the body plumage is somewhat grayer, less brownish. The sexes are alike in all plumages.

Food.—Professor Beal (1915b) examined only 41 stomachs of Townsend's solitaire, too few in his opinion to "draw general conclusions." The food was made up of 35.90 percent of animal matter and 64.10 percent of vegetable. Of the animal food, Lepidoptera in the form of caterpillars made up the largest item, 12.95 percent for the year; one stomach, taken in May, held 72 percent caterpillars. Beetles constituted the second largest item, 10.74 percent, of which 5.89 percent were the useful predatory ground beetles (Carabidae), 95 percent of the contents of one January stomach and 93 percent of the food in an October stomach consisted of Carabidae. Ants were eaten to the extent of 4.71 percent, bees and wasps amounted to less than 0.5 percent, and he found only a trace of flies (Diptera); it seems strange that a bird, supposed to take so much of its food on the wing, should have eaten so few of these flying insects. Hemiptera were found to the extent of 3.51 percent, grasshoppers amounted to less than 1 percent, and there was only a trifle of other insects. Spiders were eaten to the extent 2.94 percent, and there was one hairworm (*Gordius*).

More than half of the vegetable food was wild fruit or berries, and there was no evidence that any cultivated food had been taken. He found cedar berries in six stomachs, madrona berries in five, hackberries and rose haws in two each, and serviceberries, wild cherries, sumac berries, poison ivy, waxwork, honeysuckle berries, and elderberries in one stomach each.

Strangely enough, he does not mention mistletoe seeds, which others have referred to as a favorite food; these viscid seeds are swallowed whole and passed through the alimentary canal to adhere where they fall; thus these birds help to spread this parasite, as well as the poison-ivy. Pine seeds, pinyon seeds, and kinnikinnick berries have been mentioned by other observers.

Dr. G. F. Knowlton has sent me the following note on the contents of two stomachs: "Recognizable stomach contents consisted of one nymphal Orthoptera, six Hemiptera, one being a scutellerid and another a mirid; three adult caddisflies; 11 beetles, one being a weevil and another a click bettle; four lepidopterous larvae, apparently cutworms; two Diptera, one being a crane fly; 17 Hymenoptera, all but two of which were ants, three being carpenter ants. One stomach held four berries; the other contained plant pulp and plant fragments." The birds were taken on June 20 and July 2, in Utah.

I. McT. Cowan (1942) includes Townsend's solitaire among the birds that feed on the termite *Zootermopsis angusticollis*. And Leslie L. Haskin (1919) adds angleworms to the list, "which it secured Robin fashion, except that instead of watching for them from the ground it would drop down upon them from the lower limbs of the fruit trees, returning immediately to its perch. In fact, during the entire time I watched it, I did not see it take more than half a dozen hops along the ground." He also watched it "taking its prey in Bluebird fashion, by watching for it from fence-posts and stumps, and dropping to the ground only when an insect had been located, returning immediately to its point of observation."

Many observers have referred to the solitaire's flycatching habits, and it has been called the "flycatching thrush." Samuel F. Rathbun watched a pair thus engaged for nearly half an hour and says in his notes: "In their actions these birds were almost identical with flycatchers, sitting erect on or near the extremity of some limb, well toward the top of the tree, and from this location they would fly out and catch the passing insects. At all times the birds were perfectly silent, for during the entire time of observation neither uttered a note."

H. W. Henshaw (1875) evidently never observed this habit, for he says: "The habit of catching insects on the wing, after the manner of the Flycatchers, which is attributed to this bird, appears to be not a common one, or, as is likely the case the bird varies its habits in different localities, as, of hundreds I have seen at different seasons, none were ever thus engaged, nor have I ever seen them searching among the leaves for insects, like the thrushes. In their usual manner of procuring food, as in their habits and motions generally, they have always seemed to me nearly allied to the Bluebirds."

Behavior.—Many of the solitaire's traits have been referred to above, as well as some of the points on which it resembles other species in appearance and manners. Dawson and Bowles (1909) have summed this up very well, as follows:

Barring the matter of structure, which the scientists have now pretty well thrashed out, the bird is everything by turns. He is Flycatcher in that he delights to sit quietly on exposed limbs and watch for passing insects. These he meets in mid-air and bags with an emphatic snap of the mandibles. He is a Shrike in appearance and manner, when he takes up a station on a fence-post and studies the ground intently. When its prey is sighted at distances varying from ten to thirty feet, it dives directly to the spot, lights, snatches, and swallows, in an instant; or, if the catch is unmanageable, it returns to its post to thrash and kill and swallow at leisure. During this pouncing foray, the display of white in the Solitaire's tail reminds one of the Lark Sparrow. Like the silly Cedar-bird, the Solitaire gorges itself on fruit and berries in season. Like a Thrush, when the mood is on, the Solitaire skulks in the thickets or woodsy depths, and flies at the suggestion of approach. Upon alighting it stands quietly, in expectation that the eye of the beholder will thus lose sight of its ghostly tints among the interlacing shadows.

It is generally regarded as a solitary, quiet, retiring bird, often being seen singly, in pairs, or in family groups, but at times, mainly on the fall migration, it is sometimes seen in larger groups. Henshaw (1875) mentions such a gathering: "At the Old Crater, forty miles south of Zuni, N. Mex., they had congregated in very large numbers about a spring of fresh water, the only supply for many miles around; and hundreds were to be seen sitting on the bare volcanic rocks, apparently too timid to venture down and slake their thirst while we were camped near by."

Ridgway (1877) records the thrushlike behavior of the solitaire in the vicinity of its nest: "As we walked along the embankment of a mining-sluice it flitted before us, now and then alighting upon the ground, and, with drooping and quivering wings, running gracefully, in the manner of a Robin, then flying to a low branch, and, after facing about, repeating the same maneuvers—evidently trying to entice us away from the spot."

The flight of the solitaire is not swift, probably not over 20 miles an hour in direct flight; but the flight is usually not direct or much protracted, and is more or less erratic; Grinnell and Storer (1924) say that it reminds one of Say's phoebe, "in that the wings are widely spread and flapped rather slowly, and the flight course is irregularly circuitous."

It is ordinarily a gentle bird and not inclined to quarrel with its neighbors, but it is very solicitous in the defense of its home and will often drive away other birds from the vicinity of its nest or young.

Voice.—Much has appeared in the literature in praise of the charming song of Townsend's solitaire, and I have some interesting contributed notes on it. Aretas A. Saunders (MS.) praises it as "one of the most glorious and beautiful of bird songs" and says that it "is a rather prolonged, warblelike series of rapid notes, each note on a different pitch than the last. The notes are clear, sweet, and loud, and follow each other almost as rapidly as those of the winter wren."

Samuel F. Rathbun heard one singing in the Olympic Mountains, on July 19, 1920, and says in his notes: "While we were eating lunch the song of this bird suddenly rang out not far away. It seemed to come from near the top of one of the trees in a nearby grove of conifers. The song was most beautiful, full and clear, with sparkling, ringing notes, some of which remind one of the song of the purple finch at its best. But the solitaire's song has much more volume and is more brilliant. It was given a number of times, and well fitted its surroundings, for there was a swing to it that went with the expanse of the mountain heights."

Dr. Louis B. Bishop (1900) praises it highly in the following words, as he heard it in Alaska: "On the hot noon of June 26, while seated on the summit of a hill some 1,500 feet above Caribou Crossing, I heard the most beautiful bird song that has ever delighted my ear. It seemed to combine the strength of the robin, the joyousness and soaring quality of the bobolink, and the sweetness and purity of the wood thrush. Starting low and apparently far away, it gained in intensity and volume until it filled the air, and I looked for the singer just above my head. I finally traced the song to a Townsend solitaire that was seated on a dead tree about 150 yards away, pouring forth this volume of melody without leaving its perch."

Ralph Hoffmann (1927) says that it has "the quality of the Black-headed Grosbeak's song and a tempo between that of the Warbling Vireo and the Purple Finch." The song has been said by others to resemble the warbling of the bluebird, a decided compliment to the bluebird, and also the mockingbird, the California thrasher, and the sage thrasher, all of which seems a bit fanciful.

Forrest S. Hanford (1917) writes thus attractively of the songster in the solitude of its mountain retreat:

The little shadowy canyon wherein I rested enjoyed a hushed and solemn tranquility not diminished, but rather added to, by a drowsy murmuring from a bright brook splashing on its way to the lake. This, I thought, could be none other than the haunt of a Solitaire, and I wished that I might see the bird; and as in answer to my prayer came one, a small gray ghost of a bird that flitted silently in and out the leafy corridors of its retreat, finally resting on the limb of a pine not ten feet away. And as I watched, the feathers of his breast and throat rose with a song that softly echoed the beautiful voices of the brook, the gurgling of eddies, the silvery tinkle of tiny cascades, and the deeper medley of miniature falls. Infinitely fine and sweet was this rendering of mountain music. At times the song of the bird rose above the sound of the water in rippling cadences not shrill, but in an infinite number of runs and modulated trills, dying away again and again to low plaintive whispering notes suggestive of tender memories.

The star performance of the solitaire is its flight song, which has been referred to by only a few observers. Mr. Saunders says in his notes: "The flight that accompanies the song varies greatly. As I have observed it, the bird hovers for a long time high in the air and sings continuously while doing so." Charles L. Whittle (1922) who observed it, also in Montana, has published a diagram of the flight, and has written the following description of it:

On May 15 my attention was attracted to the Solitaires by hearing them sing as they were migrating northerly over the mountains as single birds and in pairs. They commonly flew well above the mountains so that identification was made by their songs. * * * A number of times on this date a Solitaire could be heard singing high in the air and well above us up the mountain, and sometimes it could be seen coming down the steep slope just over the trees with great velocity, alighting suddenly on a tree top, when he would again burst into song. On May 24 I witnessed the beginning of a song-flight, no doubt a courtship performance, of which the precipitate descent over the tree tops just described is the termination, although at that time the birds appeared to have mated.

I was standing on a nearly treeless ridge, at an elevation of 7,300 feet, when a Solitaire which was singing close by on a stunted pine, flew upward in two series of irregular spirals. The first series was made by circling to the left, and the second series by circling to the right, as shown diagramatically in figure 30. By this method the bird mounted to a height of perhaps 500 feet, singing at intervals. Then he started off as though to leave the vicinity, when, suddenly and with astonishing velocity, he plunged downward, apparently with set wings, in a succession of steeply-pitched zigzags, almost to the ground, and then turned abruptly upward again in a second series of spirals of the same character, which ended in another zigzag drop of at least 700 feet when he disappeared down the slope.

Authorities seem to differ greatly as to the singing season of the solitaire; several have reported its singing in fall and winter, and some state that it ceases to sing during the normal song period of other birds, late in spring and early in summer. Mr. Whittle (1922) quotes a number of observers on the subject, and then sums up the evidence, as follows: "The Solitaire is thus reported, by the combined testimony of several observers, to be in song, at least at intervals, from September to February inclusive, and by two observers to be silent during the customary singing season. Others, however, including the writer, find the species quite normal in the matter of having the usual spring singing period. It is difficult to account for the reports that this species does not sing during the courting and nesting seasons."

Mr. Saunders (MS.) says on this subject: "The season of song of the Solitaire, judged from the small amount of data I have, begins in March or April and continues to the middle of July, my earliest and latest dates being March 15, 1910, and July 20, 1911. It frequently sings in fall. In most years I heard it early in October, but in 1908 I heard it from September 7 to October 23, which are my earliest and latest dates for fall singing."

Russell K. Grater writes to me: "At Cedar Breaks National Monument, at an elevation of over 10,000 feet and in the dead of winter, I have heard solitaires singing loud and clear from the trees, while snow several feet in depth covers the ground. This song was the same familiar one heard in the warmer months."

Dr. Coues (1874) quotes T. M. Trippe as saying:

> Toward the middle and latter part of winter, as the snow begins to fall, the Flycatching Thrush delights to sing, choosing for its rostrum a pine tree in some elevated position, high up above the valleys; and not all the fields and groves, and hills and valleys of the Eastern States, can boast a more exquisite song. * * * At first it sings only on bright, clear mornings; but once fairly in the mood, it sings at all hours and during the most inclement weather. Often while travelling over the narrow, winding mountain roads, toward the close of winter, I have been overtaken and half-blinded by sudden, furious storms of wind and snow, and compelled to seek the nearest tree or projecting rock for shelter. In such situations I have frequently listened to the song of this bird, and forgot the cold and wet in its enjoyment. Toward spring, as soon as the other birds begin to sing, it becomes silent as though disdainful of joining the common chorus.

Townsend's solitaire also has some short, metallic calls or alarm notes, which have been written as *tink*, *tink*, or *clink*, or *peet*, and which are somewhat ventriloquial in effect. They suggest similar notes from some of the other thrushes. Harry S. Swarth (1922) writes, referring to the Stikine River region: "The solitaires did not sing much but the call note was uttered continually. From our rooms in town at Telegraph Creek, this was one bird note that could be heard hour after hour, monotonously repeated nearly the whole day through. To our ears it sounded so nearly like the distant barking of a California ground squirrel (*Citellus beecheyi*) that the sound would surely have been disregarded as a bird call had we been in a region where the squirrels occur."

Field marks.—The solitaire can be distinguished as a long, slim, brownish-gray bird with a long tail, a short bill, and a light eye ring. In flight the white outer tail feathers and the buff areas in the wings show conspicuously. It suggests a mockingbird, but its coloration is much duller.

Fall.—Henshaw (1875) writes: "They are quite common, in the fall, in Eastern Arizona and Western New Mexico. Having reared their young, these birds appear to forsake the pine woods, which constitute their summer abode, and appear lower down on the hill sides, covered with piñons and cedars. Their food at this season appears to consist almost exclusively of berries, particularly from the piñons and cedars, and the crops of many examined contained little else save a few insects."

Grinnell and Storer (1924) say: "The Townsend Solitaire as a species does not, in the Yosemite region, make much of a change in its haunts with the passage of the seasons. In summer the majority are to be found in and about the red fir forests of the Canadian Zone. At other times of year the birds forage and live in the western junipers which often grow close by on rocky slopes, or else they drop to the Transition Zone where mistletoe berries on the golden oaks afford bounteous forage. There are no solitaires in Yosemite Valley during the summer months, but with the coming of winter the oaks on the talus slopes become tenanted by numbers of the birds."

Frank M. Drew (1881) says that, in Colorado, "in fall the Solitaire comes out of the woods and can be found around houses, or in low bushes near water."

Winter.—Townsend's solitaire does not seem to be much affected by low temperature, its haunts and its movements in winter being dependent on the food supply in the shape of fruit and berries, in search of which it wanders about in large or small groups or in family parties. It spends the winter throughout most of its summer range, except in the most northern part of it, but at lower levels than it occupies in summer. It has been known to winter as far north as Montana, during the severest seasons, even when the thermometer is flirting with zero and winter storms are howling.

In the vicinity of his ranch, in the lowlands of Montana, E. S. Cameron (1908) records the solitaire as a winter resident, and evidently not present in summer. He says that it arrives the "second week in September and leaves middle of April. * * * A pair frequented my ranch in Dawson County during November 1904, and throughout October and November in 1905. On November 25, these were joined by two others when all four seemed to live near the water troughs and playfully chased each other round and round the cedars. They were not seen after a blizzard on Nov. 28, when the temperature fell to 14° below zero, but they are able to withstand severe cold, as a pair returned at the end of January and remained until April 14."

Frank Bond (1889) gives the following account of a great gathering of solitaires in a canyon near Cheyenne, Wyo., in winter:

On the walls of the cañon, especially in the less precipitous places, there flourishes a scattering growth of scrub cedar whose branches were well laden with the dark blue cedar berry. Living, I believe, almost entirely upon these berries, for a winter diet, were countless thousands of Townsend's Solitaire (*Myiadestes townsendii*) and Robins (*Merula migratoria propinqua*). I saw also *Sitta canadensis* and several Long-crested Jays (*Cyanocitta s. macrolopha*). Both the Solitaires and Robins were acting like school children out for a holiday. They would chase one another hither and thither, now up to the brow of the cañon 500 or 600 feet above, now back and forth across the mirrored ice of the river below, and all the while singing and chattering like mad. It warms one's heart to enter such a vale of melody in cold December." [The birds were still there up to February 7.]

In El Paso County, Colo., according to Aiken and Warren (1914), the solitaire is "a solitary bird in summer, but sometimes they congregate in flocks of 20 or more in warm, sheltered cañons and gulches in winter. Early in 1911 Solitaires were seen in the residence portion of Colorado Springs several times, which is something unusual."

DISTRIBUTION

Range.—Western North America from the Arctic Circle to central Mexico.

Breeding range.—The Townsend's solitaire breeds **north** to central eastern Alaska (Yukon River 20 miles above Circle); northwestern Yukon (Bern Creek, Selwyn River, and the Semenof Hills); and southwestern Mackenzie (Mount Tha-on-tha, at the mouth of the Nahanni River). **East** to southwestern Mackenzie (mouth of the Nahanni River); the mountains of western Alberta (Jasper Park, Banff, and Calgary); western Montana (Lake McDonald, Flathead Lake, Billings, and Kirby); northeastern Wyoming (Bear Lodge Mountains); western South Dakota (Black Hills); northwestern Nebraska (Squaw Canyon and Pine Ridge, in Sioux County); southeastern Wyoming (Wheatland and Laramie); central Colorado (Estes Park, Buffalo Creek, Manitou, and Fort Garland); and central northern New Mexico (Taos Mountains and Pecos Baldy). **South** to northern New Mexico (Pecos Baldy, Santa Fe, and Fort Wingate); central Arizona (White Mountains and Fort Whipple); and southern California (San Jacinto Mountains, San Bernardino Mountains, and

Mount Pinos); also in the Sierra Madre of Mexico from northwest Chihuahua (Colonia Garcia) to northwestern Zacatecas (Sierra Madre). **West** to southern California (Mount Pinos), the Sierra and Cascade ranges in California (King's Canyon, Yosemite, Fyffe, Butte Lake, and Salmon Mountains); Oregon (Pinehurst and Prospect); Washington (Mount St. Helens, Mount Rainier, Tacoma, and Seattle); western British Columbia (Mount Benson, Vancouver Island; Glenora, Telegraph Creek, and Atlin); southwestern Yukon (Carcross and Burward Landing); and eastern Alaska (Chitina Moraine, Joseph Village, and the Yukon River above Circle).

Winter range.—In southern British Columbia and in most of the United States range of the solitaire the migration seems to be principally altitudinal. The species winters **north** to southern British Columbia (Victoria, Sumas, the Okanagan Valley, and Arrow Lakes); northern Idaho (St. Joe National Forest near St. Maries); and central eastern Montana (Terry). **East** to eastern Montana (Terry and Kirby); eastern Wyoming (Platte Canyon and Laramie); eastern Colorado (Fort Morgan and Manitou); rare or accidental east to southeastern South Dakota (Vermillion); eastern Nebraska (Omaha and Lincoln), and Kansas (Topeka); western Texas (Palo Duro Canyon near Amarillo, Guadalupe Mountains, and Chisos Mountains; rarely to Kerrville); and southwestern Chihuahua (Maquerichic). **South** to southern Chihuahua (probably to the limit of the breeding range in Zacatecas); northern Sonora (15 miles south of Nogales); and northern Lower California (Sierra San Pedro Mártir, and Guadalupe Island, one record). **West** to northwestern Lower California (Tecate), central California and occasionally the coastal region (Indio, Claremont, Santa Barbara, the valleys of the Sierra Nevada, Berkeley, Davis, and Paynes Creek); Oregon (Klamath Basin); western Washington (Tacoma, Seattle, and Bellingham); and southwestern British Columbia (Victoria).

Migration.—Some late dates of spring departure are: Texas—Kerrville, April 17. Kansas—Hays, April 6. Nebraska—Hastings, May 26. Utah—Ogden, April 30.

Some early dates of spring arrival are: Colorado—Durango, March 25. Wyoming—Yellowstone Park, April 11. Montana—Big Sandy, March 31. Saskatchewan—Eastend, April 19. Alberta—Banff, April 20. Idaho—Rathdrum, March 4. Oregon—Corvallis, March 4. British Columbia—Chilliwack, March 29; Atlin, April 30.

Some late dates of fall departure are: British Columbia—Atlin Lake, September 9; Okanagan Lake, November 22. Washington—Pullman, October 22. Oregon—Weston, October 28. Alberta—Banff, October 20. Montana—Missoula, November 25. Wyoming—Wheatland, October 30. Colorado—Yuma, November 5.

Some early dates of fall arrival are: Montana—Terry, September 9. Colorado—Fort Morgan, September 27. Nebraska—Long Pine, October 10. Kansas—Hays, October 12. Texas—20 miles northwest of Amarillo, September 27.

The migratory movements of the solitaire seem to be rather erratic and during migration it is often found far east of its normal range; as at Lake Johnston, Saskatchewan; Stonewall, Manitoba; one banded at Wilton, N. Dak., on October 7, 1937; Falls Creek, Murray County, Okla.; and Dallas, Tex.

Casual records.—There are three records of the occurrence of the solitaire in Minnesota: a specimen collected at Collegeville, Stearns County, December 20, 1909; another taken near Fairmont, Martin County, November 30, 1916; and one at a feeding station near Groveland from January to the middle of March 1922. A specimen was collected at West Point, Wis., in February, 1910; and another in Lake County, Ill., December 16, 1875. An individual was under observation near Toledo, Ohio, from December 26, 1938 to January 14, 1939. The easternmost record is from Long Island, where a specimen was collected November 25, 1905, at Kings Park.

Egg dates.—California: 24 records, May 2 to August 7; 14 records, June 2 to June 20, indicating the height of the season.

Colorado: 20 records, May 16 to July 10; 11 records, May 30 to June 15.

New Mexico: 6 records, June 3 to July 12.

Washington: 7 records, May 22 to June 17.

Family SYLVIIDAE: Warblers, Gnatcatchers, and Kinglets

ACANTHOPNEUSTE BOREALIS KENNICOTTI (Baird)

KENNICOTT'S WILLOW-WARBLER

CONTRIBUTED BY BERNARD WILLIAM TUCKER

HABITS

Kennicott's warbler is a race of—if indeed it is really separable from—Eversmann's warbler (*A. b. borealis*) of the Old World. Ridgway (1904) stated that the Alaskan bird is smaller than the typical race, with the color of the upperparts decidedly grayer (at least in spring and summer plumage) and the underparts less strongly tinged with yellow. It must be admitted, however, that the grounds for the separation of the race *kennicotti* are not very secure. Ridgway gave the average wing measurements of seven unsexed adults as 62 millimeters against 68.5 for males and 63.7 for females in the case of birds from eastern Asia. Witherby (1938, vol. 2) gives the wing measurement for the typical race as 62–71 millimeters in males (12 measured) and 61–67 in females. This places Ridgway's measurements within the range of variation of the typical form. However, Dr. C. B. Ticehurst (1938), the leading and most recent authority on the group, gives the measurement for males of the typical form as 65–72 millimeters and of females as 62.5–66 (exceptional specimens 58.5 and 61.5), and as this is based on a large number of specimens it would seem as if an average of 62 millimeters probably does indicate a smaller race. Ticehurst himself was not able to examine sufficient material to judge the validity of the Alaskan race, but quotes the opinion of J. L. Peters, whom he consulted. Peters wrote: "This seems a rather unsatisfactory race, but I think it may be distinguished from typical *borealis* by the average smaller size and smaller bill. I compared 4 ♂♂ and 1 ♀ from Alaska with a series of Siberian breeding birds of *borealis; kennicotti* ♂ wing 63.4–67.1, ♀ 63.1 mm. The colour characters given by Ridgway do not seem to hold."

Ticehurst adds that he has seen one unsexed bird from Alaska (wing 61.5), "which is much yellower below than autumn *borealis*, and has a smaller bill than ♀ *borealis*," thus just the reverse of Ridgway's description so far as the underparts are concerned. He further notes that Friedmann (1937) also remarked on the yellowish wash on the underparts of two birds from St. Lawrence Island, Alaska, taken in July and August. Altogether one cannot but agree with Ticehurst's final conclusion that "the form requires much further study on larger material before its validity can be assured or its range demarcated."

Swarth (1934) not only queried the validity of the race *kennicotti* but even raised the question whether the species really breeds in Alaska at all. He writes: "It is an extremely rare bird on the Alaskan side, it has not been found actually nesting there, and occurrences are nearly all as in our Nunivak specimen, of migrating birds in late summer. These might be merely an overflow of migrants from the Siberian side that later retrace their course."

Reference to the original records for the localities of presumed breeding quoted in the A.O.U. Check-list shows that Swarth's comment was not unjustified. The earlier records for the coastal districts of western and southwestern Alaska include only two June dates, namely, June 14 (Grinnell, 1900) and June 19, the date of two specimens from Nushagak in the National Museum, collected by J. W. Johnson in 1884 and mentioned by Osgood (1904). Osgood himself secured two specimens near Iliamna at the base of the Alaska Peninsula on July 13 and 14, while none of the several records for St. Michael and the vicinity is earlier than July 26 (Nelson, 1887). Most other records refer to August or even September. No details are recorded about the Nushagak specimens, and those secured by Osgood were associating with other warblers, so there is little reason to think they were breeding birds. This leaves us, so far as the above records are concerned, with only the bird seen by Grinnell on June 14. This date certainly suggests breeding. The bird was hunting among some willow bushes and stunted spruces, and it was expected that it would prove to be nesting, but eventually it flew right away, so that even in this case there was no very strong evidence of breeding. Moreover, it is not irrelevant to point out that apparently not one of the earlier observers ever heard the song in Alaska, with the exception of Townsend (1887), who records that one obtained by him in a thicket far up on one of the highest hills of the middle Kowak River region was still in song on August 1.

However, the question of breeding in Alaska seems to be settled by the observations of Dixon (1938) in the Mount McKinley National Park. He states that "Kennicott's willow-warbler was a fairly common breeding bird on the upper Savage River in 1926. Here, on June 20, we found half a dozen willow warblers singing in one tract of spruce woods. * * * Three specimens were collected in June 1926, and two proved to be adult males in full breeding condition." Even Dixon, however, found no nests, and he records further that "in 1932 I repeatedly visited the exact locality where these warblers had been found in 1926, but I neither saw nor heard them. All summer a continued search was carried on in the McKinley region but not a single willow warbler could be found. The late heavy snows had apparently prevented their reaching this inland district."

Such erratic and inconstant breeding distribution from season to season, dependent mainly on weather conditions in spring, is a characteristic phenomenon with many Arctic birds: a species may be present in numbers in a given district in one year and completely absent in the next. Nevertheless, it may be suggested that the field data are not particularly favorable to the existence of a distinct Alaskan race. The writer speaks with some diffidence on this American problem, especially as there may be more recent data not accessible to him, but the evidence does seem to raise the question whether this warbler may not be still only in the process of colonizing Alaska from Asia, a question that clearly has a bearing on the status of the race *kennicotti*. Only more material from both sides of Bering Strait *and* more field study in Alaska can finally settle the problem.

Acanthopneuste, or, as the writer would prefer to call it, *Phylloscopus*, *borealis* (for Ticehurst, already quoted, has shown that the separation of *Acanthopneuste* has very little basis) is biologically noteworthy on account of its extremely interesting migration. There are well-established cases of migrants extending their winter range in relation with an extension of the breeding area, but the present species exemplifies the reverse situation, the adherence to ancestral winter quarters in spite of a great extension of summer range. There can be no reasonable doubt that the original home of the species was in the more eastern part of Siberia, with winter quarters in southeastern Asia and the Malay Archipelago. Yet the European birds have not adopted what might seem the natural course of migrating for the winter to Africa. They return annually to the much more distant ancestral winter quarters, and to do this they must first travel the long journey eastward to eastern Asia before finally turning southward through the eastern parts of China.

Throughout its summer range in the Old World this warbler is primarily a bird of the northern birch woods, though it may also be found nesting in conifer and mixed woods. In the Mount McKinley National Park, however, the typical haunts of the species are the spruce woods (Dixon, 1938). In northern Europe it is found as a rule in well-grown woodland, avoiding the mere scrub growth which satisfies its relative the willow-warbler (*Phylloscopus trochilus*), though it also appears to require a fair amount of ground cover, and it has a distinct liking for the neighborhood of streams or other water and swampy places, though in no way confined to such. In fact, it may be found in anything from dry woods to even those that are more or less flooded.

As there are next to no bionomical data on the species in America, observations in Europe will be mainly relied on in the account that follows, and as the writer has had the opportunity of studying Eversmann's warbler, as the Old World form is called, in Lapland it will be possible to draw to a considerable extent on original experience.

Migration.—The Arctic willow-warbler passes through eastern China on its spring migration during the latter half of May and the first 10 days of June. La Touche (1926) says that it "is one of the last birds seen in the spring migration, being still seen in North-East Chihli about the 10th of June. Although the song is never heard in China, it commonly utters in spring a very loud call, 'tsic-tsic.' " He says that it is "one of the earliest autumn migrants, appearing in S. E. China as early as the last days of August and in North-East China from the middle of that month."

Collett (1886) says that, in Lapland, "these birds appear to arrive rather late in the spring. Mr. Seebohm, in 1887, met with the first arrivals on June 18th, and a few days afterwards found them in considerable numbers. In 1885, at Matsjok (Tana), they could hardly have arrived before June 22nd, but two or three days afterwards they were numerous." At the other end of its summer habitat, in northeastern Siberia, Thayer and Bangs (1914) say that this species arrived at Nijni Kolymsk, May 30, 1912.

Referring to the Commander Islands, to the westward from the

Aleutians, Dr. Stejneger (1885) says: "They are not known to breed on the islands, where they have been collected only during the spring migration. In 1882 I shot only one specimen on each island, but during the 'bird-wave' of 1883, they were plentiful in the northern part of Bering Island. The birds occurring on the islands belong to the same stock as those inhabiting the mainland of Kamtschatka."

Courtship.—All the small Old World leaf-warblers have more or less well-defined postures and display actions in courtship, but those of *Acanthopneuste borealis* have yet to be adequately observed. Dixon records two birds that were "seen to perch on a limb fluttering their wings quite audibly and uttering a harsh 'chit' at frequent intervals," which sounds like a display action, though it might possibly have been aggressive, and I have observed what was evidently a sexual chase of the type met with in so many small birds, among the bushes in a Norwegian birch swamp late in June.

Nesting.—The nest is placed among vegetation on the ground and is a domed structure with entrance hole at the side, built of fine grasses with commonly some moss and dead leaves and lined with fine grass. It is characteristic that feathers are absent from the lining, though one recent author has recorded three small ones in one nest.

Collett (1886) found three nests near Matsjok, of which he says:

> The first nest I found (on July 27th) was placed at the foot of a slope thickly covered with birch trees, and was well hidden by *Cornus suecica*, halfgrown *Chamaenerion angustifolium*, *Veronica longifolia*, and *Melica nutans*. It lay under the root of a tree, which partly formed a roof to the nest. The other nest, found the same day at another slope in the wood, had no such protection; but both nests were completely domed, as is usual in those of the other *Phylloscopi*. They were most loosely constructed; the outer base was composed of some dry birch-leaves; the outside consisted of coarse straws and moss, the interior of finer straws, but without a trace of hairs or feathers. The number of young birds in the first was seven, in the other six. Each brood was about nine days old.
>
> The third nest (July 28th) also lay on a high slope covered with birch trees, protected by a thin branch of juniper and surrounded mostly by *Cornus suecica*, while the other tall forest plants here were absent. This nest was thus somewhat exposed. Like the others, it was domed and loosely put together, inside with fine straws, outside of larger, but nevertheless soft, straws, as well as a good deal of two kinds of moss which covered the ground in the immediate neighbourhood, viz. *Hylocomium splendens*, Hedw., and *Dicranum scoparium*, Hedw. The number of young was six, nearly ready to fly.

Henry Seebohm (1879) writes of its nesting in Siberia: "When I left the Arctic circle it had probably not commenced to breed; but on the 6th of July I had the good fortune to shoot a bird from its nest at Egaska, in latitude 67°. * * * The nest was built on the ground in a wood thinly scattered with trees, and was placed in a recess on the side of a tussock or little mound of grass and other plants. It was semidomed, the outside being composed of moss, and the inside of fine dry grass. There was neither feather nor hair used in the construction. I did not see this bird further north than lat. 69°."

J. H. Riley (1918) reports that Copley Amory, Jr.—

took a nest and seven slightly incubated eggs on the Kolyma, directly opposite Nijni Kolymsk, June 18, 1915. The nest was in swamp and willows on one side of a "niggerhead," with water directly below the nest and a leaning dead willow stick directly above. The nest outwardly is composed of rather coarse grass with a few pieces of sphagnum moss, loosely woven; internally of finer grass and lined with white ptarmigan feathers. The outer covering extends up over the egg cavity, forming a roof. In fact, the nest has the appearance of two nests, the outer one composed of dark-colored coarse grass and the inner of finer yellowish grass. Outwardly the nest measures about 6½ by 5 inches; the egg cavity which is rounded 2 inches. The inner nest is placed in the front of the mass that composes the outer nest.

Eggs.—The Arctic willow-warbler lays a set of five to seven eggs; probably six is the commonest number, but there are many sets of seven.

Mr. Riley (1918) describes the eggs taken by Mr. Amory as follows: "The eggs are short, ovate in shape; white, rather evenly spotted with larger and smaller spots of vinaceous russet in two tints; the spots more numerous on the larger end. They measure as follows: 16.7 by 12.6, 16.4 by 12.5, 15.5 by 12.4, 16.4 by 12.5, 16.7 by 12.7, 16.5 by 12.6, 16.2 by 12.4 mm."

Seebohm (1879) says that "the eggs are larger than those of our Willow-Warbler's, pure white, and profusely spotted all over with very small and very pale pink spots."

Young.—Only one brood is reared. There can be little doubt that only the female incubates, as in the case of its nearest relatives, though this does not seem to have been positively proved. This conclusion is supported by a statement of Robert Collett, who discovered Eversmann's warbler breeding in northern Norway and gave an excellent account of it in the Proceedings of the Zoological Society of London (1877). He mentions that a female he shot had large incubation spots but makes no mention of these in males which he obtained.

Plumages.—The plumages of *Acanthopneuste b. borealis* are described by H. F. Witherby in the "Handbook of British Birds" (1938, vol. 2) and by Ticehurst (1938, vol. 2). The nestling has not been described. The juvenal plumage is much like the adult but less greenish above, more inclined to grayish brown, and whiter below. Ticehurst states that first autumn birds incline to be brighter and greener than adults, less gray-green above and more tinged with yellow on the underparts. "Usually too they can be recognized by the unworn bar on the wings and the freshness of the flight and tail feathers, which in adults show wear, since they are not cast after breeding, but held until the early spring moult."

Food.—Collett, quoted above, states that in Norway in summer the food seems to be wholly taken from the countless myriads of mosquitoes, of which there are at least half a dozen species. All crops examined were crammed with these insects. In a later communication (1886) he mentions having on two occasions found the larvae of a *Cidaria* or other geometrid larva in the stomach, as well as other soft insects. Further exact information is scanty, but Jourdain (1938, vol. 2), summarizing the records of several other observers in addition to Collett, mentions Hymenoptera (larvae of *Tenthredo* or *Lophyrus*, ants), Coleoptera, Hemiptera, and other insects.

Behavior.—All the small leaf-warblers of the Old World have much in common in their behavior, and Eversmann's warbler differs little in this respect from its better-known European relatives. Like them it is a lively and active bird, constantly hovering and flitting about in foliage in search of insects, and sometimes it will hover for a few seconds to pick one off a leaf. It does, however, differ from some allied species in that, at any rate in the breeding season, it keeps principally in the canopy of well-grown trees, in which it chiefly feeds and sings, though it may also be seen at times in lower vegetation.

Voice.—As more or less marked geographical differences in song and notes, which may or may not coincide with accepted racial differences, are not unknown within the limits of single species of the genus, it cannot be assumed that the song and notes of Eversmann's warbler in Europe are identical with those of the same species in Alaska, though it is evident that they are similar. To an observer familiar with European birds much the best description that can be given of the song of Eversmann's warbler is that it is unmistakably like that of the cirl bunting (*Emberiza cirlus*), though it is somewhat higher pitched and less hard-sounding. To those not acquainted with that bird the best that can be said is that the song is a little rattling repetition of a single somewhat sibilant note, lasting for approximately three seconds. It is usually, though not invariably, preceded by a clicking *tzick* uttered one to three times, and when the bird is in full song the fairly regular alternation of the monotonous little song phrases and clicking *tzick* notes is very characteristic. At such times there is less than one-second interval between the end of a song and the next *tzick*. The song is generally delivered from among the foliage of trees, though at times lower down, and the bill moves rapidly during its utterance. It may be heard in Lapland from the time of the bird's arrival about mid-June until mid-July or a little later. Though in general so simple a song gives little scope for variation I have heard one variant in quicker time which was rather distinct.

Dixon (1938) states that the song of Kennicott's warbler "might well be described as intermediate between that of the orange-crowned and northern pileolated warblers." The writer is not acquainted with either of these, but judging from descriptions this would seem to suggest some little advance on the extreme simplicity and uniformity of the song of the species in Europe.

The *tzick* note is also used on the breeding ground separately from the song as a call or possibly rather as an alarm or scold. It is no doubt the note rendered by Dixon as a harsh *chit.* What is undoubtedly the same note is described by La Touche (1926) as commonly uttered by passage birds in China in spring, although the song is never heard from these migrants. A note of migrants described as a repeated husky *tswee-ep* is evidently different, and I have heard a note that could be so rendered on the breeding ground. Another quite distinct note which I have heard in the breeding season is a low rattle or churr, and during what appeared to be a courtship chase I have heard from one or both birds a kind of loud *sit-sit-sit-sit . . .*, something like the notes of the song repeated in a more staccato fashion, less sibilantly and with more emphasis.

Hamilton M. Laing (1925) writes:

For some days after our arrival at Petropavlovsk, an elusive song of good quality was heard in the birch woods, but the author could not be seen. When a breeze was blowing, the twinkling leaves of these trees made it most difficult to catch sight of birds aloft in the upper branches. On July 27, an effort was made to learn the identity of the puzzling song heard in the woods since arrival. The assumption that it was wagtail was quite wrong. It came from the treetops, and finally the bird was seen in song, even to the beak open in delivery, and then it was shot. It proved the same small warbler-like chap resembling our Tennessee Warbler, taken July 21. The song is suggestive of several others. It suggests the Northern Water-Thrush, the Oven-bird at times, and even the California Purple Finch. It might be fairly syllabized as '*Reecher! Reecher! Reecher! Reecher!*'—quite ringing and melodious.

Collett (1886) gives the following account of the song:

Whilst the females are sitting, the males have each their singing-place, which they hardly ever leave. It was on a little hill within the woods covered with larger birch and a few pine trees which towered above the others. Here the male would sit, in the top of the loftiest trees, and sing almost incessantly the whole day; it stopped only for a few moments, when it generally entirely disappeared, and sometimes it could then be seen to meet the female. Some minutes after it would perch again on the top of its tree, as a rule on the same branch, and recommence its song again. * * *

The indefatigable manner in which the male gives forth its monotonous, but nevertheless strongly sounding, song is almost incredible. The song consists, as I have previously remarked, of a single note, *zi-zi-zi-zi . . .*, repeated unusually quickly, fourteen to sixteen times in succession. After each song follows a short period of rest, which in the height of the singing-time scarcely exceeds half a minute, when it recommences its song again.

The song of the different specimens was almost precisely alike, but in some it might sound a little more or less harsh than in others. Seebohm has compared the song to the trill of the Redpole; and this seems to be a suitable description, although it appeared to me to resemble more the first quick notes of the song of *Sylvia curruca.*

In these latitudes, where the day is but little lighter than the night, the song might be heard at any hour and even at midnight. A little after the middle of July most of the males had ceased to sing, although at Matsjok once or twice I heard the song so late as the 28th of that month.

Field marks.—A small warbler something like a warbling vireo, about 4¾ inches long, with dull greenish upperparts, a prominent pale yellowish eye stripe, whitish underparts tinged with yellow and a narrow pale wing bar. In birds on the breeding ground I found this bar tolerably noticeable at close range, but not otherwise. As the season advances it may be much obscured by abrasion. In the breeding season the peculiar song and note are additional good characters.

Fall and winter.—This species leaves its breeding grounds early, and La Touche (1926) states that it appears in southeastern China as early as the last days of August and in northeastern China from the middle of that month. He states that on migration "it is quite arboreal in its habits, except perhaps while passing through poorly-wooded country, where I have seen it hunting for insects in grass and bushes. But, once in a garden, I actually saw one hopping about on an open gravel path." H. C. Robinson (1927) states:

This little willow-warbler is very common * * * throughout the Malay Peninsula, from August to the beginning of June, though very few arrive before the middle of October or remain after April. Most of our visitors seem to remain on or near the coast and do not penetrate into heavy jungle, though we have found them near the summit of Kedah Peak, at a height of 3,500 ft. in December.

Among the mangroves and Casuarinas on the coast they are often met with, in small flocks or singly, and are very active and restless, flying with a short, jerky action, and returning to their perch like a flycatcher. They seem to feed largely on very small flies and midges, though they are often seen searching the boughs like a tit.

DISTRIBUTION

Breeding range.—The principal Old World race *A. b. borealis* (Blasius) breeds in northern Europe and Asia north to the tree limit (generally around latitude 70° N., but to 75° on the Taimyr Peninsula), including north Norway, north Finland (south to about 68°), north Russia, and west Siberia south to about 61°, but in Siberia east of the Yenisei reaching much farther south and extending to the mountain regions of northern Mongolia, eastward across Siberia to Bering Sea. Allied races are found in the Kamchatka Peninsula and North Pacific Islands.

In western Alaska *A. b. kennicotti* is recorded in coastal regions **north** to the Kotzebue Sound region (Kowak River), also at Port Clarence, St. Michael and elsewhere on Norton Sound, on Nunivak Island, **south** to the Aleknagik River, Nushagak, and Iliamna at the base of the Alaska Peninsula, and inland in the Mount McKinley region, Alaska Range. Probably only accidentally (bird found dead) even as far north as Icy Cape, near Wainwright.

Winter range.—The species winters in the Philippines, French Indo-China, Siam, Tenasserim, Malay Peninsula, and East Indian Islands to the Moluccas; also Formosa, Andaman Islands, etc. Ticehurst (1938, vol. 2) states with reference to *kennicotti* that he has "seen a few very small birds from various southeastern localities in Asia in winter (Amoy; Tenasserim; Philippines; Penang; Siam) which may represent the winter-quarters, since it does not winter in the New World."

Spring migration.—Ticehurst gives a number of dates from literature and from labels of skins for the passage through eastern Asia to the breeding grounds, but the data are not extensive enough to give any exact picture of the northward spread of the species. It is apparent that the passage starts in China about the beginning of the second week of May, and the northeastern parts of the country may be reached by the end of the week, but Manchuria and Amurland are apparently not reached till the end of the month. Arrival at Nijni Kolymsk at the mouth of the Kolyma, northeastern Siberia, is recorded on May 30, and north Norway is reached about the beginning of the third week of June. For Alaska there appears to be no record earlier than June 14.

Fall migration.—The migration of the species through eastern Asia begins early. It is recorded in northeastern Mongolia by August 7 and in northeast China by August 10, and southern China (Fokien, southeastern Yunnan) may be reached by August 24. The migration continues through October and even into November. Nevertheless, not all birds leave so early as above, for specimens of *kennicotti* are recorded in Alaska from August 16 to the end of the month and even (Nunivak Island) to September 8.

Casual records.—The typical race is recorded from Great Britain, Helgoland, Italy, Holland, and near Orenburg, Russia.

LOCUSTELLA OCHOTENSIS (Middendorff)

MIDDENDORFF'S GRASSHOPPER-WARBLER

CONTRIBUTED BY WINSOR MARRETT TYLER AND BERNARD WILLIAM TUCKER

HABITS

Middendorff's grasshopper-warbler is a little bird belonging to the Old World subfamily Sylviinae. It winters in the Malay Archipelago and moves northward, by way of Japan, to its breeding grounds on the Kuril Islands, Kamchatka, and the eastern coast of Siberia. La Touche (1926) gives the bird's range in more detail, stating:

> Middendorff's Grasshopper-Warbler appears to pass in numbers at Shaweishan on its way to and from Japan and extreme North-East Asia. Dr. Hartert gives Kamtschatka, the Siberian coast of the Sea of Ochotsk, and "Schantar-Island," Kurile Islands, as the breeding range; also Behring Island. It winters in the Malay Archipelago. This bird has not often been taken on the mainland of China. I obtained it in the reed-beds and by the river-banks at the mouth of the River Min (Fohkien Coast) at the end of May and beginning of June, and have specimens from East Fohkien dated September and October. * * * There is no record of the bird having been taken in North China, and it doubtless migrates *via* Japan and Corea.

The occurrence of Middendorff's grasshopper-warbler as a North American species was established by a single specimen secured on September 15, 1927, on Nunivak Island, Alaska, an island to the north of the Aleutian Chain. Of this individual Harry S. Swarth (1928) says: "One specimen: C. A. S. No. 30760, female, bird-of-the-year, September 15. The range of Middendorff's Grasshopper-warbler includes the north-eastern coast of Siberia and the Kurile Islands, so that its occurrence in Alaska is no more extraordinary than that of some other Asiatic birds that regularly cross Behring Sea. The capture of this bird adds to our *Check-List* a species and genus in the family Sylviidae."

Swarth (1934) describes the specimen thus: "Bill, upper mandible brown, lower mandible brownish orange at base, shading through brownish yellow to a dusky tip; iris brown; tarsus purplish brown, toes rather pale brown."

The species is treated binomially in the A. O. U. Check-list, but two races are now generally recognized, *L. ochotensis ochotensis*, of Kamchatka, the Commander and Kurile Islands, Sakhalin, and the coasts of the Sea of Okhotsk, and *L. o. pleskei* Taczanowski, of Korea and some of the Japanese islands. The latter is a somewhat larger bird with a longer bill, the culmen, according to Hartert and Steinbacher (1938), measuring 18.5–22 millimeters as against 15.5–17 millimeters in the typical race. Yamashina (1931) gives the corresponding figures as 16.5–17.5 millimeters (17 specimens) and 14–16.5 millimeters (53 specimens). There are also some color differences between the races: *ochotensis* has slightly darker centers to the feathers of the upperparts, while in *pleskei* these are lacking, and the back is described as darker without the tendency to russet often shown by the typical race. Yamashina adds that the throat, breast, and belly of *pleskei* are pure whitish instead of more or less tinged with buffish or brownish. The American specimen evidently belongs to the typical race, but *L. o. pleskei* is the form whose biology is best known owing to the observations of Japanese workers, on which the account which follows is partly based.

Locustellas are extremely skulking little birds, frequenting long grass, reeds, and bushy ground, often in marshy or wet places or actually over water, and the present species is typical in this respect. It is described as frequenting willow bushes and thickets in damp places. *L. o. ochotensis* is described as the characteristic bird of the beds of reed grass at waters on Sakhalin, while in Kamchatka it was found by Stejneger (Ridgway, 1884) "among the high grass and willows which cover the swampy slopes of the mountains with a thicket almost impenetrable to foot and eye" and by Laing (1925) fairly commonly "in the low shrubbery on the more open country" at Petropavlovsk.

On the Seven Islands of Izu off the coast of Japan Yamashina (1931) describes *L. o. pleskei* as breeding in places overgrown with the small bamboo *Arundinaria simoni.*

Nesting.—B. Dybowski (1883) reports that the species (translation) "arrives the first part of June, constructs an open nest in the herbage, above the ground level, composed of dried grasses and lined with small feathers; lays five eggs and commences incubation the end of June or beginning of July."

The following is a translation of the description of the nest of *L. o. pleskei:* "The nest is built on a flat place where *Arundinaria simoni* grows thickly. It is usually built on several stems of *Arundinaria.* Occasionally this species builds a nest on stems of hydrangea, which grows among the *Arundinaria.* The nest is placed at heights of about 35–150 centimeters above the ground. It is shaped like a drinking glass; the outer part consists almost entirely of dry leaves of *Arundinaria.* On the inside are finely broken up (dunn gespaltene) pine needles, dry bents, fine fibers, etc. The outer diameter is 10–15 centimeters, the height 8–17 centimeters."

Eggs.—L. Taczanowski (1882) says of the eggs (translation): "The eggs are different from those of the other Locustellas; they have a ground of a shade almost like that of the eggs of *L. lanceolata,* but uniform, without any trace of dark spots; they have only a simple fine blackish vein, surrounding the large end completely or incompletely, otherwise, some present a crown of a little deeper shade, on which I cannot distinguish even with the aid of a lens, any trace of spots. The eggs measure 20.5 by 14.2; 20.9 by 14.8; 20.9 by 14.2; 21 by 14.2; 21 by 14.4; 21.4 by 14.2 millimeters."

According to Dresser (1902), *L. ochotensis* lays five or six eggs, which are pale rose-colored, unspotted, but marked with one or two fine blackish lines at the larger end, which sometimes form a wreath. Yamashina states that the eggs of *L. o. pleskei* have a grayish-white ground color with a violet tone, and that as regards the spotting there are two types: (1) with rather large scattered spots and short lines of a violet-blackish and gray color, and (2) with the violet-brownish and pale purple spots so close that they almost cover the ground. He adds that about two or three of type 1 are found to one of type 2. The difference in the two descriptions is sufficient to raise the question whether the eggs Dresser described were rightly identified, unless, indeed, *pleskei* is really a different species, as some have held. Moreover, Yamashina states that the nests of *pleskei* always contain four eggs. He gives the season of laying as from mid-May to mid-June. Hartert and Steinbacher (1938) give the average egg measurement for *ochotensis* as 19.44 by 14.7 and for *pleskei* as 20.96 by 15.47 millimeters. Yamashina gives the maxima for *pleskei* as 23 by 15.5 and 20.7 by 16.2 millimeters and minimum 19.5 by 15 millimeters.

Food.—Like others of its kind the species is insectivorous. Yamashina examined the stomach contents of 12 individuals of *pleskei*, which all contained the remains of Coleoptera. Three contained also the remains of small Hymenoptera, two small fruits, one larvae of Diptera and Geometridae, and one small snails.

Young.—The female incubates (Yamashina). No other details are available.

Plumage.—The plumage is described by Hartert (1910). The differences between the races *ochotensis* and *pleskei* are mentioned above.

Behavior.—Sten Bergman (1935), freely translated from the German, says of the habitat in the far north: "*Locustella ochotensis*, one of the characteristic birds of Kamchatka, arrives rather late in spring, never before June. In 1921 I saw the first individual on June 9 near Klutschi, and the last on September 5 near Ust-Kamchatka.

"The bird does not frequent the larch or coniferous forests or, customarily, the dry birch woods, but occurs in the meadows, swamps, and along the brooks, and, confining itself to the undergrowth, is seldom or never found in the high trees. It is a typical bird of the lower slopes of the mountains among the alder scrub, and does not, like *Luscinia calliope camtschatkensis*, ascend to the higher elevations."

Leonhard Stejneger (1883), writing to Professor Baird, gives a glimpse of this elusive, little known bird, breeding in a far-away corner of the world, and of the difficulty in collecting specimens.

He writes:

It is a kind of willow-warbler, common in Petropaulski, but not observed here on the islands. My only specimen is a male, shot on the 5th of July, 1882. * * * The loud song, consisting of the syllables *witshe-witshe-witshe-witsh,* and somewhat resembling the sound made by whetting a scythe, was heard, especially towards night, from all sides when walking through the high grass and willows covering the swampy slopes of the mountains with a thicket almost impenetrable both to foot and eye. You would very seldom get a glimpse of the watchful songster, when, clinging to the middle of the upright stalk of some high orchid or grass, he did his best in the singing-match with one of his own kind or a *Calliope kamtschatkensis* or a *Carpodacus.* But no sooner would you move your gun to secure the longed-for specimen than he silently disappears, as completely and suddenly as if he possessed Dr. Fortunatus's cap. The only way to obtain a specimen is to watch patiently near one of his favorite bushes, with the gun ready. For hours I have thus sat in the wet swamp, almost desperate from the bites of the numberless bloodthirsty mosquitos, which I did not dare to wipe off, fearing to drive away the silent bird, who perhaps was watching my immovable figure until he was satisfied as to his safety. Curious, but still cautious, he would come nearer, slipping between the stems and branches nearest to the ground, uttering a very low, thrush-like *tak; tak; tak; tak,* and with the tail straight upright, very much like a long-tailed *Troglodytes* both in color and conduct. And if I kept absolutely quiet he sometimes would proceed close to my feet, looking curiously at me with his pretty dark eyes. But before the challenge of a neighbor had attracted his attention and provoked his reply, which he usually began with a short trill, it would not have been advisable to move a muscle.

Then comes the time to lift your gun very slowly, stopping as often as he suspiciously stops his song, until the "crack" puts an end to it forever, and you hold in your hand a crushed specimen, unfit for preparation, when you have to shoot from too short a distance, or return without anything, while, after a longer shot, you cannot find the plain-looking little bird amidst the immense vegetation in the dim light of the vanishing day and tortured by the intolerable mosquitos. You will understand from your own rich experience how much pleasure it gave me when I, at last, obtained a tolerably good specimen.

Voice.—Bergman (1935) quotes Stejneger's rendering of the song, and adds: "The single song is sung most frequently in the evening, but during the day the bird often flies up in the air and sings while descending. Sometimes it sings almost through the whole summer night."

Yamashina (1931) described the male of *L. o. pleskei* as flying up to a height of about 2 meters and uttering a song rendered as *tschurrrr* and after a time dropping down to the same place. According to Stejneger the song of *L. o. ochotensis* somewhat resembles that made by whetting of a scythe and is rendered *witshe-witshe-witshe-witsh.* He found that it was uttered especially toward night and described the songster as clinging to the middle of an upright stalk of some high grass or orchid (not performing a song flight as described by Yamashina in the case of *pleskei*), but ready to dive into cover at the least disturbance. He describes the note as "a very low thrushlike, *tak; tak, tak; tak,*" and this was the only note heard by Laing late in July, when the birds had stopped singing.

Field marks.—All the Locustellas are small warblers with strongly graduated tails, which spend most of their time creeping or running about among vegetation near the ground and rarely show themselves at all freely except when singing. Stejneger (in Ridgway, 1884) describes Middendorff's grasshopper-warbler as "slipping between the stems and branches near the ground * * * with his tail held upright, very much in the manner of a long-tailed wren." This species has brown upperparts with only faint markings in the case of *L. o. ochotensis* or none in *L. o. pleskei.* The lateral tail feathers have a subterminal blackish band and dull white tips. A photograph of *L. o. pleskei* is given in Yamashina's paper already quoted.

Winter.—Owing to its very secretive disposition little is recorded about the habits of this species outside the breeding season, but it is evident that they are very much the same as on the nesting ground except for the absence of song. It frequents reed beds and other rank vegetation largely, if not entirely, near water or in marshy places. La Touche (1930) met with *pleskei* in numbers in mangroves at Swatow in May, "running along the banks of the lagoons on the mud under the mangrove-bushes," and no doubt it may be found under similar conditions on the fall passage.

DISTRIBUTION

Summer range.—*L. o. ochotensis:* Kamchatka, Commander Islands, coast of the Sea of Okhotsk, Sakhalin, and the Kurile Islands. *L. o. pleskei:* Korea and neighboring islands (known breeding places Dagelet, Quelpart, Hachibi, and Schichihatsu Islands), Kyushu, and the Seven Isles of Izu off Honshu. Recorded also on Honshu and Hokkaido, but breeding not proved.

Winter range.—Philippines, Borneo, Celebes, and Southwestern (Serwatti) Islands. On passage in Japan and China.

POLIOPTILA CAERULEA CAERULEA (Linnaeus

BLUE-GRAY GNATCATCHER

HABITS

Contributed by Francis Marion Weston

Our acquaintance with a new bird dates, it seems to me, not from the moment we learn to identify it in the field but rather from the first time we really have a glimpse of its "personality." Thus, my "first" blue-gray gnatcatcher was certainly not the one my ornithological mentor first pointed out to me, but another that came along months later, flitted to a bush within arm's length of where I stood and, between snatches at insects too small for me to see even at that short distance, spent several minutes looking me over.

It was upon the foundation laid in those few minutes that I have built whatever else I may have learned about the gnatcatcher. In the course of writing these pages, the memory of that first meeting has come back to me many times, almost with the clarity of a visual picture, and I feel that I am telling of the later adventures of one little bird rather than of the habits of its myriad kin.

The habitat of the blue-gray gnatcatcher evidently varies materially in different parts of its range. In the far South, where it is resident, it is common and widely distributed in the nesting season, occurring regularly even in the residential (wooded) sections of the cities and towns, as well as in all forested areas, wooded swamps, pine lands with an undergrowth of scrub oak, pecan and citrus groves—in fact, everywhere where there are trees suitable for nest sites. Farther north, it is characterized as being a bird of the watercourses and the timbered swamps, spending most of its time in the tops of the tallest trees. An interesting variation is noted from the "great open spaces" of Kansas, where N. S. Goss (1891) wrote of its being "as much at home in the shrubby bushes on the hillsides, or the mesquite growths on the plains, as within the treetops of the heavily-timbered bottom lands."

Spring.—A few years ago it would have taken a whole paragraph to describe the spring arrival of the blue-gray gnatcatcher on its breeding grounds, for its movement from winter quarters in the Southern United States and the Tropics differs from the familiar wavelike rush of the warblers. Today we need but a single word of the military parlance that has become part of our every-day speech: Infiltration. Yesterday, the gnatcatchers were not here; today, they are; and we are always a bit surprised when we discover them. Early in March they are on the move from southern Florida; by the last of the month they are halfway on their course across the country; and mid-April finds them at the northern limits of their normal breeding range.

Courtship.—For a week or two after their arrival, they are still silent and retiring in habit; then suddenly they all seem to come to life and are ready to resume that all-absorbing function of all living things—reproduction.

In common with many of the other small birds, the gnatcatcher seems to have no well-marked courtship ritual. We note the unrestrained animation of the male birds, see their frequent bloodless combats, hear their ceaseless singing and chattering—then, after a surprisingly short interval, we find that mates have been selected and the serious business of nest-building is under way.

The song of the male gnatcatcher, even at the peak of his spring animation, is scarcely louder than a whisper, and it is interesting to note the opinion of our ablest interpreter of small-bird actions, Mrs. Margaret Morse Nice (1932), who finds that the male of this species, unlike louder-voiced birds, "does not sing to proclaim his territory; perhaps the *spee* which is constantly given by both birds serves this purpose."

Nesting.—More than with most species of small birds, the attention and interest of the observer center about the nesting habits of the blue-gray gnatcatcher because of the great beauty of its nest. This nest should be even better known than its miniature counterpart, the nest of the ruby-throated hummingbird, for by virtue of its larger size and consequent better visibility it can be found much the more readily of the two—yet it seems to have been entirely overlooked by the general public and is known only to ornithologists.

The general situation of the nest in the extremes of the breeding range of the species is decidedly different. In the southern end of the range nests can be found almost everywhere that trees grow—from the residential sections of the cities to the heart of the great river swamps—but farther north they occur principally along watercourses and in timbered swamp areas.

The height of the nest above ground varies from a few feet to 70 or 80. By far the greater number—those seen by human observers, at any rate—are less than 25 feet up; but this may not represent true distribution in height for, as G. A. Petrides (MS.) points out, "the noisiness and lack of suspicion of the birds about the nest probably enabled the low ones to be located more easily." I suspect that the reported heights of the very high nests were estimated rather than measured, but there are dependable figures available for many low ones. I have found some of these low nests myself: one just 4 feet from the ground near the end of a low-sweeping branch of a pecan tree; another within camera tripod reach of the ground (not more than 5 feet up) in a small lemon tree; and several that were between 5 and 7 feet up in small scrub oaks in open pine woods. Angus McKinnon (1908) mentioned one in northern Florida that was "in a small oak only about three and one-half feet from the ground." The lowest figure of all is given in a recent letter from R. A. Hallman of Panama City, Fla., who describes a "typical Gnatcatcher nest, placed in the upright fork of a small scrub oak bush", which was "by actual measurement 38 inches * * * from the ground to the top of the nest."

The nest is usually saddled on a horizontal limb 1 to 2 inches in diameter—occasionally on a larger one—but it is often placed in a fork formed by an upright branch and a horizontal or slanting one, the lower branch furnishing the foundation and the upright lending side support. An interesting variation was reported by J. J. Murray (1934), who described a nest near Lexington, Va., that "was not saddled on a limb, but set between [*sic*] three small forks of an upright crotch, in the manner of the nests of the Yellow Warbler and Redstart."

An attempt to compile from the literature a list of the trees selected by the gnatcatcher for nesting sites resulted in a collection that reads like a catalog of the silviflora of eastern North America. It seems that this species is willing to use any tree in its habitat that provides limbs of the right size and conformation. In the North, where the bird is uncommon or rare, no generalization as to preference can be hazarded. In the South, where it is common to abundant, some local tendencies are noted but these vary widely in different sections. A. T. Wayne (1910) found that, in coastal South Carolina, the gnatcatcher prefers the live oak (*Quercus virginiana*) "because nesting material is plentiful." S. A. Grimes (1928), writing from Jacksonville, Fla., stated that "oddly, or not, the pine is the tree most commonly chosen for the nest site." In his experience "the pines (at least two varieties) have been selected * * * oftener than all other trees combined." That choice could hardly have been influenced by availability of nesting material. In extreme western Florida and southern Alabama I have found more nests in scrub oaks (principally *Quercus catesbaei* and *Q. cinerea*) growing in open pine woods than in any other trees, and I have never seen a nest in a pine. Location in a tree with lichen-covered bark provides the ultimate in concealment for a lichen-covered nest, but this correlation does not seem to be a factor in the choice of a nest site. The preference for pines in the Jacksonville area, as cited by Grimes, is a case in point, and I have seen nests in citrus trees, pecan trees, and cypresses where little or no lichen growth was present. Yet the nest, even in other than its optimum surroundings, is always difficult to see and is seldom found except as "given away" by its owners.

If it were possible to extract a composite or average of all the published descriptions of the nest of the gnatcatcher—and the wording of most of them is monotonously similar—the result would be something like this: A beautiful, cup-shaped nest, compactly built of plant down and similar materials bound together with insect silk and spider web and covered externally with bits of lichen. Materials listed seem to include every kind of soft plant fiber found in the region where the subject nest was located. Many writers use the general terms "plant down" and "fleecy plant substances," but a few particularize with "sycamore fuzz," "leaf down from the under surfaces of leaves," "dandelion and thistle down," and "dried blossoms." Fibrous materials that enter into the lining of nests include fine strips of bark, fine grasses, tendrils, feathers, and horsehair. A. T. Wayne (1910) collected several nests that were "profusely lined with feathers." W. P. Proctor (MS.) mentions having seen a gnatcatcher in northern Florida "picking the petals from dewberry blossoms [*Rubus trivialis*] for its nest."

I was interested to discover what degree of availability determines the selection of material for a particular nest. I had an old nest, a mantel decoration much the worse for dust and age, of which I knew the original location—it had been taken from a point about 20 feet up in a medium-sized live oak that grew on the edge of a highway right-of-way where it entered a wooded swamp. Pulling the nest to pieces, I found it to be composed largely of oak catkins felted together with plumed seeds and a kinky plant fiber, buff in color, that I could not name at the time. Scattered through the mass were a few small pieces of what appeared to be "sheet" spider web or fragments of cocoons, but there was no vestige (even under a strong hand lens) of the insect silk or spider web that had presumably been used to bind the outer covering of lichens to the body of the nest. Actually, there was no longer need for a mechanical binder since the lichens had attached themselves firmly to their new foundation. The inner cup of the nest was a felted structure, readily separable from the base and from the outer, lichen-covered sheathing. It had been shaped by a few stiff, wirelike grasses (unidentified) disposed through the felted material almost like the reinforcing bars in an engineer's concrete structure, and the felting was composed entirely of the plumed seeds and the kinky fiber and contained no catkins. Upon visiting the original site at the usual season of nest building, I found (as I knew I should) that the oak catkins and the lichens were obtainable in unlimited quantity within inches of the spot where the nest had been built. The plumed seeds proved to be from one of the broom-grasses (*Andropogon* sp.), a small dried patch of which, still bearing a few seeds late in the season, was within 10 yards of the base of the nest tree—and I could find no more within a hundred yards in any direction. The kinky buff fiber was discovered in inexhaustible abundance along both sides of the right-of-way: it was the stem "wool" of the cinnamon fern (*Osmunda cinnamomea*).

Measurements of two nests were kindly furnished by W. P. Proctor (MS.). One was: outside, 2 inches diameter by 2½ deep; inside, 1¼ inches diameter by 1¼ deep. The other measured: outside, 2½ inches diameter by 2¼ deep; inside, 1¼ inches diameter by 1⅜ deep. The striking feature of both these sets of figures is the difference between the inner and outer depths—1¼ inches in the first nest and ⅞ inch in the second. These differences represent the thickness of the foundations of the nests, the thickness of the pads of resilient, closely felted material between the weight of the contents and the surface of the supporting branch. It is not unlikely that the remarkable tenacity of these tiny nests under stresses of weather buffeting and family struggles resides in the elasticity of their thick, resilient foundations. A moment's consideration of the proportions of the deep cup of the nest—1¼ by 1¼ inches—and of the slim length of the little bird that must crowd itself into these narrow confines explains why the incubating gnatcatcher can assume no other posture than with bill and tail pointing straight upward.

Most observers agree that both sexes work at the construction of the nest, and my experience in northern Florida is that they share this labor fairly equally. W. E. C. Todd (1940) stated that, in Pennsylvania, the male never assists in nest building but that he "always remains near at hand and takes a great interest in the work." W. P. Proctor, reporting from southern Michigan, writes (MS.) that in one instance of the three observed by him, the female alone did the building, and that on one shared task the female did more than the male. Aretas A. Saunders, however, writing (MS.) from central Alabama, finds that where there is an unequal division of labor the male bears the heavier burden, and also that the male is a more persistent worker than his mate and that he is less sensitive to interference by human intrusion. He cites an instance of one male bird that was seen to bring nesting material five times in six minutes. Slower than this high-speed worker was a building pair reported by Mrs. Nice (1931), who counted 27 trips to the nest with material in an hour. It always gives an observer a start, when he is watching nest construction and has seen material brought in and placed in position to, have a bird come in with a seemingly empty bill, yet work as diligently at the structure as before. It is more in keeping with the known industry of this species to account for this apparently wasteful gesture as the placing of invisible lengths of spider web rather than as mere "boondoggling."

The length of time required for nest construction is subject to extreme and inexplicable variation. W. E. C. Todd (1940) stated that it requires between one and two weeks of constant labor to complete a nest. But Edward R. Ford (MS.) writes of a nest that he saw under construction from the very beginning that, after only three days, "appeared, from the ground, to be completed." Perhaps the important factor in this variation is the size of the "building crew"—one bird alone, or both birds.

A practice, apparently peculiar to the gnatcatcher, and one that has been commented upon by almost every observer familiar with its ways, is its habit of tearing up a completed or partly built nest and re-using the materials to build a new nest a short distance away. L. L. Hargrave (1933) has collected and summarized a number of published accounts of this peculiarity, and he concludes that nests are abandoned because of a change of conditions that renders the first site untenable or at least no longer desirable. He cites one case where a pair of green herons started their nest close to a still unfinished gnatcatcher's nest, and other cases where human interference was probably the determining factor in causing abandonment. A nest, once abandoned, immediately becomes the most convenient source of material for another structure. This use of an existing nest is all the more readily understandable when the extremely seasonal nature of desirable nesting material is considered. In the nest that I described earlier in this section, the predominating materials—oak catkins and "wool" from the cinnamon fern—are obtainable in quantity only over a short period. After this critical period, existing nests are the sole source of supply. Besides the nests that are moved by their constructors in extension of the original building program, I have known nests in which broods had been successfully reared to be torn up and carried off by gnatcatchers, but whether the "wrecking crew" and the original owners were the same or not, I am unable to state.

Nest-building in many parts of the gnatcatcher's range precedes egg-laying by 10 days or two weeks. C. K. Lloyd (1932) attempted to account for the need of this interval in the more northern sections by stating that the birds "nest early * * * and do not deposit eggs until the trees are well leafed out," but he gave no reason for the practice of "nesting early." In northern Florida, where many nests are built in evergreen trees and where most of the deciduous trees are in full leaf at the time of nest-building, this explanation does not hold good, yet the same length of time elapses between the completion of the nest and the laying of the first egg as in the North. It seems to me that nest-building is best accomplished at the time that the favored nesting material is obtainable in greatest abundance, even though that does not coincide with egg-laying time. Thus, in northern Florida, nest-building takes place early in April when oak catkins and fern "wool" are available with the least labor of search, but eggs are seldom laid before the third week of April.

Eggs.—[AUTHOR'S NOTE: Four or five tiny eggs usually constitute the set for the blue-gray gnatcatcher, seldom more or fewer. They are ovate or short-ovate and have little or no gloss. The ground color is pale blue or bluish white. They are rather sparingly and more or less evenly covered with small spots or fine dots of reddish brown, or darker browns; rarely there are a few very small blotches; sometimes the markings are concentrated in a ring around the large end; and very rarely an egg is almost immaculate. The measurements of 50 eggs average 14.5 by 11.2 millimeters; the eggs showing the four extremes measure **15.5** by 11.2, 15.2 by **14.7**, **13.2** by 11.2, and 15.2 by **10.7** millimeters.]

Young.—Little has been written about the nest life of the blue-gray gnatcatcher, and my own notes are peculiarly deficient in this respect. Out of the many gnatcatcher nests that I have seen, I find that I have not followed the history of a single one completely through any one phase of its development.

Incubation is said to require about thirteen days from the laying of the last egg. G. A. Petrides (MS.), writing of a nest near Washington, D. C., found "day-old young" on May 21 in a nest where the last egg of the set had been laid on May 8. William Palmer (1906) stated that, during incubation, the female parent "rarely leaves the nest" and is fed there by the male, the inference being that he does not share in the duties of incubation. On the other hand, Mrs. Nice (1932) remarked the close cooperation between the parents during incubation and stated that they relieve each other on duty at short intervals (15 to 40 minutes) and that the eggs are not left uncovered for more than a minute at a time. W. P. Proctor, of Benton Harbor, Mich., coordinates these apparently irreconcilable statements when he writes (MS.): "At some nests, both birds sit on the eggs; at others, the female alone sits. Where the birds take turns, one usually sits from 15 to 25 minutes. At a nest where the female alone sat, there was no regularity; she was off anywhere from an instant to 15 minutes, and once * * * 26 minutes."

Care of the young is characterized by the usual intense activity of this species, both parents sharing the duties of feeding and brooding the young. So unsuspicious, or so preoccupied, are they at this time that they completely ignore human proximity and fly directly to the nest with food. Added to this "dead give away," the growing young in the nest soon become very noisy, so the finding of nests at this stage of development is an easy matter for even an inexperienced observer.

In the experiment to be described later in the section "Behavior," Maurice Brooks (1933) stated that a feature of the feeding of the young "was the extreme frequency"—43 feedings in 20 minutes, 34 by the female, 9 by the male. This unequal division of labor does not obtain, I believe, under natural conditions for, although I do not have notes to verify it, my recollection is that the male visits the nest with food almost or quite as frequently as does the female.

Food brought to the young consists exclusively of animal matter, mostly insects; but so tiny are the separate items that an observer at a distance of only a few feet can seldom identify them. Sometimes larger prey is brought in, large enough, according to W. P.

Proctor (MS.), for the parent to have to "pound it on a limb" before offering it to the young.

The normal span of nest life is 10 to 12 days. S. A. Grimes (1932) wrote of a set of four eggs in the Jacksonville area that "hatched May 10 or 11, and the young left the nest on the 21st." G. A. Petrides (MS.) gives a period of 11 or 12 days for the brood that he reported as having hatched on May 20, since they were "apparently ready to leave the nest on the evening of May 31." Beryl T. Mounts (1922) reported a nest near Macon, Ga., in which the eggs hatched on or about May 16 and the young left the nest on May 26.

E. H. Forbush (1929) wrote that, in the greater part of their range, gnatcatchers rear but a single brood in a season but that two broods are normal in the far South. However, S. A. Grimes (1928), one of the most ardent and capable observers of nesting, stated that, in the Jacksonville region, this species "raises only one brood in a season." Though I cannot make a positive statement on this subject, since I have never banded or otherwise marked gnatcatchers for individual identification, I believe that some of the late nests I have found were true second nestings and not second attempts by birds that had failed the first time. Taking April 24 as a median date for complete sets of eggs in northern Florida, a nest that I found just being completed on May 25, 1930, may or may not have been a second attempt by a pair that had lost their first nest; but a nest just started on June 1, 1941, in which a brood was later successfully reared, seems to me to represent a true second nesting; and there can hardly be any doubt that a brood that I saw just out of the nest on August 8, 1926, comes in this category.

Plumages.—[AUTHOR'S NOTE: The young gnatcatcher in juvenal plumage is much like the adult female, both sexes being alike and lacking the black forehead. An incomplete postjuvenal molt occurs in July and August, which involves the contour plumage and the wing coverts, but not the rest of the wings or the tail. This produces a first winter plumage, which is similar to the previous plumage but more washed with brownish on the back and sides. The first nuptial plumage is acquired in February by a limited molt of the feathers of the forehead, throat, and chin, when the black frontal band of the male is acquired, the upperparts become bluer and the young bird is now in adult plumage. A complete postnuptial molt occurs in July and August. Young males lack the black frontal band during the first fall and winter, and the females never have it.]

Food.—In common with most of the other very small birds (though not the hummingbirds) of Eastern United States, the blue-gray gnatcatcher eats very little if any vegetable food; and, by virtue of its fondness for some of the insects most harmful to man's interests, it is considered an entirely beneficial species. A. H. Howell (1924) quoted the findings of Judd's analyses of stomach contents and cited particularly "longicorn beetles, jointworm flies, caddis-flies and several * * * unidentified Diptera." He stated also that the gnatcatcher had been seen in Alabama "feeding on cotton leaf worms." E. H. Forbush (1929) added to this, "locusts * * * gnats, * * * ants and other hymenoptera, wood-boring beetles, weevils and spiders." A. A. Allen (1929) summed up many important items of the gnatcatcher's diet under the comprehensive term, "defoliating insects." It is not unlikely that a stomach analysis of Florida and Texas specimens taken in the citrus groves, one of the favorite haunts of this species, would disclose the presence of some of the citrus pests.

Winter food of the birds that remain within our borders probably consists largely of insect eggs and pupae, the known prey of the chickadees and kinglets with which the gnatcatcher associates at that season.

Food-table offerings seem seldom to attract this species; in fact, I am able to find but a single instance of it, and that in winter. Mrs. Andrew L. Whigham, who maintains an all-year feeding station in her garden in extreme western Florida, writes (MS.): "In January and February, 1933, for six or eight weeks, two of these birds used our feed shelves. They ate the inevitable cornbread [a saltless recipe of Mrs. Whigham's, baked in quantity for the birds and proven to be as attractive to most species as cracked sunflower seeds] and the commercial mockingbird food mixed with grated carrot."

Behavior.—The gnatcatcher is a little bird of intense activity; active, not with the methodical continuity of the brown creeper, but with an irrepressible vivacity of its own in all phases of its life cycle—feeding, nesting, care of its young—at all tmes, in fact, except during the enforced inertia of incubation.

In defense of its nest, the gnatcatcher's small size places it at a disadvantage in competition with larger species, for it seems not to possess the "driving power" of the even smaller hummingbirds, though it lacks nothing in either bravery or initiative when occasion demands (see section "Enemies" for a special case of nest robbery by blue jays). Its attitude toward human invasion of the sacred precinct of the nest shows wide individual variation. On the few occasions when I have approached closely to a gnatcatcher's nest, my presence always caused great excitement, which was evidenced by noisy protests but never resulted in a direct attack. S. A. Grimes (1928) found the brooding gnatcatcher very tame, and on several occasions he "climbed to within five or six feet of a sitting bird without causing it to leave the nest, or when it did it usually returned before I could get the camera set up for photographing." In sharp contrast to this, the same writer (1932) described an attack made upon him while he was photographing a nest, when the male gnatcatcher actually "struck the writer several times on the head and once in the eye," this last blow incapacitating him completely for a time. Maurice Brooks (1933) tamed a pair of gnatcatchers by making gradual advances toward the nest during the period of incubation until, after the young had hatched, he and his family could come within 2 or 3 feet of the nest without interrupting the feeding schedule. Finally he cut off the nest branch and lowered it for easier observation, still without apparently disturbing the parent birds. His next move was to cup his hands loosely about the nest in an attempt to compel the parent birds to alight on the hands. This intimacy was more than the birds would stand and the result was surprising as the female immediately attacked viciously and repeatedly. Amicable relations were later reestablished, and the female did occasionally actually alight upon the experimenter's hands, but even then she would without warning "sometimes take time out to attack." All attacks were made by the female.

Except during courtship and in defense of its nest, the gnatcatcher has never seemed to me to be pugnacious. Certainly, in its winter association with chickadees, titmice, and kinglets, it shows no tendency to harass or tease; so Alexander F. Skutch's note (MS.) on the wintering birds in Guatemala comes as a surprise when he writes that "the adult males do not seem to get along together." He cites as an instance: "On November 12, 1934, while following a large flock of small birds through open woods near Huehuetenango, I noticed of a sudden two blue-gray gnatcatchers in the oak tree in front of me. Upon finding themselves face to face, they became excited and attempted to sing; but at this season their voices were rusty from disuse, and their notes came thin and wheezy. Flying at each other, they clashed in midair; but the momentary affray was without consequence. After the first onset, they separated. From their attempts to sing, I feel sure that these birds were males. I have witnessed similar behavior—often with singing—on the part of other small birds which are solitary during the winter, when two of the kind come together."

The flight of the gnatcatcher, as described by Dr. H. C. Oberholser (1938), is "usually quick, but the bird does not ordinarily travel far without stopping. Sometimes it flies rather high, particularly when passing from one high tree to another, but it is usually seen flitting about the underbrush." The character of the flight is somewhat undulating or wavering rather than direct, with rapid wing beats, and is similar to that of many other very small birds; but the gnatcatcher can readily be distinguished in flight by the length of its tail. Even at some distance the observer has no difficulty in recognizing a gnatcatcher in flight, while chickadees, kinglets, and small warblers pass unnamed. P. A. Taverner and B. H. Swales (1908) gave an idea of the capabilities of this species for performing sustained flights when they listed it among the migrating birds on the southward crossing from Point Pelee, Ontario. The gnatcatchers they saw were unable to face the heavy wind prevailing at the time of the observation, and had to come back ashore, tacking just as a man would do in a boat, but the inference was that, under favorable conditions, the crossing would be completed successfully.

Certainly the most expressive feature of the gnatcatcher—as of its larger counterpart, the mockingbird—is its long, ever-active tail; now up and down, now from side to side, it is never for an instant at rest. Under stress of great excitement, the bird seems to try to combine these two motions at once, and achieves a ludicrous impression of circular motion.

The gnatcatcher's manner of feeding is similar to that of many other small birds, yet it differs in some respects from the methods employed by its most frequent associates even though its food, in winter at any rate, is probably the same as theirs. In its gleaning of the twigs and leaves of trees and bushes it tends to maintain an upright position and never (as I recall) hangs beneath a twig, as is the chickadee's constant habit. Like the kinglets, it often hovers before a leaf or terminal twig to secure some morsel that cannot be reached from above, but it does not indulge in this habit with the frequency of the kinglets. Unlike the creepers and the nuthatches, it is seldom or never seen on the trunks or large branches of trees. It is adept as a catcher of flying insects (many other kinds besides gnats!) and even in winter is often seen to secure food in this way. Its darts after flying insects differ markedly from the long swings of the true flycatchers, for its forays are seldom more than five or six feet in extent and are usually vertically upward with a quick drop back to the starting point. Again, unlike the flycatchers, it does not perch motionless and wait for passing insects; and I suspect that much of its catching of flying insects is by way of retrieving prey that, disturbed by the bird's actions among the leaves, makes a sudden flight to escape capture. Rarely, when the gnatcatcher is feeding in low bushes, it drops momentarily to the ground to pick up some object that attracts its attention; but it is no sense a ground feeder for it does not search for food while it is on the ground.

Field marks.—The blue-gray gnatcatcher, one of our smallest birds, can be distinguished from all other very small birds of eastern North America by its clear blue-gray upperparts and unmarked white underparts entirely lacking in yellow or yellowish tints, especially when the coloring is noted in conjunction with the slender build, long tail and white outer tail feathers. The bird is longer, and therefore apparently larger, that the kinglets because of the length of its tail. The brown creeper, another tiny bird with a long tail, differs notably in color, shape, and habit. The parula warbler, another bluish-gray bird, lacks the long tail and always shows white wing patches and some yellowish in the plumage.

No less a writer than John Burroughs (1880), when describing the gnatcatcher, made an unfortunately inept comparison that has been copied down the years in the writings of many of his followers. He wrote: "In form and manner it seems almost a duplicate of the catbird, on a small scale. It mews like a young kitten, erects its tail, flirts, droops its wings, goes through a variety of motions when disturbed by your presence, and in many ways recalls its dusky prototype." Such a comparison would never have occurred to an observer who knew the mockingbird, for the points of similarity (except for size) between the gnatcatcher and the mockingbird are truly striking—form, proportions (even to the long, expressive, white-edged tail), color value though not color tone, many characteristic movements and attitudes, in fact in almost every feature except the lack of white in the wings of the gnatcatcher. I once knew a tyro bird-watcher who, not aware that altricial species attain full body size before leaving the nest, spoke seriously of the gnatcatcher as a tiny mockingbird.

The distinguishing mark of the male gnatcatcher in breeding plumage—the black forehead and line over the eye—is useful as a field mark only at very short distances. Many times I have tried to see it, even with binoculars and in good light, but the activity of the little birds usually defeated my efforts. Only at the nest, when an approach to within a few feet is possible, have I been able to detect it with ease and certainty. It seems to be not generally known that this distinguishing mark is not present in winter specimens.

Voice.—Unlike the winter wren and the ruby-crowned kinglet, whose bid for fame rests as much upon the surprising volume of sound as upon the beauty of their songs, the gnatcatcher does not take high rank as a singer. To an observer like myself, whose auditory nerves (with advancing years) no longer react to high-pitched sounds of small volume, the gnatcatcher must actually be seen in the act of singing before the attention can be focused sufficiently to catch the sound. Once heard, the song is appreciated as a finished performance. C. J. Maynard (1896) immortalized it in this beautiful passage:

> I heard a low warbling which sounded like the distant song of some bird I had never heard before * * *. And nothing could be more appropriate to the delicate marking and size of the tiny fairy-like bird than this silvery warble which filled the air with sweet, continuous melody. I was completely surprised, for I

never imagined that any bird was capable of producing notes so soft and so low, yet each one given with such distinctness that the ear could catch every part of the wondrous and complicated song. I watched him for some time, but he never ceased singing, save when he sprung into the air to catch some insect.

Other observers and writers, however, do not seem impressed by its beauty. F. H. Allen writes (MS.) that the song of this species is "scrappy, formless, leisurely, and faint, and is delivered somewhat in the manner of a Vireo while the bird flits about among the branches. [He] found the phrase *pirrooeet* occurring frequently in it." A. A. Saunders (MS.) regrets that he cannot describe the song in detail, since his collection of sound records "contains only a few fragments from a single bird. The song is long continued, of greatly varied rapid notes and trills, on a high pitch, and of a squeeky or nasal quality. It is more curious than beautiful."

Wells W. Cooke (1914) cited a unique variation when "one bird was heard to give a long, and beautiful and perfect trill"; and A. L. Pickens, writing (MS.) from Paducah, Ky., strikes a new note when he describes "one fact about the blue-gray gnatcatcher most observers appear to have missed. It has decided powers of mimicry. One of its most amusing performances is the apparent imitation, in its almost whispering tones, of a flock of crows, or else blue jays." He says that the first name he knew for this species was "Little Mockingbird."

The song period commences with the reanimation of the gnatcatchers about mid-March and lasts only until eggs are laid and incubation is started in mid-April. Birds heard singing later than that in the far South may be only the late nesters or those that have lost their first nests and are preparing to try again. Like many other song birds, the gnatcatcher has a mild revival of ardor in the fall, and I have a few times heard its song in October. A. F. Skutch mentions (MS.) having heard one "sing a sweet little medley in an undertone" in January in Guatemala.

The call note of the gnatcatcher is far better known, because it is more easily heard, than the song. I find it variously described by many observers and writers, most of whom use combinations of the syllable *zee* in attempting to "phonotype" it. Others liken it to "the twang of a banjo string"; and "a nasal *twee*, suggestive of the catbird's mew but thinner, shorter and fainter." Any or all of these may serve as aids to identification for one who hears the sound for the first time, but to my ear it possesses a quality that defies description in stereotyped terms. It is more long-drawn than a chirp; not as clear as (more husky or "fuzzy" than) a whistle; definitely not a trill—and there I have compared it negatively with the more usual small-bird sounds, and still I have not described it. However, the sound is characteristic of this one species and, once heard, is readily remembered and recognized.

L. A. Stimson, writing (MS.) of the gnatcatcher in its winter quarters in southern Florida, mentions another note, "a shorter, more abrupt call with less of the *zz* quality."

Enemies.—It is little short of incredible that so tiny a bird as the gnatcatcher can and does successfully fill the role of foster parent to the young of the much larger cowbird (*Molothrus ater*), but there are many instances of this on record. Dr. Herbert Friedmann (1929) recorded the gnatcatcher as "a not uncommon victim [of the cowbird] and in some places a fairly common one." It must indeed be the smallest North American species thus victimized. An extreme case is given by M. G. Vaiden (MS.), who, writing from the Yazoo-Mississippi Delta of Mississippi, states that he has examined 12 nests of the gnatcatcher since 1919 and has found only two of them without cowbirds' eggs. In one instance, on June 4, 1939, he found a nest that "contained four gnatcatcher eggs and three cowbird's eggs," implying that other parasitized nests examined by him had contained fewer than three. In another nest he found "two young gnatcatchers and two young of the cowbird." Of the two nests that had not contained cowbird's eggs, only one was definitely immune, since the gnatcatcher was incubating her own eggs when discovered. In the other, the gnatcatcher had only commenced to lay, for the nest contained but a single egg, and the observer concluded that the cowbird "just had not located the nest yet," for he is "of the opinion it later on did have cowbird eggs." Thus we have a known 83 percent and a possible 92 percent parasitization, which is, of course, too high for any species to survive if it applied to more than restricted areas. Ben J. Blincoe (1923) watched a pair of gnatcatchers attacking persistently a female cowbird and driving it away.

The gnatcatcher probably suffers to some extent from predators and nest marauders—undoubtedly a few are taken by sharp-shinned hawks and screech owls, and perhaps some others succumb to attacks by loggerhead and migrant shrikes—but there is nothing to indicate that this species is singled out, nor, on the other hand, would it be expected to enjoy greater immunity than other species of comparable size within its range. However, a surprising instance of seemingly selective predation—though this may be as localized in its application as is the gnatcatcher–cowbird relation just cited from Mississippi—was given by S. A. Grimes, of Jacksonville, Fla., who wrote (1928):

Probably the greatest enemy of the Gnatcatcher is the Florida Blue Jay [*Cyanocitta cristata florincola*]. I have seen the Jay in the act of pilfering the smaller bird's nest perhaps a score of times. One such episode remains singularly vivid in my memory * * * When the Jay alighted on the rim of the nest, the Gnatcatchers were frantic and darted wildly at him, though so far as I could see neither actually struck him. Unperturbed, the Jay * * * grasped an egg in its beak and flew to a limb some twenty feet from the nest. I watched three trips to the nest, one egg being taken each time. I am inclined to believe that the Jay did not take all the eggs, for usually the nest is pulled apart after the last egg is eaten. And on the third visit the robber appeared annoyed with the continued attacking of the owners and flew with the egg to a tree some distance away before stopping to eat it. He did not return to the nest * * *.

Fall.—In the northern part of its breeding range the gnatcatcher is one of the first species to withdraw from its summer home, and mid-August often sees the last of them there. Farther south they linger much later, temporarily joining the wandering groups of small woodland birds headed by the chickadees and titmice.

In northern Florida and southern Georgia and Alabama, where the gnatcatcher is resident, there is a gradual increase in numbers in fall as birds from the northward pass through, then an equally gradual subsidence until the small winter population becomes stabilized. It is all so quietly and unobtrusively done that, unless a constant observer actually records numbers of birds seen on each trip afield, he is not likely to realize until late in the season what has taken place before his eyes. October is the time of greatest abundance, and by mid-November only a few gnatcatchers remain.

In southern Florida, where the gnatcatcher is not normally present in summer, its fall arrival is, of course, noticeable if not conspicuous. L. A. Stimson writes (MS.) from Miami: "In fall its migration into this area is in an increasing crescendo over a short period [leading up to its extreme abundance in winter]. Its first appearance will be made by one or a very few individuals, and a week later it will be common."

From the Southern United States, the gnatcatcher's progress to its tropical wintering grounds is presumably by way of the land masses and not by direct flight across the Gulf of Mexico, for I have never found a gnatcatcher among the many specimens of known trans-Gulf migrant species killed by striking the lighthouse at Pensacola, Fla., nor can I find a record of any having been killed at any other of the Gulf-coast lights.

Winter.—The blue-gray gnatcatcher's winter home in the United States embraces the coastal regions of South Carolina and Georgia, all of Florida, and a strip of the Gulf coast from northern Florida to Texas. In the northern part of this area it is uncommon to rare, but it becomes common in extreme southern Louisiana and is abundant in southern Florida and southern Texas.

Arthur T. Wayne (1910) wrote of it in the Charleston, S. C., region: "The birds are sometimes very hard to detect during the winter, and at that season frequent the interior of large swamps where they find food and shelter." Another southern observer, S. A. Grimes, wrote (1928) from Jacksonville, Fla., that it "is easily overlooked in winter, being rather retiring and feeding mostly in the higher foliage." My experience with wintering gnatcatchers in the similar—or even colder—climate of extreme northwestern Florida and southern Alabama is widely at variance with these last two observations. Here, although the gnatcatcher is far from common, it is widely distributed, ranging low as well as high in every well-wooded habitat except pure stands of pine. Alone, and usually silent in winter, it could easily be overlooked, but I have found it almost invariably associated with Florida chickadees and tufted titmice—and what could be easier to find than a titmouse! A typical chickadee-titmouse winter group of small birds comprises half a dozen each of titmice, chickadees, ruby-crowned kinglets, and myrtle warblers, a blue-headed vireo, an orange-crowned and a yellow-throated warbler, and a gnatcatcher or two. The scolding note of the titmouse is the signal for the observer to look sharp for the rare winter visitors that, when present, attach themselves to these wandering bands of small fry, so in the course of a winter, I see many gnatcatchers.

The gnatcatcher is not susceptible to freezing temperatures and has been known to withstand successfully such extreme as 16° F., provided these frigid spells last only a day or two; but the severe and protracted freeze of January 1940, when ice formed in northern Florida every night for two weeks and on several days did not thaw all day, caused the complete disappearance of the gnatcatcher from the Pensacola region until the advent of spring migrants. During the following winter of 1940–41, few were seen, although migrants and nesting birds had seemed no less abundant than usual in season. It was not until the second winter after the "big freeze" that the gnatcatcher could again be expected with confidence in every titmouse group.

Farther south in Florida the gnatcatcher reaches its peak of winter abundance. L. A. Stimson (MS.), describing its occurrence in the extreme southern end of the State, writes: "In the winter the gnatcatcher seems to show no favoritism as to habitat. It may readily be found in the city [of Miami] in fruit, native or exotic trees; in the open country in typical hammock trees; in pine woods; in the cypress; in the mangrove, buttonwood or bay fringes of the coast or swamps; and along the Tamiami Trail it will be found in the low willows where taller trees are absent. During its stay here the gnatcatcher associates freely with other insect eaters, wintering warblers, western palm, myrtle, yellow-throated, prairie, parula, black and white, black-throated green; the vireos, white-eyed and blue-headed; and the ruby-crowned kinglet. Woodpeckers, wrens, yellowthroats, and cardinals will frequently be found in the same clump of trees. In fact its call note is often the guide to a good 'bird tree'."

Alexander Sprunt, Jr., writes (MS.) from Okeechobee, Fla.: "Gnatcatchers swarm on some days. The hammocks and canal banks in perfectly open country have hundreds of them, and the characteristic *zee-e-e-e* note sounds in one's hearing at every stop. I have seen as many as six in one small willow. They frequent the oak and cabbage palm hammocks and the willows, myrtles and other growth typical of the banks of the drainage canals. There might be a stretch of open prairie for miles about such a place, but there they are! It is certainly one of the typical passerine species of this area in mid-winter."

Beyond our limits, in Guatemala, Alexander F. Skutch (MS.) considers the gnatcatcher an abundant winter bird at middle altitudes—from 2,000 to 7,000 feet above sea level—and states that "although the records of its occurrence range from the lowlands to the summit of the Volcán de Agua (12,100 feet), it is not often seen at either extreme; and it seems likely that the birds taken at very high altitudes were migrating rather than settled in their winter home. But I found it fairly common during the winter months in the open woods of pine and oak in the lower portions of the highlands among the orchards and thickets about the shores of Lake Atitlán and among the shade trees of the great coffee plantations on the Pacific slope down to about 2,000 feet." While he finds that "the gnatcatchers may at times form small flocks of their own kind," the habits of these tropical visitors seem to conform to the social pattern of the birds that winter within our limits, for they "attach themselves singly to flocks of warblers, the Tennessee warbler in the coffee-growing districts, the Townsend warbler at higher elevations." But he suspects that "when several of the birds flock together, they are females or immature individuals, for the adult males do not seem to get along together." Length of sojourn is indicated by Skutch's "only record which would indicate the date of arrival—one from Huehuetenango for September 11, 1934," at which time he saw several individuals. He also cites Griscom's "extreme dates for the occurrence of the species in Guatemala as September 7 and March 3."

DISTRIBUTION

Range.—Southern Canada to Guatemala.

Breeding range.—Toward the limits of its range the blue-gray gnatcatcher is rather sporadic in its occurrence, nesting one year and perhaps not appearing again for several years. It has been found breeding **north** to northern California (Covelo, Baird, and probably Yreka); central Nevada (Kingston Creek and Nyala); northern Utah

(Boulton, Stansbury Island, and the Uinta Valley near Vernal); southwestern Wyoming (Green River; it has occurred at Torrington in the eastern part of the State); southern Colorado (Cortez and Pueblo; occurrences have been recorded at Two Bar Spring, Moffat County, Grand Junction, and Boulder); western and central Oklahoma (near Kenton, Oklahoma City, and Ponca); eastern Kansas (Wichita, Manhattan, and Blue Rapids); southeastern Nebraska (Lincoln and Omaha); northern Iowa (Hawarden, one record, Grinnell, and National); southeastern Minnesota (Minneapolis, Marine, and Frontenac); southern Wisconsin (Madison and Milwaukee; occasionally north to New London); southern Michigan (Grand Rapids, Lansing, Ann Arbor, and Plymouth); extreme southern Ontario (Mitchells Bay, Plover Mills, and Hamilton); western New York (Geneva); southwestern Pennsylvania (Beaver and Harmarville); and northern New Jersey (Somerset and Essex Counties). **East** to eastern New Jersey (Essex County, Sea Isle City and Cape May); and the Atlantic Coast States to central Florida (New Smyrna and Deer Park); and the Bahama Islands (Abaco and New Providence). **South** to the Bahama Islands (New Providence); central Florida (Deer Park and Braden River); Cozumel Island, Mexico, and Yucatán (Chichen Itzá); Tabasco (Balancán); Chiapas (Palenque); Oaxaca (Oaxaca); and Lower California (Cape San Lucas). **West** to Lower California (Cape San Lucas, Comondu, and the Sierra San Pedro Mártir); and western California inside the Coast Range (San Diego, Pasadena, Stockton, Napa, and Covelo).

In addition there are many records of individuals north of any known breeding localities. Many of these are late in summer or in fall and suggest the possibility that some may be individuals that have wandered north after the breeding season; some are probably migrating from nesting sites that man has not found. These records extend north to Sault Ste. Marie; Mackinac Island and Douglas Lake, Michigan; Goderich and Toronto, Ontario; Montreal and Quebec, Quebec; and the coast of New England as far north as Portland, Maine.

Winter range.—The blue-gray gnatcatcher is found in winter **north** to central California (Muir Beach, Marin County; Kettleman Hills, Death Valley, and Needles); two specimens taken at Ashland, Oreg., February 4, 1881, are in the British Museum; southern Arizona (Papago Indian Reservation, Tucson, and the Rincon Mountains); Sonora (Punta Penascosa and Tesia); Sinaloa (Caliacan); Jalisco (Ocotlan); Nuevo León (Monterey and Rodriguez); Tamaulipas (Nuevo Laredo and Camargo); southern Texas (Brownsville and rarely San Antonio); southern Louisiana (Chenier au Tigre, Lecompte, and New Orleans); Mississippi (Bay St. Louis and Biloxi); northern

Florida (Pensacola and Tallahassee); eastern Georgia (Fitzgerald and Blackbeard Island; occasionally Milledgeville and Augusta); and the coastal region of South Carolina (Charleston). It has been recorded in midwinter at Collington Island, N. C., and Washington, D. C. **East** to South Carolina (Charleston), the Bahama Islands (Abaco, Caicos, and Great Inagua). **South** to the Bahamas (Great Inagua); Cuba (Guantánamo and Isle of Pines); and Guatemala (Los Amates and Alotenango). **West** to Guatemala (Alotenango, Sacapulas, and Chanquejelve); Oaxaca (Juchitán); Guerrero (Chilpancingo); Lower California (Cape San Lucas, Cedros Island, and San Telmo); and California (Santa Catalina and Santa Cruz Islands, Santa Barbara, and Muir Beach).

The species as outlined has been divided into several subspecies or geographic races. The eastern blue-gray gnatcatcher (*P. c. caerulea*) breeds from Nebraska, Kansas, Oklahoma, and Texas eastward; the western gnatcatcher (*P. c. amoenissima*) breeds from California to Colorado and south to northern Lower California and Sonora; the San Lucas gnatcatcher (*P. c. obscura*) breeds in the southern half of Lower California.

Migration.—Late dates of spring departure are: Cuba—Habana, May 13. Texas—Somerset, May 5.

Some early dates of spring arrival are: North Carolina—Raleigh, March 16. Virginia—Lawrenceville, March 21. West Virginia—Bluefield, March 28. District of Columbia—Washington, March 6. Pennsylvania—Waynesburg, April 6. New York—Brooklyn, April 8. Arkansas—Monticello, March 16. Tennessee—Nashville, March 12. Kentucky—Madisonville, March 27. Indiana—Bloomington, March 22. Ohio—Cincinnati; March 29. Michigan—Ann Arbor, April 11. Missouri—St. Louis, March 31. Iowa—Keokuk, April 12. Wisconsin—Beloit, April 23. Minnesota—Minneapolis, April 18. Oklahoma—Copan, March 20. Kansas—Independence, March 31. Nebraska—Fairbury, April 6.

Some late dates of fall departure are: Minnesota—Minneapolis, September 7. Wisconsin—Madison, September 11. Iowa—Hillsboro, September 15. Missouri—Columbia, September 18. Michigan—Grand Rapids, October 5. Illinois—Rantoul, October 15. Ohio—Toledo, October 5. Kentucky—Eubank, September 24. Tennessee—Athens, October 6. Arkansas—Helena, October 28. New York—New York, September 10. Pennsylvania—McKeesport, September 29. District of Columbia—Washington, September 13. West Virginia—Bluefield, September 5. Virginia—Naruna, September 22. North Carolina—Chapel Hill, October 4. South Carolina—Spartanburg, October 10. Georgia—Athens, October 17. Alabama—Leighton, September 5.

Some early dates of fall arrival in the winter home are: Cuba—Habana, August 13. Guatemala—Remote, Petén, July 30, Panajachel, September 7.

Egg dates.—Lower California: 12 records, May 8 to July 23; 7 records, June 25 to July 19.

California: 108 records, April 5 to July 12; 58 records, May 16 to June 12, indicating the height of the season.

South Carolina: 11 records, April 19 to May 22; 7 records, May 7 to May 22.

Texas: 11 records, April 15 to May 25; 6 records, April 23 and April 24.

POLIOPTILA CAERULEA AMOENISSIMA Grinnell

WESTERN GNATCATCHER

HABITS

Under the name western gnatcatcher (*Polioptila caerulea obscura*) Mr. Ridgway (1904) described the blue-gray gnatcatchers of the Southwestern United States, from the interior of northern California to northern Mexico and Lower California, as far south as Cape San Lucas, as "similar to *P. c. caerulea,* but gray of upper parts slightly duller and black at base of inner web of outermost rectrix more extended, usually showing beyond tip of under tail-coverts." His type came from San José del Cabo, and he evidently thought it might be only a winter visitor, for he said of its occurrence there "in winter only?". He therefore applied the name *obscura* to all the western gnatcatchers.

Subsequently, Dr. Joseph Grinnell (1926) discovered that "the materials now accessible in sufficient amount show that there is a separately recognizable race of Blue-gray Gnatcatcher resident in the restricted faunal area known as the Cape San Lucas district of Lower California." He therefore restricted the name *obscura* to the Lower California race and proposed the name *amoenissima* for the birds of the west outside of the southern tip of Lower California. He characterized *amoenissima* as "similar to *P. c. obscura* Ridgway, of the Cape San Lucas region, but wing and tail (especially the tail) longer, bill slightly slenderer, and median lower surface less clearly white, more imbued with very pale gray."

The 1931 Check-list gives the breeding range of this form as "from northern interior California (Shasta County), southern Nevada, southern Utah, and Colorado (El Paso County) south to northern Lower California, Chihuahua, Sonora, and Colima."

In California the favorite haunts of the western gnatcatcher are the warm, dry foothills covered with chaparral, small oaks, and other small trees and underbrush, as well as the groves of cottonwoods in the river valleys. In the Lassen Peak region, according to Grinnell, Dixon, and Linsdale (1930), "willow, valley oak, manzanita, digger pine, blue oak, and buckbrush are all plants in which individul gnatcatchers were seen. However, the portion of the section most favorable for this bird, as indicated by concentration of breeding birds to cover practically the whole of such area, was the tract of blue-oak-covered hills north of Red Bluff and west of the Sacramento River. There the park-like arrangement of the trees, each with many slender branches covered with copious foliage, and with the intervening spaces between the trees of a uniformly short distance, seemed ideally to fulfill the requirements of this bird for nesting and foraging activities."

In Arizona we found this gnatcatcher only fairly common in the foothills of the Huachuca and Dragoon Mountains and in the wider, lower portions of some of the larger canyons, mainly between 5,000 and 6,000 feet, but never out on the lower plains or at higher elevations in the mountains among the coniferous forests. Some of the foothills where we found the gnatcatchers were dotted with small black-jack oaks, with more or less even, open spaces between them, covered with mountain-misery and various other shrubs. A typical canyon haunt was in Miller Canyon in the Huachucas. This was heavily wooded along the stream with a row of big sycamores, ashes, walnuts, maples, and cedars; at its wide mouth was an open, parklike forest of large black-jack oaks in which numerous Arizona jays were nesting; farther up the canyon were smaller oaks of various kinds, madrones, manzanitas, and thickets of various kinds of shrubs. The gnatcatchers seemed to prefer the smaller oaks.

Territory.—In densely populated areas western gnatcatchers seem to establish and maintain fairly definite breeding territories. Grinnell and Storer (1924) write:

During the early spring immediately after their arrival from the south, the gnatcatchers are to be seen in pairs, the male in close attendance upon the female. When the latter engages in the work of nest construction her mate remains in the vicinity, part of the time accompanying her on trips for building material or on foraging sorties. Otherwise he guards the nesting precincts against invasion of any rival male. All the while, in the heat of mid-afternoon as well as at other hours of the day, the male gnatcatcher utters his fine wheezy song at frequent intervals, and the female answers from time to time in similar tone of voice with single notes.

When settled for nesting each pair of gnatcatchers is strongly localized. Each keeps within a radius of not more than a hundred yards from the nest tree. This localization permits an observer to take a more accurate census of nesting pairs than is possible with many other birds. At Black's Creek, near Coulterville, our own counts led to an estimate of 64 breeding pairs of the Western Gnatcatcher to each square mile in that immedate district. Carrying these figures further, in consideration of the estimated area of the Upper Sonoran Zone included in our

Yosemite section, we find a total gnatcatcher population just before the appearance of the new broods, to consist of 50,000 individuals.

Nesting.—The same observers say that "the nests are of deep cup-shape, and are constructed throughout of light-weight materials. A framework of fine grass stems forms the main wall, and this is covered both inside and out with softer substances. The outside is felted with lichens such as abound on the bark of blue oaks, with a few grass seed hulls, some small oak leaves, and occasionally a feather or two, the whole being held together with spider web. The inside of the nest is lined almost entirely with feathers, laid flatways of the inner surface. Whatever the purpose of the bird in constructing such a nest, the form and outside appearance are usually such that the structure might easily be mistaken for a weathered stub or a small accumulation of débris."

They mention a typical nest in a blue oak as "situated about 10 feet above the ground near the periphery of the tree, amid small twigs and branchlets, and rested directly on a horizontal branch." Others were found "in crotches of small blue oaks, and several were found in greasewood (chamisal) bushes, at a height of not more than 3 feet from the ground."

It seems to be a characteristic habit of the western gnatcatcher to change the location of its nest once or twice before being sufficiently satisfied with the site to finish the structure. Several observers have reported that nesting material has been entirely removed from an unfinished nest and used to build a nest in another location. This may be because the birds saw that their nest had been discovered and was unsafe for eggs or young; some birds are very sensitive to human intrusion. Corydon Chamberlin (1901) says: "Of the first few nests I saw being built none were finally occupied on their original site. One pair near my camping place moved their nest and made it over three times before being satisfied to deposit eggs in it. Each time that the nest was nearly complete, the birds would discover a more suitable site and then the work of tearing down would begin and it would be moved piece-meal to the new place and until scarcely a vestige of the nest remained in the old location. The third and final resting place for this nest was in the main crotch of a small white oak bush at such a height that I could just reach the nest by standing on tip toe."

He mentions several other nests, found in Tuolumne County, Calif., one of which was "in the main crotch of an alder tree 30 feet from the ground, the tree being in a creek bed." Another, 6 feet from the ground on a small horizontal oak limb, "was made in between the stub of a small twig and a live twig carrying a bunch of leaves that hung over the nest like a parasol.

"Perhaps the most unique nesting site ever seen of this species was the top of a pine cone in a sprawling bull pine. The cone was on a lonely limb fully thirty feet above the ground at the butt of the tree, but as the tree hung over a gully the nest was double that distance vertically from the ground."

Most nests of the western gnatcatcher are more or less decorated with lichens of colors to match their surroundings, though generally not as thickly covered as the nests of the eastern race. But Mr. Chamberlin mentions two nests that were entirely devoid of lichens; these were in oaks that had been killed by fire, as the whole area had been burned over; as there were no lichens available, the nests were decorated with bits of burnt bark, which helped to make them match their surroundings in the charred trees.

Harriet W. Myers (1907) found a nest "near the top of a holly bush that had grown so tall that it was more like a tree than a shrub" and another "on the south side of a tall, straight eucalyptus tree about twenty feet from the ground, its only supports being the tiny twigs that grew out from the side of the tree."

California nests of western gnatcatchers have been recorded at heights as low as 3 feet in bushes and as high as 45 feet in oaks.

The only nest we found in Arizona was 7 feet from the ground on a horizontal limb of a black-jack oak (pl. 44), in Miller Canyon, in the Huachuca Mountains.

Eggs.—The four or five eggs that make up the usual set for the western gnatcatcher are indistinguishable from those of the eastern subspecies. The measurements of 40 eggs average 14.3 by 11.4 millimeters; the eggs showing the four extremes measure **16.2** by 11.5, 14.0 by **12.0, 13.5** by 11.5, and 13.7 by **10.7** millimeters.

Young.—The incubation, brooding, and care of the young seem to be performed by this western subspecies in the same manner as by the eastern blue-gray gnatcatcher. Mrs. Myers (1907) gives the following feeding record for one brood:

> On this same morning from 7:25 to 8:25, the birds fed thirty-five times, less than two minutes apart; the male twenty-seven times, the female fourteen. The next morning, in the hour from 6:37 to 7:37, the birds fed forty-six times, the male thirty-six and the female twenty-four times. In looking over my notes I find that the birds fed more often early in the morning than later in the day.
>
> In five hours, 6:30 to 11:30, they fed one hundred and fifty-two times, or an average of thirty-eight times an hour. Allowing sixteen hours to their day, we can estimate that they fed six hundred and eight times. The word "gnatcatcher" proved to be a misnomer, the food brought so often being small white worms.

Food.—The food of the western gnatcatcher is not essentially different from that of its eastern relative. Professor Beal (1907) makes the following general statement on the food of the California gnatcatchers: "No complaints have been made that these busy creatures ever injure fruit or other crops. Their food is composed almost exclusively of insects, which they hunt with untiring energy from morning till night. Like the titmice and kinglets, gnatcatchers are fitted by nature to perform a service which larger species are unable to accomplish. There are hosts of minute insects, individually insignificant but collectively a pest, that are too small to be attacked by ordinary birds and are to be combated by man, if at all, only at great expense. It is to so deal with such pests that they may not unduly increase that these tiny birds would seem to be especially designed."

The behavior, voice, and general habits are similar to those of the eastern bird. It can be distinguished from the other California gnatcatchers by having the outer tail feathers white for their entire length; there is much less white in the tail of the plumbeous gnatcatcher, and hardly any in the tail of the black-tailed; the gray of the upperparts is lighter in the western than in either of the others. California jays have been known to destroy the nests of this gnatcatcher as well as to rob them of eggs or young; and Mrs. Myers (1907) has seen the little birds drive away a California shrike.

Dr. Friedmann (1929) says that the western gnatcatcher is victimized by the two western races of the cowbird, *artemisiae* and *obscurus*, in much the same way as the blue-gray gnatcatcher is imposed upon by the eastern cowbird.

POLIOPTILA CAERULEA OBSCURA Ridgway

SAN LUCAS GNATCATCHER

HABITS

As explained under the previous subspecies, this race of the blue-gray gnatcatcher has been separated from the other western race on the slight characters there mentioned. It is found only in southern Lower California, from Cape San Lucas north to about latitude 28°, where it is apparently resident. It probably does not differ materially in any of its habits from the more northern form. William Brewster (1902) says of it:

> The Western Gnatcatcher is a rather common resident of the Cape Region, where it appears to be indifferent to conditions of mean temperatures or environment, for it occurs nearly everywhere from the seacoast (La Paz and San José del Cabo) to the summits of the highest mountains (Sierra de la Laguna). Mr. Frazar found it breeding at San José del Rancho in July. His first nest, discovered on the 7th, contained four eggs on the point of hatching, and was not disturbed. Two others, taken respectively on the 14th and 19th of the month, had full sets of four eggs each, all freshly laid. One of these nests, built in the fork of a bush at a height of about five feet, measures as follows: Greatest external diameter, 2.25; greatest external depth, 2.00; internal diameter at top, 1.30; internal depth, 1.10; greatest thickness of walls, .50. The exterior is composed of gray, hemp-like, vegetable fiber and narrow strips of reddish brown bark, and

is decorated with a very few lichens, all these materials being over-wrapped and kept in place by a nearly invisible tissue of spider-web. The interior is lined with fragments of silky cocoons and a few feathers. The other nest, which was placed in the fork of a small tree about ten feet above the ground, and which is essentially similar to the specimen just described, save that it has no lichens whatever, measures externally 2.15 in diameter by 2.10 in depth; internally, 1.40 in diameter by 1.50 in depth. Both nests are smaller and more compact than any nests of *P. caerulea* in my collection.

There is a set of eggs in the Doe collection, taken by J. Stuart Rowley at Miraflores on May 8, 1933; the nest was saddled on a branch of a palo blanco tree, 20 feet above ground. A nest in the Thayer collection, taken by W. W. Brown near La Paz on June 25, 1908, was placed between two upright stems of a bush, about 5 feet from the ground. The construction of this nest is similar to that of the nest described above; there are no lichens on it, but it has a decidedly gray appearance.

The eggs of the San Lucas gnatcatcher are similar to those of the species elsewhere. The measurements of 32 eggs average 14.5 by 11.2 millimeters; the eggs showing the four extremes measure **15.6** by 11.8, 14.3 by **11.9**, **12.9** by 11.2, and 13.6 by **10.6** millimeters.

POLIOPTILA MELANURA MELANURA Lawrence

PLUMBEOUS GNATCATCHER

HABITS

The type race of this species is the desert form that inhabits the arid and semiarid regions of the southeastern United States and northern Mexico. The plumbeous gnatcatcher, like many other desert forms, is much paler, especially on the dorsum, than its near relative and neighbor, the California black-tailed gnatcatcher, and it has more white in the tail and usually more on the lores and sides of the head.

The favorite haunts of this gnatcatcher are in the thorny thickets of mesquites or in the rank growths of saltbush on the southwestern deserts. Van Tyne and Sutton (1937) say that, in Brewster County, Tex., both this and the western gnatcatcher are common, "but the Plumbeous is a bird of the river banks and low country, while the Western Blue-gray is a bird of the mountains. We found the Plumbeous especially abundant in the dense thickets of mesquite in the ravines and along the Rio Grande." Mrs. Bailey (1902) says: "The small bluish figure of *plumbea* is a familiar sight in the brushy canyon mouths of the Guadalupe Mountains in Texas and in the orchard-like juniper and piñon pine tops of the mountains." We found it fairly common in the low, arid valleys and dry brushy washes in southern Arizona, nesting in the thorny bushes.

Nesting.—The nests of the plumbeous gnatcatcher are placed in low trees or small bushes at no great height from the ground, much like those of the black-tailed gnatcatcher. There are two nests of the former in the Thayer collection in Cambridge, quite similar in construction, but very different in dimensions. One taken by W. W. Brown in Sonora on April 30, 1905, measures 1¼ inches in height, 2 inches in outer diameter, 1¼ in inside diameter, and 1 inch in inner depth; it was in a mesquite. The other, a much larger nest, was taken from an *Atriplex* bush on the desert near Phoenix, Ariz., by G. F. Brenninger on April 10, 1901; it measures 2½ inches in height, 2¼ in outside and 1½ in inside diameter, and was hollowed to a depth of 1½ inches. Both nests are very neatly made of various grayish fibers, compactly woven, and are lined with pappus and other plant down; they are firmly bound with spider web, but no lichens have been used for outside decoration. Each nest held five eggs.

A nest taken by Frank C. Willard near Tombstone, Ariz., on April 22, 1897, is described in his notes as 3½ feet up in a small bush, in a fork and supported by various twigs; it was made of fine bark strips and grass and was lined with fine grass and cactus fiber, with a few small feathers woven in. A nest found by Van Tyne and Sutton (1937) in Brewster County, Tex., on April 14, 1935, was "situated about three feet from the ground in a thickly-leaved thorn bush that was growing under a huge cottonwood." Mrs. Bailey (1928) mentions a nest near Terlingua, Tex., in a fouquieria (ocotillo) bush.

J. Stuart Rowley writes to me: "I have found these birds to be very touchy about the inspection of their nests before the eggs are deposited. Many nests which I have located in the Coachella Valley of California in the process of construction were examined without being actually touched by hand, and upon returning in a week or ten days, were found to be utterly destroyed and deserted. One female sat on her eggs so closely that she was approached slowly and forcibly removed from the nest by being grasped by the bill."

Eggs.—The plumbeous gnatcatcher lays three to five eggs to the set, but four is decidedly the commonest number. The eggs of all the gnatcatchers are practically indistinguishable, though there is perhaps a tendency to be less heavily marked in eggs of this subspecies. The eggs are ovate or short-ovate and have little or no gloss. The ground color in museum specimens is very pale blue, bluish white, or nearly pure white; they are somewhat brighter, greenish blue when fresh in the nest. They are usually rather sparingly and more or less evenly covered with small spots or fine dots of reddish browns; sometimes there are a few very small blotches of various darker browns; and sometimes the markings are somewhat concentrated in an imperfect ring about the large end. Very rarely an egg is almost immaculate. The measurements of 40 eggs average 14.1 by 11.1 millimeters; the eggs showing the four extremes measure **15.2** by 11.2, 14.5 by **11.5**, **12.2** by 11.0, and 13.1 by **10.6** millimeters.

The plumbeous and the black-tailed gnatcatchers are so much alike in all their habits and characteristics, due allowance being made for the difference in environment, that the excellent life history of the latter contributed by Mr. Woods will suffice for the former.

As the plumbeous and the western gnatcatchers are sometimes both found in the same region in winter, and as the former does not show its black cap at that season, there is a chance of confusing the two species in their winter haunts. However, Ralph Hoffmann (1927) remarks: "The darker gray of the Plumbeous Gnatcatcher and the small amount of white in the tail when spread are helps to identification, but the call notes are often the only sure distinctions. The call of the Plumbeous Gnatcatcher is a series of two or three short notes, *chee chee chee*, unlike the single emphatic *pee* of the Western. The song of the Plumbeous Gnatcatcher is a slight *tsee-dee-dee-dee-dee*, suggesting a chickadee."

Dr. Friedmann (1929 and 1934) mentions several cases where this bird has been victimized by cowbirds. And Mr. Rowley tells me that "along the Colorado River area, cowbirds parasitize the nests of these birds rather abundantly," and he has "found a female sitting on three eggs of a cowbird and none of her own, with many nests containing one or two cowbird eggs."

DISTRIBUTION

Range.—Southwestern United States and northern Mexico; nonmigratory.

The gnatcatchers of the *melanura* group breed **north** to southern California (Santa Barbara, Yermo, Daggett, and Resting Spring); southern Nevada (Las Vegas and Bunkerville); northern Arizona (Beale Spring, Big Sandy Creek, Fort Whipple, and Keam Canyon); western and southeastern New Mexico (Gallup and Fort Wingate; Carlsbad and the Guadalupe Mountains); the western extension of Texas (Frijoles, Marathon, and Chisos Mountains) and the Rio Grande Valley (Lozier, Laredo, and Brownsville); with records of its occurrence at Corpus Christi and San Antonio. **East** to extreme southern Texas (Brownsville) and Tamaulipas (Jaumave). **South** to Tamaulipas (Jaumave); central Nuevo León (Monterrey); southern Coahuila (San Pedro); central Chihuahua (Santa Eulalia and Chihuahua); central Sonora (Moctezuma and Guaymas); and Lower California (Cape San Lucas). **West** to the Pacific coast of Lower California (Cape San Lucas, Todos Santos, Santa Margarita Island, San Juanica Bay, and

San Quintín); and southwestern California (San Diego, Escondido, Los Angeles, and Santa Barbara).

The range as outlined is for the entire species of which several subspecies or geographic races are recognized. The plumbeous gnatcatcher (*P. m. melanura*) occurs from southeastern California, southern Nevada, central Arizona, and the Rio Grande Valley to Chihuahua and Tamaulipas; the black-tailed gnatcatcher (*P. m. californica*) occurs in southwestern California and northwestern Lower California; the San Francisquito gnatcatcher (*P. m. pontilis*) is found in central Lower California; the Margarita gnatcatcher (*P. m. margaritae*) is found in Lower California from about 29° N. latitude southward; the Sonora gnatcatcher (*P. m. lucida*) occurs from southeastern California and southern Arizona to central Sonora.

Egg dates.—Arizona: 17 records, April 10 to July 15; 9 records, April 10 to May 4.

Lower California: 8 records, April 19 to July 5.

California: 52 records, March 18 to June 19; 26 records, April 14 to May 13, indicating the height of the season.

POLIOPTILA MELANURA MARGARITAE Ridgway

MARGARITA GNATCATCHER

HABITS

In naming this subspecies, Mr. Ridgway (1904, p. 733, footnote) wrote: "The only specimens examined from Margarita Island (two in number) very likely represent a different form; both have decidedly shorter wings and longer bills than specimens from the mainland of Lower California; they have the upper parts decidedly darker, the dull slate color of the pileum contrasting abruptly and strongly with the dull white of the loral region; there is, apparently, a distinct whitish crescentic mark immediately behind the dark grayish auricular region, a feature which I have not been able to find in any specimen of true *P. plumbea*. Both specimens were skinned from alcohol; one is an immature male, the other probably an adult female. Should the bird from Margarita Island prove to be distinct, I propose for it the name *Polioptila margaritae.*"

Santa Margarita Island lies close to the Pacific coast of Lower California, toward the southern end of the peninsula, but north of the Cape region. Although this gnatcatcher was described and named from specimens collected on this island, it seems to be well distributed on the mainland of Lower California between latitudes 24°30′ and 29°.

The only published account of its habits that I can find is the following from Griffing Bancroft (1930):

[It is] widely spread and decidedly common. There was no association and no region where the presence of a pair of these little scolds could not be expected. They were most plentiful west of José María Cañon. They seemed equally at home in the thickest brush and on the most open plains. Yet we found only five occupied nests and were able to save but one set of eggs.

The nests are cups, rather thin and quite deep. They are so extremely neat and trim and blend so well into the background that it is difficult to see them, the first time. They are usually placed in the center of some sage-like bush about three feet from the ground. They rest on both a lateral and a horizontal branch. One exceptional site was the heart of a mistletoe in a mesquite, well hidden by the parasite, at a height of twelve feet.

Fresh eggs are most numerous about the middle of May, and the season is exceptionally short. The number in a clutch, within our limited experience, was three.

There is a set of three eggs in the Doe collection, taken by Mr. Bancroft, near San Ignacio Lagoon, on June 5, 1932; the nest was three feet above the ground, well concealed near the upper center of a tall shrub, back in the hinterland and well away from the lagoon.

The eggs are evidently indistinguishable from those of the northern forms. The measurements of 6 eggs average 15.2 by 11.5 millimeters; the eggs showing the four extremes measure **15.5** by 11.5, **14.8** by **11.9**, and 15.0 by **11.0** millimeters.

POLIOPTILA MELANURA ABBREVIATA Grinnell

XANTUS'S GNATCATCHER

HABITS

This local race of the Cape district of Lower California remained for long unrecognized as distinct from the other races of the species. It remained for Dr. Joseph Grinnell (1926) to describe and name it. He says of it: "In general character similar to *Polioptila melanura melanura* (see Ridgway, 1904, p. 731, under *Polioptila plumbea*) of southeastern California and southern Arizona, but (in both sexes) tail decidedly shorter, bill somewhat larger, leaden hue of dorsum slightly deeper, and lower surface slightly more imbued with gray, not so clearly white."

Of its range he says: "So far as now definitely known, only the southern end of the Lower Californian peninsula, from San José del Cabo and Cape San Lucas north to La Paz."

Earlier writers referred the birds of the Cape region to *plumbea*, now known as *melanura*, but William Brewster (1902) remarks: "All my Lower California specimens seem to have shorter tails than the birds which inhabit Arizona and Texas." Evidently all his birds came from the Cape region, where Mr. Frazar regarded it as rather rare.

J. Stuart Rowley writes to me: "I met with these birds in the Cape region of Lower California in the lowland country. A pair were building a nest in the immediate vicinity of my base camp at Miraflores in the latter part of May 1933, but they were rather shy about divulging the actual nest site. However, after much watching, the nest was found to be in the extreme top of a tall mahogany tree, and since the birds were obviously building at this time, it was left undisturbed. On the day I broke camp, the extension ladder was set up to its fullest length of 28 feet and that distance plus my six feet, plus arm reach, fell short of reaching the nest by a good ten feet; so the nest was cut down and lowered to me, only to find it contained but one egg. I have never found a gnatcatcher nesting as high from the ground as this one did."

He took a set of three slightly incubated eggs near San José del Cabo on May 23, 1933, which is now in the Doe collection at the University of Florida. The nest was placed in a mesquite bush, 4 feet from the ground, and was made of fine fibers. These three eggs measure 15.3 by 11.9, 15.3 by 11.4, and 14.5 by 11.4 millimeters.

Since the above was written, the A. O. U. Committee has ruled in its nineteenth supplement to the Check-list (Auk, vol. 61, p. 457) that this bird is synonymous with *P. m. margaritae*, but it seems best to include the above account as a matter of historical record.

POLIOPTILA MELANURA CALIFORNICA Brewster

BLACK-TAILED GNATCATCHER

HABITS

CONTRIBUTED BY ROBERT S. WOODS

Although frequenting the environs of some of the most populous districts of the West, the black-tailed gnatcatcher remains one of the least familiar of North American birds. For some reason this species is localized and comparatively few in numbers, while the western gnatcatcher, apparently with no superior endowments, is widespread and numerous.

Formerly classified as a separate species (*Polioptila californica*), this form is now regarded as a subspecies of the plumbeous gnatcatcher, but it is easily distinguishable from the latter in the field by its decidedly darker body coloration. The dividing line between their respective territories lies across San Gorgonio Pass and along the higher mountain chain of southern California. The range of the present subspecies extends from Ventura southward into northwestern Lower California.

Semidesert valleys and lower slopes covered with a low and often scattered growth of shrubs are the home of the black-tailed gnatcatcher, and few land birds confine themselves so rigidly to their characteristic habitat. Suitable conditions are found on gravelly washes along watercourses, as around Azusa and Claremont in Los Angeles County, and in various parts of San Diego, Riverside, and San Bernardino Counties. A. C. Bent (MS.) and Dr. Louis B. Bishop also encountered this species, together with Bell's sparrow, near Norco, on "low, rolling hills covered with an open growth of low bushes, chamise, white sage, golden yarrow and wild buckwheat." Even in the most favorable situations the black-tailed gnatcatcher is not abundant, and, as it does not range far in search of food, it might be looked for in vain through a period of years in any given locality.

Measurements indicate that this species closely rivals the bushtit for the distinction of being the smallest North American passerine bird.

Nesting.—As its nesting sites, the black-tailed gnatcatcher chooses small or medium-sized bushes rather than large ones. Eight nests discovered at Azusa by the present writer, during three different years, were placed at heights of 2 to 3 feet. Four were in buckthorn bushes (*Rhamnus crocea*), two in laurel sumac (*Rhus laurina*), one in a clump of cactus and weeds, and the last in a shrubby composite (*Ericameria pinifolia*). These locations afforded varying degrees of concealment, one of the buckthorns being so densely branched that the nest could hardly be discerned from any point outside the bush, while those in the sumacs could easily be seen from a distance.

Descriptions of nesting sites in other districts differ in minor details. Wilson C. Hanna (1934) mentions a nest 4 feet from the ground, near the top of a black sagebush on a dry hillside near Riverside. Mr. Bent (MS.) saw one near Claremont "about 2 feet up in a small branching cactus growing in a clump of chamise, on a dry, dusty chamise-covered flat." Probably the earliest description of the black-tailed gnatcatcher's nest is that of Bendire (1888):

This gnat-catcher was first described by Mr. William Brewster, from specimens collected by Mr. F. Stephens near Riverside, * * * Cal., March 28, 1878. * * * *

The nest of *P. Californica*, like that of *P. plumbea* BAIRD, from Arizona Territory, differs radically in structure from that of its eastern relative, *P. caerulea* (LINN.), which is too well known to ornithologists to require description. It lacks entirely the artistic finish of the lichen-covered structure of the former, and resembles more in shape certain forms of the nest of the Summer Yellow Warbler, *Dendroica aestiva* (GML.), and the American Redstart, *Setophaga ruticilla* (LINN.)

The nest is cone-shaped, built in the forks of a small shrub, a species of mahogany, *Coleogyne ramosessima* (TORR.) I think, only 2 feet from the ground, and it is securely fastened to several of the twigs among which it is placed. Its walls are about half an inch in thickness. The material of which the nest is composed, is well quilted together and makes a compact and solid structure.

Externally the nest is composed principally of hemp-like vegetable fiber mixed with small curled-up leaves of the white sage, *Eurotia lanata*, plant-down, and fragments of spiders' webs. Inside the nest is lined with the same hemp-like fiber, only much finer, and a few feathers. The cavity of the nest is cup-shaped and rather deep. Externally the nest measures 2½ inches in diameter by 3¼ inches in depth. The inner diameter is 1½ inches by 1¾ inches in depth. Compared with a nest of *Polioptila plumbea* Baird, now before me, from Arizona Territory, it seems much better constructed and also somewhat larger.

The nests at Azusa were deeply cup-shaped, sometimes slightly constricted at the top, and were compactly and neatly constructed of small pieces of grass, bark, fiber, paper, cloth, string, etc., and lined with small feathers, rabbit fur, and soft cottony material. Small bits of newspaper entered largely into the construction of one of them. The interior measurements of three of these nests were 1½ inches in diameter by 1¼ inches in depth. Ordinarily only the top of the head and the tail of the incubating bird project above the rim.

The birds being permanently mated, the nest-building urge sometimes seems to develop long before the time for egg-laying. About March 1, 1921, a pair of black-tailed gnatcatchers started a nest near the top of a low sumac bush about 2½ feet from the ground. The work proceeded rapidly at first, most of it done by the male, and then gradually slowed up. By the middle of the month the nest, which was ready for the lining, seemed to be deserted, and a week later was found overturned and partially destroyed. This suggested a search for a new nest, which was discovered a day or two later about 125 feet away in a clump of cactus and weeds, and about 2 feet from the ground. It was complete except for part of the lining, which was added to from time to time until finished, after which the birds showed no further interest in it.

On April 22, by following one of the gnatcatchers, I found a third nest containing three eggs in a buckthorn about 250 feet from the previous nest. The site was about 2½ feet from the ground in the midst of an unusually dense mass of twiggs. The young having left this nest on May 12, an inspection on June 8 showed that a section of it had been removed, and it was found that only a small portion of the second unused nest remained. A search revealed a nest with three eggs, about 50 feet away, and situated very similarly to the last nest, but less thickly surrounded by foliage. This nest appeared a little looser and bulkier than those built of new materials.

In 1927 a nest was discovered in a buckthorn bush, at a height of about 3 feet, the set of four eggs being completed April 7, two days after the male was seen incubating two eggs. On April 26 the four young, which had hatched within less than 24 hours on April 20 or 21, were found dead in and around the nest, though the parents remained in the vicinity. The next nest was about 200 feet distant, and at a height of about 3 feet in a small laurel sumac. The four eggs were hatched on or about May 21 and the young left on June 5. A third nest with four well-grown young, undoubtedly belonging to the same pair, was encountered on July 12, about 100 yards from the second and at the same height.

The duty of incubation is divided between the two parents, the eggs seldom being left uncovered even for a moment; the anxiety of one of the males to resume his duties would cause him sometimes to almost shove his mate from the nest. While sitting, this same individual maintained a vigilant watch, frequently peering over the edge and closely scrutinizing the ground beneath. In another case it was found that the male occupied the nest more than half the time during the day, being relieved by his mate at intervals of approximately one hour. Soon after sunset, however, the female would take her place there and presently, with head tucked in, would be settled for the night.

The male, or less frequently the female, would fiercely attack other birds, including gnatcatchers, which strayed within perhaps 50 feet of the nest, darting at them repeatedly with rapidly snapping bill. A flock of bushtits offered greater difficulties, as each one attacked merely dodged back into the foliage, to emerge serenely as soon as the assault was transferred to another member of the flock.

Eggs.—The normal set of eggs consists of four. Any larger number is at least very rare, but some nests contain only three. It appears that the laying of sets of three may be an individual tendency, as the female observed in 1921 laid two sets of three, while that of 1927 laid three sets of four. Mr. Hanna (1934) gives the average weight of 38 eggs as 0.99 gram, with a maximum of 1.11 grams and a minimum of 0.82 gram. Comparing the eggs of the black-tailed and western gnatcatchers, Harry H. Dunn (1906) says: "Several writers, and even as good an authority as Davie, claim that there is a difference in the eggs of the two species, but from a long string of sets which have come under my observation I am unable to agree to this. To my mind they are absolutely indistinguishable, especially a few days after being blown, when the eggs of both species fade considerably, becoming a pale, washed-out blue, instead of the deep green they possess when freshly laid."

The measurements of 36 eggs average 14.4 by 11.5; the eggs showing the four extremes measure **15.1** by 11.5, 14.5 by **12.0**, **12.9** by 10.8, and 13.0 by **10.6** millimeters.

Young.—The incubation period, which I ascertained in only one case, proved to be rather longer than expected, considering the small size of the eggs, being 14 days from the laying of the last egg. Although the eggs were laid on successive days and incubation was apparently begun with the first, or at least the second egg, all four were hatched within less than 24 hours.

The food brought to the young consists of a large variety of small insects and spiders, unidentifiable from a distance. Larger insects are thoroughly beaten against a branch before being offered. The largest noted was a walking-stick nearly as long as one of the young birds, and was swallowed only after some difficulty. The nest and its surroundings are kept scrupulously clean at all times.

On May 31, 1927, a record was made of visits to the nest by both parents between 4:12 and 5:55 P. M. During this period of 1 hour and 43 minutes, the female brought food 37 times, at intervals ranging from one to nine minutes, and averaging slightly less than three minutes. The longer intervals occurred previous to 4:46 P. M., but no indication of periodicity could be detected at any time. The record of the male during this time probably was hardly representative: he brought food only five times in the first 69 minutes, and nine times in the last 32 minutes.

Whenever the partly feathered young would otherwise be exposed to bright sunshine in the middle of the day, one of the parents is accustomed to stand over them with wings extended sufficiently to shade the interior of the nest. The same attitude is assumed when incubating on hot days. With four young in the nest, the quarters become somewhat cramped as the brood approaches full growth, and the young birds are sometimes disposed in two tiers, with one or both of those in the lower layer almost concealed from sight. Presumably they change places from time to time, as their rate of development always seemed remarkably uniform, with the birds leaving the nest almost simultaneously. As with some other passerine birds, the fledging period shows wide variations. One brood of three left the nest in nine days, and a second in 10 or 11 days, while a family of four, in another year, remained 14 or 15 days.

After leaving the nest, the fledglings show little activity for a few days, sitting quietly together and moving from bush to bush only at the urging of the parents. To find their progeny, the old birds must depend on their memories, as the young at this stage remain practically mute. On two occasions I saw a mother bring food to a bush which the young had left some time previously, and after searching for a while and then calling without answer from the young ones which sat stolidly in the interior of another bush, she appeared to recollect and flew directly to them. The young are fed for about three weeks after leaving the nest; thereafter the birds of the first brood are not allowed to loiter near the scene of the second nesting.

Plumages.—The bodies of the young are entirely bare until the primaries begin to appear, three or four days after hatching. When feathered, their coloration is similar to that of the adult female, but until the fall molt takes place they can be distinguished by the greater amount of white on their tails. The general molt occurs only in the fall, so that by summer the white edgings of the outer tail feathers of the adults have by abrasion become much reduced or altogether lacking.

As observed by the present writer, the black cap of the male, a seasonal character, is assumed in February, black patches appearing on the crown and quickly spreading over the entire top of the head. The reverse change in the fall takes place much more slowly, in the form of a gradual obscuring and replacing of the glossy black by gray. The first signs of gray could be detected about the middle of July, or early in August in another year, and it required approximately a month and a half for all traces of darker color, with the exception of a permanent blackish streak above the eye, to disappear. The change appeared to be complete before the new tail feathers were entirely grown out. Concerning variations in the molt, H. S. Swarth (1902) writes:

> I have taken many specimens between August and March showing no black on the head, with the exception of the almost invisible black streak over the eye, which is, I believe, always present in the male; and others during March and April undergoing moult over the entire crown; so I was the more surprised on taking on December 13, 1901, a male bird with the black cap nearly complete, though not quite as extensive as in most spring specimens, and with the black feathers tipped with the blue-gray of the rest of the upper parts, so that the black was not apparent unless the feathers were ruffled. It would have taken but little abrasion of the tips of these feathers to have rendered this bird indistinguishable from specimens taken during April and May. On December 19 I secured another, almost a counterpart of the one described above. * * *
>
> The time for the spring moult seems to be extremely variable; I have specimens taken at the end of February, with no trace of the black crown, and not yet commencing to moult; while on January 20, 1902 I took one with many pin feathers on the head and the black cap nearly complete. Usually, I think that the change of plumage is not finished before the first week in April. No part of the bird but the crown seems to be affected by the moult.

I am unable to confirm a further statement by Mr. Swarth that the winter plumage is tinged with brown. However, one female which was watched during the nesting season was notable for the decidedly brown tone of the back and wings, in contrast to the clear gray of the male. I have noticed no brownish tendencies in any other individuals at any time of year.

Food.—Prof. F. E. L. Beal (1907) reports as follows on the food of the gnatcatchers:

> Only 30 stomachs of *P. c. obscura* and the same number of *P. californica* have been examined, and their contents were so similar that they may be treated as from a single species.

Vegetable food.—Of the 60 stomachs three only contained any vegetable food whatever, and in only one did it amount to a respectable percentage. This one held 92 percent of seeds of some species of Rhus; another contained 8 percent of unknown seeds, and the third a few bits of rubbish, which amounted to only 2 percent of the whole contents. The total vegetable matter in the 60 stomachs aggregated less than 2 percent of the entire food.

Animal food.—The remainder of the food, over 98 percent, is made up of beetles, wasps, bugs, and caterpillars, with a few flies, grasshoppers, and spiders. Bugs (Hemiptera) constitute more than half of the whole food, 64 percent. * * * In one stomach were 20 percent of black olive scales (*Saissetia oleae*). All of these are harmful to trees and other plants. Wasps and a few ants (Hymenoptera) are next in importance as an element of the gnatcatcher's food, and amount to over 16 percent of the whole. * * * The only decidedly useful insects in any of the stomachs were 2 ladybird beetles (*Coccinella t. californica*), which had been eaten by *P. californica.*

Observation of the foraging birds gives the impression that small moths (not mentioned in the report) must make up a large portion of the food.

Behavior.—Though Professor Beal found the food of the two species of *Polioptila* to be essentially the same, their methods of obtaining it are not identical. The aerial activities which have earned for *caerulea* its appropriate name "gnatcatcher" were rarely practiced by the various individuals of *californica* which I have watched. If they ever obtained any of their food on the wing, it was usually while hovering, like a kinglet, to pick off an insect from a leaf or terminal twig. A sound of the snapping of the bill would occasionally give the impression of flycatching at other times were the birds not actually in sight. Practically all their foraging consists of systematic search through the branches of shrubs. They do not seem to care for water, either for drinking or bathing.

A peculiarity of these birds is their reluctance to leave their accustomed surroundings. Neither orchard nor garden seems to offer inducement to exploration; when they reach the boundaries of their arid, brushy habitat, they seldom go farther. Their feeding territory covers but a few acres, and throughout the year the pair may ordinarily be found at any time with very little search. However, I have not known them to remain more than about a year, after which they moved on to parts unknown.

These gnatcatchers showed very noticeable individual differences in temperament. The 1921 male, notable for his solicitousness and watchfulness, was also most fearless and trustful in his attitude toward the spectator and the camera. Others evinced varying degrees of timidity and suspicion, but none showed any sign of hostility or resentment at these intrusions upon their family affairs, in marked contrast to their actions toward trespassing birds or toward cats, which they would scold vigorously whenever encountered at any time of year. The anxiety of some of the birds increased as the time approached for the young to leave the next, though their fear seemed to be for their offspring rather than for themselves. The female of 1921, though sometimes distrustful, occasionally indicated a certain curiosity by approaching silently to within arm's length or hovering close above one's head. In foraging, when not engaged in nesting, a pair usually pursued independent courses through the bushes, though remaining not far apart; one male, however, would always closely follow his mate from bush to bush.

Voice.—The black-tailed gnatcatcher is less inclined to song than is the western; in fact, I have never heard anything which could be definitely so called. The ordinary call note resembles that of *caerulea*, but it can be recognized by a rather querulous, mewing tone; that of the female is especially thin and plaintive. On rare occasions, the male has been heard to utter a short, harsh note like that of the plumbeous gnatcatcher.

Field marks.—Within the territory of this subspecies, it could readily be confused with only two other species, the western gnatcatcher and the bushtit, both of which resemble it in size, length of tail, and gray color. From the former it can be distinguished by its darker general coloration and narrower white tail edgings. From the bushtit it differs in its longer bill, more slender form, white tail edgings (except in very worn plumage), and more restless movements.

Enemies.—Mr. Hanna (1934) says: "It was not until May 25, 1933, that I found a nest of the Black-tailed Gnatcatcher * * * in this vicinity parasitized by the Dwarf Cowbird," and adds: "It must be that the Black-tailed Gnatcatcher has not suffered from the cowbirds to any great extent in any locality, as Friedmann in his book 'The Cowbirds' fails to mention them as victims. In this locality one of the probable reasons for their escape has been due to their usual habitat being on the dry bush-covered hillsides or dry washes between 700 and 2,000 feet elevation above sea level." On one occasion at Azusa I saw a nearly grown cowbird following and being fed by a black-tailed gnatcatcher.

Upon visiting one nest which then held eggs, I found a good-sized alligator lizard (*Gerrhonotus scincicauda*) resting immediately beside the nest while the male bird incubated, apparently unperturbed. The only inference I could draw was that the reptile was waiting until the eggs should be left unguarded. Probably with the return of the female bird, one would have driven the lizard away while the other protected the eggs; however, I removed the intruder to prevent any untoward happenings.

In one case, three out of a brood of four disappeared within a few days after leaving the nest. The presence of a shrike in the vicinity at that time furnished a plausible reason for their disappearance.

POLIOPTILA MELANURA LUCIDA van Rossem

SONORA GNATCATCHER

The gnatcatchers of central Sonora, southern Arizona, and southeastern California have been given the above name by A. J. van Rossem (1931). In describing it he says: "Compared with *Polioptila melanura melanura* of the Lower Rio Grande Valley, *lucida* is slightly smaller in all dimensions and the bill is notably smaller; in color it is decidedly paler below (often nearly white medially) and the flanks are very much paler and less extensively gray. Females and young males of *lucida* also lack much of the brown wash seen in the corresponding plumages of *melanura*."

It is not known to differ in its habits from other races of the species.

POLIOPTILA MELANURA PONTILIS van Rossem

SAN FRANCISQUITO GNATCATCHER

This seems to be an intermediate form, found only in central Lower California, originally described by Mr. van Rossem (1931) under the name *P. m. nelsoni*. In his description of it he says: "In color and size *nelsoni* is intermediate between *margaritae* of the Cape Region and *californica* of southwestern California and northwestern Lower California."

As the name *nelsoni* was found to be preoccupied, Mr. van Rossem gave the poor bird the new name *pontilis* in the Proceedings of the Biological Society of Washington, vol. 44, July 15, 1931, page 99. It would seem to the author of these bulletins that the naming of what appears to be a strictly intermediate form is undesirable and confusing, as it increases the number of intermediate forms.

REGULUS SATRAPA SATRAPA Lichtenstein

EASTERN GOLDEN-CROWNED KINGLET

HABITS

Many years ago, a boy found on the doorstep the body of a tiny feathered gem. Perhaps the cat had left it there, but, as it was a bitter, cold morning in midwinter, it is more likely that it had perished with the cold and hunger. He picked it up and was entranced with the delicate beauty of its soft olive colors and with its crown of brilliant orange and gold, which glowed like a ball of fire. In his eagerness to preserve it, he attempted to make his first birdskin. It made a sorry-looking specimen, but it was the beginning of a life-long interest in birds, which has lasted for over a half century. Since then many a winter landscape in southern New England has been enlivened by the cheery little groups of kinglets, wandering through our evergreen woods, bravely facing winter's storms and cold, for it is only at that season that we are likely to see them south of the Canadian Zone.

The summer home is in the coniferous forests of the northern tier of States and in the southern Provinces of Canada. Ora W. Knight (1908) says that, in Maine, "pine, fir, spruce and hemlock woods, or mixed growth in which these trees predominate are their preference." Most observers say that they prefer the spruces. William Brewster (1888) found them breeding in Winchendon, Mass., in dense woods of white pine and spruce. Based on my limited experience, golden-crowned kinglets seem to prefer the more open forests of more or less scattered, second-growth spruces, rather than the dense forests of mature growth. In these more open forests there are often a few balsam firs or white birches scattered through the spruces, but the presence of spruces seems to be necessary for nesting purposes.

In the Adirondack Mountains of New York, according to Aretas A. Saunders (1929a), this kinglet "lives in the coniferous forests, especially in the tops of tall spruces. Spruce, hemlock, balsam, and tamarack all attract it, and it is seldom seen in summer in the hardwoods, and then only where spruces are near. On the Avalanche Pass Trail I found it in second growth spruce, where the trees were dense but only ten or fifteen feet high."

The golden-crowned kinglet is found in similar situations in the mountains of western Massachusetts, in places where the spruces have not been cut off. And Prof. Maurice Brooks writes to me: "This is a permanent resident in the Appalachian spruce forests, the most notable thing about it being its extraordinary abundance, especially late in summer. I recall one 10-day period spent in the Cheat Mountains when it seemed that kinglets were around us during almost every daylight minute. The spruce tops swarmed with them, parent birds and young of the year. In the same area, during subzero January weather, the birds were still abundant, although I do not know that the same individuals occurred."

Referring to northern Minnesota, Dr. Thomas S. Roberts (1932) says: "In nesting-time the Golden-crown makes its home in the dense spruce and arbor vitae bogs so numerous in the northern woods."

Spring.—As some golden-crowned kinglets spend the winter well up toward the northern limits of their breeding range, the spring migration is seldom conspicuous and is not easily traced. Robie W. Tufts tells me that it is normally resident throughout the year near Wolfville, Nova Scotia, and begins "nest-building with great regularity about April 15th." But in some seasons it seems to be conspicuous by its absence, for he says: "In the spring of 1918 none of these birds was seen about their favourite haunts near Wolfville, in spite of the fact that a diligent search was made." On migration in New England it is not confined to the coniferous woods, but may be found wherever there are trees and bushes, in the undergrowth in deciduous woods, in brushy thickets, in sprout lands, and even in orchards or the shrubbery in our gardens.

Milton B. Trautman (1940) records a well-marked migration around Buckeye Lake, Ohio, saying: "The first spring arrivals occasionally appeared in the first week in March, but usually they did not arrive until March 15 to 23, and it was not until after March 27 that the species could be daily encountered in small numbers. The daily numbers rapidly increased after April 3, and at the height of spring abundance, between April 9 and 21, between 25 and 150 birds were recorded daily. During migrations the majority of individuals inhabited the brushier portions of woodlands, brushy thickets, weedy fence rows, and thickets of hawthorn and wild plum."

Nesting.—Henry D. Minot (1877) was the first ornithologist to discover the nest of the golden-crowned kinglet, on July 16, 1875, "in a forest of the White Mountains [New Hampshire], which consisted chiefly of evergreens and white birches." The nest "hung four feet above the ground, from a spreading hemlock-bough, to the twigs of which it was firmly fastened; it was globular, with an entrance in the upper part, and was composed of hanging moss, ornamented with bits of dead leaves, and lined chiefly with feathers. It contained six young birds, but much to my regret no eggs."

The most elaborate account of the nesting of this species is that given by Mr. Brewster (1888), describing the three nests that he secured, near Winchendon, Mass., during that season. The first nest, taken June 29, "was placed in a tall, slender spruce (*A. nigra*), on the south side, within about two feet from the top of the tree, and at least sixty feet above the ground, suspended among the pendant twigs about two inches directly below a short horizontal branch, some twelve inches out from the main stem, and an equal distance from the end of the branch. The tree stood near the upper edge of a narrow strip of dry, rather open woods bordered on one side by a road, on the other by an extensive sphagnum swamp." Externally, the nest varies in depth from 3.60 to 2.70, and in diameter from 4.20 to 3.00 inches, being irregular in outline.

Brewster says:

The top of the nest is open, but the rim is slightly contracted or arched on every side over the deep hollow which contained the eggs. * * * The cavity is oblong, not round. The walls vary in thickness from 1.35 to .40. Outwardly they are composed chiefly of green mosses [five species of *Hypnum* and one of

Frullania, added in footnote] prettily diversified with grayith lichens and *Usnea*, the general tone of the coloring, however, matching closely that of the surrounding spruce foliage. The interior at the bottom is lined with exceedingly delicate strips of soft inner bark and fine black rootlets similar to, if not identical with, those which invariably form the lining of the nest of the Black-and-yellow Warbler. Near the top are rather numerous feathers of the Ruffed Grouse, Hermit Thrush, and Oven-Bird, arranged with the points of the quills down, the tips rising to, or slightly above, the rim and arching inward over the cavity, forming a screen that partially concealed the eggs.

The second nest, taken the same day, was in—

a lonely glen on high land between two ridges. The ridges were covered with young white pines. The prevailing growth in the glen was spruce and hemlock, the trees of large size and standing so thickly together as to shut out nearly all sunlight from the ground beneath. The nest was on the west side of a sturdy, heavily limbed spruce (*A. nigra*) about fifty feet above the ground, twenty feet below the top of the tree, six feet out from the trunk, and two and a half feet from the end of the branch, in a dense cluster of stiff, radiating (not pendant) twigs, the top of the nest being only an inch below, but the whole structure slightly on one side of the branch from which its supports sprang. Above and on every side it was so perfectly concealed by the dense flakelike masses of spruce foliage that it was impossible to see it from any direction except by parting the surrounding twigs with the hand. From directly below, however, a small portion of the bottom was visible, even from the ground. The foliage immediately over the top was particularly dense, forming a canopy which must have been quite impervious to the sun's rays, and a fairly good protection from rain also. Beneath this canopy there was barely sufficient room for the birds to enter.

This nest is similar to the other, though somewhat smaller and rounder, and the lining "is wholly of the downy under feathers of the Ruffed Grouse. These are used so lavishly that, radiating inward from every side, they nearly fill the interior and almost perfectly conceal its contents."

Referring to the third nest, he says: "The position of the third nest is different from that of either of the others. Placed nearly midway between two stout branches which in reality are forks of the same branch, one above the other, and at the point in question about six inches apart, it is attached by the sides and upper edges to the twigs which depend from the branch above, while its bottom rests firmly on a bristling platform of stems which rise from the branch below."

Mr. Brewster's lowest nest, the third, was 30 feet from the ground. Owen Durfee's experience, near Lancaster, N. H., was quite different; he says in his notes on nine nests: "The nests were all, with exception of two, in small spruces, most of our hunting being done in what we called 'pasture spruces'—really a second growth." Only one of the nests was up in the air, the average of the other eight being only 14 feet. His highest nest was 46 feet from the ground, "in a 12 inch spruce, in tall, hard woods growth, with a few scattered evergreens."

His lowest nest was only 8 feet from the ground. The only nest that was not in a spruce was 18 feet up in a balsam fir.

Ora W. Knight (1908) mentions a nest found near Bangor, Maine, that was only 6 feet from the ground, and says that most of those located by him in inland localities were "nearer forty to fifty feet in elevation." Miss Cordelia J. Stanwood has sent me some voluminous notes on the home life of the golden-crowned kinglet, near Ellsworth, Maine, where she finds them nesting in both black and white spruces. They begin nest-building in April, in spite of occasional snowstorms at that season, and she has found a nest about half finished on April 25. It requires about a month to complete the nest, in which the female apparently gathers the material and does all the building, while the male accompanies her and encourages her with song. She describes the building process, as follows: "The kinglets selected for the roof of their cradle a heavy spruce limb with a dense tip; and the female, hopping down through the branch from twig to twig, attached her pensile nest to the sprays.

"The bird wove her spherical structure about herself much as the caterpillar of the luna or cecropia moth weaves its cocoon about itself, except that the kinglet had to gather her materials. The bird stood on a twig on one side of the space she had chosen for her nest and measured off her length, as far as the situation of the twigs would permit, by attaching bits of spider's silk and moss to the twigs. Thus she laid off the points for the approximate circle for the top of the nest. Then she spanned the space through the center of the circle, roughly speaking from north to south, with spider's silk and moss, forming a sort of cable, which later assumed the appearance of a hammock. After a time, when the bird came with moss or silk, she would fly down upon the hammock as if to test its strength and lengthen it. At all times, however, she worked all over the nest from left to right, moving her beak back and forth as she secured the silk and moss and stretched the web from one point of attachment to another. As soon as the hammock would support the bird, she stood in the center and walked around from left to right. When the hammock was wide enough to admit of her sitting down, she modeled the center of the suspended band by burrowing against it with her breast, and making a kicking motion with her feet. Gradually she embodied some of the twigs in the structure, as if for ribs, and occasionally she snipped off a spruce twig to use in shaping the globular nest. At last the bottom, or basketlike part, arose to meet the top of the nest and the industrious gold-crest was hidden from sight as she labored.

"The creation was really a silken cocoon, in the walls of which was suspended enough moss, hair, and feathers to render it a nonconductor of heat, cold, and moisture. This primitive incubator was made of the same fine, dark yellow-green moss, *Hypnum uncinatum*, that seems characteristic of the habitations of the golden-crowned kinglet in this locality, *Usnea longissima*, a long, fringelike lichen, and animal silk. More of the gray-green *Usnea* lichen was used in the hammock-like band around the middle of the nest than in other parts of the well-made structure. The lining consisted of rabbit hair, I think, and partridge feathers. The wall of the abode was all of an inch and a half thick, and the window in the roof measured an inch and a half in diameter."

Nest-building starts early in Nova Scotia; Mr. Tufts tells me that he found two nests just started on April 10, 1921. In order to determine how many nests the kinglets would build and how many eggs they would lay, if the nests were destroyed, he tried the experiment of taking three nests from each of two pairs in isolated groves. He took the three sets from one pair on May 26, June 11, and June 30, 1915; and the other pair was robbed on May 27, June 15, and June 29, 1917. Each pair laid nine eggs each in the first two nests and eight in the third. The third nest was a flimsy affair. The birds must have worked fast to have built these nests and laid the large sets of eggs in such short intervals.

S. F. Blake (1916) found an interesting nest, in an unusual situation, near Stoughton, Mass., of which he says:

My attention was first attracted by the familiar call-notes of the birds coming from the edge of a rather close growth of Red Cedar (*Juniperus virginiana*) and deciduous trees at the base of a low hill close to a little-travelled wood-road. Pushing in among the trees, I soon caught a glimpse of the female Kinglet being pursued by a Black-and-White Warbler. The male soon came into view, and very soon the female disappeared in the top of a red cedar about twenty feet high. After a few minutes' wait I climbed a nearby tree and found her sitting on the nest. This was placed 18 feet 10 inches above the ground on the upper side of a small branch about a foot long, near the trunk and about a foot and a half from the top of the tree, rather firmly fastened and requiring some effort to dislodge.

Eggs.—The golden-crowned kinglet lays large sets of its tiny eggs, from five to ten in number, perhaps most often eight or nine. The nest is so small that they have to be deposited in two layers, probably five in the lower and four in the upper layer in a set of nine; that was the arrangement in one of Mr. Brewster's (1888) nests. His description of the eggs is so good that I cannot improve upon it; of the 18 eggs, he says:

The majority are more or less regularly ovate, but several are elliptical-ovate while two are very nearly perfectly elliptical-oval. The ground color varies from creamy white to exceedingly deep, often somewhat muddy, cream color. Over this light ground are sprinkled numerous markings of pale wood-brown, while at least three specimens have a few spots and blotches of faint lavender.

The brown markings vary in size from the finest possible dots to rather large blotches. In most of the specimens they are distributed pretty thickly over the entire shell, but in nearly all they are most numerous about the larger ends where they form a more or less distinct wreath pattern, while in four or five (and these have the lightest ground color) they are nearly confined to the larger ends, the remainder of the egg being sparsely marked. * * * In both sets the whitest, most sparsely spotted eggs were the freshest, showing that they were the last ones laid.

The measurements of 50 eggs average 13.3 by 10.4 millimeters; the eggs showing the four extremes measure **15.0** by 10.5, 14.4 by **10.7, 11.9** by 9.8, and 14.7 by **9.7** millimeters.

Young.—Miss Stanwood (MS.) writes: "The young kinglets are about as large as bumble bees when they come from the shell. They are blind and almost naked, save for a few tufts of fine, gray down. At the approach of the parent birds, they raise their little, palpitating bodies and open wide their tiny, orange-red mouths for food. These mouths are about the color of the meat of a peach around the stone. The veins showing through the thin skin give the bodies much the same tone. At first the young are fed by regurgitating partly digested food; later moths, caterpillars, and other insects furnish their diet. They are very fond of spruce bud moths and caterpillars. A beautiful triple spruce was attacked by these pests and almost denuded of its foliage. I noticed the kinglets frequenting this tree a great deal. In a season or two, the foliage was as luxuriant as it had been in the past. Such are the good offices performed by the golden-crowned kinglets and their young. The feet of the young are large and strong for the size of their bodies. If a person attempts to lift one from the nest, the little fellow will tear the lining out before he will release his hold. Just before the feathers appear the young begin to preen, and after that spend much of the remainder of their time in the nest smoothing and oiling their plumage. The parent birds remove all waste, depositing it far away from the little home, which is kept clean and sweet.

"I have seen kinglets feeding young in the nest as late as the last of June, but by the eighteenth or twentieth day of June, goldcrest families are usually foraging in the trees. As late as the middle of September in 1912, I saw mature kinglets industriously feeding a large family of young birds in a seedling grove."

I can find no reference anywhere to the period of incubation or to the duration of life in the nest.

Plumages.—Miss Stanwood (MS.) says that the small nestlings have "a few tufts of fine, gray down." The sexes are alike in the juvenal plumage, which Ridgway (1904) describes as follows: "Pileum brownish gray or grayish olive, margined laterally with a rather indistinct line of black; otherwise similar to adults, but hindneck concolor with back, etc., the color more brownish olive, and texture of plumage much looser." There is no orange or yellow in the crown of either sex.

The first winter plumage is acquired by a partial postjuvenal molt, involving all the contour feathers and the lesser wing coverts, but not the rest of the wings or the tail. The molt begins early in August, and after its completion the young birds are practically indistinguishable from the adults of their respective sexes. The young male has acquired the orange and yellow crown, bordered with black, and the young female has the yellow crown patch.

There is apparently no spring molt, and wear is not very conspicuous until late in the season. Year-old birds and adults have a complete postnuptial molt beginning in July. The fall and winter plumage is more brightly colored than the worn summer plumage, the upperparts being more decidedly olivaceous, and the underparts are strongly suffused with pale buffy-olive.

Food.—No comprehensive analysis of the food of the golden-crowned kinglet seems to have been made, but it apparently consists almost entirely of insects, their larvae and eggs, and other forms of minute animal life.

These items are obtained in various ways from different sources, but mainly from trees and shrubs. The kinglet feeds largely on bark beetles, scale insects, and the eggs of injurious moths and plant lice, which it obtains from the trunks, branches, and twigs of trees and bushes, mainly the coniferous trees.

Edward H. Forbush (1907) writes: "At Wareham, on Dec. 25, 1905, I watched the Gold-crest hunting its insect food amid the pines. The birds were fluttering about among the trees. Each one would hover for a moment before a tuft of pine 'needles,' and then either alight upon it and feed, or pass on to another. I examined the 'needles' after the kinglets had left them, and could find nothing on them; but when a bird was disturbed before it had finished feeding, the spray from which it had been driven was invariably found to be infested with numerous black specks, the eggs of plant lice. Evidently the birds were cleaning each spray thoroughly, as far as they went." Again, he saw kinglets feeding in the pines near his home, mainly on the trunks and the larger branches; they were feeding on the eggs of the aphids, which "were deposited in masses on the bark of the pines from a point near the ground up to a height of thirty-five feet. The trees must have been infested with countless thousands of these eggs, for the band of Kinglets remained there until March 25, almost three months later, apparently feeding most of the time on these eggs. When they had cleared the branches the little birds fluttered about the trunks, hanging poised on busy wing, like Hummingbirds before a flower, meanwhile rapidly pecking the clinging eggs from the bark."

W. L. McAtee (1926) says: "If we may apply to eastern conditions the findings of a study of the species in California, we may be sure that the Kinglet consumes little if any vegetable food, and that it gets numerous spiders as well as a variety of small insects principally of the hymenoptera, beetles, bugs, and flies. Moths, caterpillars, and small grasshoppers also are devoured. Forest pests taken are leaf beetles, leaf hoppers, plant lice, and scale insects." F. H. King (1883) says of nine specimens examined in Wisconsin, "two had eaten twelve small *diptera*; three, nine small beetles; one, five caterpillars; one a small chrysalid, and three, very small bits of insects, too fine to be identified." Junius Henderson (1927) says that it has been "seen feeding on locusts in Nebraska." Miss Stanwood mentioned in her notes that the kinglets, old and young, are very fond of the spruce bud moths and caterpillars, which are so destructive to the spruces in Maine.

Kinglets are expert flycatchers, taking small flying insects readily on the wing. Some observers have expressed surprise at seeing kinglets feeding on the ground, but it is not a rare occurrence. Francis H. Allen tells me that, when feeding on the ground, it progresses by surprisingly long hops. Miss Stanwood says in her notes that "the kinglet in winter finds considerable of his food on the snow under the trees; he even went under branches partly submerged by the snow and fed on the melted places close to the base of the trunk."

The golden-crowned kinglet has been observed apparently drinking the sap that flows from the fresh drillings of sap-sucking woodpeckers, but it may be that the birds are after the insects that are also attracted to such places. Francis Zirrer, of Hayward, Wis., has sent me the following note on the subject: "During the flow of maple sap the woodpeckers, especially the hairy, occasionally tap a tree. On a warm day, especially toward the end of the flow, sap thickens, ferments, and attracts many insects, mostly flies and small beetles, of which many stick to the syrupy fluid. Noticing a number of small fluttering forms in front of a tree trunk some 30 feet from the ground, I walked closer to investigate. To my surprise, there was a small flock of kinglets picking insects from the bark of the tree. In the course of the same afternoon and the following days, I found many more birds taking advantage of the bountiful supply; besides the two kinglets, woodpeckers, chickadees, nuthatches, and a phoebe."

Milton P. Skinner (1928), referring to the winter food of this kinglet in North Carolina, writes: "Sometimes they hunt the opening blossoms of trees and shrubs to prey on the small insects attracted by the flowers, and quite often they look over the bases of the bunches of loblolly and long-leaf needles for the tiny insects that hide there.

In spite of their almost universal insect hunt in winter, I noticed one Golden-crowned Kinglet fly over and take two bites from each one of two persimmon fruits on January 1, 1927."

This is the only reference I can find to indicate that either this kinglet or its western race ever eats any vegetable food; this is strange, as the ruby-crowned kinglet takes a small amount of fruit and seeds. I have often seen golden-crowned kinglets foraging in the Japanese barberry bushes about my house; the bushes were full of bright red berries, but I could not see that the kinglets ever touched any of them. They were probably feeding on some form of insect life, too small for me to see. Incidentally, I have noticed that none of the birds seem to like these barberries, though the common, wild barberry is very popular.

Behavior.—Golden-crowned kinglets are tame and confiding little creatures. They pay but little attention to the close presence of humans, and even come flitting about on the low branches or in the bushes near us, with beady little eyes glistening below their glowing crowns, and frequently opening and closing their little wings with their characteristic quivering motion.

Two quotations will suffice to illustrate their tameness and friendliness. A. H. Wood (1884) relates this experience with them while he was on a boat in Michigan: "One morning we found our boat invaded by eight or ten of these birds. It was not long before they found their way into the cabin, attracted there by the large number of flies, and at dinner time they caused no little amusement and some annoyance by perching on the heads of the passengers and on the various dishes which covered the table. I caught flies, which they would readily take from my hand with a quick flutter. I caught several, and even when in my hand, they manifested no fear, but lay quiet and passive." Cynthia Church (1927) found them very friendly in her garden; she writes: "On October 15, Golden-crowns became so tame that when I followed them quietly they allowed me to approach them and even to stroke them. Even when I patted and stroked their beautiful crest or parted their wings, they showed no fear. They even sat on my hands or lit on my coat. They were incredibly friendly."

Voice.—The golden-crowned kinglet is no such brilliant singer as the ruby-crowned, but it has a pretty little song at times. Aretas A. Saunders has given me the following description of it: "The song of the golden-crowned kinglet is much less musical and pleasing than that of the ruby-crowned, yet it bears a certain resemblance. The song is in two parts. The first part is a series of rather long, squeaky, very high-pitched notes, either all on the same pitch, or the pitch gradually rising. It is similar to the beginning of the ruby-crowned song, but higher pitched and with longer notes. My records show from two to nine notes in this part of the song. The second part is a series of very rapid, loud, harsh notes, descending in pitch, so different from the first part that it hardly seems to belong to the same song or bird. There are from four to nine notes in this part of the song, and the drop in pitch to the last note is sometimes more than an octave. A fairly typical song would be *eeee, teeee, teeee, teeee, teeee, chĭtĭtătătŭtŭp*. The pitch of fourteen records in my collection varies from F'''' to D''. Individual songs vary considerably, especially in the last part.

"This song is rather rarely heard in the spring migration in April, but is commonly heard in June, or early July, on the breeding grounds. Twelve of my 14 records come from breeding birds in the Adirondacks, and the other two from migrating birds in Connecticut. In winter the common call is like the first part of the song, but the notes are shorter and fainter, and so high-pitched that the sound is difficult for many people to hear."

Francis H. Allen refers to the song in his notes as "a pleasing performance, beginning with a number of fine, high notes and containing a lower-pitched and mellow *willy, willy, willy* that is quite charming." On April 20, 1900, when my hearing was good, on the coast of Maine, I recorded in my notes a song of nine notes, of which I wrote that "the first three notes are the same as their winter notes, rather faint and lisping, uttered slowly; the second three are on a higher key, louder and fuller toned; the last three notes are on the descending scale, with increasing rapidity, but decreasing in volume, suggesting the last part of the chickadee's song." Miss Stanwood (MS.) puts the song partly into words, which are rather expressive, "*zee, zee, zee, zee, zee, why do you shilly-shally.*"

Her notes record the kinglets in song, occasionally as early as March 15, regularly from the middle of April, on through the breeding season, once as late as August 26, and occasionally in fall, September 26 and October 12. Professor Brooks tells me that, curiously enough, he has never heard the golden-crowned kinglet in full song in West Virginia, in spite of the fact that it breeds there abundantly.

Field marks.—The kinglet is one of our smallest birds, a tiny ball of fluffy plumage, olive and buffy-gray in color. The orange-and-yellow crown of the male and the yellow crown of the female, bordered with black, are quite distinctive. The orange center in the male's crest does not always show, but flashes out under excitement. Young birds of both sexes have no orange or yellow in the crown, and might be mistaken for ruby-crowned kinglets, but the ruby-crowned has a conspicuous light eye ring which the young golden-crowned lacks.

Enemies.—Probably only the smaller hawks and owls, such as the sharp-shinned hawk and the screech owl, would be likely to bother with such small fry as kinglets. The cowbird does not seem to have found access to its well-concealed nest but once (Friedmann, 1934), and it has no competitors for its nesting site. Harold S. Peters (1936) lists one louse, *Philopterus incisus*, and one fly, *Ornithoica confluenta*, as external parasites on the eastern golden-crowned kinglet.

James G. Needham (1909) shows some photographs of a number of golden-crowned kinglets that had become entangled in the hooks of the ripening heads on several clumps of burdocks; he says:

> They were visible in all directions, scores of them sticking to the tops of the clumps on the most exposed clusters of heads. The struggle had ended fatally for all that I saw, and its severity was evidenced by the attitudes of their bodies and the disheveled condition of their plumage.
>
> I examined a number of the burdock heads to determine what attraction had brought the Kinglets within range of the hooks, and found insect larvae of two species present in considerable abundance. Most abundant were the seed-eating larvae of an obscure little moth (*Metzgeria lapella*), but the larvae of the well-known burdock weevil were also present in some numbers. Doubtless, it was in attempting to get these larvae that the Kinglets (mostly young birds) were captured.

Winter.—In spite of its diminutive size, the golden-crowned kinglet is a hardy little mite and spends the winter in much of its summer range, though in reduced numbers, even as far north as Maine and Nova Scotia. Miss Stanwood says in her notes: "The kinglets were abundant during the severe winter of 1906 and 1907. When I went to distribute my food supply for the birds near the boiling spring in the woods, they followed me to the spring and back, sometimes gleaning from tree to tree, or hopping and running ahead of me over the snow. Undoubtedly, in very cold weather many of the kinglets perish for lack of sufficient food to keep the vital fires burning. The winter of 1906 and 1907 was a cruel winter for the birds."

With us, in Massachusetts, these little feathered gems are among our most charming winter visitors, sometimes abundant but often scarce or entirely absent. We usually find them in the evergreen woods, pines or hemlocks, or in the cedar swamps where they find more protection from the cold winds. We see them flitting through the woods, gleaning from the lower branches, or hovering close to the tree trunks in search of food; sometimes we catch a glimpse of the golden crown, as the bird forages upon the ground among the pine needles. Often they form jolly little roving bands, with chickadees, a brown creeper or two, and perhaps a downy woodpecker, adding cheer to the dark and dreary winter woods. But they are not always confined to the coniferous woods; they frequent mixed woods and open woods, where birches grow along the woodland paths, and are often seen in orchards or in the shrubbery of our home grounds and gardens. Wherever they are found they are always a welcome addition to our winter bird life.

DISTRIBUTION

Range.—From southern Alaska and central Canada to Guatemala.

Breeding range.—In so far as our present knowledge goes the breeding range of the golden-crowned kinglet is discontinuous. The western breeding range is **north** to southern Alaska (Kodiak, Kenai Peninsula, Montague Island, Chitina Moraine, and Glacier); northern British Columbia (the Atlin region probably, and Fort Nelson); central Alberta (Grand Prairie and Glenevis; single birds have been recorded from 40 miles below Athabaska Landing and Point La Brie on Lake Athabaska). **East** to central Alberta (Glenevis); western Montana (Flathead Lake, Anaconda, Mystic Lake, and the mountains near Kirby); western Wyoming (Yellowstone National Park and Sheep Mountains); western Colorado (Walden, Montrose, and San Juan County); and central northern New Mexico (Cowles and Pecos Baldy). **South** to northern New Mexico (Pecos Baldy); southeastern Arizona (White Mountains, Graham Mountains, and Santa Catalina Mountains); and southern California (San Bernardino and San Jacinto Mountains). **West** to California (San Jacinto Mountains) and north through the Sierra Nevada (Mount Breckenridge, Trout Meadows, and Yosemite Valley) and the Coast Range (Santa Cruz Mountains, Mount Sanhedrin, and Trinity Mountains); the coast forests of Oregon (Yaquina Bay and Garabaldi); Washington (Point Chehalis, Ozette Lake, and Neah Bay); coastal region and islands of British Columbia (Nootka Sound, Vancouver Island; Queen Charlotte Islands, and Porcher Island); and the coastal region of southern Alaska (Forrester Island, Sitka, Yakutat Bay, and Kodiak Island).

The eastern range is **north** probably to central or northern Manitoba, since the species is a rare but fairly regular migrant through the Red River Valley and southern Manitoba; it has been reported to breed at Aweme, is a regular migrant but not breeding at Lake St. Martin, has been found in August and September at The Pas; a specimen was collected in June on the lower Echimamish River and it has been found in June of three different years near Churchill. It breeds north to central Ontario (Lake Nipigon, Abitibi Lake, Algonquin Park, and Ottawa); southern Quebec (St. Margaret, Kamouraska, Esquimaux Point, and Anticosti Island), and central Newfoundland (South Brook and Brigus). **East** to Newfoundland (Brigus and Placentia); Nova Scotia (Halifax and Barrington); Maine (Machias and Ellsworth); and rarely eastern Massachusetts (Lynn and Stoughton). **South** to Massachusetts, rarely (Stoughton); New York (Catskill Mountains); northern Pennsylvania (Pocono Mountains and Warren);

southern Ontario (Port Dover and Lucknow); northern Michigan (Mackinac Island, Blaney, and Porcupine Mountains); possibly northern Wisconsin (recorded from Door and Vilas Counties in early July); and central Minnesota (St. Cloud). **West** to central Minnesota (St. Cloud, Gull Lake, Cass County, and Island Lake); and Manitoba (Aweme and The Pas, probably).

Resident races occur from central Mexico to Guatemala.

Winter range.—The golden-crowned kinglet is found in winter **north** to southeastern Alaska (Sitka and Juneau); southern British Columbia (Comox, Vancouver, and Vernon); central Alberta (Glenevis and the Battle River Valley near Camrose); southern Saskatchewan (Nipawin); southeastern Wyoming, casually (Cheyenne); eastern Colorado (Boulder, Denver, and Colorado Springs); central Iowa (Ames); occasionally north to southeastern South Dakota (Yankton) and central eastern Minnesota (Minneapolis and Cambridge); southern Wisconsin (Madison and Milwaukee); southern Michigan (Ann Arbor and Detroit); rarely southern Ontario (Harlow, Guelph, and Toronto); central New York (Rochester and Geneva); eastern Massachusetts (Boston); southern Maine (Auburn, Waterville, Dover-Foxcroft, and Calais); southern New Brunswick (Scotch Lake and St. John); and central Newfoundland (Grand Falls). **East** to Newfoundland (Grand Falls, Brigus, and St. John's); southern Nova Scotia (Yarmouth); southern Maine (Bath); and the Atlantic Coast States to northern Florida (Jacksonville). **South** to northern Florida (Jacksonville, Tallahassee, Apalachicola, and Pensacola); the Gulf coast of Alabama, Mississippi, and Louisiana to southern Texas (Houston, San Antonio, San Angelo, and the Guadalupe Mountains); northern Mexico (Casas Grandes, Chihuahua, and Hermosillo, Sonora); and southern California (Los Angeles and Santa Barbara). **West** to the valleys and coast of California (Santa Barbara, Fresno, San Francisco, and Seiad Valley); Oregon (Fort Klamath, Yaquima Bay, and Portland); Washington (Grays Harbor, Tacoma, and Seattle); southwestern British Columbia (Victoria and Comox) and southeastern Alaska (Craig and Sitka).

The breeding ranges as outlined apply to the whole species, which has been divided into three subspecies or geographic races. The typical race, the eastern golden-crowned kinglet (*R. s. satrapa*) breeds from Manitoba and Minnesota eastward; the western golden-crowned kinglet (*R. s. olivaceus*) breeds from Alaska, British Columbia, and Alberta south to southern California, Utah, and Colorado; the Arizona golden-crowned kinglet (*R. s. apache*) breeds in Arizona and New Mexico.

Migration.—Some late dates of spring departure from its wintering grounds are: Florida—Pensacola, March 30. Alabama—Anniston,

March 26. Georgia—Athens, April 12. South Carolina—Spartanburg, April 23. Virginia—Lynchburg, April 17. District of Columbia—Washington, April 27. Pennsylvania—Pittsburgh, April 28. Massachusetts—Holyoke, May 9. Mississippi—Biloxi, April 5. Louisiana—New Orleans, March 26. Arkansas—Rogers, April 18. Missouri—St. Louis, May 6. Kentucky—Bowling Green, April 21. Illinois—Quincy, May 3. Ohio—Cleveland, May 12.

Some early dates of spring arrival are: Vermont—Rutland, March 6. Quebec—Montreal, March 27. Illinois—Chicago, March 21. Michigan—Battle Creek, March 7. Minnesota—Northfield, March 20. South Dakota—Brookings, April 5. North Dakota—Argusville, April 4. Manitoba—Winnipeg, April 14. Montana—Columbia Falls, March 21. Washington—Pullman, March 20.

Some late dates of fall departure are: Alberta—Belvedere, October 22. Washington—Yakima, November 14. Montana—Columbia Falls, November 20. Wyoming—Laramie, November 12. Manitoba—Aweme, November 16. North Dakota—Fargo, November 7. South Dakota—Faulkton, November 1. Nebraska—Lincoln, November 27. Minnesota—St. Paul, November 22. Wisconsin—Ripon, November 30. Ontario—Galt, November 21. Quebec—Quebec, November 23. Prince Edward Island—October 25. Newfoundland—St. Anthony, October 17. Vermont—Wells River, November 8. Maine—Ellsworth, November 25.

Some early dates of fall arrival are: Wyoming—Wheatland, September 15. North Dakota—Fargo, September 27. Nebraska—Omaha, October 7. Oklahoma—Tulsa, October 29. Texas—Corpus Christi, October 5. Wisconsin—North Freedom, September 19. Michigan—Grand Rapids, September 2. Indiana—Fort Wayne, September 22. Ohio—Columbus, September 27. Kentucky—Lexington, September 13. Tennessee—Nashville, October 7. Arkansas—Jonesboro, October 4. Louisiana—Covington, October 18. Mississippi—Bay St. Louis, October 14. Massachusetts—Danvers, September 16. Pennsylvania—Berwyn, September 24. District of Columbia—Washington, September 6; a casual record, July 25, 1932. Virginia—Lexington, October 10. North Carolina—Louisburg, October 10. South Carolina—Charleston, October 14. Georgia—Atlanta, October 9. Florida—New Smyrna, October 18.

Few kinglets are banded but one record is available to indicate migration. A bird banded at Elmhurst, Long Island, N. Y., on November 5, 1931, was found dead at Suwanee, Ga., about January 9, 1933.

Casual records—In the Bermudas, one was collected in the spring of 1883; and on November 5, 1928, one was noted on shipboard 140 miles east of Cape Charles, Va.

Egg dates.—California: 8 records, April 9 to June 19.

Labrador: 3 records, June 8 to June 18.

New Brunswick: 14 records, May 21 to June 17; 9 records, May 21 to May 27.

Washington: 17 records, April 15 to August 1; 9 records, May 1 to May 20, indicating the height of the season.

REGULUS SATRAPA OLIVACEUS Baird

WESTERN GOLDEN-CROWNED KINGLET

HABITS

This subspecies occupies a wide breeding range, from the Rocky Mountains to the Pacific coast and from southern Alaska, south of the peninsula, to southern California and northern New Mexico. It differs but slightly in appearance from the eastern bird, being more brightly colored above, more greenish or more olivaceous, and having shorter wings and tail and a more slender bill. It seems to be confined, in the breeding season at least, to the coniferous forests of tall firs and Douglas spruce. Samuel F. Rathbun tells me that he finds it an abundant bird throughout western Washington, from the Cascade Mountains to the Pacific, and "from tidewater up to an altitude of 4,500 feet in the mountains." Farther south it breeds at higher elevations as far south as the Boreal Zone extends, as in the San Jacinto Mountains in southern California. In northern New Mexico, according to Mrs. Bailey (1928), it breeds in the Sangre de Cristo Mountains at elevations from 9,800 to 11,500 feet.

While out with Mr. Rathbun near South Tacoma, Wash., D. E. Brown showed us some attractive country, where he and J. H. Bowles had been most successful in finding the nests of the western golden-crowned kinglet. It was smooth, level land, with a fine parklike growth of firs and cedars scattered about; the two or three local species of firs were the most abundant trees, growing to perfection in the open, where they were well branched down to the ground; the largest firs were magnificent specimens, reminding me of our eastern Norway spruces with their downward-sweeping branches; it is in these larger trees that the kinglets prefer to build their nests in the pendant sprays.

Aretas A. Saunders tells me that in the mountains of Montana there are three types of forest, spruce, fir, and lodgepole pine, and that the golden-crowned kinglets are confined to the spruce forests and the ruby-crowned kinglets to the Douglas firs.

Nesting.—Mr. Rathbun (MS.) writes, referring to the vicinity of Seattle: "In its time for nesting this kinglet appears to have quite an extended period. We have seen it carrying material for its nest as early as April 4; on this occasion the bird was gathering bits of green moss from a decayed log, which would indicate the earlier stage of construction; and we have a record of unfledged young in the nest on May 17. We also have a record of a nest with eggs as late as August 1; but we are of the opinion that the great majority of the birds nest, in this locality, between the middle of May and the middle of July; and there is little doubt that many raise two broods of young during the season.

"It is partial to certain types of coniferous trees in which to place its nest, those whose limbs have a decidedly drooping foliage; evidently this is not only for concealment, but it affords protection from the elements. The nest may be located at various heights, the nests that I have found ranging from 9 to 45 feet above ground; and we have no doubt that many are placed even higher, in which case they would be difficult to detect. In the larger trees the nest will be placed quite near the extremity of the limb, making it very difficult to secure; but on occasions a nest is found in some tree of small size, not far from the trunk and at but little height above ground."

He describes a nest found on June 8, 1912: "This nest was at a height of 14 feet, on the under side and near the extremity of one of the lower limbs of a good-sized fir tree, situated on the edge of a large open space in the forest. It was attached to four very small drooping twigs that grew from one of the laterals of the main branch, the sides of the nest being firmly bound thereto. In shape it is somewhat round, excepting its upper surface, and is made outwardly of small pieces of green moss and lichens, much of the latter being used in the bottom; with both of these materials are interwoven bits of fine dead grasses and fir and hemlock needles, all firmly bound by plant fibers of a soft character and some spider webs; the walls are three-quarters of an inch in thickness; the top of the nest at the opening is somewhat arched over or constricted, as if its edges were drawn together. The interior has a thick lining of the softest of vegetable fibers, into which are neatly woven many downy feathers of small size. Outside height 3 inches, outside diameter 3½ inches; inside diameter 1¾ inches, diameter at the opening 1 inch, and depth inside 1½ inches."

Nests described by others were, in the main, similar to the above. They have been found at lesser and at greater heights from the ground than the figures given by Mr. Rathbun; Dawson and Bowles (1909) say "from eye-level to fifty feet." What few nests I have seen are less globular than those of the eastern bird, flatter on top and more open at the top. Some nests are lined with cow's or deer's hair, or with the feathers of various small birds or of grouse or even poultry.

Eggs.—The set of eggs laid by the western golden-crowned kinglet may consist of anywhere from five to eleven, but eight and nine are the commonest numbers; sets of four are probably incomplete, and any numbers larger than nine are uncommon or very rare. The eggs have very frail shells. They are indistinguishable from those of the eastern race. The measurements of 40 eggs average 13.5 by 10.5 millimeters; the eggs showing the four extremes measure **14.8** by 11.0, 14.5 by **11.4**, and **12.5** by **10.0** millimeters.

Food.—Professor Beal (1907) has this to say about the food of the western golden-crowned kinglet in California: "Only 9 stomachs have been examined, but these in the nature of their contents are so similar to those of the ruby-crown that statements applicable to the latter are almost certain to apply as well to this species. No vegetable matter was found in any of the 9 stomachs, and the insects belong to the same orders and were taken in essentially the same proportions as by the other species."

The behavior, voice, and other habits of the western bird seem to be similar to those of the eastern golden-crowned kinglet. Mr. Rathbun tells me that, in western Washington, after the breeding season has passed, the adults with their broods wander over almost the entire region. During the autumn months there is a movement from the higher altitudes to the lower, and the birds become very common throughout the lowlands, always found associated with individuals of their own species, and occasionally with some of the small arboreal species that likewise are winter residents.

REGULUS SATRAPA APACHE Jenks

ARIZONA GOLDEN-CROWNED KINGLET

HABITS

Randolph Jenks (1936) has given the above name to the kinglets that breed in the White Mountains and adjacent ranges in east-central Arizona. After considerable discussion of its distribution and habitat, and a detailed description of it, he sums up its characters as follows:

"It differs from *R. r. olivaceus*, its closest geographical counterpart, by being a grayer bird, having a noticeable, wide gray, dorsal nuchal band, having decidedly longer wings, tail, and bill, and having a deeper and more richly colored central pileum. It differs from *R. r. satrapa* in having a wide gray dorsal nuchal band, a much longer, more slender bill, and a much more deeply colored central pileum. Finally, it differs from *R. r. clarus* by being a considerably grayer bird, having a wide gray dorsal nuchal band, and having longer wings, tail, and bill."

Probably it does not differ materially in its haunts and habits from other adjacent races of the species.

REGULUS CALENDULA CALENDULA (Linnaeus)

EASTERN RUBY-CROWNED KINGLET

HABITS

The ruby-crowned kinglet is not brilliantly colored, for it is clad in soft olive and gray, but it is a dainty little bird with attractive manners; only when it shows its red crown-patch under excitement is there any brilliancy in its plumage, but when it bursts into its marvelous song it ranks as one of our most brilliant songsters. What it lacks in color it makes up for in music.

It ranges much farther north in summer and goes farther south in winter than the golden-crowned kinglet, breeding from northwestern Alaska down through the Rocky Mountains to Arizona and New Mexico, and in eastern Canada as far south as Nova Scotia. There are also breeding records for Michigan and Maine and indicated breeding in Massachusetts.

I found the ruby-crowned kinglet breeding in some very attractive hillside woods back of Bay of Islands, Newfoundland. It was rather an open tract of mixed woods consisting mostly of fir balsams, with some red and white spruces, larches, and white pines and a sprinkling of canoe birches, black birches, and mountain-ash. *Amelanchier* and *Rhodora* were in bloom, and the blossoms on the *Arbutus* were larger and whiter than we see them at home. The kinglets were in full song, adding much to the beauty of their surroundings.

Dr. Paul Harrington writes to me that, in the vicinity of Toronto, Ontario, "the ruby-crowned kinglet is a typical bird of the black spruce bogs, and it is only on rare occasions that this bird is found out of these regions in the breeding season."

Bagg and Eliot (1937) record the almost certain breeding of this kinglet at Savoy, Mass. William J. Cartwright had seen the kinglets feeding their young in a grove of spruce at the top of a hill on July 5, 1915; there was a flock of about 20, including old and young, evidently two families. On July 19, 1920, he again found six in the same grove of spruces. And Mr. Bagg adds: "Visiting this hill-top with Mr. Cartwright early on July 3, 1932, Mr. H. E. Woods and I both had the thrill of seeing a Ruby-top feed an out-of-the-nest fledgling and in the act erect its crown-spot. None, however, could be found on June 11, 1933, and the old spruces are dying and fast being removed by the State Forest authorities." (See also Bagg, 1932, p. 486.)

The ruby-crowned kinglet undoubtedly breeds more or less rarely in Maine, where the dense woods of mixed spruces and fir balsam often extend quite down to the coast. Ora W. Knight (1908) mentions watching a pair building a nest in deep woods of this type near Orono.

Aretas A. Saunders tells me that, in Montana, the ruby-crowned kinglet is confined to the forests of Douglas fir, while the golden-crowned kinglet inhabits the spruce forests.

Spring.—The spring migration of the ruby-crowned is more conspicuous than that of the golden-crowned kinglet. It has much farther to go, as it winters farther south and breeds farther north. It sings on migration, not the full, rich song that one hears on its northern breeding grounds, but pleasing enough to attract attention. It travels singly or in small groups, sometimes as many as 20 or more, and is often associated with migrating warblers. Throughout most of the United States it occurs mainly as a migrant in spring and fall. In Massachusetts it passes through in April and the first half of May. Then we need not look for it in coniferous woods only, for it is likely to be found almost anywhere; its favorite haunts are the swampy thickets along streams, or around ponds or bogs; but it is sometimes seen in the trees and shrubbery about our houses and gardens; and when we see it pouring out its charming song among the apple blossoms in the orchard, then we enjoy one of the greatest delights of the spring migration.

In Ohio the first arrivals come early in April, but Milton B. Trautman (1940) says that, at Buckeye Lake, "the species remained uncommon until mid-April, when the numbers rapidly increased, and from April 20 to May 10 the greatest daily numbers, 15 to 40, were attained. As many as 60 a day were noted during large flights. The last transients were seen between May 14 and 18. * * * As with the Golden-crowned Kinglet, this species chiefly inhabited woodlands, thickets, and brushy fence rows, and in such situations was found most frequently where there were dense clumps of hawthorn, wild plum, honey locust, and osage orange. The bird appeared to be less numerous in this area than it was in other localities of similar size in central Ohio; it was decidedly less numerous than it was in localities which contained many conifers."

Courtship.—This seems to consist mainly of the display between rival males of the flaming red crest, which is usually partially concealed or at least restricted by the surrounding dull feathers of the crown, but which can be uncovered or perhaps erected in the ardor of courtship or in the anger of combat. John Burroughs (Far and Near, pp. 178–179) thus describes such rivalry between two males: "They behaved exactly as if they were comparing crowns, and each extolling his own. Their heads were bent forward, the red crown patch uncovered and showing as a large, brilliant cap, their tails spread out, and the side feathers below the wings were fluffed out. They did not come to blows, but followed each other about amid the branches, uttering their thin, shrill notes, and displaying their ruby crowns to the utmost."

It would not be surprising if rivalry in song were also one of the features in the contest for supremacy.

Nesting.—So far as I can learn, the nests of the ruby-crowned kinglet are always built in coniferous trees, generally in spruces, sometimes in firs, and more rarely in some of the western pines. They are placed at various heights from the ground, from 2 to 100 feet; a number of nests have been found at 50 or 60 feet, many between 15 and 30, and comparatively few below 15 feet. Winton Weydemeyer (1923) reports a Montana nest in an unusual location; it "was about fifty feet from the base of a partly fallen spruce, * * * fourteen feet from the ground, and eighteen inches from the end of a seven-foot branch extending downward from the trunk."

Most of the nests reported have been attached to the pendant twigs beneath the branch of a spruce, well concealed among the twigs, partially or wholly pensile, and usually near the end of the branch where the foliage is thickest; but very rarely the nest may be placed *on* a branch; W. L. Sclater (1912) states, probably on the authority of Denis Gale, who found a number of nests in the mountains of Colorado, that the nests are "sometimes simply saddled on a horizontal bough." The nest that I found in Newfoundland was only 8 feet from the ground in a spruce, suspended between two drooping branches, or rather large twigs; the tree stood in a rather open situation; it contained no eggs on June 5, but Edward Arnold collected a set from it later. The lowest nest reported was found by Maj. Allan Brooks (1903) in the Cariboo District of British Columbia; it was "in a small spruce not four feet high; the nest was close to the stem and about two feet from the ground; it was a very deep cup, almost a vertical cylinder." And at the other extreme, Dr. Mearns (1890) records a nest in the Mogollon Mountains in Arizona that was "attached to the end of a horizontal branch upwards of a hundred feet above the ground," in a spruce. John Swinburne (1890) found a nest in the White Mountains of Arizona, at an altitude of "about 8500 to 9000 feet," that "was placed in a bunch of cones at the end of a small branch, in a spruce-fir tree, at an altitude of about sixty or seventy feet from the ground. * * * The nest was completely hidden by the fir cones surrounding it, and was placed about four feet out from the stem of the tree."

Dr. Paul Harrington, of Toronto, writes to me that he found a nest of the ruby-crowned kinglet, at Chapleau on June 10, 1937, 30 feet from the ground in a black spruce on the border of a bog, and says: "I was first attracted to the spot by the agitated male, which sang from close range. Whenever I came near the nesting tree the bird became agitated, even at a distance of 100 yards. I found the bird to act in this manner near another nest, which contained nine incubated eggs. On numerous occasions I have observed the male to act in this manner and believe that it is a regular nesting characteristic. The nest found June 10 was a perfect example of the utmost in bird architecture, a compact structure of moss, lined thickly with rabbit fur and feathers, and, although globular in shape, was in no way semi-pensile, but really a deep, cup-shaped structure. All the nests I have seen have been placed near the top clump of needles, straddled on small branches adjoining the trunk and exceptionally well concealed from the ground."

Most descriptions of nests give the impression (and what few nests I have seen confirm it) that they are made mainly of green mosses, such as *Cladonia*, *Hypnum*, and *Parmelia*, gathered from fallen logs and trees, mixed with the long, green tree-lichen, *Usnea longissima*. But careful studies of nests have shown that much other material is often, if not regularly, used. Mr. Weydemeyer (1923), for example gives the following good description of a nest in Montana:

In color, the nest looked much like the surrounding spruce foliage. In general appearance, it resembled an elongated Wright Flycatcher's nest constricted at the top. The cup was between four and four and one-half inches deep, and two and one-half inches wide at the center, narrowing toward the top to form a circular opening not more than an inch and a quarter in diameter.

* * * Thistle down, cotton from the catkins of the aspen, and small feathers made up a large part of the body of the nest. The outside was thickly covered with finely shredded inner bark of aspen, a few blades of dry grass, and ground and tree mosses, with a surface covering of grayish lichens and a few small spruce twigs.

The interior of the nest was thickly lined with feathers. The sides were covered with body feathers of the Canadian Ruffed Grouse, arranged with the points of the quills down and covered by the tips of the feathers below. The tips of the uppermost feathers curved slightly inward just below the opening of the nest. At the bottom was a thick covering of breast feathers of the female mallard.

With the exception of the feathers forming the inner lining, the various materials composing the nest were strongly bound together by an intricate and extensive network of extremely fine fibers from insect cocoons. The coarser material on the outside of the nest was also held together by stiff porcupine hairs, while the bottom was further strengthened with several long horsehairs. Thus, though the nest was unusually soft and quite yielding to the touch, it was nevertheless strongly held in shape.

The nests are almost always pensile, or semipensile, and usually partly attached to surrounding twigs. They are generally roughly globular in shape, though somewhat flattened on the top, measuring 3 or 4 inches in both height and diameter, but sometimes elongated downward to 5 or 6 inches. The opening above seems to vary from 1¼ to 2 inches in diameter; the internal depth varies from 1½ to 3 or 4 inches, but the cavity is usually deep enough completely to conceal the incubating bird.

A. D. Henderson tells me that the ruby-crowned kinglet is an abundant breeder at Belvedere, Alberta, "building its nest usually in a slim spruce in a muskeg. The nest is very difficult to see, and is usually found by watching the birds go back to it, when the female comes off to feed. If the exact location of the nest is not seen, but its presence is suspected, every nearby spruce is rapped with a stick; and when the one with the nest is struck, the sitting female drops like a bullet to within a few feet from the ground. I have seen one nest of the ruby-crowned kinglet only 5 feet from the ground and another 45 feet up; ten other nests ranged from 7 to 25 feet up."

Eggs.—The ruby-crowned kinglet lays 5 to 11 tiny eggs, closely packed in its little nest; sets of four are probably incomplete, from seven to nine are the commonest numbers found, and any larger numbers are uncommon or very rare. The eggs are so much like those of the golden-crowned kinglet that the two are practically indistinguishable. They vary in shape from ovate to oval or rounded oval. The ground color is pale buffy white, dirty white, or clear white. The egg is more or less evenly covered with very fine dots or small spots of reddish brown or dull brown; sometimes these markings are concentrated around the larger end, and sometimes they are so faint that the eggs appear immaculate. The measurements of 40 eggs average 13.7 by 10.8 millimeters; the eggs showing the four extremes measure **14.8** by 10.9, 14.0 by **11.4, 11.9** by 10.6, and 13.5 by **9.8** millimeters.

Young.—The period of incubation does not seem to have been learned, but it is probably the same as that of the European goldcrest, 12 days, and the young probably remain in the nest for about the same length of time under normal circumstances. Incubation is apparently performed wholly by the female, but both parents feed the young in the nest, and for some time after they leave it. J. Dewey Soper (1920) watched a pair of these kinglets feeding their young, while he was in the tree near the nest and again from the top of a ladder within 3 feet of it; he writes:

> During the half hour which I clung to the tree the male visited the nest with food three times and the female twice. The former upon deposition of the food vacated the nest promptly but the female on the contrary, often remained with the young until the return of her mate, when she then slipped quietly away. In this manner the young were left alone for certain periods but sheltered again for longer ones when the female returned. * * *
>
> The detention of the female at the nest I observed, was due to her habit of regularly cleaning the nest of all the sac-like excrement; due to the rapid digestion of the hungry infants, her obligations in this respect seemed never to cease. The matter was probed for with scrupulous care, some consumed by her, and the remainder dropped overboard at some distance from the nest. In this the male

never assisted. Candor bids me remark however, that his tireless assiduity in harvesting for the young more than offset this disparity. * * *

With my face only a couple of feet distant from the nest the pair continued their work scarcely conscious of my presence. True, at first they hovered above me with sweet queries in their throats and entered the nest from the opposite side of the bough but soon this discretion was forsaken for perfect freedom. Twice, the male warbling an undertone alighted within two feet of my hand on the supporting guy rope of the ladder. A pretty performance and employed only by the male was to flit from the nest and become suspended on whirring wings before me, like a hummingbird before a flower.

Plumages.—Young ruby-crowned kinglets, in juvenal plumage, look very much like young goldencrowns, but their coloration is darker and they show the light eye ring instead of the superciliary stripes. Ridgway (1904) described the young bird, in first plumage, as "similar to the adult female, but upper parts browner (nearly hair brown), wing-bands tinged with brownish buffy, under parts less yellowish, and texture of plumage more lax."

The postjuvenal molt, accomplished before the bird migrates, is incomplete, involving the contour plumage and the wing coverts, but not the rest of the wings or the tail. This produces a first winter plumage which is practically indistinguishable from that of the adult in each sex. The young males usually assume the scarlet vermilion crown patch, but in some cases it is more nearly orange in color. Some males in full song in spring have no vestige of the crown patch.

The spring plumage is acquired by wear, with sometimes the renewal of a few feathers. A complete postnuptial molt occurs late in summer. Fall birds, in fresh plumage, are more brightly colored than spring birds, more olive above and more buffy below.

There has been some discussion in the past as to whether the female ever has a red or an orange crown patch, but I believe it is now agreed that she does not. Specimens labeled as females may have been wrongly sexed. But it may be that, as in some other species, a very old female may assume, at least partially the plumage of the male.

The reported orange or yellow crown patch in the young male seems to be very rare, and it has been suggested that this color may change later to the usual bright red, this suggestion is strengthened by the fact that the specimens showing this yellowish crown were taken in fall; however, this matter still remains to be settled.

Food.—Professor Beal's (1907) analysis of the contents of the stomachs of 294 ruby-crowned kinglets, although taken in California, will probably give a very fair idea of the food of the species elsewhere. The food consisted of 94 percent animal matter, "insects, spiders, and pseudoscorpions—minute creatures resembling microscopic lobsters," and 6 percent vegetable matter, fruit and seeds. "Hymenoptera, in the shape of wasps, and a few ants appear to be the favorite food, as they aggregate over 32 percent of the whole." Hemiptera make up 26 percent of the diet, including assassin bugs, lace bugs, leafbugs, leafhoppers, jumping plant-lice, plant-lice, and scale insects. Beetles were eaten to the extent of 13 percent, only 2 percent of which were the useful ladybirds and the remainder all more or less harmful. Butterflies, moths, and caterpillars were eaten rather sparingly, aggregating only 3 percent of the whole. Flies amounted to 17 percent, and spiders and pseudoscorpions made up an additional 2 percent.

The small amount of vegetable food was divided as follows: Fruit, principally elderberries, less than 1 percent, weed seeds 0.01 percent, and miscellaneous matter, including seeds of poison-oak and leaf galls, over 4 percent.

Dr. George F. Knowlton (MS.) lists the ruby-crowned kinglet among the birds that eat the beet leafhopper in Utah; five birds examined had eaten two nymphs and two adults.

Milton P. Skinner (1928) gives the following account of the feeding habits of this kinglet in North Carolina in winter:

> During the winter, they depend largely on small insects for food. At times they are on the ground amid the fallen leaves, searching herbaceous plants less than a foot high, or on the twigs of low bushes or shrub oaks, but often on the three species of pines searching the trunks, limbs, twigs, and the bunches of needles. When hunting the clusters of pine needles, the kinglets search carefully at the base of each needle and in the pockets between the needles, frequently swinging back down below the clusters, and sometimes hovering in mid-air on fast-beating wings before the clusters. One kinglet that searched the tufts of needles appeared to catch an insect every five or six seconds as long as I watched it, and another one found something to eat on every four inches of pine limb that it searched. Sometimes the Ruby-crowned Kinglets hunt insects in the cedars, hollies, gums and dogwoods. In this limb and twig hunting, they depend chiefly on picking insects from the bark, or on catching those that fly from the bark. But many of these birds perch on limbs and dart on insects that attempt to fly past them. Sometimes the Ruby-crowns collect dogwood berries from the ground and eat them, but reject the seeds probably, and occasionally they take a few sumac berries. More often they consume cedar berries, both pulp and seeds, and some of the pulp from wild persimmons.

Behavior.—The above account of the feeding habits of this kinglet by Mr. Skinner gives a good idea of its behavior anywhere, for the constant search for food is always the main activity of this busy little bird. Mr. Skinner adds: "When they fly, these kinglets show a peculiar, jerky, undulating flight that is more or less characteristic of them."

Except during the breeding season, the ruby-crowned kinglet is a sociable bird, being seen on migrations and in winter loosely associated with various other birds, such as warblers, bluebirds, titmice, nuthatches, creepers, golden-crowned kinglets, as well as with individuals of its own species. It is probably some community of interest or some similarity in foraging ground, rather than any special attachment for each other or desire for company, that brings together these loose associations of very different birds. There is no apparent flocking instinct among them; each species, and in fact each individual, acts independently in pursuit of its special line of activity. This kinglet is also a tame and unsuspicious little bird, not easily frightened and easily approached.

W. E. Clyde Todd (1940) calls attention to certain differences in the behavior of our two kinglets: "The Ruby-crown is by all odds the more active, nervous, and irascible of the two, as it is also the more musical. It does not manifest the same partiality for conifers, and it also tends to keep nearer the ground. It has a characteristic way of flirting its wings with a sudden jerking motion; otherwise its actions while exploring the trees and bushes for its minute insect food are warbler-like."

Voice.—For its remarkable song the ruby-crowned kinglet is justly famous. Those who have not heard it in its full richness on the breeding grounds cannot appreciate it, for we seldom hear the full song on migration even in spring. The remarkable part of the song is the great volume of sound that issues from the tiny throat in the latter part of the performance, much greater than would seem possible from such a small bird. Much has been written in praise of it. Bradford Torrey (1885) says: "The song is marvelous,—a prolonged and varied warble, introduced and often broken into, with delightful effect, by a wrennish chatter. For fluency, smoothness, and ease, and especially for purity and sweetness of tone, I have never heard any bird-song that seemed to me more nearly perfect."

Aretas A. Saunders has sent me the following description of the song: "The song of the ruby-crowned kinglet is of three distinct parts. It begins with four to eight high-pitched, rather squeaky notes. This is followed by a rapid chatter of five to ten notes, often a full octave lower than the first notes, and usually rising slightly in pitch. The third part is loudest and most musical. It consists of a 3- or 4-note phrase repeated two to seven (commonly three or four) times. In this phrase the last note is highest, loudest, and strongly accented. The whole song then is like *eee-tee-tee-tee-too-too-tu-tu-ti-ta-tidaweét-tidaweét-tidaweét.*

"I have 22 records of this song. In 16 of them the drop between the first and second parts is exactly one octave. Most of the songs begin on C'''', but some on B''' or C#''''. The complete range in pitch is from C#'''' to B'', one tone more than an octave.

"It is quite common to hear the bird sing this song through two or three times without a pause. At other times it may sing the last part of the song only."

He says that all the breeding birds of this species that he heard in Montana sang a somewhat different song from that of the eastern birds: "The songs of eastern and western birds are alike in the first two parts, but in the third and loudest part they are very different, the western bird singing *whάytay, whάtay, whάytay, whάtay*, with the accent strongly on the first note, rather than the last. I have heard songs like those of the eastern bird from migrating individuals in Montana, and Weydemeyer has reported a number in northwestern Montana with the eastern song."

Weydemeyer (1923) describes a very elaborate and probably a very unusual song, as follows:

> The first two parts were the same as in the usual song, but the final notes were quite different and much more pleasing. The song sounded something like this: *Kezee kezee, zeek, zeek, eek, eek, eek, eek, chiva, chiva, chiva, chiva, chiva, chiva, chiva, chiva-lete! te-telete! te-telete! te-telete! te-telete!* Nearly every day that summer and fall, except during the molting season, this song, or a portion of it, was heard in the flat. As the nesting season approached, the song was not so often heard, and usually when it was, only the last part was given. During August it was seldom heard; by September, the last part was heard occasionally; and by the middle of that month the song was again given as in the spring.

In 1909 in southern Labrador, and again in 1912 in Newfoundland, when my hearing was only fairly good, I evidently missed the high-pitched first part of the song, and wrote down the other, louder parts as *toot, toot, toot, peabody, peabody, peabody*, or the latter part as *liberty, liberty, liberty*, the phrases often repeated more than three times; again I wrote the latter part in French as *toute suite, toute suite, toute suite!* I also recorded an alarm note, a loud *peu, peu*, almost as loud as the similar note of some thrushes.

Dr. Harrison F. Lewis (1920) recorded five types of song-endings, as heard from migrating birds near Quebec during the season of 1920, which he classified as follows:

1. *wud-a-weét, wud-a-weét*, etc. (3 syllables, accent on third), 1 record.
2. *pul-é-cho, pul-é-cho*, etc. (3 syllables, accent on second), 2 records.
3. *jim-in-y, jim-in-y*, etc. (3 syllables, accent on first), 50 records.
4. *you-eét, you-eét*, etc. (2 syllables, accent on second), 1 record.
5. *pé-to, pé-to*, etc. (2 syllables, accent on first), 9 records.

The third song-ending seems to have been by far the commonest, and agrees very closely with what I heard on the breeding grounds. The fifth is much like the common call of the tufted titmouse. Francis H. Allen's notes for August 17, 1928, at Matamek, Quebec, state: "I heard a puzzling incoherent song which I soon traced to a young ruby-crowned kinglet. The song was a long-continued succession of short phrases, resembling somewhat the fall song of a young song sparrow, but having a suggestion of the full song of its own species."

The ruby-crowned kinglet's song period is spread out, more or less continuously, from early in spring until quite late in fall, with some cessation during the period of greatest nesting activity and during the molting season. It sings during both migrations, but much more frequently and more fully in spring; the fall songs are not so regularly heard and are more fragmentary. In Frederic H. Kennard's notes, I find records of the song as early as March 27 and as late as October 16 in Massachusetts. Arthur T. Wayne (1910) says that the song period in South Carolina begins early in April, and that "when engaged in singing, the males display the vermilion patch on the crown." Mr. Saunders tells me that he has heard the song from migrating birds only in April or early May, or more rarely in October or November. Mr. Todd (1940) says, referring to western Pennsylvania: "After the first few days of May this song is seldom heard, since the later migrants are all females. These have only an odd chattering, snapping, scolding note, which, once learned, will always serve to distinguish this kinglet from the Golden-crowned species."

Field marks.—If the bright-red crown patch of the male can be seen, it is a positive field mark; but it is often partially, or wholly, concealed by the surrounding plumage; and it is not present at all on the female or the young bird. If no crown patch is seen, and the bird is a kinglet, it is a ruby crown, for the crown patch shows conspicuously in both adult sexes of the golden-crowned kinglet. Moreover, the ruby-crowned kinglet has a conspicuous white eye ring, which the other species lacks. An eye ring is somewhat in evidence in some of the small flycatchers, but these are mostly larger than kinglets and behave differently. If a bird is seen sitting quietly in an upright position, it is a flycatcher and not a kinglet; if flitting actively about, almost constantly in motion, it is more likely to be a kinglet; kinglets are tiny, plump little birds clad in olive and buffy gray plumage.

Enemies.—The ruby-crowned kinglet is a very rare victim of the cowbird; Dr. Friedmann (1929) could find but one record. Harold S. Peters (1936) records one fly, *Ornithomyia confluenta* Say, and one tick, *Haemaphysalis leporis-palustris* Packard, as external parasites on the eastern ruby-crowned kinglet.

Winter.—Although most of the ruby-crowned kinglets go far south in winter, ranging as far as Mexico and Guatemala, some spend the winter as far north as southern British Columbia, Iowa, and Virginia more or less regularly; there are a number of records for Massachusetts, and Mr. Tufts tells me that several came to a feeding station at Digby, Nova Scotia, in January 1941.

This kinglet winters abundantly all through the Southern States, where it is much commoner than the goldencrown. C. J. Maynard (1896) writes:

The Ruby-crowned Kinglets are the most common birds of Florida during winter, arriving from the North about the first of December, scattering through the hammocks all over the state, even as far south as Key West, and they may occasionally be found in company with other birds, but are generally independent; indeed, I think they seldom pay any attention to the movements of even their own companions; each pursues a course agreeable to itself. They can therefore hardly be called gregarious at this season, being equally numerous in every wooded locality, unless we choose to consider all which are in Florida as constituting one vast flock. They move about among the luxuriant growth of trees and shrubs in a manner which plainly indicates that they are at home. They seem to be always busily engaged in searching for insects upon the branches, yet will pause to gaze inquisitively at a stranger. They are not noisy at such times, and although very abundant, one who is not a naturalist would scarcely notice them, for they come without bustle, remain in the seclusion offered by the hammocks, quietly pursuing their avocations, then, by the middle of March, retire northward as silently as they came.

DISTRIBUTION

Range.—North America from northern Canada to southern Mexico; occasionally to Guatemala.

Breeding range.—The ruby-crowned kinglet breeds **north** to northern Alaska (Kobuk River; reported to breed to the edge of the willows a few miles south of Point Barrow; specimens from Point Barrow and Cape Halkett); northwestern Mackenzie (Mackenzie River, 100 miles below Fort Good Hope; Grandin River, and Fort Resolution); northern Saskatchewan (north shore of Lake Athabaska, 8 miles northeast of Moose Island, and the Churchill River); northern Manitoba (Reindeer Lake, Oxford House, and Churchill); northern Ontario (Moose Factory); central Quebec (Fort George, Lake Mistassini, Mingan, and Little Mecatina); and eastern Labrador (Makkovik). **East** to eastern Labrador (Makkovik, Rigolet, and Paradise River); Newfoundland (St. Anthony, Twillingate, and White Bear River); and Nova Scotia (Baddeck). **South** to Nova Scotia (Baddeck, Halifax, and Yarmouth); Maine (Calais, possibly Ship Harbor, and Scarboro Beach); possibly New Hampshire (Holderness); northwestern Massachusetts (Savoy); probably northern New York (Mount Whiteface); southern Ontario (Guelph, Sault Ste. Marie, Port Arthur, and Kenora); northern Michigan (Mackinac Island, Newberry, and Iron County); southern Manitoba (Winnipeg and Aweme); central Saskatchewan (Hudson Bay Junction and Big River); central southern Montana (Fort Custer and the Big Horn Mountains); central Wyoming (Sheridan, Parco, and Laramie); the eastern slope of the Rocky Mountains in Colorado (Estes Park,

Idaho Springs, and Fort Garland); central northern and southwestern New Mexico (Lost Trail Creek, Pecos Baldy, and Black Range); southeastern to central northern Arizona (Tombstone, Santa Catalina Mountains, Mogollon Mountains, San Francisco Mountain, and the north rim of Grand Canyon); southwestern Utah (Cedar Breaks); southern Nevada (Charleston Mountains); and the mountains of southern California (San Jacinto Mountains, San Bernardino Mountains, and Mount Wilson). **West** to southern California and the Sierra Nevada (Mount Wilson, Yosemite Valley, Pyramid Peak, and Mount Shasta); western Oregon (Fort Klamath, Coos Bay, Corvallis, and Newport); the Cascades of Washington (Mount Rainier and Bumping Lake); western British Columbia (Cape Scott, Vancouver Island; Bella Coola, and Graham Island, Queen Charlotte Islands); and western Alaska (Sitka, Yakutat Bay, Kenai, Nushagak, Nulato, and Kobuk River).

Winter range.—The ruby-crowned kinglet is found in winter **north** to southwestern British Columbia (Comox and Victoria, Vancouver Island, and occasionally at Okanagan Lake); western Washington (Bellingham, Everett, and Mount Rainier National Park); western Oregon (Portland, Salem, and Corvallis); eastern California (Susanville, Yosemite Valley, and Death Valley); southern Nevada (Colorado River opposite Fort Mojave); occasionally to the Ogden Valley of Utah; central Arizona (Prescott, Camp Verde, and the Salt River Wildlife Refuge); southern New Mexico (Silver City, San Antonio, Tularosa, and Carlsbad); central to northeastern Texas (San Angelo, Gainesville, and Texarkana); central Arkansas (Hot Springs and Little Rock); southern Missouri (Ozark Region and occasionally St. Louis); southern Illinois (Odin and Mount Carmel, occasionally to Chicago); southern Indiana (Bicknell and Richmond); and southern Virginia (Blacksburg and Lynchburg); rare or occasional north to North Platte, Nebr.; Washington, D. C.; Easton, Pa.; Demarest, N. J.; Hartford, Conn., and Falmouth, Maine. **East** to southern Maine (Falmouth); Connecticut (Hartford and New Haven); Long Island (Orient); eastern Pennsylvania (Easton); District of Columbia (Washington); the Atlantic coast from North Carolina (Cape Hatteras) to southern Florida (Royal Palm State Park). **South** to southern Florida (Royal Palm Hammock) and the Gulf coast of Alabama (Mobile); Mississippi (Biloxi and Gulfport); Louisiana (New Orleans, New Iberia, and Chenier au Tigre); and Texas (Houston and Brownsville); through Tamaulipas (Camargo and Victoria); Puebla (Tziutlan and Puebla); to Oaxaca (Parada); occasionally to Guatemala, since specimens are in existence from that country taken, probably, in the Department of Vera Paz previous to 1859. **South** to Oaxaca and occasionally to Guatemala. **West** to Oaxaca (Parada); Guerrero (Chalpancingo and Taxco); Michoacán (Nahuatzen); Durango (Chacala); Lower California (Victoria Mountains, Cape Region, San Telma, and Las Cruces, and formerly resident on Guadalupe Island); the Pacific Coast of California (Santa Catalina and Santa Cruz Islands, Santa Barbara, Watsonville, and San Francisco); Oregon (Coos Bay); Washington (Cape Disappointment and Tacoma); and southern Vancouver Island, British Columbia (Victoria and Comox).

The ranges as outlined apply to the species as a whole, of which four subspecies or geographic races are recognized. The eastern ruby-crowned kinglet (*R. c. calendula*) breeds from northwestern Mackenzie, Alberta, and central southern Montana eastward; the western ruby-crowned kinglet (*R. c. cinerasceus*) breeds from northwestern Alaska, Yukon, and northeastern British Columbia, south through the Rocky Mountains to New Mexico, Arizona, and southern California, and west to the Cascade Mountains; the Sitka kinglet (*R. c. grinnelli*) breeds in the coastal belt from Kenai, Alaska, to Washington; the dusky kinglet (*R. c. obscurus*) was resident on Guadalupe Island, Lower California, but is probably now extinct.

Migration.—Some late dates of spring departure are: Nuevo León—San Pedro Mines, May 8. Florida—Gainesville, April 29. Georgia—Athens, May 10. South Carolina—Columbia, May 8. North Carolina—Raleigh, May 10. District of Columbia—Washington, May 17. Pennsylvania—Beaver, May 11. New York—Watertown, May 14. Massachusetts—Northampton, May 21. Mississippi—Rodney, April 13. Louisiana—Lobdell, April 25. Arkansas—Helena, April 19. Kentucky—Bardstown, May 4. Ohio—Columbus, May 23. Illinois—Lake Forest, May 12. Texas—Corpus Christi, May 10. Oklahoma—Norman, May 6. South Dakota—Sioux Falls, May 21.

Some early dates of spring arrival are: West Virginia—French Creek, March 26. New York—Rochester, March 29. Massachusetts—Boston, April 8. Vermont—Bennington, March 31. Maine—Auburn, April 5. Nova Scotia—Wolfville, April 21. New Brunswick—Scotch Lake, April 18. Quebec—Kamouraska, April 22. Newfoundland—St. John's, April 16. Ohio—Columbus, March 8. Michigan—Grand Rapids, March 31. Ontario—London, April 8. Missouri—Kansas City, March 24. Iowa—Keokuk, March 24. Wisconsin—Milwaukee, March 18. Minnesota—Redwing, March 21. South Dakota—Brookings, April 5. North Dakota—Fargo, April 5. Manitoba—Pilot Mound, April 11. Saskatchewan—Regina, April 25. Colorado—Grand Junction, April 18. Wyoming—Laramie, April 16. Montana—Fortine, March 30. Alberta—Banff, April 16. Mackenzie—Willow River, near Providence, May 2. Utah—

Corrine, April 3, Idaho—Coeur d'Alene, April 7. Alaska—Craig, April 13.

Some late dates of fall departure are: Yukon, Carcross, September 24. Alberta—Glenevis, October 6. Idaho—Meridian, December 23. Montana—Columbia Falls, October 12. Wyoming—Cheyenne, October 25. Colorado—Yuma, October 29. Saskatchewan—Eastend, October 24. Manitoba—Aweme, October 14. North Dakota—Fargo, October 27. South Dakota—Faulkton, November 1. Minnesota—Minneapolis, November 1. Iowa—Sigourney, November 10. Wisconsin—Unity, October 25. Michigan—Ann Arbor, November 13. Ontario—Ottawa, November 10. Ohio—Oberlin, November 23. Indiana—Notre Dame, November 14. Kentucky—Bowling Green, November 14. Newfoundland—St. Anthony, October 3. Quebec—Montreal, November 3. New Brunswick—Scotch Lake, October 29. Nova Scotia—Yarmouth, October 12. Maine—Phillips, October 22. New Hampshire—Jefferson, October 16. Massachusetts—Harvard, December 11. New York—Brooklyn, November 13. Pennsylvania—Pittsburgh, November 2. West Virginia—French Creek, November 4.

Banding records.—In the banding files are several records of the return of ruby-crowned kinglets to the station where banded one or two years after banding. One banded at Waukegan, Ill., on April 18, 1937, was found 10 days later at Green Lake, Wis.

Casual records.—In 1852 an individual was found at Loch Lomond, Scotland; a specimen was collected at Nenortalik, Greenland, in 1859; and one was recorded April 13 and 21, 1909, in the Bermudas.

Egg dates.—Alberta: 13 records, June 2 to June 24.

California: 65 records, May 30 to July 17; 40 records, June 11 to June 25, indicating the height of the season.

Colorado: 10 records, June 3 to July 9.

New Brunswick: 5 records, June 14 to July 5.

REGULUS CALENDULA CINERACEUS Grinnell

WESTERN RUBY-CROWNED KINGLET

HABITS

In the mountain ranges of California, we find this larger race of the ruby-crowned kinglet, paler and grayer, less yellowish, throughout than the eastern form.

Like the eastern subspecies, this kinglet seems to be confined in the breeding season to the coniferous forests of the higher mountains, mainly above 5,000 feet in the Lassen Peak region, and from 7,000 to 8,500 feet in the San Bernardinos. Referring to the former region, Grinnell, Dixon, and Linsdale (1930) write: "In the summer when the kinglets were in the mountains they lived among the tops of the coniferous trees, especially white fir, lodgepole pine, red fir, and hemlock. On June 14, 1925, a kinglet was watched as it foraged about the ends of the branches close to the summit of a fir fully fifty meters tall."

Courtship.—Howard L. Cogswell sends me the following note: "In the spring, of course, the males often sing and display before the females, but twice during the fall of 1942, late October and November, I saw a male with red crest raised to its fullest extent over the top of his head, posturing before a female and singing a somewhat wheezy and subdued song, though of characteristic ruby-crown pattern."

Nesting.—J. Stuart Rowley (1939) found three nests of the western ruby-crowned kinglet in Mono County, Calif., of which he says:

Around Virginia Lakes, the breeding range of the Ruby-crowned Kinglet is limited to lodgepole pine stands, above 8,500 feet elevation. Over the five seasons of search for nests, only three were found. The first was some sixty feet up in a lodgepole pine, well concealed in the needles. This nest, on July 7, 1927, contained only one fresh egg although the female was flushed from the nest in midday. A second nest was found the next day, containing six heavily-incubated eggs; it was placed not more than twenty feet from the ground. The third nest found on July 6, 1930, was about forty feet up in a lodgepole pine and contained seven heavily-incubated eggs.

Each was discovered by patiently watching and following females at feeding time early in the morning or late in the evening. At each location, the male kept a vigilant guard against intruding birds of other species, making furious darts at casual passing robins, warblers, and the like. By locating a singing male, one could assume that a nest was near, but to find it was another matter.

All three nests were made of lichens and pieces of bark, tied together with cobwebs. The linings were chiefly of feathers. The persistence of incubating females in remaining on the nest is quite remarkable for such a shy nester. In our experience, the females left the nest reluctantly, one remaining until I was a foot or so from the nest. None of the three females flew farther away from their nests than twenty feet when inspection was going on.

Most other observers have reported nests in spruces, which probably are the trees most often chosen for nesting sites. Grinnell and Storer (1924) mention a nest in an incense cedar, near the Sentinel Hotel Annex in Yosemite Valley, Calif.

Eggs.—The eggs of the western ruby-crowned kinglet are, apparently, indistinguishable in every way from those of the eastern race. The measurements of 30 eggs in the United States National Museum average 14.0 to 10.9 millimeters; the eggs showing the four extremes measure **14.7** by **13.2, 12.7** by 11.4, and 14.2 by **10.2** millimeters.

Food.—Professor Beal's (1907) report on the food of this species is quoted under the eastern race, as it is the only comprehensive analysis we have for the species; but, as it was based mainly on specimens collected in California, it might just as well have been included here.

Behavior.—Grinnell and Storer (1924) give such a good account of the activities of this species that it is worth quoting here:

Both of our kinglets are busy birds at all times, but the Ruby-crown shows even more activity than does its relative. Its temperament is of the high-strung or nervous sort, which keeps the bird constantly on the go—in decided contrast to the phlegmatic behavior of, for instance, the Hutton Vireo. The kinglet has relatively long legs, and standing up on these its body is kept well clear of any perch so that the bird can hop or turn readily in any direction. Such twists and jumps are often assisted by fluttering movements of the wings. Not infrequently a Ruby-crowned Kinglet will poise on rapidly moving wings while it picks off an insect from some leaf not to be reached from a foothold. In routine foraging the bird moves through the foliage rapidly, peering this way and that as it goes, spending but a moment in any one spot or pose.

The Ruby-crowned Kinglet lacks the sociable attribute of the Golden-crown. During the nesting season the pairs give close attention to the rearing of their broods, but as soon as the young are able to live independently the families break up and each individual takes up a separate existence. While in the foothill and valley country, the Ruby-crowns are to be seen singly, each keeping to a particular forage area and usually resisting approach by another of the same species. When something excites one of their kind, however, other individuals are quick to gather and all unite in a community of effort until the object of their concern has disappeared. Then each kinglet goes its way alone once more. * * *

At about nine o'clock in the morning one of our party noticed a remarkable assemblage of Ruby-crowned Kinglets about the foliage of a certain tree. Fifteen or more of the birds were buzzing about as actively and excitedly as bees, and each kinglet was uttering its "ratchet call" with vigorous persistence. A couple of Plain Titmouses joined the group while it was being watched. The cause of the excitement became apparent when a pigmy owl flew out from the foliage of the tree. As the owl made off the crowd of excited kinglets followed in his wake.

In the nesting season Ruby-crowned Kinglets often give warning of the insidious activities of Blue-fronted Jays. On one occasion, at Chinquapin, on June 14, 1915, one of our party followed up a kinglet which was giving its *yer-rup, yer-rup,* over and over again in low but insistent tones. The cause of concern proved to be a pair of silent jays one of which was shot—to the seeming satisfaction of the kinglet, which immediately sang.

The voice of the western bird seems to be similar to that of the eastern race, with the same variations and with equal charm. It can be recognized by the same field marks and by similar behavior and voice as its eastern relative. Mr. Cogswell has sent me the following note on this subject:

"I find beginning students in field identification have difficulty separating ruby-crowned kinglets from Hutton's vireos, unless a very close view is obtained. To me, however, the chief distinguishing feature is the much slenderer head and especially the thinner bill of the rubycrown, as opposed to the thicker-billed, bull-headed appearance of the Hutton's. Call notes are more positive yet, when one knows them; the short, grating *jzidit,* or *tchidit,* of the kinglet is absolutely distinctive. The wing-flitting habit of the kinglet, given by some as an identification aid, is also indulged in by the vireo to some extent, though the kinglet, nearly always once a second or *oftener*, opens his wings and shuts them again all in a flash, whereas the vireo does it only occasionally, or at the most once every 2 or 3 seconds."

Ralph Hoffmann (1927) adds to the comparison: "The Vireo is a stockier bird and much more deliberate in its movements. It drops lazily from one twig to the next, and often stays for some seconds motionless or with only a slight movement of the head."

Winter.—In the fall these kinglets move down from their breeding grounds in the mountains and spread out in scattering groups over the foothills and valleys. Mr. Cogswell tells me that it is common, at times exceedingly so, throughout the lower areas in winter, from late September to early April. "In the oak regions of the foothills and the willow regions along the lowland streams, as many as 35 to 40 individual kinglets can be counted on a forenoon's bird walk of 3 or 4 miles."

Referring to the Lassen Peak region, Grinnell, Dixon, and Linsdale (1930) write: "During the winter, ruby-crowned kinglets foraged among the branches and foliage of tall shrubs and trees. Individuals were observed, at that season, about the following kinds of plants: digger pine, yellow pine, live oak, blue oak, buck-brush, clumps of mistletoe in cottonwood, willow, and cat-tail. Almost any sort of twiggery, whether leafy or not, where small insects might be found seemed to be a suitable winter forage place."

REGULUS CALENDULA GRINNELLI Palmer

SITKA KINGLET

HABITS

The type of this dark-colored race of the ruby-crowned kinglet was collected by Dr. Joseph Grinnell, at Sitka, Alaska, on June 23, 1896, and named in his honor by William Palmer (1897), who says: "The Sitkan Kinglet is a smaller and darker bird than its near relative *R. calendula*, approaching closer, except in the coloring of its crown patch, to *R. obscurus* of Guadalupe Island. It lacks the grayness and paleness above and on the sides of the head and neck characteristic of *calendula*. The bill is larger and differently shaped. The wing is much darker, nearly black in places, and the anterior bar especially is narrower."

Its breeding range is on the humid Northwest coast, from Prince William Sound and Skagway, Alaska, to British Columbia. As this region has produced so many dark-colored subspecies, it is interesting to consider the type of gloomy and humid weather that has helped to produce them. Mr. Palmer quotes the following from a circular of the United States Weather Bureau:

The fringe of islands that separates the mainland from the Pacific Ocean from Dixon Sound northward, and also a strip of the mainland for possibly 20 miles back from the sea, following the sweep of the coast as it curves to the northwestward to the western extremity of Alaska, form a distinct climatic division which may be termed temperate Alaska. The temperature rarely falls to zero; winter does not set in until about December 1, and by the last of May the snow has disappeared except on the mountains. The mean winter temperature of Sitka is 32.50, but little less than that of Washington, D. C. * * * The rainfall of temperate Alaska is notorious the world over not only as regards the quantity that falls, but also as to the manner of its falling, viz.: In long and incessant rains and drizzles. Cloud and fog naturally abound, there being on an average but 66 clear days in the year.

The nesting and all other habits of the Sitka kinglet seem to be similar to those of the species elsewhere, and need not be referred to further here. I have been unable to locate any eggs of this subspecies.

Richard C. Harlow has sent me a nest of this kinglet, taken by C. DeB. Greene on Porcher Island, one of the Queen Charlottes, in June 1921. At the time it held broken eggshells. It was apparently located like the nests of other kinglets, in the pendant twigs of some species of fir or spruce, pieces of such twigs still adhering to it. But no further data are available. In its present condition it is a rather large, cup-shaped ball, open at the top, measuring about 4 by 5 inches in outside diameter and about 2 inches in height; the inner cavity probably measured less than 2 inches in diameter and about an inch and a half in depth. It is composed mainly of a mass of mosses and lichens, much of which was probably green when used, reinforced with many fine strips of shredded weed stems and many fine, white, threadlike fibers. It was originally lined with feathers, but these have been eaten out by moths and no trace of them remains.

REGULUS CALENDULA OBSCURUS Ridgway

DUSKY KINGLET

HABITS

This kinglet, found only on Guadalupe Island, is darker than even the Sitka kinglet, with a shorter wing and larger bill and feet, and the crown patch is more pinkish red. In fact, all the birds peculiar to Guadalupe Island are darker than their nearest relatives on the mainland and have shorter wings and larger bills and feet.

Practically all we know about this insular form of the ruby-crowned kinglet, which was originally described as a full species, comes from Walter E. Bryant's (1887) report on the birds of that remote island. Of the haunts of these birds, he says: "Frequenting more numerously the large cypress grove, they are nevertheless found in the smaller grove, and also among the pines. In the former and latter places they are positively known to breed, and there is but little doubt that they also nest in the small grove."

They were evidently very numerous on the island, for he collected a series of ten males and three females.

Nesting.—Bryant says: "As early as the middle of February nest-building was in order, the birds selecting the topmost foliage of a cypress, and sometimes the very outer extremity of a horizontal branch," and continues:

As the result of many days' diligent search, three nests came under my observation, and these were detected only by watching the birds as they collected building material, or by tracing to its source a peculiar, low song, which the male sometimes sings when close to the nest.

These nests were all found over twenty feet high, and only one could be seen from the ground, and that merely during the intervals when the wind parted the branches. They were placed in the midst of a thick bunch of foliage, and but lightly secured to the twigs. Compact, though not very smooth in structure, they were composed of soft strips of bark intermingled with feathers, bits of moss, fine grass and cocoons. Additional warmth is secured by a quantity either of goat's hair or feathers, and, lastly, a thin lining of goat's hair. Their external measurement is about 70 mm. in height by 90 mm. in diameter, while the internal depth is about 45 mm., and diameter from 35 mm. to 45 mm. The mouth of the opening is smaller than immediately below.

Eggs.—Quoting further from Mr. Bryant's account:

In color the eggs are white, with a dense wreath of pale yellowish-brown spots encircling the larger end. In some places, these spots appear to be laid over a pale lavender washing, and in one specimen, these fine, almost indistinct dots extend sparingly over the entire surface. They measure in millimeters 14 x 11 and 15 x 11.

T. E. McMullen has sent me the following measurements of five eggs of the dusky kinglet in his collection: 14.7 by 11.4, 15.2 by 11.4, 14.7 by 11.2, 14.5 by 11.2, and 15.2 by 11.4 millimeters.

Voice.—Mr. Bryant continues:

In December I found them in full song and as common as in April. * * * Their song is indescribably sweet and musical, and of wonderful power for so small a bird, commencing with a few low, quick notes, as though the singer were merely trying his voice, then bursting into a full animated warble, it ends in a dissyllabic measure, accented on the first syllable, and usually repeated from three to six times. One remarkably fine songster repeated the final dissyllable eight or ten times. Only once did I hear the metallic click, so common with the Oakland birds in winter, but even then it flowed immediately into song.

LITERATURE CITED

ABEL, ARTHUR R.
1914. Notes on a northern robin roost. Wilson Bull., vol. 26, pp. 165–172.
ADAIR, WARD W.
1920. A railroad robin. Bird-Lore, vol. 22, pp. 289–290.
AIKEN, CHARLES EDWARD HOWARD, and WARREN, EDWARD ROYAL.
1914. Birds of El Paso County, Colorado. Colorado College Publ., gen. ser., No. 74 (sci. ser., vol. 12, No. 13, pt. 2), pp. 497–603.
ALDRICH, JOHN WARREN.
1939. Geographic variation of the veery. Auk, vol. 56, pp. 338–340.
1945. Additional breeding and migration records of the black-backed robin. Auk, vol. 62, pp. 310–311.
ALDRICH, JOHN WARREN, and NUTT, DAVID CLARK.
1939. Birds of eastern Newfoundland. Sci. Publ. Cleveland Mus. Nat. Hist., vol. 4, pp. 13–42.
ALEXANDER, BOYD.
1908. The Victoria history of the county of Kent. Birds, pp. 267–301.
ALEXANDER, HORACE GUNDRY.
1927. A list of the birds of Latium, Italy, between June 1911 and February 1916. Compiled from the notes and letters of the late C. J. Alexander. Ibis, ser. 12, vol. 3, pp. 659–691.
ALLEN, ARTHUR AUGUSTUS.
1929. Blue-gray gnatcatcher. Bird-Lore, vol. 31, p. 222.
1934. The veery and some of his family. Bird-Lore, vol. 36, pp. 68–78.
ALLEN, JOEL ASAPH.
1879. Odd behavior of a robin and a yellow warbler. Bull. Nuttall Orn. Club, vol. 4, pp. 178–182.
ALLEN, FRANCIS HENRY.
1913. More notes on the morning awakening. Auk, vol. 30, pp. 229–235.
ALLEN, GLOVER MORRILL.
1902. The birds of New Hampshire. Proc. Manchester Inst. Sci., vol. 4, pp. 23–205.
AMERICAN ORNITHOLOGISTS' UNION.
1931. Check-list of North American birds, ed. 4.
ANDRESEN, C.
1942. Townsend solitaire uses camp table for nest site. Condor, vol. 44, p. 284.
ANTHONY, ALFRED WEBSTER.
1889. New birds from Lower California, Mexico. Proc. California Acad. Sci., ser. 2, vol. 2, pp. 73–82.
1903. Nesting of the Townsend solitaire. Condor, vol. 5, pp. 10–12.
ANTONIUS, OTTO.
1937. Begattung der Amsel. Beitr. Fortpflanzungsbiologie Vögel, Jahrg. 13, p. 150.
APLIN, OLIVER VERNON.
1903. Letter to the Ibis, pp. 132–133.

ARNOLD, CLARENCE M.
1907. Robins and sparrows. Bird-Lore, vol. 9, p. 84.
ARRIGONI DEGLI ODDI, ETTORE.
1929. Ornitologia Italiana.
AUDUBON, JOHN JAMES.
1840. The birds of America, vol. 1.
1841. The birds of America, vol. 3.
AUSTIN, OLIVER LUTHER, JR.
1932. The birds of Newfoundland Labrador. Mem. Nuttall Orn. Club, No. 7.
BAGG, AARON CLARK.
1932. Ruby-crowned kinglet feeding young in Massachusetts. Auk, vol. 49, p. 486.
BAGG, A. C., and ELIOT, SAMUEL ATKINS, JR.
1937. Birds of the Connecticut Valley in Massachusetts.
BAILEY, ALFRED MARSHALL.
1926. A report on the birds of northwestern Alaska and regions adjacent to Bering Strait, pt. 10. Condor, vol. 28, pp. 165-170.
1927. Notes on the birds of southeastern Alaska. Auk, vol. 44, pp. 351-367.
BAILEY, A. M., and NIEDRACH, ROBERT JAMES.
1936. Community nesting of western robins and house finches. Condor, vol. 38, p. 214.
BAILEY, FLORENCE MERRIAM.
1902. Handbook of birds of the western United States.
1928. Birds of New Mexico.
BAIRD, SPENCER FULLERTON.
1864. Review of American birds, in the museum of the Smithsonian Institution, pt. 1: North and Middle America. Smithsonian Misc. Coll., No. 181, vol. 12.
BAIRD, S. F.; BREWER, THOMAS MAYO; and RIDGWAY, ROBERT.
1874. A history of North American birds: Land birds, vol. 1.
BANCROFT, GRIFFING.
1930. The breeding birds of central Lower California. Condor, vol. 32, pp. 20-49.
BANGS, OUTRAM.
1898. Some new races of birds from eastern North America. Auk, vol. 15, pp. 173-183.
BANGS, OUTRAM, and PENARD, THOMAS GILBERT.
1921. The name of the eastern hermit thrush. Auk, vol. 38, pp. 432-434.
BATCHELDER, CHARLES FOSTER.
1900. An undescribed robin. Proc. New England Zool. Club, vol. 1, pp. 103-106.
BAXTER, EVELYN VIDA, and RINTOUL, LEONORA JEFFREY.
1914. Birds singing while on migration. Scottish Nat., 1914, pp. 188-189.
BEAL, FOSTER ELLENBOROUGH LASCELLES.
1907. Birds of California in relation to the fruit industry, pt. 1. Biol. Surv. Bull. 30.
1915a. Food of the robins and bluebirds of the United States. U. S. Dept. Agr. Dept. Bull. 171.
1915b. Food habits of the thrushes of the United States. U. S. Dept. Agr. Dept. Bull. 280.

BEAN, TARLETON HOFFMAN.
1883. Notes on birds collected during the summer of 1880 in Alaska and Siberia. Proc. U. S. Nat. Mus., vol. 5, pp. 143–173.

BECK, ROLLO HOWARD.
1900. An unusually high nest of Audubon's hermit thrush. Condor, vol. 2, p. 19.

BELDING, LYMAN.
1884. Second catalogue of a collection of birds made near the southern extremity of Lower California. Proc. U. S. Nat. Mus., vol. 6, pp. 344–352.
1889a. Description of a new thrush from Calaveras County, California. Proc. California Acad. Sci., ser. 2, vol. 2, pp. 18–19.
1889b. The small thrushes of California. Proc. California Acad. Sci., ser. 2, vol. 2, pp. 57–72.

BENDIRE, CHARLES EMIL.
1888. Description of the nest and eggs of the California black-capped gnatcatcher (*Polioptila californica* Brewster). Proc. U. S. Nat. Mus., vol. 10, pp. 549–550.

BERGMAN, STEN.
1935. Zur Kenntnis nordostasiatischer Vögel.

BICKNELL, EUGENE PINTARD.
1884. A study of the singing of our birds. Auk, vol. 1, pp. 126–140.

BINGER, HARRY F.
1932. A robin and snake story. Bird-Lore, vol. 34, pp. 390–391.

BIRD, CHARLES GODFREY, and BIRD, EDWARD GODFREY.
1941. The birds of North-east Greenland. Ibis, 1941, pp. 118–161.

BISHOP, LOUIS BENNETT.
1900. Birds of the Yukon region. North American Fauna, No. 19.

BLAIR, HELEN.
1935. Key for identifying birds' nests, Pittsburg, Pa.

BLAIR, HUGH MOVAY SUTHERLAND.
1936. On the birds of east Finmark. Ibis, 1936, pp. 280–308, 429–459, 651–674.

BLAIR, RICHARD HENRY, and TUCKER, BERNARD WILLIAM.
1941. Nest sanitation, by R. H. Blair. With additions from published sources, by B. W. Tucker. British Birds, vol. 34, pp. 206–215, 226–235, 250–255.

BLAKE, SIDNEY FAY.
1916. Breeding of the golden-crowned kinglet in Norfolk County, Massachusetts. Auk, vol. 33, pp. 326–327.

BLINCOE, BENEDICT JOSEPH.
1923. Gnatcatchers attacked cowbird. Bird-Lore, vol. 25, pp. 253–254.

BOLANDER, L. PH., JR.
1932. A robin roost in Oakland, California. Condor, vol. 34, pp. 142–143.

BOLLES, FRANK.
1891. Land of the lingering snow.

BOND, FRANK.
1889. *Myiadestes townsendii* apparently wintering in Wyoming. Auk, vol. 6, pp. 193–194.

BORASTON, JOHN MACLAIR.
1905. Nature tones and undertones.

Bowdish, Beecher Scoville.
1890. Sialia-Mus. Oologist, vol. 7, p. 140.

Bowles John Hooper.
1927. Nesting of the western robin. Murrelet, vol. 8, p. 74.

Boyd, Arnold Whitworth.
1941. Display of blackbirds. British Birds, vol. 35, p. 157.

Brackbill, Hervey.
——. A wood thrush study by color-banding. (Unpublished manuscript.)

Bralliar, Floyd.
1922. Knowing birds through stories.

Brand, Albert Rich.
1938. Vibration frequencies of passerine bird song. Auk, vol. 55, pp. 263–268.

Brandt, Herbert.
1943. Alaska bird trails.

Brannon, Peter A.
1921. Notes on Alabama birds. Auk, vol. 38, pp. 463–464.

Brewer, Thomas Mayo.
1878. Wilson's thrush, with spotted eggs and nesting on a tree. Bull. Nutthall Orn. Club, vol. 3, p. 193.

Brewster, William.
1885. Preliminary notes on some birds obtained in Arizona by Mr. F. Stephens in 1884. Auk, vol. 2, pp. 84–85.
1888. Breeding of the golden-crested kinglet (*Regulus satrapa*) in Worcester County, Massachusetts, with a description of its nest and eggs. Auk, vol. 5, pp. 337–344.
1890. Summer robin roosts. Auk, vol. 7, pp. 360–373.
1902. Birds of the Cape region of Lower California. Bull. Mus. Comp. Zool., vol. 41, No. 1, pp. 1–241.
1906. The birds of the Cambridge region of Massachusetts. Mem. Nuttall Orn. Club, No. 4.
1936. October Farm.
1937. Concord River.
1938. The birds of the Lake Umbagog region of Maine, pt. 4. Bull. Mus. Comp. Zool., vol. 66, pt. 4, pp. 525–620.

Briggs, Guy H.
1902. The bluebird. Journ. Maine Orn. Soc., vol. 4, pp. 16–17.

Brooks, Allan.
1903. Notes on the birds of the Cariboo District, British Columbia. Auk, vol. 20, pp. 277–284.
1905. Notes on the nesting of the varied thrush. Auk, vol. 22, pp. 214–215.

Brooks, Maurice.
1933. Taming the blue-gray gnatcatchers. Bird-Lore, vol. 35, pp. 90–93.

Brown, Nathan Clifford.
1906. A great flight of robins and cedar-birds. Auk, vol. 23, pp. 342–343.
1911. A remarkable number of robins in Maine in winter. Auk, vol. 28, pp. 270–272.

Bryant, Walter (Pierc)E.
1887. Additions to the ornithology of Guadalupe Island. Bull. California Acad. Sci., vol. 2, pp. 269–318.

Burleigh, Thomas Dearborn.
1923. Notes on the breeding birds of Clark's Fork, Bonner County, Idaho. Auk, vol. 40, pp. 653–665.

1931. Notes on the breeding birds of State College, Centre County, Pennsylvania. Wilson Bull., vol. 43, pp. 37–54.
1941. Bird life on Mt. Mitchell. Auk, vol. 58, pp. 334–345.

BURNS, FRANKLIN LORENZO.
1915. Comparative periods of deposition and incubation of some North American birds. Wilson Bull., vol. 27, pp. 275–286.
1919. The ornithology of Chester County, Pennsylvania.
1921. Comparative periods of nestling life of some North American nidicolae. Wilson Bull., vol. 33, pp. 4–15.

BURROUGHS, JOHN.
1880. Wake-robin.
1894. Riverby.

BUTLER, AMOS WILLIAM.
1898. The birds of Indiana. Indiana Dept. Geol. and Nat. Resources, 22d Ann. Rep., 1897, pp. 515–1197, pls. 21–25.

CAMERON, EWEN SOMERLED.
1908. The birds of Custer and Dawson Counties, Montana. Auk, vol. 25, pp. 39–56.

CAREY, HENRY REGINALD.
1925. How a family of hermit thrushes came to camp. Bird-Lore, vol. 27, pp. 225–228.

CHAMBERLAIN, MONTAGUE.
1882. The wood thrushes (*Hylocichla*) of New Brunswick. Ornithologist and Oologist, vol. 7, pp. 185–187.
1889. Some accounts of the birds of southern Greenland, from the manuscripts of A. Hagerup. Auk, vol. 6, pp. 291–297.

CHAMBERLIN, CORYDON.
1901. Some architectural traits of the western gnatcatcher. Condor, vol. 3, pp. 33–36.

CHILDS, JOHN LEWIS.
1913. Destruction of robins in a storm. Auk, vol. 30, p. 590.

CHISLETT, RALPH.
1933. Northward Ho!—for birds.

CHURCH, CYNTHIA.
1927. Friendly kinglets. Bird-Lore, vol. 29, p. 342.

CLARK, AUSTIN HOBART.
1910. The birds collected and observed during the cruise of the United States Fisheries steamer *Albatross* in the North Pacific Ocean, and in the Bering, Okhotsk, Japan, and eastern seas, from April to December, 1906. Proc. U. S. Nat. Mus., vol. 38, pp. 25–74.
1945. Animal life of the Aleutian Islands. Smithsonian Inst. War Background Studies No. 21, pp. 31–61.

CLARKE, MRS. J. FREDERICK.
1930. A robin roost close to a house. Wilson Bull., vol. 42, p. 291.

CLEVELAND, LOTTA A.
1923. Robins and cicadas. Bird-Lore, vol. 25, p. 254.

COCHRAN, MRS. ARCH.
1935. Two migrating returns of olive-backed thrushes. Bird-Banding, vol. 6, p. 66.

COKER, COIT M.
1931. Hermit thrush feeding on salamanders. Auk, vol. 48, p. 277.

COLLETT, ROBERT.

1877. On *Phylloscopus borealis* and its occurrence in Norway. Proc. Zool. Soc. London, 1877, pp. 43–47.

1886. Further notes on *Phylloscopus borealis* in Norway. Ibis, 1886, pp. 217–223.

COLLINGE, WALTER EDWARD.

1927. The food of some British wild birds, 2d rev. ed.

COMEAU, NAPOLEON ALEXANDER.

1890. Additional notes on the probable breeding of *Saxicola oenanthe* near Godbout, Province of Quebec, Canada. Auk, vol. 7, p. 294.

1891. The robin wintering at Godbout, Quebec. Auk, vol. 8, pp. 317–318.

COOKE, MAY THACHER.

1937. Some longevity records of wild birds. Bird-Banding, vol. 8, pp. 52–65.

COOKE, WELLS WOODBRIDGE.

1914. Some winter birds of Oklahoma. Auk, vol. 31, pp. 473–493.

CORRINGTON, JULIAN DANA.

1922. The winter birds of Biloxi, Mississippi, region. Auk, vol. 39, pp. 530–556.

COTTAM, CLARENCE, and KNAPPEN, PHOEBE.

1939. Food of some uncommon North American birds. Auk, vol. 56, pp. 138–169.

COUES, ELLIOTT.

1874. Birds of the Northwest.

COWAN, IAN MCTAGGART.

1942. Termite-eating birds in British Columbia. Auk, vol. 59, p. 451.

COWARD, THOMAS ALFRED.

1920. The birds of the British Isles and their eggs.

CRIDDLE, NORMAN.

1922. A calendar of bird migration. Auk, vol. 39, pp. 41–49.

1927. Habits of the mountain bluebird in Manitoba. Can. Field Nat., vol. 41, pp. 40–44.

DALL, WILLIAM HEALEY, and BANNISTER, HENRY MARTYN.

1869. List of the birds of Alaska, with biographical notes. Trans. Chicago Acad. Sci., vol. 1, pp. 267–310, pls. 27–34.

DARCUS, S. J.

1930. Notes on birds of the northern part of the Queen Charlotte Islands in 1927. Can. Field Nat., vol. 44, pp. 45–49.

DAUKES, A. H.

1932. Breeding of the redwing in Scotland. British Birds, vol. 26, pp. 132–134.

DAVID, ARMAND, and OUSTALET, ÉMILE.

1877. Les oiseaux de la Chine.

DAWSON, WILLIAM LEON.

1923. The birds of California, vol. 2.

DAWSON, W. L., and BOWLES, JOHN HOOPER.

1909. The birds of Washington, vol. 1.

DEANE, RUTHVEN.

1878. Deadly combat between an albino robin and a mole. Bull. Nuttall Orn. Club, vol. 3, p. 104.

DEMERITTE, EDWIN.

1920. Peculiar nesting of hermit thrushes. Auk, vol. 37, pp. 138–140.

DICKEY, DONALD RYDER, and VAN ROSSEM, ADRIAAN JOSEPH.

1938. The birds of El Salvador. Publ. Field Mus. Nat. Hist., zool. ser., vol. 23, pp. 1–635.

DIXON, JOSEPH SCATTERGOOD.
1938. Birds and mammals of Mount McKinley National Park, Alaska. Nat. Park Serv. Faunal Ser., No. 3.
DOMBROWSKI, ROBERT RITTER VON.
1903. Materialen zu einer Ornis Rümaniens. Bull. Soc. Sci. Bucharest, Ann. 12, Nos. 3, 4.
DRESSER, HENRY EELES.
1902. A manual of Palaearctic birds.
DREW, FRANK MAYO.
1881. Field notes on the birds of San Juan County, Colorado. Bull. Nuttall Orn. Club, vol. 6, pp. 85–91
DUNN, HARRY H.
1906. The gnatcatchers of southern California. Warbler, ser. 2, vol. 2, pp. 60–61.
DWIGHT, JONATHAN JR.
1900. The sequence of plumages and moults of the passerine birds of New York. Ann. New York Acad. Sci., vol. 13, pp. 73–360, pls. 1–7.
DYBOWSKI, BÉNOÎT.
1883. Remarques sur les oiseaux du Kamtschatka et des îles Comandores Bull. Soc. Zool. France, vol. 8, pp. 351–370.
FLEMING, JAMES HENRY.
1907. Birds of Toronto, Canada. Auk, vol. 24, pp. 71–89.
FORBES, JOHN RIPLEY.
1938. Recent observations on the Greenland wheatear. Auk, vol. 55, pp. 492–495.
FORBES, STEPHEN ALFRED.
1880. The food-habits of thrushes. Amer. Ent., new ser., vol. 1, pp. 12–13.
FORBUSH, EDWARD HOWE.
1907. Useful birds and their protection.
1929. Birds of Massachusetts and other New England States, vol. 3. Land birds from sparrows to thrushes.
FRIEDMANN, HERBERT.
1929. The cowbirds: A study in the biology of social parasitism.
1934. Further additions to the list of birds victimized by the cowbird. Wilson Bull., vol. 46, pp. 25–36.
1937. Further additions to the known avifauna of St. Lawrence Island, Alaska. Condor, vol. 39, p. 91.
1938. Additional hosts of the parasitic cowbirds. Auk, vol. 55, pp. 41–50.
GÄTKE, HEINRICH.
1895. Heligoland as an ornithological observatory.
GARNIER, E.
1934. Zu "Paarung bei *Turdus merula.*" Orn. Monatsb., vol. 42, pp. 54–55.
GEYR, H. BARON.
1933. Paarung bei *Turdus merula.* Orn. Monatsb., vol. 41, p. 119.
GILLESPIE, JOHN ARTHUR.
1927. Singing by migrant gray-cheeked thrush. Auk, vol. 44, p. 112.
GOSS, NATHANIEL STICKNEY.
1891. History of the birds of Kansas.
GRIMES, SAMUEL ANDREW.
1928. The blue-gray gnatcatcher. Florida Nat., new ser., vol. 2, pp. 25–27.
1932. Notes on the 1931 nesting season in the Jacksonville region.—III. Florida Nat., vol. 6, pp. 8–13.
GRINNELL, JOSEPH.
1898. Summer birds of Sitka, Alaska. Auk, vol. 15, pp. 122–131.

1900. Birds of the Kotzebue Sound region, Alaska. Pacific Coast Avifauna, No. 1.
1901. Two races of the varied thrush. Auk, vol. 18, pp. 142–145.
1904. Midwinter birds at Palm Springs, California. Condor, vol. 6, p. 44.
1908. The biota of the San Bernardino Mountains. Univ. California Publ. Zool., vol. 5, pp. 1–170.
1909. Birds and mammals of the 1907 Alexander expedition to southeastern Alaska. The birds. Univ. California Publ. Zool., vol. 5, pp. 181–244.
1918. Seven new or noteworthy birds from east-central California. Condor, vol. 20, pp. 86–90.
1926. A critical inspection of the gnatcatchers of the Californias. Proc. California Acad. Sci., ser. 4, vol. 15, pp. 493–500.

GRINNELL, JOSEPH; DIXON, JOSEPH; and LINSDALE, JEAN MYRON.
1930. Vertebrate natural history of a section of northern California through the Lassen Peak region. Univ. California Publ. Zool., vol. 35.

GRINNELL, JOSEPH, and LINSDALE, JEAN MYRON.
1936. Vertebrate animals of Point Lobos Reserve, 1934–35. Carnegie Inst. Washington Publ. No. 481.

GRINNELL, JOSEPH, and STORER, TRACY IRWIN.
1924. Animal life in the Yosemite.

GRINNELL, JOSEPH, and WYTHE, MARGARET WILHELMINA.
1927. Directory to the bird-life of the San Francisco Bay region. Pacific Coast Avifauna, No. 18.

GRISCOM, LUDLOW.
1926. Notes on the summer birds of the west coast of Newfoundland. Ibis, 1926, pp. 656–684.
1932. The distribution of bird-life in Guatemala. Bull. Amer. Mus. Nat. Hist., vol. 64.

HAMILTON, WILLIAM JOHN, JR.
1935. Notes on nestling robins. Wilson Bull., vol. 47, pp. 109–111.
1943. Spring food of the robin in central New York. Auk, vol. 60, p. 273.

HANFORD, FORREST S.
1917. The Townsend solitaire. Condor, vol. 19, pp. 13–15.

HANNA, WILSON CREAL.
1934. The black-tailed gnatcatcher and the dwarf cowbird. Condor, vol. 36, p. 89.

HANTZSCH, BERNHARD.
1905. Beitrag zur kenntnis der vogelwelt Islands.

HARBAUM, FRANK.
1921. A family of wood thrushes. Bird-Lore, vol. 23, pp. 140–141.

HARGRAVE, LYNDON L.
1933. The western gnatcatcher also moves its nest. Wilson Bull., vol. 45, pp. 30–31.

HARPER, W. T.
1926. A bluebird's nest. Bird-Lore, vol. 28, pp. 187–190.

HARTERT, ERNST.
1910. Die Vögel der paläarktischen Fauna, vol. 1.

HARTERT, ERNST, and STEINBACHER, FRIEDRICH.
1938. Die Vögel der paläarktischen Fauna. Ergänzungsband.

HASKIN, LESLIE L.
1919. Townsend's solitaire. Bird-Lore, vol. 21, pp. 242–243.

HAZEN, HENRY HONEYMAN.
1928. Nocturnal song of migrants. Auk, vol. 45, p. 230.

HEINROTH, OSCAR and MAGDALENA.
1926. Die Vogel Mitteleuropas, vol. 1.
HELMS, OTTO.
1926. The birds of Angmagsalik. Medd. om Grønland, vol. 58, pp. 205–274.
HELMS, OTTO, and SCHIØLER, EILER THEODOR LEHN.
1917. Om nogle for Grønlands Øst og Westkyst nye og sjaeldeme Arter. Dansk Orn. Foren. Tidsskr., Aarg. 11, pp. 172–175.
HENDERSON, JUNIUS.
1927. The practical value of birds.
HENSHAW, HENRY WETHERBEE.
1875. Report upon ornithological collections made in portions of Nevada, Utah, California, Colorado, New Mexico, and Arizona during the years 1871, 1872, 1873, and 1874. (Wheeler Survey.)
HODGE, CLIFTON FREMONT.
1904. A summer with the bluebirds. Bird-Lore, vol. 6, pp. 41–46.
HOFFMANN, RALPH.
1927. Birds of the Pacific States.
HOWE, REGINALD HEBER, JR.
1898. Breeding habits of the American robin (*Merula migratoria*) in eastern Massachusetts. Auk, vol. 15, pp. 162–167.
1900. A new subspecies of the genus *Hylocichla*. Auk, vol. 17, pp. 270–271.
HOWELL, ARTHUR HOLMES.
1924. Birds of Alabama.
1932. Florida bird life.
HOWELL, JOSEPH CORWIN.
1940. Spring roosts of the robin. Wilson Bull., vol. 52, pp. 19–23.
HOWES, PAUL GRISWOLD.
1914. The migration of the olive-backed thrush, 1912. Oologist, vol. 31, pp. 162–166.
IVOR, HANCE ROY.
1941. Observations on "anting" by birds. Auk, vol. 58, pp. 415–416.
1943. Further studies of anting by birds. Auk, vol. 60, pp. 51–55.
JENKS, RANDOLPH.
1936. A new race of golden-crowned kinglet from Arizona. Condor, vol. 38, pp. 239–244.
JENSEN, JENS KNUDSON.
1923. Notes on the nesting birds of northern Santa Fe County, New Mexico. Auk, vol. 40, pp. 452–469.
1925. English sparrows and robins. Auk, vol. 42, p. 591.
JEWETT, STANLEY GORDON.
1928. Assistant parentage among birds. Condor, vol. 30, pp. 127–128.
JOHNSTON, VERNA RUTH.
1943. An ecological study of nesting birds in the vicinity of Boulder, Colorado. Condor, vol. 45, pp. 61–68.
JONES, LYNDS.
1910. The birds of Cedar Point and vicinity. Wilson Bull., vol. 22, pp. 172–182.
JOURDAIN, FRANCIS CHARLES ROBERT.
1938. The handbook of British birds, vols. 1, 2. (By H. F. Witherby, F. C. R. Jourdain, Norman F. Ticehurst, and Bernard W. Tucker.)
KALMBACH, EDWIN RICHARD.
1914. Birds in relation to the alfalfa weevil. U. S. Dept. Agr. Dept. Bull. 107.

KANE, SUSAN M.
1924. Battle between a western robin and Steller's jay, witnessed on the campus at the University of Washington. Murrelet, vol. 5, No. 2, p. 9.
KELLY, BERNERS B.
1913. The building of a robin's nest. Bird-Lore, vol. 15, pp. 310–311.
KELSO, LEON HUGH.
1935. Bird notes from Fall River, Larimer County, Colorado. Oologist, vol. 52, pp. 14–19.
KING, FRANKLIN HIRAM.
1883. Economic relations of Wisconsin birds. Geology of Wisconsin, vol. 1, pp. 441–610.
KIRKMAN, FREDERICK BERNUF BEVER.
1911. The thrush family. *In*, "The British Bird Book," vol. 1.
KNIGHT, ORA WILLIS.
1908. The birds of Maine.
KÖNIG, D.
1938. Paarung der amsel. Beitr. Fortpflanzungsbiologie Vögel, Jahrg. 14, pp. 69–70.
LACK, DAVID.
1942. Ecological features of the bird faunas of British small islands. Journ. Animal Ecol., vol. 11, pp. 9–36.
1943. The age of the blackbird. British Birds, vol. 36, pp. 166–175.
LACK, DAVID, and LIGHT, WILLIAM.
1941. Notes on the spring territory of the blackbird. British Birds, vol. 35, pp. 47–53.
LACK, H. LAMBERT.
1941. Display in blackbirds. British Birds, vol. 35, pp. 54–57.
LAING, HAMILTON MACK.
1925. Birds collected and observed during the cruise of the *Thiepval* in the North Pacific, 1924. Victoria Mem. Mus. Bull. 40.
LANGILLE, JAMES HIBBERT.
1884. Our birds in their haunts.
LASKEY, AMELIA RUDOLPH (MRS. F. C. LASKEY).
1939. A study of nesting eastern bluebirds. Bird-Banding, vol. 10, pp. 23–32.
1942. Nesting observations for 1942 a la bird-banding. Migrant, vol. 13, pp. 36–38.
LA TOUCHE, JOHN DAVID DIGUES DE.
1920. Notes on the birds of North-East Chihli, in North China. Ibis, 1920, pp. 629–671.
1925. A handbook of the birds of eastern China, vol. 1, pt. 2.
1926. A handbook of the birds of eastern China, vol. 1, pt. 3.
LEWIS, HARRISON FLINT.
1919. Winter robins in Nova Scotia. Auk, vol. 36, pp. 205–217.
1920. The singing of the ruby-crowned kinglet (*Regulus c. calendula*). Auk, vol. 37, pp. 594–596.
1928. Notes on the birds of the Labrador peninsula in 1928. Can. Field Nat., vol. 42, pp. 191–194.
LINCOLN, FREDERICK CHARLES.
1935. The migration of North American birds. U. S. Dept. Agr. Circ. 363.
1939. The migration of American birds.

LINSDALE, JEAN MYRON.
1938. Environmental responses of vertebrates in the Great Basin. Amer. Midl. Nat., vol. 19, pp. 1–206.

LLOYD, BERTRAM.
1933. The courtship and display of wheatears. Trans. Hertfordshire Nat. Hist. Soc., vol. 19, pp. 135–139.

LLOYD, CLARK K.
1932. The blue-gray gnatcatcher moves its nest. Wilson Bull., vol. 44, p. 185.

LONGSTAFF, TOM GEORGE.
1932. An ecological reconnaissance in west Greenland. Journ. Animal Ecol., vol. 1, pp. 119–142.

LOW, SETH HASKELL.
1934. Bluebird studies on Cape Cod. Bird-Banding, vol. 5, pp. 39–41.

LYNES, HUBERT.
1925. On the birds of north and central Darfur with notes on the west-central Kordofan and North Nuba Provinces of British Sudan (part 3). Ibis, 1925, pp. 71–131.

MACKAY, GEORGE HENRY.
1897. A great flight of robins in Florida. Auk, vol. 14, p. 325.

MACOUN, JOHN, and MACOUN, JAMES MELVILLE.
1909. Catalogue of Canadian birds, ed. 2.

MAILLIARD, JOSEPH.
1908. A migration wave of varied thrushes. Condor, vol. 10, pp. 118–119.
1930. Happenings in a robin household. Condor, vol. 32, pp. 77–80.

MAILLIARD, JOSEPH, and HANNA, G. DALLAS.
1921. New bird records for North America, with notes on the Pribilof Island list. Condor, vol. 23, pp. 93–95.

MARSHALL, W. A.
1921. Robin and snake. Bird-Lore, vol. 23, p. 304.

MAYNARD, CHARLES JOHNSON.
1896. The birds of eastern North America, ed. 2.
1910. Records of walks and talks with nature, vol. 3.

MCATEE WALDO LEE.
1926. The relation of birds to woodlots in New York State. Roosevelt Wild-life Bull., vol. 4, pp. 7–152.

MCCLINTOCK, NORMAN.
1910. A hermit thrush study. Auk, vol. 27, pp. 409–418.

MCINTOSH, FRANKLIN GRAY.
1922. Robin and snake. Bird-Lore, vol. 24, p. 152.

MCKINNON, ANGUS.
1908. A pair of blue-gray gnatcatchers that moved their nest. Bird-Lore, vol. 10, p. 173.

MEARNS, EDGAR ALEXANDER.
1890. Observations on the avifauna of portions of Arizona. Auk, vol. 7, pp. 251–264.

MEIKLEJOHN, M. F. M.
1941. In "Additions to the Handbook of British Birds," vol. 3, by Witherby, Jourdain, Ticehurst, and Tucker.

MEINERTZHAGEN, RICHARD.
1922. Notes on some birds from the near east and from tropical East Africa. Ibis, 1922, pp. 1–74.
1930. In "Birds of Egypt" by Nicoll.

MERRIAM, CLINTON HART.
1882. Late breeding of the hermit thrush in northern New York. Ornithologist and Oologist, vol. 7, p. 171.
MICHAEL, CHARLES WILSON.
1934. Unusual behavior of the western robin. Condor, vol. 36, pp. 33–34.
MILLER, ALDEN HOLMES.
1935. Some breeding birds of the Pine Forest Mountains, Nevada. Auk, vol. 52, pp. 467–468.
MILLER, WALDRON DEWITT.
1911. The notes of the hermit thrush. Bird-Lore, vol. 13, p. 99.
MINOT, HENRY DAVIS.
1877. The land birds and game birds of New England.
1895. The land birds and game birds of New England, ed. 2.
MOREAU, REGINALD ERNEST.
1937. Migrant birds in Tanganyika Territory. Tanganyika Notes and Records, No. 4, pp. 1–34.
MOREAU, R. E., AND MOREAU, W. M.
1928. Some notes on the habits of Palaearctic migrants while in Egypt. Ibis, 1928, pp. 233–252.
MORLEY, AVERIL.
1937. Some activities of resident blackbirds in winter. British Birds, vol. 31, pp. 34–41.
1938. (See under Witherby, Jourdain, Ticehurst, and Tucker.)
MOUNTS, BERYL T.
1922. A blue-gray gnatcatcher's nest. Wilson Bull., vol. 34, pp. 116–117.
MOUSLEY, HENRY.
1916. Five years personal notes and observations on the birds of Hatley, Stanstead County, Quebec—1911–1915. Auk, vol. 33, pp. 168–186.
MURDOCH, JOHN.
1885. Report of the International Polar Expedition to Point Barrow, Alaska. Part 4, Natural History, pp. 91–135.
MURRAY, JOSEPH JAMES.
1934. The blue-gray gnatcatcher moving its nest. Wilson Bull., vol. 46, p. 128.
MUSSELMAN, THOMAS EDGAR.
1935. Three years of eastern bluebird banding and study. Bird-Banding, vol. 6, pp. 117–125.
1939. The effect of cold snaps upon the nesting of the eastern bluebird (*Sialia sialis sialis*). Bird-Banding, vol. 10, pp. 33–35.
1942. A probable case of parasitism in the starling. Auk, vol. 59, pp. 589–590.
MYERS, HARRIET WILLIAMS.
1907. Nesting ways of the western gnatcatcher. Condor, vol. 9, pp. 48–51.
1912. Nesting habits of the western bluebird. Condor, vol. 14, pp. 221–222.
NATORP, OTTO.
1928. Blaukelchens Stern. Beitr. Fortpflanzungsbiologie Vögel, vol. 4, pp. 136–139.
NEEDHAM, JAMES GEORGE.
1909. Kinglets captured by burdocks. Bird-Lore, vol. 11, pp. 261–262.
NELSON, EDWARD WILLIAM.
1883. Birds of Bering Sea and the Arctic Ocean. Cruise of the Revenue-Steamer *Corwin* in Alaska and the N. W. Arctic Ocean in 1881, pp. 56–118.
1887. Report upon natural history collections made in Alaska between the years 1877 and 1881. U. S. Signal Service Arctic Ser., No. 3.

Nethersole-Thompson, Caroline and Desmond.
1942. Egg-shell disposal by birds. British Birds, vol. 35, pp. 162–169, 190–200, 214–223, 241–250.
1943. Nest-site selection by birds. British Birds, vol. 37, pp. 70–74, 88–94, 108–113.

Nice, Margaret Morse.
1930a. Do birds usually change mates for the second brood? Bird-Banding, vol. 1, pp. 70–72.
1930b. A list of the birds of the campus of the University of Oklahoma. Publ. Univ. Oklahoma Biol. Surv., vol. 2, pp. 195–207.
1931. The birds of Oklahoma, rev. ed.
1932. Observations on the nesting of the blue-gray gnatcatcher. Wilson Bull., vol. 44, p. 54.
1933. Robins and Carolina chickadees remating. Bird-Banding, vol. 4, p. 157.

Nicholson, Edward Max, and Koch, Ludwig.
1936. Songs of wild birds.

Niethammer, Günther.
1937. Handbuch der deutschen Vogelkunde, vol. 1.

Oberholser, Harry Church.
1898. Description of a new North American thrush. Auk, vol. 15, pp. 303–306.
1917. Description of a new *Sialia* from Mexico. Proc. Biol. Soc. Washington, vol. 30, pp. 27–28.
1938. The bird life of Louisiana. Louisiana Dept. Conserv., Bull. 28.

Odum, Eugene Pleasants, and Burleigh, Thomas Dearborn.
1946. Southward invasion in Georgia. Auk, vol. 63, pp. 388–401.

Oldendow, K.
1933. Fugleliv in Grønland. Grønl. Selsk. Aarsskr., 1932–33, pp. 17–224.

Oldys, Henry.
1913. A remarkable hermit thrush song. Auk, vol. 30, pp. 538–541.
1916. Rhythmical singing of veeries. Auk, vol. 33, pp. 17–21.

Ortega, James L.
1926. Robin carrying water to young. Condor, vol. 28, p. 244.

Osgood, Wilfred Hudson.
1904. A biological reconnaissance of the base of the Alaska Peninsula. North Amer. Fauna, No. 24.
1909. Biological investigations in Alaska and Yukon Territory. North Amer. Fauna, No. 30.

Owen, Daniel Edward.
1897. Notes on a captive hermit thrush. Auk, vol. 14, pp. 1–8.

Palmer, William.
1897. The Sitkan kinglet. Auk, vol. 14, pp. 399–401.
1906. The blue-gray gnatcatcher. Osprey, vol. 5, pp. 86–88.

Parker, Harry G.
1887. Notes on the eggs of the thrushes and thrashers. Ornithologist and Oologist, vol. 12, pp. 69–71.

Parsons, Katharine S.
1906. Our robin's nest. Bird-Lore, vol. 8, pp. 66–67.

Pearson, Thomas Gilbert.
1910. The robin. Bird-Lore, vol. 12, pp. 206–209.
1916. The veery. Bird-Lore, vol. 18, pp. 270–273.
1917. The bird study book.

PEDERSON, ALWIN.
1930. Fortgesetzte beitrage zur kenntnis der säugetier—und Vogelfauna der Ostküste Grönlands. Medd. om Grønland, vol. 77, pp. 341–507.
PEET, MAX MINOR.
1908. Annotated list of the birds of Isle Royale, Michigan. Ecology of Isle Royale. Michigan Survey, 1908, pp. 337–386.
PERRIOR, A. W.
1899. Food of the robin. Auk, vol. 16, p. 284.
PERRY, GEORGE P.
1908. Nest of wood thrush into which a cowbird had deposited five eggs. Bird-Lore, vol. 10, pp. 126–127.
PETERS, HAROLD SEYMORE.
1933. External parasites collected from banded birds. Bird-Banding, vol. 4, pp. 68–75.
1936. A list of external parasites from birds of the eastern part of the United States. Bird-Banding, vol. 7, pp. 9–27.
PETTINGILL, OLIN SEWALL, JR.
1930. Observations on the nesting activities of the hermit thrush. Bird-Banding, vol. 1, pp. 72–77.
PHILLIPS, JOHN CHARLES.
1927. Catbirds and robins as fish-eaters. Bird-Lore, vol. 29, pp. 342–343.
PITELKA, FRANK ALOIS.
1941. Foraging behavior in the western bluebird. Condor, vol. 43, pp. 198–199.
PRICE, JOHN BASYE.
1933. Winter behavior of two semi-albino western robins. Condor, vol. 35, pp. 52–54.
PROVOST, MAURICE W.
1939. Altitudinal distribution of birds in the White Mountains and its application. Bull. Audubon Soc. New Hampshire, vol. 14, pp. 3–7.
PUTNAM, F. W. (editor), and WHEATLAND, H. (secretary).
1866. [Note on bluebirds nesting.] Proc. Essex Inst., vol. 4, p. cxlix.
RACEY, KENNETH.
1939. Western bluebird notes. Can. Field-Nat., vol. 53, p. 30.
RANEY, EDWARD C.
1939. Robin and mourning dove use the same nest. Auk, vol. 56, pp. 337–338.
REINHARDT, JOHANNES.
1861. List of the birds hitherto observed in Greenland. Ibis, vol. 3, pp. 1–19.
REISER, OTMAR.
1894, 1896, 1905. Materialen zu einer Ornis Balcanica, vols. 2–3.
RIDGWAY, ROBERT.
1877. United States geological exploration of the fortieth parallel, pt. 3: Ornithology.
1894. On geographic variation in *Sialia mexicana* Swainson. Auk, vol. 11, pp. 145–160.
1904. The birds of North and Middle America. U. S. Nat. Mus. Bull. 50, pt. 3.
1907. The birds of North and Middle America. U. S. Nat. Mus. Bull. 50, pt. 4.
1912. Color standards and color nomenclature.

RILEY, JOSEPH HARVEY.
1918. Annotated catalogue of a collection of birds made by Mr. Copley Amory Jr., in northeastern Siberia. Proc. U. S. Nat. Mus., vol. 54, pp. 607–626.
ROBERTS, THOMAS SADLER.
1936. The birds of Minnesota, ed. 2, vol. 2.
ROBINSON, HERBERT CHRISTOPHER.
1927. The birds of the Malay Peninsula, vol. 1.
ROHRBACH, J. H.
1915. Plaster for the robin's nest. Bird-Lore, vol. 17, pp. 212–213.
ROOT, OSCAR MITCHELL.
1942. Wood thrush nesting in the coniferous bogs of Canadian Zone. Auk, vol. 59, pp. 113–114.
ROWLEY, JOHN STUART.
1939. Breeding birds of Mono County, California. Condor, vol. 41, pp. 247–254.
RYVES, B. H.
1943. An investigation into the roles of males in relation to incubation. British Birds, vol. 37, pp. 10–16.
SAGE, JOHN HALL.
1885. Return of robins to the same nesting-places. Auk, vol. 2, p. 304.
1886. A partial albino hermit thrush. Auk, vol. 3, p. 282.
SALVIN, OSBERT.
1888. A list of the birds of the islands of the coast of Yucatan and of the Bay of Honduras. Ibis, 1888, pp. 241–265.
SAUNDERS, ARETAS ANDREWS.
1916. A note on the food of the western robin. Condor, vol. 18, p. 81.
1921a. The song periods of individual birds. Auk, vol. 38, pp. 283–284.
1921b. A distributional list of the birds of Montana. Pacific Coast Avifauna, No. 14.
1924. Recognizing individual birds by song. Auk, vol. 41, pp. 242–259.
1929a. The summer birds of the northern Adirondack Mountains. Roosevelt Wild Life Bull., vol. 5, pp. 323–490, figs. 93–160.
1929b. Bird song. New York State Mus. Handbook, No. 7.
1938. Studies of breeding birds in the Allegany State Park. New York State Mus. Bull. No. 318.
SAUNDERS, WILLIAM EDWIN.
1887. The bluebird as a mimic. Ornithologist and Oologist, vol. 12, pp. 61–62.
1907. A migration disaster in western Ontario. Auk, vol. 24, pp. 108–110.
SAXBY, HENRY LINCKMYER.
1874. The birds of Shetland.
SCHANTZ, WILLIAM EDWARD.
1939. A detailed study of a family of robins. Wilson Bull., vol. 51, pp. 157–169.
SCHEEL, H.
1927. Nathejre, *Nycticorax griseus*, og sjagger, *Turdus pilaris*, nue for Grønland. Dansk Orn. Foren. Tidskr., Aagr. 21, pp. 76–81.
SCHIØLER, EILER THEODOR LEHN.
1926. Danmarks Fugle.
SCLATER, WILLIAM LUTLEY.
1912. A history of the birds of Colorado.
SCOTT, WILLIAM EARL DODGE.
1888. On the avi-fauna of Pinal County, with remarks on some birds of Pima and Gila Counties, Arizona. Auk, vol. 5, pp. 159–168.

SEEBOHM, HENRY.
1879. Contributions to the ornithology of Siberia (third part). Ibis, 1879, pp. 1-18.
1881. Catalogue of the birds in the British Museum, vol. 5.
1901. The birds of Siberia.
SEEBOHM, HENRY, and HARVIE-BROWN, JOHN ALEXANDER.
1876. Notes on the birds of the Lower Petchora. Ibis, 1876, pp. 105-126.
SELOUS, EDMUND.
1901. Bird-watching.
SETON, ERNEST THOMPSON.
1911. The Arctic prairies.
SEVERSON, H. P.
1921. Robin's nest on a trolley wire. Bird-Lore, vol. 23, p. 249.
SHAW, TSEN-HWANG.
1936. The birds of Hopei Province. Zoologica Sinica, ser. b. The vertebrates of China, vol. 15, fasc. 2, pp. 529-974, figs. 304-506, pls. 14-25.
SHELDON, HARRY H.
1908. Three nests of note from northern California. Condor, vol. 10, pp. 120-124.
SHELLEY, LEWIS ORMAN.
1930. Notes on the feeding reactions of some spring birds during a late snow storm. Auk, vol. 47, pp. 100-101.
SHERMAN, ALTHEA ROSINA.
1912. Robin (*Planesticus migratorius migratorius*). Wilson Bull., vol. 24, pp. 50-51.
SILLOWAY, PERLEY MILTON.
1901. Summer birds of Flathead Lake. Bull. Univ. Montana, No. 3.
SIMMONS, GEORGE FINLAY.
1925. Birds of the Austin region.
SKINNER, MILTON PHILO.
1928. A guide to the winter birds of the North Carolina sandhills.
SMITH, WENDELL PHILLIPS.
1937. Some bluebird observations. Bird-Banding, vol. 8, pp. 25-30.
SMITH, WILBUR FRANKLIN.
1920. Some robins' nests. Bird-Lore, vol. 22, pp. 147-150.
SOPER, JOSEPH DEWEY.
1920. Nesting of the ruby-crowned kinglet at Guelph, Ontario. Can. Field-Nat., vol. 34, pp. 72-73.
SPINDLER, ETHEL M.
1933. A true snake-story. Bird-Lore, vol. 35, p. 323.
SPROT, G. D.
1926. Mouse eaten by a robin. Murrelet, vol. 7, No. 3, p. 65.
STANWOOD, CORDELIA JOHNSON.
1910. The hermit thrush; the voice of the northern woods. Bird-Lore, vol. 12, pp. 100-103.
1913. The olive-backed thrush (*Hylocichla ustulata swainsoni*) at his summer home. Wilson Bull., vol. 25, pp. 118-137.
STATON JACK.
1941. Display in blackbirds. British Birds, vol. 35, p. 107.
STEJNEGER, LEONHARD.
1883. Contributions to the history of the Commander Islands. No. 1.—Notes on the natural history, including descriptions of new cetaceans. Proc. U. S. Nat. Mus., vol. 6, pp. 58-89.

1885. Results of ornithological explorations in the Commander Islands and Kamtschatka. U. S. Nat. Mus. Bull. 29.

1901. On the wheatears (*Saxicola*) occurring in North America. Proc. U. S. Nat. Mus., vol. 23, pp. 473–481.

STOCKARD, CHARLES RUPERT.

1905. Nesting habits of birds in Mississippi. Auk, vol. 22, pp. 273–285.

STONE, D. D.

1884. Colorado notes. Ornithologist and Oologist, vol. 9, pp. 20–21.

STONER, DAYTON.

1932. Ornithology of the Oneida Lake region: With reference to the late spring and summer seasons. Roosevelt Wild Life Ann., vol. 2, Nos. 3, 4.

STORER, TRACY IRWIN.

1923. English blackbird in California. Condor, vol. 25, pp. 67–68.

1926. Range extensions by the western robin in California. Condor, vol. 28, pp. 264–267.

STOVER, A. J.

1912. A robin's roost. Wilson Bull., vol. 24, pp. 169–171.

STRICKLAND, LAURA RAYMOND.

1934. A timely intervention. Bird-Lore, vol. 36, p. 239.

SUCKLEY, GEORGE, and COOPER, JAMES GRAHAM.

1860. The natural history of Washington Territory and Oregon.

SWAINSON, WILLIAM, and RICHARDSON, JOHN.

1831. Fauna Boreali-Americana, vol. 2. The birds.

SWARTH, HARRY SCHELWALD.

1902. Winter plumage of the black-tailed gnatcatcher. Condor, vol. 4, pp. 86–87.

1904. Birds of the Huachuca Mountains, Arizona. Pacific Coast Avifauna, No. 4.

1914. A distributional list of the birds of Arizona. Pacific Coast Avifauna, No. 10.

1922. Birds and mammals of the Stikine River region of northern British Columbia and Southeastern Alaska. Univ. California Publ. Zool., vol. 24, pp. 125–314.

SWARTH, HARRY SCHELWARD—Continued

1926. Report on a collection of birds and mammals from the Atlin region, northern British Columbia. Univ. California Publ. Zool., vol. 30, pp. 51–162.

1928. Occurrence of some Asiatic birds in Alaska. Proc. California Acad. Sci., ser. 4, vol. 17, pp. 247–251.

1934. Birds of Nunivak Island, Alaska. Pacific Coast Avifauna, No. 22.

SWEET, DANA W.

1906. Bicknell's thrush on Mt. Abraham. Journ. Maine Orn. Soc., vol. 3, pp. 81–84.

SWINBURNE, JOHN.

1890. The nest and eggs of *Regulus calendula*. Auk, vol. 7, pp. 97–98.

TACZANOWSKI, LADISLAS.

1872–1873. Bericht über die ornithologischen Untersuchungen der Dr. Dybowski in Ost-Siberien. Journ. für Orn., Jahrg. 20, pp. 340–366, 433–454; Jahrg. 21, pp. 81–119.

1882. Liste des oiseaux recueillis par le Dr. Dybowski au Kamtschatka et dans les îles Comandores.

1891. Faune ornithologique de la Sibérie Orientale. Mém. Acad. Impériale Sci. St. Pétersbourg, ser. 7, vol. 39.

TATE, RALPH C.

1926. Some materials used in nest construction by certain birds of the Oklahoma Panhandle. Univ. Oklahoma Bull., vol. 5, pp. 103–104.

TAVERNER, PERCY ALGERNON, and SWALES, BRADSHAW HALL.

1908. The birds of Point Pelee. Wilson Bull., vol. 20, pp. 107–129.

1940. Fieldfare, an addition to the American list, and some Arctic notes. Auk, vol. 57, p. 119.

TAYLOR, WALTER PENN.

1912. Field notes on amphibians, reptiles and birds of northern Humboldt County, Nevada, with a discussion of some of the faunal features of the region. Univ. California Publ. Zool., vol. 7, pp. 319–436.

TAYLOR, W. P., and SHAW, WILLIAM THOMAS.

1927. Mammals and birds of Mount Rainier National Park.

TESS, STANLEY.

1926. A plucky robin. Bird-Lore, vol. 28, pp. 202–203.

THAYER, JOHN ELIOT.

1911. Eggs and nest of the San Lucas robin. Oologist, vol. 28, p. 77.

THAYER, J. E., and BANGS, OUTRAM.

1914. Notes on the birds and mammals of the Arctic coast of east Siberia. Birds. Proc. New England Zool. Club, vol. 5, pp. 1–48.

THOMAS, RUTH HARRIS.

1946. A study of eastern bluebirds in Arkansas. Wilson Bull., vol. 58 pp. 143–183.

THOMS, CRAIG S.

1929. A robin's strange nest. Bird-Lore, vol. 31, pp. 118–119.

TICEHURST, CLAUD BUCHANAN.

1925. [Exhibition and remarks on redwings from Iceland.] Bull. British Orn. Club, vol. 45, pp. 90–91.

1938. A systematic review of the genus *Phylloscopus*.

TICEHURST, NORMAN FREDERIC.

1938. The handbook of British birds, vol. 1, 2. (By H. F. Witherby, F. C. R. Jourdain, Norman F. Ticehurst, and Bernard W. Tucker.)

TIMMERMANN, GÜNTER.

1934. Die rotdrossel (*Turdus musicus coburni* Sharpe) als Stadtvogel in Südwest-Island. Journ. für Orn., Jahrg. 82, pp. 319–324.

TODD, WALTER EDMOND CLYDE.

1940. Birds of western Pennsylvania.

TORREY, BRADFORD.

1885. Birds in the bush.

1892. The foot-path way.

TOWNSEND, CHARLES HASKINS.

1887. Notes on the natural history and ethnology of northern Alaska. Report of the cruise of the revenue marine steamer *Corwin* in the Arctic Ocean in the year 1885.

TOWNSEND, CHARLES WENDELL.

1905. The birds of Essex County, Massachusetts. Mem. Nuttall Orn. Club, No. 3.

1909. Robins used same nest six seasons. Journ. Maine Orn. Soc., vol. 11 pp. 30–31.

1923. Thrush killed by red-squirrel. Auk, vol. 40, p. 135.

1924. Mimicry of voice in birds. Auk, vol. 41, pp. 541–552.

Townsend, Charles Wendell, and Allen, Glover Morrill.
1907. Birds of Labrador. Proc. Boston Soc. Nat. Hist., vol. 33, pp. 277–428.
Trafton, Gilbert Haven.
1907. Robins nesting in bird-houses. Bird-Lore, vol. 9, p. 270.
Trautman, Milton Bernard.
1940. The birds of Buckeye Lake, Ohio. Mus. Zool. Univ. Michigan Misc. Publ. 44.
Tucker, Bernard William.
1938. The handbook of British birds, vol. 2. (By H. F. Witherby, F. C. R. Jourdain, Norman F. Ticehurst, and B. W. Tucker.)
Turner, Emma Louisa.
1911. The wheatear. In "The British Bird Book," vol. 1, pp. 398–407.
Tyler, John Gripper.
1913. Some birds of the Fresno District, California. Pacific Coast Avifauna, No. 9.
Tyler, Winsor Marrett.
1913. A successful pair of robins. Auk. vol, 30, pp. 394–398.
1916. The call-notes of some nocturnal migrating birds. Auk, vol. 33, pp. 132–141.
Vaiden, Meredith Gordon.
1940. Interesting Mississippi birds. Migrant, vol. 11, pp. 66–68.
van Rossem, Adriaan Joseph.
1931. Concerning some western races of *Polioptila melanura*. Condor, vol. 33, pp. 35–36.
1942. A note on the western robin in the Pasadena area. Condor, vol. 44, p. 130.
Van Tyne, Josselyn, and Sutton, George Miksch.
1937. The birds of Brewster County, Texas. Mus. Zool. Univ. Michigan Misc. Publ. 37.
Vaughan, Robert E., and Jones, Kenneth Hurlstone.
1913. The birds of Hong Kong, Macao and the West River or Si Kiang in South-East China, with special reference to their nidification and seasonal movements. Ibis, ser. 10, vol. 1, pp. 17–76.
Wallace, George John.
1939. Bicknell's thrush, its taxonomy, distribution, and life history. Proc. Boston Soc. Nat. Hist., vol. 41, pp. 211–402.
Walpole-Bond, John.
1938. A history of Sussex birds, vol. 2.
Walton, Herbert A.
1903. Notes on the birds of Peking. Ibis, ser. 8, vol. 3, pp. 19–35.
Wayne, Arthur Trezevant.
1910. Birds of South Carolina. Contr. Charleston Mus., No. 1.
Weston, Francis Marion.
1935. Pensacola (Florida) region. The season. Bird-Lore, vol. 37, pp. 467–468.
Wetmore, Alexander.
1936. The number of contour feathers in passeriform and related birds. Auk, vol. 53, pp. 159–169.
1937. Observations on the birds of West Virginia. Proc. U. S. Nat. Mus., vol. 84, pp. 401–441.
1940. Notes on the birds of Kentucky, Proc. U. S. Nat. Mus., vol. 88, pp. 529–574.

WEYDEMEYER, WINTON.
1923. Notes on the song and the nest of the ruby-crowned kinglet. Condor, vol. 25, pp. 117–118.
1934a. The song of the mountain bluebird. Condor, vol. 36, p. 164.
1934b. Singing of the mountain bluebird and the western bluebird. Condor, vol. 36, pp. 249–251.
WHARTON, WILLIAM PICKMAN.
1941. Twelve years of banding at Summerville, S. C. Bird-Banding, vol. 12, pp. 137–147.
WHEATLAND, H. (See under F. W. Putnam.)
WHEELOCK, IRENE GROSVENOR.
1904. Birds of California.
WHEELWRIGHT, HORACE WILLIAM ("An Old Bushman").
1871. A spring and summer in Lapland.
WHISTLER, HUGH.
1928. Popular handbook of Indian birds.
WHITE, FRANCIS BEACH.
1937. Local notes on the birds at Concord, New Hampshire.
WHITE, STEWART EDWARD.
1893. Birds observed on Mackinac Island, Michigan, during the summers of 1889, 1890, and 1891. Auk, vol. 10, pp. 221–230.
WHITEHEAD, JOHN.
1899. Field-notes on birds collected in the Philippine Islands in 1893–6. Ibis, ser. 7, vol. 5, pp. 81–111.
WHITTLE, CHARLES LIVY.
1922. Miscellaneous bird notes from Montana. Condor, vol. 24, pp. 73–81.
WIDMAN, OTTO.
1895. A winter robin roost in Missouri, and other ornithological notes. Auk, vol. 12, pp. 1–11.
1907. A preliminary catalogue of the birds of Missouri.
WILSON, ALEXANDER, and BONAPARTE, CHARLES LUCIAN.
1832. American ornithology, vol. 1.
WITHERBY, HARRY FORBES.
1920. A practical handbook of British birds, vol. 1.
1938. The handbook of British birds, vols. 1, 2. (By H. F. Witherby, F. C. R. Jourdain, Norman F. Ticehurst, and Bernard W. Tucker.)
WOOD, A. H.
1884. Tameness of the golden-crested kinglet (*Regulus satrapa*). Ornithologist and Oologist, vol. 9, p. 62.
WOOD, HAROLD BACON.
1937. Robin strangled by a grass snake. Bird-Lore, vol. 39, p. 311.
WRIGHT, HORACE WINSLOW.
1909. Birds of the Boston Public Garden.
1912. Morning awakening and even-song. Auk, vol. 29, pp. 307–327.
1920. Hermit thrush's nest in unusual location. Auk, vol. 37, p. 138.
YAMASHINA, MARQUIS YOSHIMARO.
1931. Die Vögel der Kurilen. Journ. für Orn., Jahrg. 79, pp. 491–541.
YARRELL, WILLIAM.
1874. A history of British birds, ed. 4, vol. 1. Edited by Alfred Newton.
YOUNGWORTH, WILLIAM.
1929. A city robin roost. Wilson Bull., vol. 41, pp. 105–106.
ZIEMER, EWALD.
1887. Ornithologische Beobachtungen. Monatsschrift deutschen Vereins Schutze Vogelwelt. Vol. 12, pp. 297–300.

INDEX

○

Norton, Mass., May 5, 1944. Grice & Grice.

EASTERN ROBIN AT ITS NEST.

Near Toronto, Ontario. H. M. Halliday.

Nest Used for 6 Years, Two Broods a Year.

EASTERN ROBIN.

Saskatchewan, Canada. H. O. Todd, Jr.

NEST OF EASTERN ROBIN.

Marthas Vineyard, Mass., May 26, 1901. Owen Durfee.

Nest in an Eel Trap.

Fall River, Mass., May 23, 1900. Owen Durfee.

Very Low Nest.

NESTING OF EASTERN ROBIN.

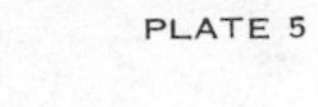

PLATE 5

Toronto, Ontario. H. M. Halliday.

AN ALBINO ROBIN AND ITS MATE.

Coeur d'Alene, Idaho, May 29, 1919. H. J. Rust.

Near Spencer, Idaho, June 11, 1915. H. J. Rust.

YOUNG WESTERN ROBIN AND NEST.

Mulino, Oreg., May 4, 1912. Alexander Walker.

Arizona. F. C. Willard.

NESTS OF WESTERN ROBINS.

W. L. Finley.

Wintering in Oregon.

S. F. Rathbun.

Haunts in Western Washington.

PACIFIC VARIED THRUSH.

Duval County, Fla., May 1942. S. A. Grimes.

WOOD-THRUSH NEST IN POISON-IVY VINE.

Hennepin County, Minn., June 12, 1940. A. D. DuBois.

PAIR OF WOOD THRUSHES.

Duval County, Fla., April 27, 1933. S. A. Grimes.

NEST OF WOOD THRUSH.

Erie County, N. Y., June 1928. S. A. Grimes.

NEST OF WOOD THRUSH.

Mono County, Calif., July 4, 1939. J. S. Rowley. Washington. W. L. Finley.

SIERRA HERMIT THRUSH AND NESTS.

Near Toronto, Ontario. H. M. Halliday.

EASTERN HERMIT THRUSH.

Ellsworth Maine. Cordelia J. Stanwood

Douglas Lake, Mich., June 30, 1928. A. O. Gross.

NESTS OF EASTERN HERMIT THRUSH.

Douglas Lake, Mich., July 13, 1928. A. O. Gross.

Brunswick, Maine, May 31, 1920. A. O. Gross.

EASTERN HERMIT THRUSH

Douglas Lake, Mich., July 19, 1928. A. O. Gross.

Young Cowbird 7 Days Old; Young Thrush 6 Days Old.

Ellsworth, Maine. Cordelia J. Stanwood.

EASTERN HERMIT THRUSH.

Santa Clara County, Calif. Gayle Pickwell.

San Mateo, Calif., May 17· 1931 J. S. Rowley.

RUSSET-BACKED THRUSH AND NESTS

Oregon. W. L. Finley.

Tillamook, Oreg., July 4, 1915. Alexander Walker.

NESTS OF RUSSET-BACKED THRUSH.

Hog Island, Maine.

A. D. Cruickshank.

OLIVE-BACKED THRUSH.

Ellsworth, Maine — Cordelia J. Stanwood.

OLIVE-BACKED THRUSH.

Ellsworth, Maine. Cordelia J. Stanwood.

Price County, Wis., June 19, 1912. A. D. DuBois.

OLIVE-BACKED THRUSH.

Askinuk Mountains, Alaska, June 27, 1924. Herbert Brandt.

NEST OF GRAY-CHEEKED THRUSH.

Mount Washington, N. H., June 26, 1908. Owen Durfee.

Seal Island, Nova Scotia, July 6, 1904. A. C. Bent.

NESTING SITES OF BICKNELL'S THRUSH.

Nest in Blackberry Bush.

Erie County, N. Y., June 10, 1927. S. A. Grimes.

Nest on Ground.

VEERY.

Asquam Lake, N. H., June 26, 1927 A. C. Bent.

Young.

Allen Frost

Adult at Nest.

VEERY.

Hennepin County, Minn., June 11, 1935. A. D. DuBois.

WILLOW THRUSH.

Hennepin County, Minn., June 7, 1935. A. D. DuBois.

NEST OF WILLOW THRUSH.

Toronto, Ontario. H. M. Halliday

Young. Adult.

EASTERN BLUEBIRD.

Logan County, Ill., June 19, 1913. A. D. DuBois.

Nest in a Fence Post.

Hennepin County, Minn., June 25, 1936. A. D. DuBois.

Nest in a Box.

EASTERN BLUEBIRD.

Duval County, Fla., June 23, 1932. S. A. Grimes.

NEST OF FLORIDA BLUEBIRD

Clinton, Iowa, June 1937. M. L. Miles.

EASTERN BLUEBIRD AND NESTING BOX.

Los Angeles County, Calif., June 16, 1942. James Murdock.

WESTERN BLUEBIRD AND NESTING SITE IN AN OLD LOG.

Near Spencer, Idaho. H. J. Rust.

NESTING SITES OF MOUNTAIN BLUEBIRD.

Near Spencer, Idaho. H. J. Rust.

YOUNG MOUNTAIN BLUEBIRDS.

Altitude 6,000 feet. E. N. Harrison.

San Bernardino Mountains, Calif., June 20, 1916. W. M. Pierce.

NESTS OF TOWNSEND'S SOLITAIRE.

Pinehurst, Oreg., May 12, 1924.

J. E. Patterson.

NEST OF TOWNSEND'S SOLITAIRE.

Sequoia National Park, Calif. Gayle Pickwell.

TOWNSEND'S SOLITAIRE.

Duval County, Fla., May 1931. S. A. Grimes.

BLUE-GRAY GNATCATCHERS.

Duval County, Fla., May 1930. S. A. Grimes.

BLUE-GRAY GNATCATCHERS.

Duval County Fla., May 12, 1933. S. A. Grimes.

Washington, D. C., May 8, 1938 G. A. Petrides.

NESTS OF BLUE-GRAY GNATCATCHER.

Calhoun County, Mich., June 18, 1933.

L. H. Walkinshaw.

BLUE-GRAY GNATCATCHER AND NEST.

Near Seneca, Md., June 1941. R. E. Lawrence.

YOUNG BLUE-GRAY GNATCATCHERS.

Huachuca Mountains, Ariz., May 14, 1922. A. C. Bent.

NEST OF WESTERN GNATCATCHER.

Arizona. F. C. Willard.

Yuma County, Ariz., May 30, 1942. J. S. Rowley.

NESTS OF PLUMBEOUS GNATCATCHER.

Azusa, Calif., June 30, 1921. R. S. Woods.

Habitat on San Gabriel Wash.

Imperial County, Calif., April 11, 1942. E. N. Harrison.

NESTING OF BLACK-TAILED GNATCATCHER.

Ellsworth, Maine. Cordelia J. Stanwood.

NEST OF EASTERN GOLDEN-CROWNED KINGLET.

Ellsworth, Maine. Cordelia J. Stanwood.

YOUNG AND NEST OF EASTERN GOLDEN-CROWNED KINGLET.

St. Marys River, Mich., July 5, 1925. K. Christofferson.

Nest and Young.

Jacksonville, Fla., March 7, 1943. S. A. Grimes.

Drinking Sap.

EASTERN RUBY-CROWNED KINGLET.

Coeur d'Alene, Idaho. H. J. Rust.

NESTING HABITAT OF WESTERN RUBY-CROWNED KINGLET.

Coeur d'Alene, Idaho. H. J. Rust.

NEST OF WESTERN RUBY-CROWNED KINGLET.

A CATALOGUE OF SELECTED DOVER BOOKS IN ALL FIELDS OF INTEREST

A CATALOGUE OF SELECTED DOVER BOOKS IN ALL FIELDS OF INTEREST

YUCATAN BEFORE AND AFTER THE CONQUEST, Diego de Landa. First English translation of basic book in Maya studies, the only significant account of Yucatan written in the early post-Conquest era. Translated by distinguished Maya scholar William Gates. Appendices, introduction, 4 maps and over 120 illustrations added by translator. 162pp. 5⅜ x 8½.
23622-6 Pa. $3.00

THE MALAY ARCHIPELAGO, Alfred R. Wallace. Spirited travel account by one of founders of modern biology. Touches on zoology, botany, ethnography, geography, and geology. 62 illustrations, maps. 515pp. 5⅜ x 8½.
20187-2 Pa. $6.95

THE DISCOVERY OF THE TOMB OF TUTANKHAMEN, Howard Carter, A. C. Mace. Accompany Carter in the thrill of discovery, as ruined passage suddenly reveals unique, untouched, fabulously rich tomb. Fascinating account, with 106 illustrations. New introduction by J. M. White. Total of 382pp. 5⅜ x 8½. (Available in U.S. only) 23500-9 Pa. $4.00

THE WORLD'S GREATEST SPEECHES, edited by Lewis Copeland and Lawrence W. Lamm. Vast collection of 278 speeches from Greeks up to present. Powerful and effective models; unique look at history. Revised to 1970. Indices. 842pp. 5⅜ x 8½. 20468-5 Pa. $8.95

THE 100 GREATEST ADVERTISEMENTS, Julian Watkins. The priceless ingredient; His master's voice; 99 44/100% pure; over 100 others. How they were written, their impact, etc. Remarkable record. 130 illustrations. 233pp. 7⅞ x 10 3/5. 20540-1 Pa. $5.00

CRUICKSHANK PRINTS FOR HAND COLORING, George Cruickshank. 18 illustrations, one side of a page, on fine-quality paper suitable for watercolors. Caricatures of people in society (c. 1820) full of trenchant wit. Very large format. 32pp. 11 x 16. 23684-6 Pa. $5.00

THIRTY-TWO COLOR POSTCARDS OF TWENTIETH-CENTURY AMERICAN ART, Whitney Museum of American Art. Reproduced in full color in postcard form are 31 art works and one shot of the museum. Calder, Hopper, Rauschenberg, others. Detachable. 16pp. 8¼ x 11.
23629-3 Pa. $2.50

MUSIC OF THE SPHERES: THE MATERIAL UNIVERSE FROM ATOM TO QUASAR SIMPLY EXPLAINED, Guy Murchie. Planets, stars, geology, atoms, radiation, relativity, quantum theory, light, antimatter, similar topics. 319 figures. 664pp. 5⅜ x 8½.
21809-0, 21810-4 Pa., Two-vol. set $10.00

EINSTEIN'S THEORY OF RELATIVITY, Max Born. Finest semi-technical account; covers Einstein, Lorentz, Minkowski, and others, with much detail, much explanation of ideas and math not readily available elsewhere on this level. For student, non-specialist. 376pp. 5⅜ x 8½.
60769-0 Pa. $4.00

THE COMPLETE BOOK OF DOLL MAKING AND COLLECTING, Catherine Christopher. Instructions, patterns for dozens of dolls, from rag doll on up to elaborate, historically accurate figures. Mould faces, sew clothing, make doll houses, etc. Also collecting information. Many illustrations. 288pp. 6 x 9. 22066-4 Pa. $4.00

THE DAGUERREOTYPE IN AMERICA, Beaumont Newhall. Wonderful portraits, 1850's townscapes, landscapes; full text plus 104 photographs. The basic book. Enlarged 1976 edition. 272pp. 8¼ x 11¼. 23322-7 Pa. $6.00

CRAFTSMAN HOMES, Gustav Stickley. 296 architectural drawings, floor plans, and photographs illustrate 40 different kinds of "Mission-style" homes from *The Craftsman* (1901-16), voice of American style of simplicity and organic harmony. Thorough coverage of Craftsman idea in text and picture, now collector's item. 224pp. 8⅛ x 11. 23791-5 Pa. $6.00

PEWTER-WORKING: INSTRUCTIONS AND PROJECTS, Burl N. Osborn. & Gordon O. Wilber. Introduction to pewter-working for amateur craftsman. History and characteristics of pewter; tools, materials, step-by-step instructions. Photos, line drawings, diagrams. Total of 160pp. 7⅞ x 10¾. 23786-9 Pa. $3.50

THE GREAT CHICAGO FIRE, edited by David Lowe. 10 dramatic, eyewitness accounts of the 1871 disaster, including one of the aftermath and rebuilding, plus 70 contemporary photographs and illustrations of the ruins—courthouse, Palmer House, Great Central Depot, etc. Introduction by David Lowe. 87pp. 8¼ x 11. 23771-0 Pa. $4.00

SILHOUETTES: A PICTORIAL ARCHIVE OF VARIED ILLUSTRATIONS, edited by Carol Belanger Grafton. Over 600 silhouettes from the 18th to 20th centuries include profiles and full figures of men and women, children, birds and animals, groups and scenes, nature, ships, an alphabet. Dozens of uses for commercial artists and craftspeople. 144pp. 8⅜ x 11¼. 23781-8 Pa. $4.00

ANIMALS: 1,419 COPYRIGHT-FREE ILLUSTRATIONS OF MAMMALS, BIRDS, FISH, INSECTS, ETC., edited by Jim Harter. Clear wood engravings present, in extremely lifelike poses, over 1,000 species of animals. One of the most extensive copyright-free pictorial sourcebooks of its kind. Captions. Index. 284pp. 9 x 12. 23766-4 Pa. $7.50

INDIAN DESIGNS FROM ANCIENT ECUADOR, Frederick W. Shaffer. 282 original designs by pre-Columbian Indians of Ecuador (500-1500 A.D.). Designs include people, mammals, birds, reptiles, fish, plants, heads, geometric designs. Use as is or alter for advertising, textiles, leathercraft, etc. Introduction. 95pp. 8¾ x 11¼. 23764-8 Pa. $3.50

SZIGETI ON THE VIOLIN, Joseph Szigeti. Genial, loosely structured tour by premier violinist, featuring a pleasant mixture of reminiscenes, insights into great music and musicians, innumerable tips for practicing violinists. 385 musical passages. 256pp. 5⅝ x 8¼. 23763-X Pa. $3.50

TONE POEMS, SERIES II: TILL EULENSPIEGELS LUSTIGE STREICHE, ALSO SPRACH ZARATHUSTRA, AND EIN HELDENLEBEN, Richard Strauss. Three important orchestral works, including very popular *Till Eulenspiegel's Marry Pranks,* reproduced in full score from original editions. Study score. 315pp. 9⅜ x 12¼. (Available in U.S. only)
23755-9 Pa. $7.50

TONE POEMS, SERIES I: DON JUAN, TOD UND VERKLARUNG AND DON QUIXOTE, Richard Strauss. Three of the most often performed and recorded works in entire orchestral repertoire, reproduced in full score from original editions. Study score. 286pp. 9⅜ x 12¼. (Available in U.S. only)
23754-0 Pa. $7.50

11 LATE STRING QUARTETS, Franz Joseph Haydn. The form which Haydn defined and "brought to perfection." *(Grove's)*. 11 string quartets in complete score, his last and his best. The first in a projected series of the complete Haydn string quartets. Reliable modern Eulenberg edition, otherwise difficult to obtain. 320pp. 8⅜ x 11¼. (Available in U.S. only)
23753-2 Pa. $6.95

FOURTH, FIFTH AND SIXTH SYMPHONIES IN FULL SCORE, Peter Ilyitch Tchaikovsky. Complete orchestral scores of Symphony No. 4 in F Minor, Op. 36; Symphony No. 5 in E Minor, Op. 64; Symphony No. 6 in B Minor, "Pathetique," Op. 74. Bretikopf & Hartel eds. Study score. 480pp. 9⅜ x 12¼.
23861-X Pa. $10.95

THE MARRIAGE OF FIGARO: COMPLETE SCORE, Wolfgang A. Mozart. Finest comic opera ever written. Full score, not to be confused with piano renderings. Peters edition. Study score. 448pp. 9⅜ x 12¼. (Available in U.S. only)
23751-6 Pa. $11.95

"IMAGE" ON THE ART AND EVOLUTION OF THE FILM, edited by Marshall Deutelbaum. Pioneering book brings together for first time 38 groundbreaking articles on early silent films from *Image* and 263 illustrations newly shot from rare prints in the collection of the International Museum of Photography. A landmark work. Index. 256pp. 8¼ x 11.
23777-X Pa. $8.95

AROUND-THE-WORLD COOKY BOOK, Lois Lintner Sumption and Marguerite Lintner Ashbrook. 373 cooky and frosting recipes from 28 countries (America, Austria, China, Russia, Italy, etc.) include Viennese kisses, rice wafers, London strips, lady fingers, hony, sugar spice, maple cookies, etc. Clear instructions. All tested. 38 drawings. 182pp. 5⅜ x 8.
23802-4 Pa. $2.50

THE ART NOUVEAU STYLE, edited by Roberta Waddell. 579 rare photographs, not available elsewhere, of works in jewelry, metalwork, glass, ceramics, textiles, architecture and furniture by 175 artists—Mucha, Seguy, Lalique, Tiffany, Gaudin, Hohlwein, Saarinen, and many others. 288pp. 8⅜ x 11¼.
23515-7 Pa. $6.95

THE AMERICAN SENATOR, Anthony Trollope. Little known, long unavailable Trollope novel on a grand scale. Here are humorous comment on American vs. English culture, and stunning portrayal of a heroine/villainess. Superb evocation of Victorian village life. 561pp. 5⅜ x 8½.
23801-6 Pa. $6.00

WAS IT MURDER? James Hilton. The author of *Lost Horizon* and *Goodbye, Mr. Chips* wrote one detective novel (under a pen-name) which was quickly forgotten and virtually lost, even at the height of Hilton's fame. This edition brings it back—a finely crafted public school puzzle resplendent with Hilton's stylish atmosphere. A thoroughly English thriller by the creator of Shangri-la. 252pp. 5⅜ x 8. (Available in U.S. only)
23774-5 Pa. $3.00

CENTRAL PARK: A PHOTOGRAPHIC GUIDE, Victor Laredo and Henry Hope Reed. 121 superb photographs show dramatic views of Central Park: Bethesda Fountain, Cleopatra's Needle, Sheep Meadow, the Blockhouse, plus people engaged in many park activities: ice skating, bike riding, etc. Captions by former Curator of Central Park, Henry Hope Reed, provide historical view, changes, etc. Also photos of N.Y. landmarks on park's periphery. 96pp. 8½ x 11. 23750-8 Pa. $4.50

NANTUCKET IN THE NINETEENTH CENTURY, Clay Lancaster. 180 rare photographs, stereographs, maps, drawings and floor plans recreate unique American island society. Authentic scenes of shipwreck, lighthouses, streets, homes are arranged in geographic sequence to provide walking-tour guide to old Nantucket existing today. Introduction, captions. 160pp. 8⅞ x 11¾. 23747-8 Pa. $6.95

STONE AND MAN: A PHOTOGRAPHIC EXPLORATION, Andreas Feininger. 106 photographs by *Life* photographer Feininger portray man's deep passion for stone through the ages. Stonehenge-like megaliths, fortified towns, sculpted marble and crumbling tenements show textures, beauties, fascination. 128pp. 9¼ x 10¾. 23756-7 Pa. $5.95

CIRCLES, A MATHEMATICAL VIEW, D. Pedoe. Fundamental aspects of college geometry, non-Euclidean geometry, and other branches of mathematics: representing circle by point. Poincare model, isoperimetric property, etc. Stimulating recreational reading. 66 figures. 96pp. 5⅝ x 8¼.
63698-4 Pa. $2.75

THE DISCOVERY OF NEPTUNE, Morton Grosser. Dramatic scientific history of the investigations leading up to the actual discovery of the eighth planet of our solar system. Lucid, well-researched book by well-known historian of science. 172pp. 5⅜ x 8½. 23726-5 Pa. $3.00

THE DEVIL'S DICTIONARY. Ambrose Bierce. Barbed, bitter, brilliant witticisms in the form of a dictionary. Best, most ferocious satire America has produced. 145pp. 5⅜ x 8½. 20487-1 Pa. $1.75

HISTORY OF BACTERIOLOGY, William Bulloch. The only comprehensive history of bacteriology from the beginnings through the 19th century. Special emphasis is given to biography-Leeuwenhoek, etc. Brief accounts of 350 bacteriologists form a separate section. No clearer, fuller study, suitable to scientists and general readers, has yet been written. 52 illustrations. 448pp. 5⅝ x 8¼. 23761-3 Pa. $6.50

THE COMPLETE NONSENSE OF EDWARD LEAR, Edward Lear. All nonsense limericks, zany alphabets, Owl and Pussycat, songs, nonsense botany, etc., illustrated by Lear. Total of 321pp. 5⅜ x 8½. (Available in U.S. only) 20167-8 Pa. $3.00

INGENIOUS MATHEMATICAL PROBLEMS AND METHODS, Louis A. Graham. Sophisticated material from Graham *Dial,* applied and pure; stresses solution methods. Logic, number theory, networks, inversions, etc. 237pp. 5⅜ x 8½. 20545-2 Pa. $3.50

BEST MATHEMATICAL PUZZLES OF SAM LOYD, edited by Martin Gardner. Bizarre, original, whimsical puzzles by America's greatest puzzler. From fabulously rare *Cyclopedia,* including famous 14-15 puzzles, the Horse of a Different Color, 115 more. Elementary math. 150 illustrations. 167pp. 5⅜ x 8½. 20498-7 Pa. $2.50

THE BASIS OF COMBINATION IN CHESS, J. du Mont. Easy-to-follow, instructive book on elements of combination play, with chapters on each piece and every powerful combination team—two knights, bishop and knight, rook and bishop, etc. 250 diagrams. 218pp. 5⅜ x 8½. (Available in U.S. only) 23644-7 Pa. $3.50

MODERN CHESS STRATEGY, Ludek Pachman. The use of the queen, the active king, exchanges, pawn play, the center, weak squares, etc. Section on rook alone worth price of the book. Stress on the moderns. Often considered the most important book on strategy. 314pp. 5⅜ x 8½. 20290-9 Pa. $3.50

LASKER'S MANUAL OF CHESS, Dr. Emanuel Lasker. Great world champion offers very thorough coverage of all aspects of chess. Combinations, position play, openings, end game, aesthetics of chess, philosophy of struggle, much more. Filled with analyzed games. 390pp. 5⅜ x 8½. 20640-8 Pa. $4.00

500 MASTER GAMES OF CHESS, S. Tartakower, J. du Mont. Vast collection of great chess games from 1798-1938, with much material nowhere else readily available. Fully annoted, arranged by opening for easier study. 664pp. 5⅜ x 8½. 23208-5 Pa. $6.00

A GUIDE TO CHESS ENDINGS, Dr. Max Euwe, David Hooper. One of the finest modern works on chess endings. Thorough analysis of the most frequently encountered endings by former world champion. 331 examples, each with diagram. 248pp. 5⅜ x 8½. 23332-4 Pa. $3.50

SECOND PIATIGORSKY CUP, edited by Isaac Kashdan. One of the greatest tournament books ever produced in the English language. All 90 games of the 1966 tournament, annotated by players, most annotated by both players. Features Petrosian, Spassky, Fischer, Larsen, six others. 228pp. 5⅜ x 8½. 23572-6 Pa. $3.50

ENCYCLOPEDIA OF CARD TRICKS, revised and edited by Jean Hugard. How to perform over 600 card tricks, devised by the world's greatest magicians: impromptus, spelling tricks, key cards, using special packs, much, much more. Additional chapter on card technique. 66 illustrations. 402pp. 5⅜ x 8½. (Available in U.S. only) 21252-1 Pa. $3.95

MAGIC: STAGE ILLUSIONS, SPECIAL EFFECTS AND TRICK PHOTOGRAPHY, Albert A. Hopkins, Henry R. Evans. One of the great classics; fullest, most authorative explanation of vanishing lady, levitations, scores of other great stage effects. Also small magic, automata, stunts. 446 illustrations. 556pp. 5⅜ x 8½. 23344-8 Pa. $5.00

THE SECRETS OF HOUDINI, J. C. Cannell. Classic study of Houdini's incredible magic, exposing closely-kept professional secrets and revealing, in general terms, the whole art of stage magic. 67 illustrations. 279pp. 5⅜ x 8½. 22913-0 Pa. $3.00

HOFFMANN'S MODERN MAGIC, Professor Hoffmann. One of the best, and best-known, magicians' manuals of the past century. Hundreds of tricks from card tricks and simple sleight of hand to elaborate illusions involving construction of complicated machinery. 332 illustrations. 563pp. 5⅜ x 8½. 23623-4 Pa. $6.00

MADAME PRUNIER'S FISH COOKERY BOOK, Mme. S. B. Prunier. More than 1000 recipes from world famous Prunier's of Paris and London, specially adapted here for American kitchen. Grilled tournedos with anchovy butter, Lobster a la Bordelaise, Prunier's prized desserts, more. Glossary. 340pp. 5⅜ x 8½. (Available in U.S. only) 22679-4 Pa. $3.00

FRENCH COUNTRY COOKING FOR AMERICANS, Louis Diat. 500 easy-to-make, authentic provincial recipes compiled by former head chef at New York's Fitz-Carlton Hotel: onion soup, lamb stew, potato pie, more. 309pp. 5⅜ x 8½. 23665-X Pa. $3.95

SAUCES, FRENCH AND FAMOUS, Louis Diat. Complete book gives over 200 specific recipes: bechamel, Bordelaise, hollandaise, Cumberland, apricot, etc. Author was one of this century's finest chefs, originator of vichyssoise and many other dishes. Index. 156pp. 5⅜ x 8. 23663-3 Pa. $2.50

TOLL HOUSE TRIED AND TRUE RECIPES, Ruth Graves Wakefield. Authentic recipes from the famous Mass. restaurant: popovers, veal and ham loaf, Toll House baked beans, chocolate cake crumb pudding, much more. Many helpful hints. Nearly 700 recipes. Index. 376pp. 5⅜ x 8½. 23560-2 Pa. $4.00

"OSCAR" OF THE WALDORF'S COOKBOOK, Oscar Tschirky. Famous American chef reveals 3455 recipes that made Waldorf great; cream of French, German, American cooking, in all categories. Full instructions, easy home use. 1896 edition. 907pp. 6⅝ x 9⅜. 20790-0 Clothbd. $15.00

COOKING WITH BEER, Carole Fahy. Beer has as superb an effect on food as wine, and at fraction of cost. Over 250 recipes for appetizers, soups, main dishes, desserts, breads, etc. Index. 144pp. 5⅜ x 8½. (Available in U.S. only) 23661-7 Pa. $2.50

STEWS AND RAGOUTS, Kay Shaw Nelson. This international cookbook offers wide range of 108 recipes perfect for everyday, special occasions, meals-in-themselves, main dishes. Economical, nutritious, easy-to-prepare: goulash, Irish stew, boeuf bourguignon, etc. Index. 134pp. 5⅜ x 8½. 23662-5 Pa. $2.50

DELICIOUS MAIN COURSE DISHES, Marian Tracy. Main courses are the most important part of any meal. These 200 nutritious, economical recipes from around the world make every meal a delight. "I . . . have found it so useful in my own household,"—*N.Y. Times*. Index. 219pp. 5⅜ x 8½. 23664-1 Pa. $3.00

FIVE ACRES AND INDEPENDENCE, Maurice G. Kains. Great back-to-the-land classic explains basics of self-sufficient farming: economics, plants, crops, animals, orchards, soils, land selection, host of other necessary things. Do not confuse with skimpy faddist literature; Kains was one of America's greatest agriculturalists. 95 illustrations. 397pp. 5⅜ x 8½. 20974-1 Pa.$3.95

A PRACTICAL GUIDE FOR THE BEGINNING FARMER, Herbert Jacobs. Basic, extremely useful first book for anyone thinking about moving to the country and starting a farm. Simpler than Kains, with greater emphasis on country living in general. 246pp. 5⅜ x 8½. 23675-7 Pa. $3.50

A GARDEN OF PLEASANT FLOWERS (PARADISI IN SOLE: PARADISUS TERRESTRIS), John Parkinson. Complete, unabridged reprint of first (1629) edition of earliest great English book on gardens and gardening. More than 1000 plants & flowers of Elizabethan, Jacobean garden fully described, most with woodcut illustrations. Botanically very reliable, a "speaking garden" of exceeding charm. 812 illustrations. 628pp. 8½ x 12¼. 23392-8 Clothbd. $25.00

ACKERMANN'S COSTUME PLATES, Rudolph Ackermann. Selection of 96 plates from the *Repository of Arts,* best published source of costume for English fashion during the early 19th century. 12 plates also in color. Captions, glossary and introduction by editor Stella Blum. Total of 120pp. 8⅜ x 11¼. 23690-0 Pa. $4.50

MUSHROOMS, EDIBLE AND OTHERWISE, Miron E. Hard. Profusely illustrated, very useful guide to over 500 species of mushrooms growing in the Midwest and East. Nomenclature updated to 1976. 505 illustrations. 628pp. 6½ x 9¼. 23309-X Pa. $7.95

AN ILLUSTRATED FLORA OF THE NORTHERN UNITED STATES AND CANADA, Nathaniel L. Britton, Addison Brown. Encyclopedic work covers 4666 species, ferns on up. Everything. Full botanical information, illustration for each. This earlier edition is preferred by many to more recent revisions. 1913 edition. Over 4000 illustrations, total of 2087pp. 6⅛ x 9¼. 22642-5, 22643-3, 22644-1 Pa., Three-vol. set $24.00

MANUAL OF THE GRASSES OF THE UNITED STATES, A. S. Hitchcock, U.S. Dept. of Agriculture. The basic study of American grasses, both indigenous and escapes, cultivated and wild. Over 1400 species. Full descriptions, information. Over 1100 maps, illustrations. Total of 1051pp. 5⅜ x 8½. 22717-0, 22718-9 Pa., Two-vol. set $12.00

THE CACTACEAE,, Nathaniel L. Britton, John N. Rose. Exhaustive, definitive. Every cactus in the world. Full botanical descriptions. Thorough statement of nomenclatures, habitat, detailed finding keys. The one book needed by every cactus enthusiast. Over 1275 illustrations. Total of 1080pp. 8 x 10¼. 21191-6, 21192-4 Clothbd., Two-vol. set $35.00

AMERICAN MEDICINAL PLANTS, Charles F. Millspaugh. Full descriptions, 180 plants covered: history; physical description; methods of preparation with all chemical constituents extracted; all claimed curative or adverse effects. 180 full-page plates. Classification table. 804pp. 6½ x 9¼. 23034-1 Pa. $10.00

A MODERN HERBAL, Margaret Grieve. Much the fullest, most exact, most useful compilation of herbal material. Gigantic alphabetical encyclopedia, from aconite to zedoary, gives botanical information, medical properties, folklore, economic uses, and much else. Indispensable to serious reader. 161 illustrations. 888pp. 6½ x 9¼. (Available in U.S. only) 22798-7, 22799-5 Pa., Two-vol. set $11.00

THE HERBAL or GENERAL HISTORY OF PLANTS, John Gerard. The 1633 edition revised and enlarged by Thomas Johnson. Containing almost 2850 plant descriptions and 2705 superb illustrations, Gerard's *Herbal* is a monumental work, the book all modern English herbals are derived from, the one herbal every serious enthusiast should have in its entirety. Original editions are worth perhaps $750. 1678pp. 8½ x 12¼. 23147-X Clothbd. $50.00

MANUAL OF THE TREES OF NORTH AMERICA, Charles S. Sargent. The basic survey of every native tree and tree-like shrub, 717 species in all. Extremely full descriptions, information on habitat, growth, locales, economics, etc. Necessary to every serious tree lover. Over 100 finding keys. 783 illustrations. Total of 986pp. 5⅜ x 8½. 20277-1, 20278-X Pa., Two-vol. set $10.00

AMERICAN BIRD ENGRAVINGS, Alexander Wilson et al. All 76 plates. from Wilson's *American Ornithology* (1808-14), most important ornithological work before Audubon, plus 27 plates from the supplement (1825-33) by Charles Bonaparte. Over 250 birds portrayed. 8 plates also reproduced in full color. 111pp. 9⅜ x 12½. 23195-X Pa. $6.00

CRUICKSHANK'S PHOTOGRAPHS OF BIRDS OF AMERICA, Allan D. Cruickshank. Great ornithologist, photographer presents 177 closeups, groupings, panoramas, flightings, etc., of about 150 different birds. Expanded *Wings in the Wilderness.* Introduction by Helen G. Cruickshank. 191pp. 8¼ x 11. 23497-5 Pa. $6.00

AMERICAN WILDLIFE AND PLANTS, A. C. Martin, et al. Describes food habits of more than 1000 species of mammals, birds, fish. Special treatment of important food plants. Over 300 illustrations. 500pp. 5⅜ x 8½. 20793-5 Pa. $4.95

THE PEOPLE CALLED SHAKERS, Edward D. Andrews. Lifetime of research, definitive study of Shakers: origins, beliefs, practices, dances, social organization, furniture and crafts, impact on 19th-century USA, present heritage. Indispensable to student of American history, collector. 33 illustrations. 351pp. 5⅜ x 8½. 21081-2 Pa. $4.00

OLD NEW YORK IN EARLY PHOTOGRAPHS, Mary Black. New York City as it was in 1853-1901, through 196 wonderful photographs from N.-Y. Historical Society. Great Blizzard, Lincoln's funeral procession, great buildings. 228pp. 9 x 12. 22907-6 Pa. $7.95

MR. LINCOLN'S CAMERA MAN: MATHEW BRADY, Roy Meredith. Over 300 Brady photos reproduced directly from original negatives, photos. Jackson, Webster, Grant, Lee, Carnegie, Barnum; Lincoln; Battle Smoke, Death of Rebel Sniper, Atlanta Just After Capture. Lively commentary. 368pp. 8⅜ x 11¼. 23021-X Pa. $8.95

TRAVELS OF WILLIAM BARTRAM, William Bartram. From 1773-8, Bartram explored Northern Florida, Georgia, Carolinas, and reported on wild life, plants, Indians, early settlers. Basic account for period, entertaining reading. Edited by Mark Van Doren. 13 illustrations. 141pp. 5⅜ x 8½. 20013-2 Pa. $4.50

THE GENTLEMAN AND CABINET MAKER'S DIRECTOR, Thomas Chippendale. Full reprint, 1762 style book, most influential of all time; chairs, tables, sofas, mirrors, cabinets, etc. 200 plates, plus 24 photographs of surviving pieces. 249pp. 9⅞ x 12¾. 21601-2 Pa. $6.50

AMERICAN CARRIAGES, SLEIGHS, SULKIES AND CARTS, edited by Don H. Berkebile. 168 Victorian illustrations from catalogues, trade journals, fully captioned. Useful for artists. Author is Assoc. Curator, Div. of Transportation of Smithsonian Institution. 168pp. 8½ x 9½. 23328-6 Pa. $5.00

CATALOGUE OF DOVER BOOKS

THE STANDARD BOOK OF QUILT MAKING AND COLLECTING, Marguerite Ickis. Full information, full-sized patterns for making 46 traditional quilts, also 150 other patterns. Quilted cloths, lame, satin quilts, etc. 483 illustrations. 273pp. 6⅞ x 9⅝. 20582-7 Pa. $4.50

ENCYCLOPEDIA OF VICTORIAN NEEDLEWORK, S. Caulfield, Blanche Saward. Simply inexhaustible gigantic alphabetical coverage of every traditional needlecraft—stitches, materials, methods, tools, types of work; definitions, many projects to be made. 1200 illustrations; double-columned text. 697pp. 8⅛ x 11. 22800-2, 22801-0 Pa., Two-vol. set $12.00

MECHANICK EXERCISES ON THE WHOLE ART OF PRINTING, Joseph Moxon. First complete book (1683-4) ever written about typography, a compendium of everything known about printing at the latter part of 17th century. Reprint of 2nd (1962) Oxford Univ. Press edition. 74 illustrations. Total of 550pp. 6⅛ x 9¼. 23617-X Pa. $7.95

PAPERMAKING, Dard Hunter. Definitive book on the subject by the foremost authority in the field. Chapters dealing with every aspect of history of craft in every part of the world. Over 320 illustrations. 2nd, revised and enlarged (1947) edition. 672pp. 5⅜ x 8½. 23619-6 Pa. $7.95

THE ART DECO STYLE, edited by Theodore Menten. Furniture, jewelry, metalwork, ceramics, fabrics, lighting fixtures, interior decors, exteriors, graphics from pure French sources. Best sampling around. Over 400 photographs. 183pp. 8⅜ x 11¼. 22824-X Pa. $5.00

Prices subject to change without notice.

Available at your book dealer or write for free catalogue to Dept. GI, Dover Publications, Inc., 180 Varick St., N.Y., N.Y. 10014. Dover publishes more than 175 books each year on science, elementary and advanced mathematics, biology, music, art, literary history, social sciences and other areas.